Sepsis and Non-infectious Systemic Inflammation

Edited by
Jean-Marc Cavaillon and
Christophe Adire

Related Titles

Meyers, R. A. (ed.)

Immunology

From Cell Biology to Disease

2007

ISBN: 978-3-527-31770-7

Deretic, V. (ed.)

Autophagy in Immunity and Infection

A Novel Immune Effector

2006

ISBN: 978-3-527-31450-8

Lutz, M. B., Romani, N., Steinkasserer A. (eds.)

Handbook of Dendritic Cells

Biology, Diseases, and Therapies

2006

ISBN: 978-3-527-31109-5

Frelinger, J. A. (ed.)

Immunodominance

The choice of the Immune System

2006

ISBN: 978-3-527-31274-0

Pollard, K. M. (ed.)

Autoantibodies and Autoimmunity

Molecular Mechanisms in Health and Disease

2006

ISBN: 978-3-527-31141-5

Sepsis and Non-infectious Systemic Inflammation

From Biology to Critical Care

Edited by
Jean-Marc Cavaillon and
Christophe Adire

WILEY-BLACKWELL

WILEY-VCH Verlag GmbH & Co. KGaA

The Editors

Prof. Dr. Jean-Marc Cavaillon
Unit Cytokines et Inflammation
Institut Pasteur
28, rue Dr. Roux
75015 Paris
France

Dr. Christophe Adrie
Hôpital Delafontaine
Réanimation Polyvalente
93205 Saint-Denis
France

Library of Congress Card No.: applied for

British Library Cataloguing-in-Publication Data
A catalogue record for this book is available from the British Library.

Bibliographic information published by the Deutsche Nationalbibliothek
The Deutsche Nationalbibliothek lists this publication in the Deutsche Nationalbibliografie; detailed bibliographic data are available in the Internet at ⟨http://dnb.d-nb.de⟩.

Typesetting Laserwords Private Ltd, Chennai, India
Printing Strauss GmbH, Mörlenbach
Binding Litges & Dopf GmbH, Heppenheim
Cover Design Adam-Design, Weinheim

Printed in the Federal Republic of Germany
Printed on acid-free paper

ISBN: 978-3-527-31935-0

Contents

Preface *XV*

List of Contributors *XVII*

Part I Clinical Aspects of Sepsis and SIRS

1 Definition of Sepsis and Non-infectious SIRS *3*
Jean-Louis Vincent
1.1 Introduction *3*
1.2 Sepsis Syndrome *3*
1.3 Systemic Inflammatory Response Syndrome *3*
1.4 Sepsis Markers *5*
1.5 Inflammation versus Infection *7*
1.6 Conclusion *9*
References *10*

2 Epidemiology of Sepsis and Non-infectious SIRS *13*
Michael C. Reade and Derek C. Angus
2.1 Introduction *13*
2.2 Epidemiology of SIRS: Incidence and Prevalence *13*
2.3 Epidemiology of Sepsis: Incidence and Prevalence *13*
2.4 Epidemiology of SIRS and Sepsis: Outcome *14*
2.5 Explanations for the Variability in Estimates of Incidence, Prevalence and Mortality *14*
2.5.1 Problems with the Definitions of Sepsis *14*
2.5.2 Differences in Study Design *20*
2.5.3 Lack of Consistency in Reported Indices *21*
2.5.4 'Hidden' Sepsis *21*
2.5.5 Poor and Missing Data *22*
2.5.6 Real Differences in Regional Epidemiology of Sepsis *22*
2.6 Conclusions Based on Available Epidemiological Data *23*
2.6.1 Trends over Time: Incidence, Mortality *23*

Sepsis and Non-infectious Systemic Inflammation. Edited by J.-M. Cavaillon and C. Adrie

ISBN: 978-3-527-31935-0

2.6.2 Disparities in the Burden of Sepsis: Race and Gender *24*
2.6.3 Site of Primary Infection *27*
2.6.4 Organism *27*
2.6.5 Environmental Factors in Risk *28*
2.7 Special Populations *29*
2.7.1 Malignancy *29*
2.7.2 HIV *29*
2.7.3 Children *30*
2.8 Resource Utilization *30*
2.9 Prevention, Early Intervention, and Systems of Care *31*
2.10 Outcome other than Mortality *33*
2.11 Implications for Planning of Health Services *34*
References *34*

3 Similarities of the Clinical Aspects of Sepsis and Non-infectious SIRS *41*
Anne Claire Lukaszewicz and Didier Payen
3.1 Introduction *41*
3.2 Case Report *43*
3.3 Epidemiology of SIRS *43*
3.4 SIRS Criteria reflect Systemic Inflammation *44*
3.4.1 Fever *44*
3.4.2 Tachycardia *45*
3.4.3 Leucocytosis *45*
3.4.4 Tachypnea *47*
3.5 Phenomenon Triggering Inflammation *47*
3.5.1 Infectious Etiology *47*
3.5.2 Non-infectious Etiology *48*
3.6 Circulating Markers for Systemic Inflammation Characterization *49*
3.7 Conclusion *50*
Acknowledgments *50*
References *51*

4 Organ Dysfunctions during Severe Sepsis and Septic-like Syndromes: Epidemiology, Classification, and Mechanisms *57*
Olfa Hamzaoui and Jean Carlet
4.1 Definitions *57*
4.2 Classification of Organ Dysfunctions *57*
4.3 Epidemiology *59*
4.4 Pathophysiology of Organ Dysfunctions in Sepsis *60*
4.4.1 Microvascular Dysfunction *60*
4.4.2 Endothelial Injury and Coagulopathy *61*
4.4.3 Endothelial Activation *62*
4.4.4 Endothelial Dysfunction *62*

4.4.5 Cytopathic Hypoxia 62
4.4.5.1 Inhibition of Pyruvate Dehydrogenase 63
4.4.5.2 Inhibition of Cytochrome a_1a_3 63
4.4.5.3 Inhibition of Mitochondrial Enzyme under the Influence of Peroxynitrite ($ONOO^-$) 63
4.4.5.4 The Poly (ADP-ribose) Polymerase (PARP-1) Hypothesis 63
4.4.6 Inflammatory Cytokines 63
4.5 Main Organ Dysfunctions 64
4.5.1 Cardiocirculatory Dysfunction 64
4.5.1.1 Cardiac Dysfunction 64
4.5.1.2 Vascular Dysfunction 65
4.5.2 Renal Dysfunction 66
4.5.3 Respiratory Dysfunction 66
4.5.4 Brain Dysfunction 67
4.5.5 Hematologic Dysfunction 68
4.5.6 Hepatic Dysfunction 69
4.6 Critique of the Organ Dysfunction Concept and of the Scores used in ICU Patients 69
4.7 Organ Failure: Severity Marker or Protective Phenomenon? 70
4.8 Conclusion 71
References 71

Part II Pathogens

5 Pathogens in Sepsis: Gram-negative Bacterial PAMPs and PRRs 79
Andra Schromm, Christian Alexander, Thomas Gutsmann, Jörg Andrä and Cordula Stamme
5.1 Introduction 79
5.2 Pathogen-associated Molecular Pattern (PAMP) Molecules of GNB 80
5.2.1 Lipopolysaccharides 81
5.2.2 Lipoproteins of GNB 84
5.2.3 Flagellins 84
5.2.4 m-DAP-type PG 86
5.3 The LPS-activated TLR4/MD-2-signaling system: Central PRR Signaling Pathways and Negative Regulatory Factors 87
5.4 Endotoxic Activity of Enterobacterial LPS is determined by Physico-chemical Properties 90
5.5 Modulation of Biological Activity of LPS 92
5.5.1 Antimicrobial Peptides and Proteins 92
5.5.2 Mechanisms of LPS Neutralization by Cationic Peptides 94
5.5.3 Pulmonary Collectins 95
5.6 Therapeutic Approaches 96

Acknowledgments 96
References 97

6 Pathogens in Sepsis: Gram-positive Bacterial PAMPs, PRRs and Superantigens 109
Christian Alexander, Romney Haylett, Andra Schromm, Ulrich Zähringer and Lothar Rink
6.1 Introduction 109
6.2 Pathogen-associated Molecular Pattern (PAMP) Molecules of Gram-positive Bacteria 110
6.2.1 Lipoproteins of Gram-positive Bacteria 113
6.2.2 Peptidoglycans of Gram-positive Bacteria 114
6.2.3 Lipoteichoic Acids and Wall Teichoic Acids 117
6.3 Activation of TLR2-dependent Signaling by Lipoproteins of Gram-positive Bacteria: Central Signaling Pathways and Major Negative Regulatory Factors 120
6.4 Superantigens: A Specific Class of Gram-positive Pathogenicity Factors 122
6.4.1 Superantigen Structure 123
6.4.2 Gene Localization, Superantigen Groups, and Allelic Variation 128
6.4.3 Superantigen Involvement in Sepsis 129
6.4.3.1 Toxic Shock Syndrome (TSS) 129
6.4.3.2 Streptococcal Toxic Shock Syndrome (STSS) 131
6.5 Novel Therapeutic Concepts in Gram-positive Septic Diseases 132
Acknowledgments 133
References 133

7 Pathogens in Sepsis: Fungi, Parasites 145
Frédéric Méchaï, Fabrice Bruneel, Fanny Lanternier, Stephane Legriel, Jean-Pierre Bédos and Olivier Lortholary
7.1 Fungi 145
7.1.1 Introduction 145
7.1.2 Epidemiology of Candidemia in Intensive Care Unit 145
7.1.3 Pathogenesis of Candida Sepsis 149
7.1.4 Diagnosis of Candidemia 150
7.1.5 Complications and Prognosis of Candidemia 151
7.1.6 Treatment of Candidemia 152
7.1.7 Conclusion 156
7.2 Severe Falciparum Malaria: An Interesting Example of Parasite-induced Sepsis 156
7.2.1 Introduction 156
7.2.2 Epidemiology of Falciparum Malaria 157
7.2.3 Definition of Severe Falciparum Malaria and Relevance to Severe Imported Malaria 157
7.2.4 Pathophysiology of Severe Falciparum Malaria 158

7.2.4.1 The Sequestration Hypothesis *159*
7.2.4.2 The Immunological (or Inflammation, or Humoral) Hypothesis *159*
7.2.4.3 The Role of Hemostasis and Platelets *160*
7.2.4.4 Genetic Polymorphisms *160*
7.2.4.5 Other Factors *161*
7.2.5 Important Clinical Issues *161*
7.2.5.1 From Uncomplicated to Severe Malaria *161*
7.2.5.2 Populations at Risk *161*
7.2.5.3 Clinical Aspects of SFM *161*
7.2.6 The Management of Severe Falciparum Malaria *163*
7.2.6.1 Specific Antimalarial Treatments *163*
7.2.6.2 Continuing Supportive Care *164*
7.2.6.3 Other Adjuvant Treatments *165*
7.2.7 Conclusion *166*
References *166*

Part III Physiopathology of Sepsis and SIRS

8 Inflammatory Mediators *177*
Francois Philippart, Jean-Marc Cavaillon and Charles Dinarello
8.1 Inflammatory Mediators in Sepsis *177*
8.1.1 Cytokines, Chemokines, and Growth Factors *177*
8.1.2 Complement System *182*
8.1.3 Coagulation and Fibrinolysis *186*
8.1.4 Proteases *187*
8.1.5 Lipid Mediators *188*
8.1.6 Nitric Oxide *189*
8.1.7 Cellular Markers of Stress and Soluble Cell Surface Markers *189*
8.1.8 Lipoproteins and Lipopolysaccharide Inhibitors *190*
8.1.9 Anti-inflammatory Mediators *190*
8.1.10 Neuromediators *192*
8.2 Inflammatory Mediators in Non-infectious SIRS *192*
8.2.1 Sepsis and Non-infectious SIRS share Similar Inflammatory Mediators *192*
8.2.2 Are Inflammatory Mediators Markers of Evolution and/or Infection? *194*
8.3 Conclusion *195*
References *195*

9 Apoptosis: A Potential Therapeutic Target in Sepsis and Non-Infectious Systemic Inflammation *205*
Chun-Shiang Chung, Ryan Swan, Mario Perl, Doreen E. Wesche-Soldato and Alfred Ayala
9.1 Introduction *205*

9.2 Apoptosis 206
9.2.1 Overview 206
9.2.2 Pathways of Apoptosis 207
9.2.3 Biochemistry of Apoptosis 209
9.3 Pathological Role of Apoptotic Cell Death in Sepsis 210
9.4 Potential Anti-apoptotic Therapies for the Treatment of Sepsis 213
9.4.1 Recombinant Human Activated Protein C 213
9.4.2 Targeting the Extrinsic Apoptotic Pathway 214
9.4.2.1 Targeting Fas 214
9.4.2.2 Targets Downstream of Fas 216
9.4.3 Targeting the Intrinsic Apoptosis Pathway 217
9.4.3.1 Targeting the BCL-2 Family 217
9.4.4 Targeting Both Pathways 218
9.4.4.1 Caspase Inhibitors 218
9.4.4.2 Bid Protein 219
9.4.5 Other Potential Anti-apoptotic Therapies 219
9.4.5.1 Protease Inhibitors in Sepsis 219
9.4.5.2 Akt 219
9.4.5.3 Adrenomedullin 220
9.4.5.4 Clearance of Apoptotic Cells 220
9.4.5.5 Milk Fat Globule Epidermal Growth Factor 8 (MFG-E8) 221
9.5 Conclusions 222
Acknowledgments 222
References 222

10 The Role of Endothelium 231
Siew-Yean Lau and Fumito Ichinose
10.1 Introduction 231
10.2 Endothelial Cell Functions and their Changes in SIRS and Sepsis 231
10.2.1 Regulation of Coagulation 232
10.2.1.1 Regulation of Coagulation in Normal Endothelium 232
10.2.1.2 Procoagulant Properties of Endothelium in Severe Sepsis 232
10.2.2 Regulation of Leukocyte Adhesion/Migration and Vascular Permeability 233
10.2.3 Regulation of the Microcirculation/Vasomotor Properties 234
10.2.3.1 Nitric Oxide 234
10.2.3.2 Prostacyclin (PGI_2) and Thromboxane A_2 (TXA_2) 236
10.2.3.3 Endothelin (ET) 238
10.3 Endothelium-centered Therapeutic Approaches in Severe Sepsis 239
10.3.1 Soluble Inflammatory Mediators as Therapeutic Targets 240
10.3.2 Therapeutic Strategies Targeting Coagulation/Anti-coagulation Balance 240
10.3.3 Nitric Oxide 241

10.3.4 Statins *242*
10.4 Role of Endothelium and Future Therapeutic Challenges in Sepsis and SIRS *243*
References *243*

11 Coagulation *251*
Marcel Levi
11.1 Introduction *251*
11.2 Relevance of Coagulation Abnormalities in Patients with Sepsis *251*
11.3 Incidence of Coagulation Abnormalities in Sepsis *253*
11.4 Pathogenetic Pathways in the Coagulopathy of Sepsis *254*
11.4.1 Initiation of Coagulation Activation *254*
11.4.2 Impairment of the Antithrombin, Protein C, and TFPI Anticoagulant Pathways in Sepsis *255*
11.4.3 Inhibition of Endogenous Fibrinolysis in Sepsis *258*
11.4.4 Regulatory Role of Cytokines in the Coagulopathy of Sepsis *259*
11.4.5 Cross-talk between Coagulation and Inflammation in Sepsis *260*
11.5 Diagnostic Approach to Coagulation Abnormalities in Sepsis *261*
11.5.1 Tests for Intravascular Fibrin Formation and Fibrin Degradation Products *262*
11.5.2 Markers for Thrombin Generation and Coagulation Activation *263*
11.5.3 Platelet Count and Coagulation Factors in Patients with Sepsis *263*
11.5.4 Diagnostic Management in Clinical Practice *264*
11.6 Supportive Treatment of Coagulation Abnormalities in Sepsis *265*
11.6.1 Plasma and Platelet Substitution Therapy *266*
11.6.2 Anticoagulants *266*
11.6.3 Restoration of Anticoagulant Pathways *267*
References *268*

12 Epithelial Barrier Dysfunction as a Mechanism in the Pathogenesis of Multiple Organ Dysfunction *279*
Mitchell P. Fink, Antonella Cotoia, Donna Beer-Stolz, Shiguang Liu, Penny L. Sappington and Russell L. Delude
12.1 Introduction *279*
12.2 Is "Cytopathic Hypoxia" the underlying Mechanism Responsible for MODS? *279*
12.3 What is the Role of Cellular Apoptosis in the Pathogenesis of MODS? *280*
12.4 Tight Junctions maintain Epithelial Polarity and Barrier Function *281*
12.5 Epithelial Dysfunction as a Pathogenic Mechanism underlying MODS *281*
12.6 The Role of Nitric Oxide (NO) and Peroxynitrite ($ONOO^-$) in the Regulation of TJ Protein Expression and Function *283*

12.7 Functional iNOS Expression is essential for LPS-induced Alterations in Intestinal Barrier Function in Mice *285*
12.8 LPS impairs Hepatobiliary Barrier Function via an iNOS-dependent Mechanism *286*
12.9 A Pro-inflammatory Milieu decreases Pulmonary Epithelial Barrier Function *287*
12.10 HMGB1 as a Mediator of Epithelial Dysfunction *288*
12.11 Cross-talk between Gut Epithelial Cells and Intestinal Bacteria *289*
References *290*

13 Neural Regulation of Cytokines *299*
Jared M. Huston and Kevin J. Tracey
13.1 Introduction *299*
13.2 Cytokine Stimulation of the Nervous System *300*
13.3 Neuro-Endocrine Mediation of Cytokines *301*
13.3.1 Hypothalamic–Pituitary–Adrenal (HPA) Axis *301*
13.3.2 Hypothalamic–Pituitary–Gonadal (HPG) Axis *302*
13.3.3 Hypothalamic–Growth Hormone Axis *303*
13.3.4 Hypothalamic–Pituitary–Thyroid (HPT) Axis *303*
13.4 Direct Neural Control of Cytokines *304*
13.4.1 The Peripheral Nervous System *304*
13.4.2 The Sympathetic Nervous System *304*
13.5 Neural Control of Cytokines via the Inflammatory Reflex *306*
13.5.1 The Afferent Arm of the Inflammatory Reflex *306*
13.5.2 The Cholinergic Anti-inflammatory Pathway *306*
13.5.3 Clinical Implications of the Cholinergic Anti-inflammatory Pathway and Future Directions *309*
References *310*

14 Genetic Polymorphisms and Other Parameters that Affect Predisposal to Infection and Outcome *315*
Jean-Marc Cavaillon and Christophe Adrie
14.1 Non-genetic Parameters that Affect the Occurrence of Infection, Organ Failure and Outcome *315*
14.1.1 The Infectious Agent and Site of Infection *315*
14.1.2 Parameters Linked to the Patients *317*
14.1.2.1 Stochastic Events *317*
14.1.2.2 Underlying Diseases *318*
14.1.2.3 Age *318*
14.1.2.4 Gender *319*
14.1.2.5 Lifestyle *320*
14.1.2.6 Asymptomatic Viral Infection *321*
14.1.2.7 Ethnic Group *321*
14.2 Parameters Linked to Health Care *322*
14.2.1 Antibiotherapy *322*

14.2.2 Iatrogenic Events *322*
14.2.3 Quality of Health Care *323*
14.3 Genetic Polymorphisms *323*
14.3.1 Cytokines *324*
14.3.2 Innate Immune Receptors *328*
14.3.3 Coagulation Factors *328*
14.3.4 Others *332*
References *335*

15 Altered Immune Status and Leukocytes Reprogramming *347*
Jean-Marc Cavaillon, Minou Adib-Conquy and Christophe Adrie
15.1 Introduction *347*
15.2 Leukocytes Surface Markers *348*
15.3 Altered Function of Circulating Leukocytes *350*
15.4 Leukocytes Apoptosis *350*
15.5 Reduced *"Ex Vivo"* Production of Cytokines *351*
15.5.1 Lymphocytes *351*
15.5.2 Neutrophils *352*
15.5.3 Monocytes *352*
15.5.4 Dendritic Cells *353*
15.6 Altered *Ex Vivo* Cytokine Production is not a Generalized Phenomenon *354*
15.6.1 Nature of the Insult *354*
15.6.2 Nature of the Cell Culture *354*
15.6.3 Nature of the Cytokines Produced *355*
15.6.4 Nature of the Activating Agent: Response to Whole Bacteria *355*
15.7 Does Altered *Ex Vivo* Cytokine Production Correlate with Severity? *356*
15.8 Molecular Mechanisms of Altered TLR-Dependent TNF Production *357*
15.9 Desensitizing Agents in Plasma *358*
15.10 Mediators and Cells that are Responsible for Leukocyte Reprogramming and Altered Immune Status *358*
15.11 Compartmentalization *360*
15.12 Reversal of Reprogramming *361*
15.13 Clinical Meaning *362*
References *362*

Part IV Experimental Models and Therapies

16 Experimental Models of Sepsis and Non-infectious SIRS *375*
Raghavan Raju, William J. Hubbard and Irshad H. Chaudry
16.1 Introduction *375*
16.2 Definition of Sepsis, Endotoxemia and Bacteremia *376*

16.2.1 Endotoxin Models *376*
16.2.2 Bacteremia Models *378*
16.3 Exogenous Peritonitis Models *379*
16.3.1 Cecal Ligation and Perforation Models *380*
16.3.1.1 The CLP Model *381*
16.3.1.2 CLP as a Model of Sepsis *381*
16.3.1.3 Potential Limitations of the CLP Model *383*
16.4 Animal Models versus Human Sepsis *383*
Acknowledgments *384*
References *384*

17 The Past, Present and Future of Therapies for Sepsis and Non-infectious Systemic Inflammation *391*
Nitin Seam and Anthony F. Suffredini
17.1 Introduction *391*
17.2 Historical Elements in the Treatment of Infection *391*
17.3 Historical Elements in the Treatment of Inflammation Associated with Infection or Injury *393*
17.4 Historical Developments in the Treatment of Shock *394*
17.5 Contemporary Developments in Sepsis and Non-infectious Systemic Inflammation: the Molecular Basis of Inflammation associated with Infection and Shock *397*
17.6 Endotoxin as a Therapeutic Target *398*
17.7 Immunotherapy Revisited *399*
17.8 Anti-inflammatory Approaches to Sepsis and Systemic Inflammation *400*
17.9 Resurrecting the Use of Corticosteroids in Sepsis *400*
17.10 The Role of Anti-coagulant Therapy in Sepsis *401*
17.11 Immunostimulatory Therapies for Sepsis and Shock *403*
17.12 Contemporary Therapy for Infection and Shock *404*
17.13 The Future of Therapies for Sepsis and Non-infectious Systemic Inflammation *405*
References *408*

Index *417*

Preface

In 1519 when Lucrece Borgia succumbed to puerperal septicemia while she was giving birth to her seventh child, probably no doctor could explain the events. It was only in 1847 that Ignaz Semmelweis understood the possible reasons for the high percentage of deaths due to puerperal septicemia, and he was the first to define the antiseptic methods required to reduce mortality in his hospital in Vienna. In 1879, Louis Pasteur further promoted antiseptic methods, and identified the presence of common bacteria in the bloodstream of these patients. Circa 1904, Sir William Osler offered a quite provocative definition of sepsis when, including the potential deleterious effects of the inflammatory response, he wrote: 'Except on few occasions, the patient appears to die from the body's response to infection rather than from it'. More recently, Roger Bone redefined sepsis, and with others, introduced the concept of 'systemic inflammatory response syndrome' (SIRS), a clinical setting that mimics many pathophysiologic observations made in sepsis but in the absence of infection. Nowadays, the discovery of endogenous 'alarmins' or 'danger associated molecular patterns' (DAMPs) that share similar receptors with exogenous 'pathogen associated molecular patterns' (PAMPs) partially explains the similarity between sepsis and non-infectious SIRS. Bone also introduced the concept of 'compensatory anti-inflammatory response syndrome' (CARS) to explain the consequences of the altered immune status observed among sepsis and non-infectious SIRS patients.

Interestingly, systemic inflammatory responses share many similarities whatever their infectious or non-infectious origin. Despite many years of intensive basic and clinical research which has increased our understanding of the ongoing processes, establishing better therapeutic strategies appeared to be the main approach to improving survival rather than the development of specific new treatments targeted at an intrinsic mechanism. We felt that this was an appropriate time to offer an overview of all the new insights into this syndrome, regardless of whether it has originated from an infectious process or from any other cause. It therefore seemed logical to report on the state-of-the art in this increasingly frequent clinical presentation from an epidemiologic, mechaninistic, and predisposal standpoint, and on the experimental models that will help to further decipher, and hopefully define new treatments for this

Sepsis and Non-infectious Systemic Inflammation. Edited by J.-M. Cavaillon and C. Adrie

ISBN: 978-3-527-31935-0

dreadful syndrome. Accordingly, we have asked the most distinguished and prominent doctors and scientists in the field to contribute their expertise to the discussion of many of the important aspects and treatment of sepsis and SIRS.

Paris and Saint-Denis
July 2008

Jean-Marc Cavaillon
Christophe Adrie

List of Contributors

Minou Adib-Conquy
Institut Pasteur
Unit Cytokines & Inflammation
Paris
France

Christophe Adrie
Delafontaine Hospital
Intensive Care Unit
Saint Denis
France

Christian Alexander
Leibniz-Center for Medicine and Biosciences
Research Center Borstel
Borstel
Germany

Jörg Andrä
Leibniz-Center for Medicine and Biosciences
Research Center Borstel
Borstel
Germany

Derek C. Angus
University of Pittsburgh
Department of Critical Care Medicine
CRISMA (Clinical Research, Investigation, and Systems Modeling of Acute Illness) Laboratory
Pittsburgh
USA

Alfred Ayala
Rhode Island Hospital
Division of Surgical Research
Providence
USA

Jean-Pierre Bédos
Centre Hospitalier de Versailles
Hôpital André Mignot
Intensive Care Unit
Le Chesnay
France

Donna Beer-Stolz
University of Pittsburgh
School of Medicine
Department of Cell Biology and Physiology
Pittsburgh
USA

Sepsis and Non-infectious Systemic Inflammation. Edited by J.-M. Cavaillon and C. Adrie

ISBN: 978-3-527-31935-0

Fabrice Bruneel
Centre Hospitalier de Versailles
Hôpital André Mignot
Intensive Care Unit
Le Chesnay
France

Jean Carlet
Groupe Hospitalier Paris Saint-Joseph
Intensive Care Unit
Paris
France

Jean-Marc Cavaillon
Institut Pasteur
Unit Cytokines and Inflammation
Paris
France

Ishrad H. Chaudry
University of Alabama at Birmingham
Center for Surgical Research
Birmingham
USA

Chun-Shiang Chung
Rhode Island Hospital/Brown
University
School of Medicine
Department of Surgery
Providence
USA

Antonella Cotoia
University of Pittsburgh
School of Medicine
Department of Critical
Care Medicine
Pittsburgh
USA

Russell L. Delude
University of Pittsburgh
School of Medicine
Departments of Critical
Care Medicine and Pathology
Pittsburgh
USA

Charles Dinarello
University of Colorado
Health Sciences Center
Department of Medicine
Denver
USA

Mitchell P. Fink
University of Pittsburgh
School of Medicine
Department of Critical
Care Medicine
Pittsburgh
USA

Thomas Gutsmann
Leibniz-Center for Medicine and
Biosciences
Research Center Borstel
Borstel
Germany

Olfa Hamzaoui
Groupe Hospitalier Paris Saint-Joseph
Intensive Care Unit
Paris
France

Romney Haylett
Leibniz-Center for Medicine and
Biosciences
Research Center Borstel
Borstel
Germany

William J. Hubbard
University of Alabama at Birmingham
Center for Surgical Research and
Department of Surgery
Birmingham
USA

Jared M. Huston
North Shore-LIJ Research Institute
Manhassset
USA

Fumito Ichinose
Massachusetts General
Hospital/Harvard Medical School
Anesthesia Center for Critical
Care Research
Charlestown
USA

Fanny Lanternier
Université Paris 5
Centre d'Infectiologie Necker-Pasteur
CHU Necker-Enfants maladies
Paris
France

Siew-Yean Lau
Harvard Medical School
Massachusetts General Hospital
Anesthesia Center for Critical
Care Research
Charlestown
USA

Stephane Legriel
Centre Hospitalier de Versailles
Hôpital André Mignot
Intensive Care Unit
Le Chesnay
France

Marcel Levi
University of Amsterdam
Academic Medical Center
Department of Medicine
Amsterdam
The Netherlands

Shiguang Liu
University of Pittsburgh
School of Medicine
Department of Critical
Care Medicine
Pittsburgh
USA

Olivier Lortholary
Université Paris 5
Centre d'Infectiologie Necker-Pasteur
CHU Necker-Enfants maladies
Paris
France
and
Institut Pasteur
Centre National de Référence
Mycologie et Antifongiques
Paris
France

Anne Claire Lukaszewicz
Hôpital Lariboisiére
Department of Anesthesiology &
Critical Care Medicine
and
University Diderot
Paris
France

Frédéric Méchaï
Université Paris 5
Centre d'Infectiologie Necker-Pasteur
CHU Necker-Enfants maladies
Paris
France

Didier Payen
Hôpital Lariboisière
Department of Anesthesiology & Critical Care
Paris
and
University Diderot
Paris
France

Mario Perl
Rhode Island Hospital/Brown University
School of Medicine
Department of Surgery
Providence
USA

François Philippart
Institut Pasteur
Unit Cytokines and Inflammation
Paris
France
and
Groupe Hospitalier Paris Saint-Joseph
Paris
France

Raghavan Raju
University of Alabama at Birmingham
Center for Surgical Research and Department of Surgery
Birmingham
USA

Michael C. Reade
Austin Hospital Intensive Care Unit
University of Melbourne
Melbourne
Australia

Lothar Rink
Leibniz-Center for Medicine and Biosciences
Research Center Borstel
Borstel
Germany

Penny L. Sappington
University of Pittsburgh
School of Medicine
Department of Critical Care Medicine
Pittsburgh
USA

Andra Schromm
Leibniz-Center for Medicine and Biosciences
Research Center Borstel
Borstel
Germany

Nitin Seam
National Institutes of Health
Critical Care Medicine Department
Bethesda
USA

Cordula Stamme
Leibniz-Center for Medicine and Biosciences
Research Center Borstel
Borstel
Germany

Anthony F. Suffredini
National Institutes of Health
Critical Care Medicine Department
Bethesda
USA

Ryan Swan
Rhode Island Hospital/Brown University
School of Medicine
Department of Surgery
Providence
USA

Kevin J. Tracey
North Shore-LIJ Research Institute
Manhassset
USA

Jean-Louis Vincent
Erasme University Hospital
Department of Intensive Care
Brussels
Belgium

Doreen E. Wesche-Soldato
Rhode Island Hospital/Brown University
School of Medicine
Department of Surgery
Providence
USA

Ulrich Zähringer
Leibniz-Center for Medicine and Biosciences
Research Center Borstel
Borstel
Germany

PART I
Clinical Aspects of Sepsis and SIRS

Sepsis and Non-infectious Systemic Inflammation. Edited by J.-M. Cavaillon and C. Adrie

ISBN: 978-3-527-31935-0

1
Definition of Sepsis and Non-infectious SIRS

Jean-Louis Vincent

1.1
Introduction

The word "sepsis" has its origins in the word "σήψις", which is the original Greek word for decomposition or putrefaction, and has been used in that context since before Hippocrates [1, 2]. However, although the word, sepsis, has been used for more than 2700 years, it is only relatively recently that we have begun to understand the pathophysiology of sepsis in any depth [3]. With this new insight into the mechanisms underlying sepsis has come the potential for new and improved therapeutic interventions, and simultaneously a realization that the available terminology and definitions of sepsis were confusing and inadequate. In this chapter, I will outline progress in the field of sepsis definitions, and discuss possible approaches for the future.

1.2
Sepsis Syndrome

In 1989, Roger Bone [4] proposed the term "sepsis syndrome", defining it as hypothermia (temperature less than 96 °F (35.5 °C)) or hyperthermia (greater than 101 °F (38.3 °C)), tachycardia (greater than 90 beat/min), tachypnea (greater than 20 breaths/min), clinical evidence of an infection site, and at least one end-organ demonstrating inadequate perfusion or dysfunction. This terminology was somewhat redundant as sepsis was already a known syndrome, and is no longer used, having being replaced by the term "severe sepsis".

1.3
Systemic Inflammatory Response Syndrome

In 1991, a Consensus Conference was held by the American College of Chest Physicians (ACCP) and the Society of Critical Care Medicine (SCCM) to

Sepsis and Non-infectious Systemic Inflammation. Edited by J.-M. Cavaillon and C. Adrie

ISBN: 978-3-527-31935-0

create a "set of definitions that could be applied to patients with sepsis and its sequelae" [5]. The goal of the conference was to provide a "framework" to define the systemic inflammatory response to infection, and by so doing to improve the early diagnosis of sepsis, thus allowing earlier therapeutic intervention. It was realized that the lack of a single definition for sepsis created difficulties in identifying patients, particularly for clinical trials, and it was believed that having a single, universally accepted definition would facilitate ongoing research in this field.

It had been recognized that the same systemic response seen in patients with severe infections, could also occur in patients without infection but with other inflammatory processes, e.g. pancreatitis, multiple trauma, ischemia, burns, etc. and the consensus conference believed it was necessary to introduce new terminology to define such patients. The key aspect of the consensus conference definitions was, therefore, the introduction of the term Systemic Inflammatory Response Syndrome or SIRS to define this phenomenon. SIRS was defined as being the presence of more than one of four clinical criteria:

1. Body temperature greater than 38 °C or less than 36 °C
2. Heart rate greater than 90 beats/min
3. Respiratory rate greater than 20 breaths/min or hyperventilation with a $PaCO_2$ less than 32 mmHg
4. White blood cell count $>12000/mm^3$, $<4000/mm^3$, or with $>10\%$ immature neutrophils.

The combination of SIRS with a confirmed infectious process was then called sepsis. Severe sepsis was defined as sepsis associated with organ dysfunction, hypoperfusion abnormality, or sepsis-induced hypotension, and septic shock was defined as severe sepsis with sepsis-induced hypotension persisting despite adequate fluid resuscitation.

The SIRS approach was rapidly adopted by many and has been widely used to define populations of patients for inclusion in clinical trials. However, not all have considered the SIRS criteria useful, arguing that they are too sensitive and non-specific to be of any real use in clinical diagnosis or in the clinical trial setting [6]. Indeed, most intensive care unit (ICU) patients and many general ward patients meet the SIRS criteria [7–11]; in the recent Sepsis Occurrence in Acutely ill Patients (SOAP) study, 93% of ICU admissions had at least two SIRS criteria at some point during their ICU stay [11]. In addition, a "diagnosis" of SIRS provides no real information regarding the underlying disease process; each of the SIRS criteria can be present in many conditions. For example, fever can be present in sepsis, after myocardial infarction or pulmonary embolism, or post-operatively. Similarly tachycardia and tachypnea may be present in sepsis, but also in heart failure, anemia, respiratory failure, hypovolemia, etc. A raised white blood cell count can be present in many other diseases encountered in ICU patients, including trauma, heart failure, pancreatitis, hemorrhage, and pulmonary edema. The presence of the SIRS criteria generally reflects an appropriate adaptative response to a physiologic insult rather than an abnormality and certainly does not constitute a separate disease entity [12].

1.4 Sepsis Markers

Despite the 1991 ACCP/SCCM consensus conference definitions, a survey of 1058 ICU physicians in 2000 reported that many of the participants remained concerned about the lack of a common definition, with only 22% of intensivists giving the consensus conference definition when asked to define sepsis [13]. With continuing advances in our understanding of sepsis pathophysiology, identification of various proposed sepsis markers, and persistent uncertainty and disagreement about the usefulness of the SIRS criteria, a Sepsis Definitions conference was convened in 2001 to re-evaluate and update definitions (Table 1.1) [14]. The conference included 29 participants from Europe and North America, and was sponsored by several leading intensive care societies.

The participants at this conference concluded that the SIRS criteria were indeed too sensitive and non-specific and that, in preference to the SIRS criteria, an expanded list of signs and symptoms of sepsis should be used to reflect the clinical response to infection (Figure 1.1). However, unfortunately, no marker is 100% specific for sepsis and diagnosis must, at present, rely on the presence of a combination of clinical symptoms and signs and available markers. Various markers have been proposed over the years. Cytokine levels may seem an obvious choice as cytokines are key mediators of the inflammatory response to sepsis. Raised levels of certain cytokines have been well documented in patients with sepsis and some have been correlated with outcome [15–18]. However, no cytokine is specific for sepsis, and not all cytokine levels are raised at all time points during the course of the disease. For example, tumor

Table 1.1 Current definitions of infection and sepsis [14].

Infection	A pathologic process caused by the invasion of normally sterile tissue or fluid by pathogenic or potentially pathogenic microorganisms.
Sepsis	The presence of infection, documented or strongly suspected, with a systemic inflammatory response, as indicated by the presence of some of the features in Figure 1.2.
Severe sepsis	Sepsis complicated by organ dysfunction.
Septic shock	Severe sepsis complicated by acute circulatory failure characterized by persistent arterial hypotension, despite adequate volume resuscitation, and unexplained by other causes.

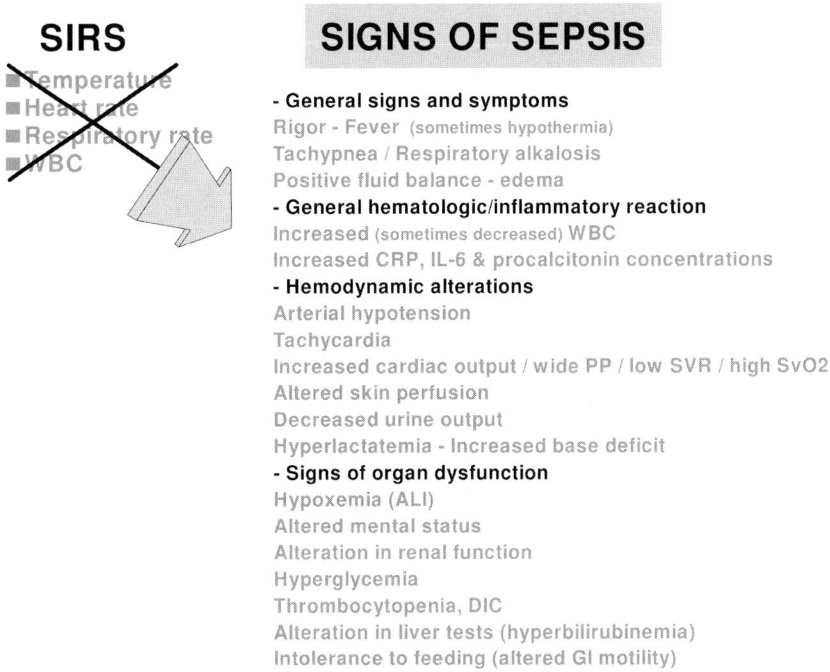

Figure 1.1 Proposed change from SIRS to a longer list of signs for the diagnosis of sepsis [14].

necrosis factor (TNF) levels are raised early in the course of sepsis, but raised levels are also found in other conditions including acute pancreatitis [19], trauma [20], myocardial infarction [21], and heart failure [22], and later in the disease process levels may fall. The same is seen with other cytokines including interleukin (IL)-6, although this is generally the cytokine whose levels are most consistently raised in sepsis. Other markers of inflammation have also been suggested as being of use in the diagnosis of sepsis and some, such as C-reactive protein (CRP), are in common use, particularly in Europe. CRP has been shown to be a useful indicator of the presence of sepsis [23], and more indicative of infection than the white cell count or fever [24]. CRP levels >17 mg/dl have been suggested as providing a means of separating patients with sepsis from those with a non-septic inflammatory response due to trauma [25]. More recently, procalcitonin has been proposed as a marker of infection [26–28], but may be more useful as an indicator of the severity of infection rather than as a marker of the presence of infection *per se* [29]. Procalcitonin levels have been used to guide therapy in patients with lower respiratory tract infections, community-acquired pneumonia, and exacerbations of chronic obstructive pulmonary disease [30–32]; whether they could also be used to guide therapy in general populations of patients with sepsis remains to be determined.

Many other molecules have been suggested as markers of sepsis, but again all are markers of inflammation rather than infection, and none are specific for sepsis. The development of multiplex technology now allows the presence of multiple markers to be detected from a single blood sample, enabling a so-called "sepsis profile" to be constructed for individual patients. Bozza and colleagues recently reported the results of a preliminary study in 60 patients with severe sepsis using a multiplex assay that included the levels of 17 cytokines [33]. These authors were able to identify cytokine profiles associated with an increased risk of early and late death and with evolution of organ dysfunction. Clearly further study is needed but multiplex profiling of ICU patients may offer an effective means of monitoring the severity of sepsis.

Importantly, the list of suggested markers proposed by the participants in the Definitions Conference should be considered as a guide to diagnosis – not all patients with sepsis will have all the markers included on the list, but their presence should be used to raise suspicion of sepsis and to encourage a continued, repeated, or more thorough search for an infectious focus. As research continues and more potential markers are identified, this list may be adapted and expanded.

By these new definitions [14], sepsis is thus defined as the presence of infection plus some of the listed signs and symptoms of sepsis (Table 1.1). Severe sepsis is defined as sepsis complicated by organ dysfunction, and septic shock is severe sepsis with acute circulatory failure characterized by persistent arterial hypotension unexplained by other causes.

1.5 Inflammation versus Infection

Importantly, sepsis, whichever official definition is used, represents the body's systemic response to severe infection, and thus true sepsis can only be present if there is an associated infection (Figure 1.2). It is important to differentiate patients with sepsis from those with SIRS, as treatments may be very different, e.g. antibiotics should not be given to all patients with SIRS, but only to those with clinical and/or bacteriological evidence of infection. However, infection can be difficult to diagnose, particularly in the ICU patient who has already received, or is receiving, antibiotics, and has multiple risk factors for infection and often several concurrent disease processes. In the recent SOAP study, 40% of patients classified as having sepsis had negative culture results [34]. Nevertheless, negative cultures do not necessarily mean that no infection is present, but may just reflect our inability to detect or locate it. Hence, sepsis may be defined as being strongly suspected without microbiological confirmation [14].

Various attempts have been made to improve the diagnosis of infection. The biphasic activated partial thromboplastin time (aPTT) may provide a means of differentiating sepsis from systemic inflammation. Chopin *et al.* [35] recently reported that the biphasic waveform had 90% sensitivity and 92% negative

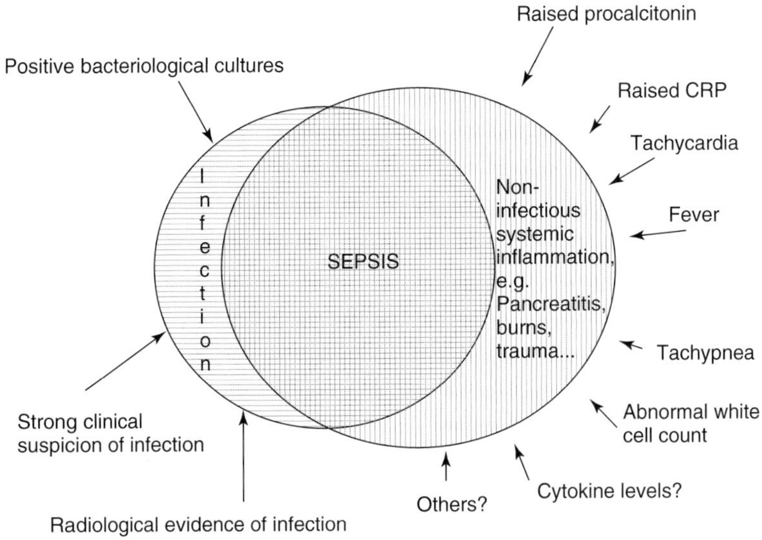

Figure 1.2 Sepsis represents the presence of infection accompanied by a systemic inflammatory response.

predictive value for differentiating severe sepsis and septic shock from SIRS or sepsis. Others have suggested that gene profiling or protein profiling could be used to distinguish SIRS due to infection from non-infectious SIRS. Using mass spectrometry techniques, Lissauer *et al.* recently reported that there were 134 different unique proteins between patients with non-infectious SIRS and those with sepsis [36]. Over 20% of these proteins were related to the complement or coagulation systems. The same group also reported that patients with sepsis had different gene expression profiles compared to patients with non-infectious SIRS; of the 54 613 gene probes evaluated, there were 459 unique gene differences between the two groups of patients, functionally involved in four major areas: innate immunity, cytokine receptors, T cell differentiation, and protein synthesis regulation [37]. Importantly, in both these studies [36, 37], the differences between patients with non-infectious SIRS and those with sepsis were already present before sepsis was clinically apparent.

An alternative approach is the development of scores that combine several variables. Peres Bota *et al.* [38] developed an infection probability score, which uses six variables (body temperature, heart rate, respiratory rate, white blood cell, CRP, and SOFA), to assess the likelihood of infection, resulting in a score from 0–26 (Table 1.2). Infection was absent in 90% of ICU patients when the score was below a cut-off value of 14. A recent prospective study supported the good predictive value of the IPS (Infection Probability Score) for a diagnosis of infection, and suggested that changes in IPS over time may be useful in following the response to antimicrobial therapy [39].

Table 1.2 Infection probability score (IPS) [38].

	IPS points						
	0	**1**	**2**	**3**	**6**	**8**	**12**
BT (°C)	≤37.5		>37.5				
HR (beats/min)	≤81					81–140	>140
RR (breaths/min)	≤25	>25					
WBC ($\times 10^3$/mm^3)	5–12	>12		<5			
CRP (mg/dl)	≤6				>6		
SOFA score	≤5		>5				

BT, body temperature; HR, heart rate; RR, respiratory rate; WBC, white blood cell; CRP, C-reactive protein; SOFA, sequential organ failure assessment.

In addition to improving the diagnosis of infection in general, a consensus conference was organized by the International Sepsis Forum to provide definitions for specific infections [40]. Definitions were developed for the six most frequent causes of infections in septic patients: pneumonia, bloodstream infections (including infective endocarditis), intravascular catheter-related sepsis, intra-abdominal infections, urosepsis, and surgical wound infections. The main aim of these definitions is to facilitate patient selection for inclusion in clinical trials; by classifying patients into prospectively defined infection categories, treatments could be more specifically targeted. However, such definitions could also potentially be used as a framework for guiding diagnostic or therapeutic decisions.

1.6 Conclusion

Clear definitions of sepsis are important clinically in facilitating accurate diagnosis and appropriate treatment. Clear definitions are also important for the purposes of clinical trials, insuring that only patients who do have sepsis are enrolled, reducing the risks of misclassification bias. New markers specific for infection rather than inflammation need to be developed, but currently physicians must rely on the presence of a number of signs and markers of sepsis in their diagnosis; no one variable alone is sufficient. Multiplex technology or combinations of elements in scoring systems are two approaches that can improve the diagnosis of sepsis, but further research is needed in this field. Current definitions may need to be adapted as new techniques or markers are identified that can usefully differentiate between sepsis and non-infectious inflammation.

References

1. Sullivan, R. (**1996**) Thales to Galen: a brief journey through rational medical philosophy in ancient Greece. Part I: pre-Hippocratic medicine. *Proc. R. Coll. Physicians Edinb.*, **26**, 135–142.
2. Geroulanos, S. and Douka, E.T. (**2006**) Historical perspective of the word "sepsis". *Intensive Care Med.*, **32**, 2077.
3. Vincent, J.L. and Abraham, E. (**2006**) The last 100 years of sepsis. *Am. J. Respir. Crit. Care Med.*, **173**, 256–263.
4. Bone, R.C., Fisher, C.J., Clemmer, T.P., Slotman, G.J., Metz, C.A. and Balk, R.A. (**1989**) Sepsis syndrome: A valid clinical entity. *Crit. Care Med.*, **17**, 389–393.
5. ACCP-SCCM Consensus Conference (**1992**) Definitions of sepsis and multiple organ failure and guidelines for the use of innovative therapies in sepsis. *Crit. Care Med.*, **20**, 864–874.
6. Vincent, J.L. (**1997**) Dear Sirs, I'm sorry to say that I don't like you. *Crit. Care Med.*, **25**, 372–374.
7. Pittet, D., Rangel-Frausto, S. and Li, N., *et al.* (**1995**) Systemic inflammatory response syndrome, sepsis, severe sepsis and septic shock: incidence, morbidities and outcomes in surgical ICU patients. *Intensive Care Med.*, **21**, 302–309.
8. Rangel Frausto, M.S., Pittet, D., Costigan, M., Hwang, T., Davis, C.S. and Wenzel, R.P. (**1995**) The natural history of the systemic inflammatory response syndrome (SIRS). A prospective study. *JAMA*, **273**, 117–123.
9. Salvo, I., de Cian, W. and Musicco, M., *et al.* (**1995**) The Italian SEPSIS study: Preliminary results on the incidence and evolution of SIRS, sepsis, severe sepsis and septic shock. *Intensive Care Med.*, **21**, S244–S249.
10. Bossink, A.W., Groeneveld, J., Hack, C.E. and Thijs, L.G. (**1998**) Prediction of mortality in febrile medical patients: How useful are systemic inflammatory response syndrome and sepsis criteria? *Chest*, **113**, 1533–1541.
11. Sprung, C.L., Sakr, Y., Vincent, J.L., *et al.* (**2006**) An evaluation of systemic inflammatory response syndrome signs in the Sepsis Occurrence in Acutely ill Patients (SOAP) study. *Intensive Care Med.*, **32**, 421–427.
12. Marshall, J. (**1997**) Both the disposition and the means of cure: "Severe SIRS," "sterile shock," and the ongoing challenge of description. *Crit. Care Med.*, **25**, 1765–1766.
13. Poeze, M., Ramsay, G., Gerlach, H., Rubulotta, F. and Levy, M. (**2004**) An international sepsis survey: a study of doctors' knowledge and perception about sepsis. *Crit. Care*, **8**, R409–R413.
14. Levy, M.M., Fink, M.P., Marshall, J.C., *et al.* (**2003**) 2001 SCCM/ESICM/ACCP/ATS/SIS International Sepsis Definitions Conference. *Crit. Care Med.*, **31**, 1250–1256.
15. Calandra, T., Baumgartner, J.D., Grau, D.G., *et al.* (**1990**) Prognostic values of tumor necrosis factor/cachectin, interleukin-1, interferon-alpha, and interferon-gamma in the serum of patients with septic shock. *J. Infect. Dis.*, **161**, 982–987.
16. Pinsky, M.R., Vincent, J.L. and Deviere, J. (**1993**) Serum cytokine levels in human septic shock: Relation to multiple-systems organ failure and mortality. *Chest*, **103**, 565–575.
17. Terregino, C.A., Lopez, B.L., Karras, D.J., Killian, A.J. and Arnold, G.K. (**2000**) Endogenous mediators in emergency department patients with presumed sepsis: are levels associated with progression to severe sepsis and death? *Ann. Emerg. Med.*, **35**, 26–34.
18. Gogos, C.A., Drosou, E., Bassaris, H.P. and Skoutelis, A. (**2000**) Pro- versus anti-inflammatory cytokine profile in patients with severe sepsis: A marker for prognosis and future therapeutic options. *J. Infect. Dis.*, **181**, 176–180.
19. Brivet, F.G., Emilie, D. and Galanaud, P. (**1999**) Pro- and anti-inflammatory cytokines during

acute severe pancreatitis: an early and sustained response, although unpredictable of death. Parisian Study Group on Acute Pancreatitis. *Crit. Care Med.*, **27**, 749–755.

20. Ferguson, K.L., Taheri, P., Rodriguez, J., Tonapi, V., Cardellio, A. and Dechert, R. (**1997**) Tumor necrosis factor activity increases in the early response to trauma. *Acad. Emerg. Med.*, **4**, 1035–1040.
21. Neumann, F.J., Ott, I., Gawaz, M., *et al.* (**1995**) Cardiac release of cytokines and inflammatory responses in acute myocardial infarction. *Circulation*, **92**, 748–755.
22. Vonhof, S., Brost, B., Stille-Siegener, M., Grumbach, I.M., Kreuzer, H. and Figulla, H.R. (**1998**) Monocyte activation in congestive heart failure due to coronary artery disease and idiopathic dilated cardiomyopathy. *Int. J. Cardiol.*, **63**, 237–244.
23. Matson, A., Soni, N. and Sheldon, J. (**1991**) C-reactive protein as a diagnostic test of sepsis in the critically ill. *Anaesth. Intensive Care*, **19**, 182–186.
24. Póvoa, P., Almeida, E., Moreira, P., *et al.* (**1998**) C-reactive protein as an indicator of sepsis. *Intensive Care Med.*, **24**, 1052–1056.
25. Miller, P.R., Munn, D.D., Meredith, J.W. and Chang, M.C. (**1999**) Systemic inflammatory response syndrome in the trauma intensive care unit: who is infected? *J. Trauma*, **47**, 1004–1008.
26. Assicot, M., Gendrel, D., Carsin, H., Raymond, J., Guilbaud, J. and Bohoun, C. (**1993**) High serum procalcitonin concentration in patients with sepsis and infection. *Lancet*, **341**, 515–518.
27. Al-Nawas, B., Krammer, I. and Shah, P.M. (**1996**) Procalcitonin in diagnosis of severe infections. *Eur. J. Med. Res.*, **1**, 331–333.
28. Brunkhorst, F.M., Eberhard, O.K. and Brunkhorst, R. (**1999**) Discrimination of infectious and noninfectious causes of early acute respiratory distress syndrome by procalcitonin. *Crit. Care Med.*, **27**, 2172–2176.
29. Ugarte, H., Silva, E., Mercan, D., de Mendonça, A. and Vincent, J.L. (**1999**) Procalcitonin as a marker of infection in the intensive care unit. *Crit. Care Med.*, **27**, 498–504.
30. Christ-Crain, M., Jaccard-Stolz, D., Bingisser, R., *et al.* (**2004**) Effect of procalcitonin-guided treatment on antibiotic use and outcome in lower respiratory tract infections: cluster-randomised, single-blinded intervention trial. *Lancet*, **363**, 600–607.
31. Christ-Crain, M., Stolz, D., Bingisser, R., *et al.* (**2006**) Procalcitonin-guidance of antibiotic therapy in community-acquired pneumonia: A randomized trial. *Am. J. Respir. Crit. Care Med.*, **174**, 84–93.
32. Stolz, D., Christ-Crain, M., Bingisser, R., *et al.* (**2007**) Antibiotic treatment of exacerbations of COPD: a randomized, controlled trial comparing procalcitonin-guidance with standard therapy. *Chest*, **131**, 9–19.
33. Bozza, F.A., Salluh, J.I., Japiassu, A.M., *et al.* (**2007**) Cytokine profiles as markers of disease severity in sepsis: a multiplex analysis. *Crit. Care*, **11**, R49.
34. Vincent, J.L., Sakr, Y., Sprung, C.L., *et al.* (**2006**) Sepsis in European intensive care units: results of the SOAP study. *Crit. Care Med.*, **34**, 344–353.
35. Chopin, N., Floccard, B., Sobas, F., *et al.* (**2006**) Activated partial thromboplastin time waveform analysis: a new tool to detect infection? *Crit. Care Med.*, **34**, 1654–1660.
36. Lissauer, M.E., Johnson, S.B., Siuzdak, G., *et al.* (**2007**) Coagulation and complement protein differences between septic and uninfected systemic inflammatory response syndrome patients. *J. Trauma*, **62**, 1082–1094.
37. Johnson, S.B., Lissauer, M., Bochicchio, G.V., Moore, R.,

Cross, A.S. and Scalea, T.M. (**2007**) Gene expression profiles differentiate between sterile SIRS and early sepsis. *Ann. Surg.*, **245**, 611–621.

38. Peres-Bota, D., Melot, C. and Lopes Ferreira, F. and Vincent, J.L. (**2003**) Infection probability score (IPS): A simple method to help assess the probability of infection in critically patients. *Crit. Care Med.*, **31**, 2579–2584.
39. Martin, A., Gottin, L., Mélot, C. and Vincent, J.L. (**2008**) A prospective evaluation of the Infection Probability Score (IPS) in the intensive care unit. *J. Infect.*, **56**, 313–318.
40. Calandra, T. and Cohen, J. (**2005**) The international sepsis forum consensus conference on definitions of infection in the intensive care unit. *Crit. Care Med.*, **33**, 1538–1548.

2
Epidemiology of Sepsis and Non-infectious SIRS

Michael C. Reade and Derek C. Angus

2.1
Introduction

Sepsis is a common condition, associated with high morbidity and mortality. This chapter examines published attempts to quantify 'common' and 'high', highlighting the methodological difficulties that make definitive estimates elusive. Incidence, prevalence and mortality estimates vary greatly between studies. Reasons for these differences will be explored following a summary of the information available.

2.2
Epidemiology of SIRS: Incidence and Prevalence

The epidemiology of SIRS is poorly understood, principally because lack of specificity in its definition reduces the value of even an accurate estimation. As pointed out in an often-quoted editorial, moderate exercise in a healthy person will precipitate SIRS [1]. A review of published cohort studies found one-third of all hospital patients, more than half all ICU patients, and more than 80% of surgical ICU patients met two or more SIRS criteria [2]. Point prevalence studies or cohort studies that rely on diagnosis at admission may underestimate incidence, as SIRS may only develop after a period of time in the ICU. A study in 99 Italian ICUs found only 52% of 1101 patients had SIRS on admission, but that 85% had at least SIRS at some point during their ICU stay [3]. Studies of the incidence and prevalence of SIRS are summarized in Table 2.1.

2.3
Epidemiology of Sepsis: Incidence and Prevalence

While problems remain with the precise definition of sepsis and in the identification of cases, a large number of epidemiological studies have been

Sepsis and Non-infectious Systemic Inflammation. Edited by J.-M. Cavaillon and C. Adrie

ISBN: 978-3-527-31935-0

published. Summarized in Table 2.1, these fall into three main categories: prospective single- or multi-center observational cohort studies; point prevalence studies; and retrospective analyses of administrative databases. Even allowing for methodological problems, it is apparent that various authors report different statistics. Possible reasons for this are explored below.

2.4 Epidemiology of SIRS and Sepsis: Outcome

Many patients initially diagnosed with sepsis or non-infectious SIRS subsequently progress to septic shock. For example, of patients with two, three and four SIRS criteria, 32, 36 and 45% developed sepsis by day 14 [4]. If attributing mortality to the various grades of systemic inflammatory response, it is important to follow patients over time.

The best measure of mortality in sepsis remains open to question. Most studies quote ICU, hospital or 28-day mortality, principally because these figures are easiest to obtain. Increasingly it has been recognized that attributable mortality in sepsis often occurs long after 28 days [5]. In a prospective observational study of 103 septic patients, only half of those who died from sepsis did so in the first month [6]. In a cohort of 153 patients admitted with sepsis, 30-day mortality was 40.3%, but this had increased to 71.9% at 1 year [7].

Comorbidities increase the risk of developing sepsis [8], and are themselves associated with substantial mortality. For example, a retrospective cohort study in six French ICUs found hospital mortality was greater if severe sepsis developed in the ICU (59%) than if it was present on admission (29%) [9]. The higher mortality of ICU-acquired severe sepsis probably reflects the additional adverse effects of the condition that precipitated admission to the ICU. Many patients may therefore die *with* sepsis rather than *of* sepsis, but this is rarely accounted for in quoted statistics. Crude mortality rates in sepsis ignore this effect, but measures of attributable mortality are rarely made and are at best rough estimates.

2.5 Explanations for the Variability in Estimates of Incidence, Prevalence and Mortality

2.5.1 Problems with the Definitions of Sepsis

Consensus definitions proposed in 1992 [10] have been useful in both clinical trials and patient management, but their ambiguity has hampered epidemiological studies. The original 1992 definitions required that infection

Table 2.1 Studies of the epidemiology of sepsis.

Period of data collection	Region	Type of study	No. of patients[a] or admissions[b] screened	ICU incidence SIRS/sepsis	ICU incidence severe sepsis/septic shock	Mortality SIRS /sepsis (usually 28-day or hospital)	Mortality severe sepsis/ septic shock (usually 28-day or hospital)	Population incidence sepsis	Population incidence severe sepsis	Population mortality sepsis	Population mortality severe sepsis
Apr 92	USA [84]	Prospective cohort	170[a]	93%/49%	16%/7.1%	4.7%/NA	35%/58%	NA	NA	NA	NA
Jan–Feb 93	France [85]	Prospective cohort	11 828[b]	NA/NA	9% of admissions; this represents 742 patients, but the total number of patients screened is not stated/NA	NA/NA	56%/NA	NA	NA	NA	NA
Aug 92–Apr 93	USA [4]	Prospective cohort	3708[a]	68%/26%, plus a further 35% who were culture negative	18% + 21%/4% + 3% (culture positive + culture negative)	10%/16% (culture positive only)	Hospital mortality 16% (culture negative) 20% (culture positive)/46% (culture positive and culture negative combined)	NA	NA	NA	NA

(continued overleaf)

Table 2.1 *(continued)*.

Period of data collection	Region	Type of study	No. of patients[a] or admissions[b] screened	ICU incidence SIRS/ sepsis	ICU incidence severe sepsis/septic shock	Mortality SIRS /sepsis (usually 28-day or hospital)	Mortality severe sepsis/ septic shock (usually 28-day or hospital)	Population incidence sepsis	Population incidence severe sepsis	Population mortality sepsis	Population mortality severe sepsis
Apr 93–Mar 94	Italy [3]	Prospective cohort	1101[a]	61.6%/9.6% on admission; 85.9%/ 27.9% at any time	5.1%/3% on admission; 11.6%/6.1% at any time	26.5%/36%	52.2%/ 81.8%	NA	NA	NA	NA
Jan 93–Apr 94	USA [22]	Prospective cohort	9763[a]	40%/20%	10%/NA	NA/NA	34%/NA	NA	NA	NA	NA
May 97–May 98	Europe, Canada, Israel [62]	Prospective cohort	14 364[b]	NA/37.7% of admissions >24 h	25.4%/14.1% of admissions >24 h	NA/17.0–38.7%	25.5–47.8%/45.7–66.8%	NA	NA	NA	NA
1998–1999	USA [86]	Retrospective database análisis	21 480[a]	NA/NA	11.3%/NA	NA/NA	36%/NA	NA	NA	NA	NA
1993–2000	France [87]	Retrospective database analysis	100 554[b]	NA/NA	8.2%/NA (7.0% in 1993 to 9.7% in 2000)	NA/NA	61.2%/NA	NA	NA	NA	NA
Dec 95–Feb 00	UK excl. Scotland [60]	Retrospective database analysis	56 673[b]	NA/NA	27.1%/NA	NA/NA	47.3%/NA	NA	51 per 100 000	NA	NA

May 99–Jul 99	Australia and New Zealand [13]	Prospective cohort	5878[b]	NA/NA	11.8%/NA	NA/NA	37.5%/NA	NA	77 per 100 000	NA	NA
Nov–Dec 2001	France [35]	Prospective cohort	3738[a]	NA/NA	16.6%/NA	NA/NA	41.9%/NA	NA	95 per 100 000	NA	NA
May 01–Jan 02	Brazil [30]	Prospective cohort	1383[a]; of whom 884 with ICU LOS >24 h	88.8%/ 46.9%	27.3%/23%	24.2%/ 33.9%	46.9%/ 52.2%	NA	NA	NA	NA
Jul–Dec 02	Slovak Republic [88]	Prospective cohort	121[a]	NA/NA	7.9%/NA	NA/NA	51.2%/NA	NA	80–90 per 100 000	NA	NA
Dec 95–Jan 05	UK excluding Scotland [34]	Retrospective database analysis	343 860[b]	NA/NA	27% (23.5% in 1996 to 28.7% in 2004)/NA	NA/NA	48.3% in 1996 to 44.7% in 2004/NA	NA	46 per 100 000 in 1996 to 66 per 100 000 in 2003		23 per 100 000 in 1996 to 30 per 100 000 in 2003
May 2002	24 European countries [14]	Prospective cohort	3147[a]	NA/37%	30%/NA	NA/27%	32.2%/ 54.1%	NA	NA	NA	NA

(continued overleaf)

Table 2.1 *(continued)*.

Period of data collection	Region	Type of study	No of patients[a] or admissions[b] screened	ICU incidence SIRS/sepsis	ICU incidence severe sepsis/septic shock	Mortality SIRS/sepsis (usually 28-day or hospital)	Mortality severe sepsis/septic shock (usually 28-day or hospital)	Population incidence sepsis	Population incidence severe sepsis	Population mortality sepsis	Population mortality severe sepsis
March 1995	Mexico [20]	Cross sectional study	895[a]	49.4%/32.8%	16.5%/6.0%	31.1%/17.8%	45.7%/55.6%	NA	NA	NA	NA
Single day, Dec 2001	Netherlands [21]	Cross sectional study	455[a]	NA/NA	29.5%/11.6%	NA/NA		60 per 100 000	54 per 100 000	NA	NA
Jan–Dec 95	USA [18]	Retro-spective database analysis	6 621 559[b]	NA/NA	Hospital incidence 2.9% ICU incidence 11.2%/NA	NA/NA	28.6%/NA	NA	30 per 100 000	NA	NA
1979–2000	USA [17]	Retro-spective database analysis	750 million[b]	NA/NA	NA/NA	27.8% in 1979–1984 to 17.9% in 1995–2000	15% in those with no organ failure; 70% with 3+ failing organs	82.7 per 100 000 in 1979 to 240.4 per 100 000 in 2000	NA	21.9 per 100 000 in 1979 to 43.9 per 100 000 in 2000	NA
1999	Norway [89]	Retro-spective database analysis	700 107[b]	NA/NA	NA/NA	NA/7.1%	27%/29.3%	149 per 100 000	47 per 100 000	13.5%	27%

				4544 patients had sepsis without organ failure	2121 patients had severe sepsis, and 1562 of these had septic shock						
Jul 99–Jun 2003	Australia [90]	Retro-spective database analysis	3 122 515[b]	NA/Hospital incidence 1.1%, 23.8% of whom received care in an ICU	39% of patients with sepsis had severe sepsis/NA	NA/10.2%	31.1%/NA	166 per 100 000 in 1999–2000 to 194 per 100 000 in 2002–2003	65 per 100 000 in 1999 to 76 per 100 000 in 2002	NA	NA
1993–2003	USA [32]	Retro-spective database analysis	391 571 824[b]	NA/NA	NA/NA	NA/NA	Case fatality rate 45% in 1993 to 37.7% in 2003/NA	NA	64.7 per 100 000 in 1993 to 134.6 per 100 000 in 2003	NA	30.3 per 100 000 in 1993 to 49.7 per 100 000 in 2003

[a,b] In all these figures, classification as having 'sepsis' implies having 'at least sepsis'. i.e. the number quoted includes patients who had severe sepsis and septic shock. This is the approach taken in many, but not all, of the published papers and so has been calculated from the available data where necessary.

be ‘confirmed’ to allow the diagnosis of sepsis, but how this was to be done was not specified [10]. Following this guideline, some studies only included patients with microbiologically documented infection [11]. However a minority of patients with sepsis will have positive cultures [12–14]. The 2001 update to the definitions [15] recognized this problem, broadening ‘sepsis’ to incorporate infection that is ‘strongly suspected, without being microbiologically confirmed’. Obviously the degree of suspicion required might be variously interpreted.

The 1992 definition of sepsis required the presence of one of four SIRS criteria. This requirement was relaxed in 2001, recognizing that the signs of sepsis are protean and that the original criteria were insufficiently sensitive. ‘Severe sepsis’ in the new definition remains sepsis with organ dysfunction; however organ dysfunction can be variably interpreted.

Imprecise definitions are particularly problematic in studies using administrative databases, which may underreport the diagnosis of sepsis by up to 25% [16]. In the United States and elsewhere, administrative databases use the International Classification of Disease, most commonly the 9th revision modified for clinical use (ICD-9 CM). Published epidemiological studies of administrative datasets often rely on imprecise definitions such as ICD-9 CM codes for ‘septicemia’ and ‘bacteremia’ along with separate codes for organ dysfunction [17]. Diagnosis of severe sepsis can be made more sensitive by combining codes for various infections (pneumonia, etc.) and acute organ system dysfunctions [18], but this introduces the possibility of reduced specificity (for example, potentially including patients with mild pneumonia and transient organic psychosis). Unfortunately the 1992 consensus definition adopted in clinical research was imperfectly coded into ICD-9 CM. Sepsis was coded 038.x and severe sepsis 995.92 + 038.x. Septic shock was grouped with other causes of non-traumatic shock into 785.59, but it accounts for only 50% of the cases in this category. Prior to 1992, reimbursement for sepsis care in the United States was less dependent on accurate coding of diagnoses. There is suspicion that the rapid increase in the incidence of severe sepsis after this date may reflect financial motivation to more accurately assign codes. Only in October 2003 were specific codes for sepsis (995.91), severe sepsis (995.92) and septic shock (785.52) introduced. Use of updated codes for severe sepsis has increased rapidly (5324 US Medicare patients in 2003, 75 098 in 2004 and 127 511 in 2005) but the true incidence is still thought to be underreported [19].

2.5.2
Differences in Study Design

Point prevalence studies are the simplest approach to describing the epidemiology of sepsis. One such example is that which found 32.8% of 895 patients in 254 Mexican ICUs had sepsis on a single day in 1995 [20]. Extrapolation of such data to population estimates, such as was undertaken in the Netherlands [21], assumes all patients with sepsis will be in an ICU. Even in the most advanced health care systems this is unlikely to be the case [18].

Prevalence may increase if illness duration increases with better survival, even if incidence falls. Data from point prevalence studies have been used to estimate population incidence [21], but without information on illness duration, these figures are difficult to interpret. Prospective cohort studies in which incidence is directly observed are potentially more accurate. A cohort study of sufficient duration may also overcome problems of seasonal variation. However, extrapolating ICU incidence to population incidence remains flawed if not all patients with sepsis are treated in an ICU. Furthermore, an unrepresentative sample of ICUs may bias the result.

These problems are only overcome by using administrative databases that record data from an entire population or correctly weighted samples thereof. Such an approach relies on accurate coding of disease by personnel entering data for another purpose, usually reimbursement. Problems of definition are particularly important when using administrative databases, as described above. In summary, there are both strengths and weakness to each epidemiological method of quantifying the incidence and prevalence of sepsis, but none is ideal.

2.5.3 Lack of Consistency in Reported Indices

Different studies report different epidemiological indices of sepsis, making comparisons difficult. Some studies quote rates per patient and some per admission, which might include multiple observations on the same patient. Some studies report incidence at the time of ICU admission, some overall incidence at any point in the ICU course, and some both. Only two studies [22, 23] report incidence outside the ICU, one [22] in only a select subset of the hospital population. Unlike the other methodological problems listed in this chapter, there is no technical reason why inconsistent reporting should not be easily overcome.

2.5.4 'Hidden' Sepsis

Many patients dying of infection will never be counted as having 'sepsis'. A palliative approach rather than aggressive treatment is often pursued for many patients with infection, due to the presence of other factors – for example, advanced malignancy or extreme frailty in old age. While such patients have 'sepsis' as defined, they are unlikely to be present in an ICU or perhaps even an acute care hospital. A discussion of the epidemiology of sepsis is therefore really one of 'treated sepsis' [24]. The threshold of eligibility for treatment is almost certainly different in different historical periods and in different countries, with different cultural approaches to end-of-life care, different availability of acute hospital and ICU beds, varying levels of universal health insurance, and other cultural and economic factors [25]. For example, in Spain in 2003 only 32% of patients with severe sepsis were admitted to the ICU [26], compared to 51.1% in the United States [18]. The increasing incidence of sepsis in the United States (described later in this chapter) almost certainly in part reflects the incorporation into the statistics of cases that would previously have been 'hidden', as the resources and expectations of physicians and patients for aggressive management have increased.

Most countries have only quantified the epidemiology of sepsis in their intensive care populations. This must be greatly influenced by the provision in ICU beds in each country. It has been postulated that the high ICU incidence of sepsis in countries such as the UK (27.1%) and Brazil (27.3%) reflects a scarcity of ICU beds, as only the most unwell patients can be admitted [24]. There are 8.6 ICU beds per 100 000 population in the UK compared to 38.4 and 30.5 per 100 000 in France and the United States [27], where the mean ICU incidence of sepsis is 12.4 and 12.6%, respectively. However, Australia, with only 5.2 ICU beds per 100 000 population [28] appears to have a much lower ICU incidence of sepsis (11.8%) than this hypothesis would predict.

2.5.5 Poor and Missing Data

Considerable resources are required to conduct prospective observational cohort studies and to create reliable administrative databases. This is likely to disproportionately affect studies in the developing world, as has been found by reviewers attempting to consolidate available data [29]. Two studies (one from Mexico [20], the other from Brazil [30]) appear to have overcome these difficulties, but represent essentially the only published data which exists outside North America, Europe and Australasia. People die of infection much more in the developing than the developed world [31]. Almost all of these patients will have progressed through the stages of sepsis prior to their deaths. Existing estimates of the epidemiology of sepsis ignore these patients.

2.5.6 Real Differences in Regional Epidemiology of Sepsis

Although the factors listed above appear more than sufficient to explain the wide variations in sepsis incidence, prevalence and mortality, there are probably also real differences in the risk of sepsis in different populations. The SOAP study found dramatic differences in the ICU prevalence (18–73%) and ICU mortality (10–35%) of sepsis in various European countries (Figure 2.1) [14].

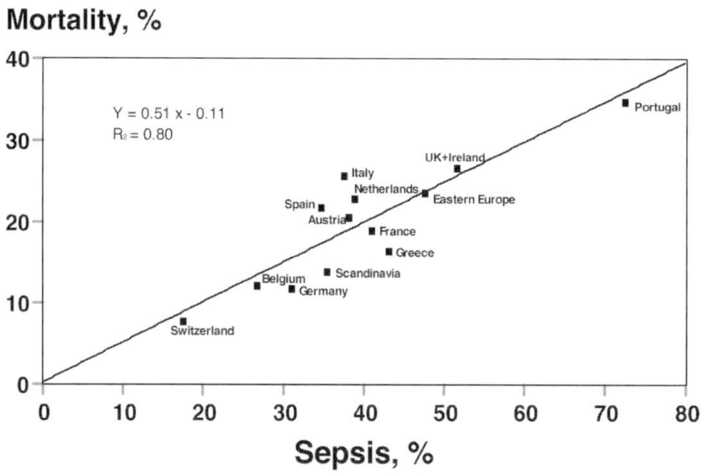

Figure 2.1 Relationship between intensive care unit mortality rates for all patients and frequency of sepsis in the various European countries [14].

The methodology of this single study was consistent across all of the countries. The casemix in the national samples of ICUs studied may have been different, but it is likely that much of the observed variation is real. Presumably much of the variability observed is attributable to different characteristics of the various health services. The risk of nosocomial infection will depend on patterns of hospitalization and utilization of surgery. There may also be differences in susceptibility to sepsis (due to genetic and environmental factors, population age structure and patterns of comorbidity) and types of infectious agents in the various countries studied. The relative importance of such factors has never been explored.

2.6 Conclusions Based on Available Epidemiological Data

Despite the many concerns listed above, some general conclusions can be drawn from the available data.

2.6.1 Trends over Time: Incidence, Mortality

Analysis of the United States Nationwide Inpatient Sample database [32] found that the age-adjusted population incidence of hospitalization for septicemia or SIRS due to an infectious process combined with organ dysfunction (using these ICD-9 CM codes to estimate severe sepsis) increased from 66.0 ± 0.16 to

132.0 ± 0.21 per 100 000 population between 1993 and 2003. The age-adjusted population mortality rate increased from 30.3 ± 0.11% to 49.7 ± 0.13% in the same period. In contrast, the hospital case fatality rate fell from 45.8 ± 0.17 to 37.8 ± 0.10%, in agreement with an earlier comprehensive review of the literature back to 1958 [33]. Of all patients hospitalized with sepsis, the proportion with severe sepsis increased from 25.6% in 1993 to 43.8% in 2003. These trends are summarized in Figure 2.2. Similar trends were observed in England, Wales and Northern Ireland [34] and in France [35].

It is reasonable to infer from these studies that the effectiveness of treatment once in hospital is improving, such that hospital case fatality rate is falling despite an increase in the severity of illness. The population incidence of sepsis is increasing, so despite falling case fatality rates more people are dying of sepsis. Noting age as a major risk factor for developing sepsis, a database study in the United States predicted that, based on the disproportionate growth in the number of elderly people within the population, the rate of hospitalization for severe sepsis would increase by 1.5% per year from 1995 [18]. More recent data [32] therefore reflects growth five times faster than predicted based purely on demographic changes. This may be due to a number of factors. The elderly are increasingly surviving with comorbid disease that in previous times would have been fatal, and comorbidity independently influences susceptibility to sepsis (Figure 2.3) [18]. For example, age-specific hospitalization for pneumonia increased 20% over the 15 years to 2002 [36], which in patients aged <85 years was thought to reflect the higher prevalence of comorbid disease. The prevalence of comorbidities not necessarily related to age, such as HIV infection [37], organ transplantation and aggressive chemotherapy for advanced malignancy [38], has also increased. The reducing proportion of sepsis that is 'hidden' must also account for part of the increased incidence.

A number of hypotheses have been advanced to explain falling case fatality rates. Delivery of critical care services seems to have improved over time, although this has proved difficult to quantify. Monitoring technology continues to improve and the range of available antibiotics is larger, although neither has a clear effect on mortality. Many evidence-based therapeutic interventions have recently been introduced but the trends observed predate these. In the United States at least, the proportion of sepsis that is 'hidden' is probably falling with time, as more patients are treated with curative intent despite advanced age and comorbid disease. At the same time, expansion of ICU bed availability may also lead to admission of less unwell patients, so the overall effect on the average case fatality rate is unclear. Unfortunately there is little published data to support or refute these anecdotal impressions.

2.6.2
Disparities in the Burden of Sepsis: Race and Gender

Males appear to be at greater risk of developing sepsis than females (Figure 2.2). Analysis of the United States National Hospital Discharge Survey found males

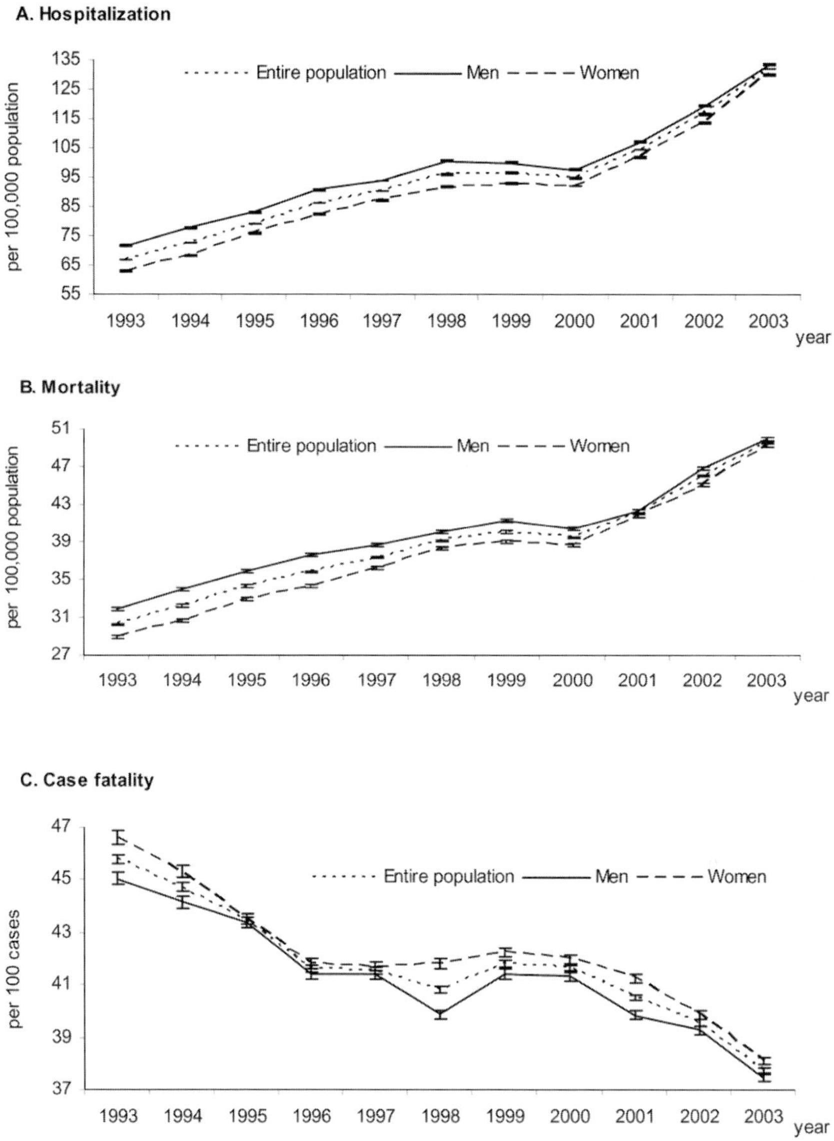

Figure 2.2 Age-adjusted hospitalization (A), mortality (B), and case fatality (C) rates for septicemia or 'SIRS plus infection' and organ dysfunction (the ICD-9 CM codes most closely approximating severe sepsis) in the United States in 1993–2003. Error bars represents the standard error of the age-adjusted rate [32].

had 1.28 times the risk of developing septicemia than females [17]. Significantly more males than females in the United States National Inpatient Sample 1993–2003 had septicemia and 'SIRS due to infection' [32]. Whether the

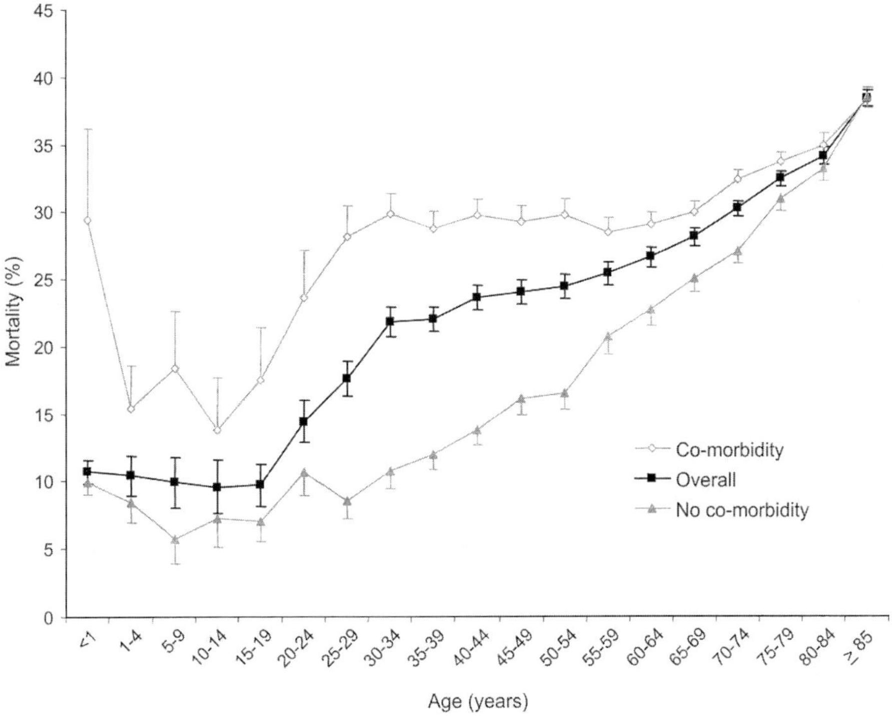

Figure 2.3 Unites States national age-specific mortality rates for all cases of severe sepsis and for those with and without underlying comorbidity. Comorbidity is defined as a Charlson-Deyo score > 0. National estimates are generated from a seven-state cohort using state and national age- and gender-specific population estimates from the National Center for Health Statistics and the U.S. Census. Error bars represent 95% confidence intervals [18].

greater male risk of developing sepsis reflects an increased risk of developing infection, or of progressing to sepsis in the presence of infection, is not known. More males than females in the population died of sepsis, as would be expected from their higher incidence. However case fatality rates were 0.2 to 1.9% greater for females [32]. Female trauma patients in an 18 133-patient trauma registry study were less likely to develop infection, but more likely to die if they did [39]. It should be noted that this conclusion may be misleading if the patient and injury characteristics of the men and women in this study were different, as is likely to be the case. A large single center prospective study found a similar trend to increased mortality in women with infection, which was significant for infections of lung and soft tissue [40]. Other studies have found no sex difference in mortality [17, 41, 42]. Postulated explanations for observed gender differences include both biological causes (for example, differences in inflammatory response [43, 44]) and cultural factors (for example, differences in the provision of critical care services [45] and in the likelihood of having a 'do not resuscitate' order written [46]), but the difficulty in observational studies of properly matching men and women on confounding factors is also likely to be highly influential.

Almost all of the data concerning racial variation in sepsis comes from the United States. Blacks and other non-white races have the highest incidence of septicemia (relative risk compared to whites 1.89 and 1.90) [17]. Case fatality is highest in blacks (22.8%), but interestingly not significantly higher than in whites (22.3%); other races have a significantly lower case fatality rate than both these groups (18.8%) [17]. In 2002, sepsis was one of the 10 leading causes of death for blacks, but not whites, in the United States [47]. Analogous disparities appear to exist elsewhere. The incidence of sepsis in Aboriginal Australians admitted to intensive care was 14.5%, compared to 4.0% for non-indigenous Australians, although adjusted for chronic health status and age, outcome was similar [48].

At least part of the higher incidence of sepsis in blacks in the United States and in Australian Aborigines is likely to be due to the confounding effect of economic disadvantage and differences in the prevalence of comorbid disease. For example, in the 2002 New Jersey state database study, blacks were 3.28 times less likely to have health insurance than whites, and 19.6 times more likely to be infected with HIV [49]. In contrast to what might be expected from this data, however, a study in the United States found whites had slightly inferior quality of care indices than blacks or Hispanics [50]. There is emerging evidence that genetic polymorphisms of inflammatory molecules (for example capsase-12) confer increased risk in populations of African descent [51], suggesting biological as well as social factors may underpin these disparities.

2.6.3 Site of Primary Infection

Pneumonia is the commonest cause of sepsis in almost all studies, and is associated with the highest mortality [42]. The relative importance of pneumonia has increased over time [33]. Women are more likely than men to have genitourinary infections [18, 42], and men are particularly likely to develop pneumonia [42], as are alcoholics [52]. Blacks in the United States are more likely to have Gram-positive infections than whites, but there are no significant racial differences in the sites of infection [42].

2.6.4 Organism

The type of organism responsible for sepsis is also an important determinant of outcome. In one study, Candida sp. and Enterococcus sp. had the highest attributable mortality, and coagulase-negative Staphylococcus the lowest [53]. A very inclusive systematic review of 510 published reports correlated infectious agents with mortality according to site of infection [54]. Overall in this study, Gram-negative bacteremia was associated with a higher mortality than that due to Gram-positive organisms. The commonest bloodstream infections were due to coagulase-negative Staphylococcus and *E. coli*, but these were associated with a relatively low mortality (20 and 19%) compared to Candida (43%) and Acinetobacter (40%) species. Gram-positive pneumonia due to *Staphylococcus aureus* had a higher mortality (41%) than that due to the commonest Gram-negative (*Streptomyces pneumoniae*, 13%), but another Gram-negative, *Pseudomonas aeruginosa*, had the highest mortality of all (77%). This study clearly demonstrated the interaction of organism and site of infection in determining mortality, and called for this to be incorporated into the risk stratification of clinical trials. Such a recommendation must be tempered by the knowledge that a significant proportion (estimated at 47% [23] and 42% [13]) of septic patients never have an infecting organism identified by culture [4]. Before ascribing causative risk to a particular organism, it is also necessary to take into account the confounding effect of the context in which the organism most commonly develops. For example, the association of Acinetobacter with high mortality probably reflects the tendency of Acinetobacter to develop as a nosocomial infection after a prolonged ICU course in patients with many comorbidities. These factors, rather than the pathogenic ability of the organism, may be the explanation for the high associated mortality.

Gram-positive organisms as a cause of sepsis have increased in frequency over time, such that they are now more common than Gram-negative infection [13, 17, 23]. This may be due to greater use of invasive procedures and the increasing proportion of hospital acquired infection [33]. Bacterial resistance to antibiotics has increased over time [55], for reasons that have been recently reviewed [56], including more frequent use of broad-spectrum antibiotics in increasingly unwell patients who remain in the ICU for more prolonged periods. Antibiotic resistance is problematic, prolonging length of stay and duration of mechanical ventilation, although the effect on mortality is uncertain [57–59]. International variations in the implementation of the two main strategies to control resistance (the more rational use of antibiotics and the prevention of cross infection between patients) may explain the dramatically different rates in different countries [56].

2.6.5
Environmental Factors in Risk

Severe sepsis is more common in colder months, both in the United Kingdom (winter 35% higher than summer) [60] and the United States (autumn 17.7% higher than summer) [61]. The case fatality rate for sepsis is also higher in winter, despite similar severity of illness. Respiratory infections have the greatest seasonal change, with their highest incidence in colder months, whereas genitourinary infections are significantly more frequent in summer. That this seasonal variation relates to climate is reflected by the regional differences within the United States: incidence variation is highest in the northeast and lowest in the south.

2.7 Special Populations

Many comorbidities influence susceptibility to and outcome from sepsis. However, some patient populations are worthy of special mention.

2.7.1 Malignancy

Cancer is the most common comorbid condition amongst patients with sepsis [17, 62]. Clinicians are sometimes reluctant to treat aggressively patients with advanced malignancy who develop sepsis, with the thought that recovery from the acute illness will only briefly delay an inevitable early death [63]. Such thinking may be unjustified. Sixty of 124 patients with hematological malignancy admitted to intensive care (mainly for respiratory failure and sepsis) and categorized as 'low risk' on the basis of a low blood urea had a 75% 30-day and 55% 6-month survival, which is not substantially worse than that in other groups [64]. While the risk of developing severe sepsis was 8.7 times higher in hematological malignancy compared with solid tumors, the in-hospital mortality from severe sepsis was similar in each group [65]. Analysis of the subgroup of patients with cancer in the 1979–2001 National Hospital Discharge Survey found cancer of all types increased the risk of developing sepsis 9.8 times. Malignancy increased the risk of sepsis more than any other comorbidity: incidence per 100 000 in patients with cancer was 1075.9, compared to HIV/AIDS 1051.9 [66]. The source of infection was related to the type of cancer; for example lung cancer patients were particularly likely to develop pneumonia. Sepsis contributed to 30% of all hospitalized cancer deaths. Cancer increased the case fatality rate of sepsis by 55%. However this is declining with time (cancer-associated sepsis case fatality rates fell from 44.7% in 1979 to 23.8% in 2001), perhaps due to safer chemotherapy, or maybe just in parallel with the overall improvement in sepsis treatment.

2.7.2 HIV

The epidemiology of sepsis in patients with HIV is changing dramatically with improvements in highly active antiretroviral therapy and *Pneumocystis carinii* prophylaxis. A 2.5-year single center study in the United States

(1993–1996) [67] found 79% of patients with AIDS were admitted for infection, of which 45% was bacterial. This is in contrast to earlier data showing 67–90% of ICU admissions were for respiratory failure, usually due to *Pneumocystis carinii* infection. Sepsis in AIDS patients is increasingly due to multiresistant organisms [68]. A reluctance to admit patients with AIDS to intensive care has been observed [69], presumably with similar justification to that used in malignancy. However, although at increased risk of death from sepsis, patients with AIDS who survive their acute septic episode now have a relatively good prognosis [7].

2.7.3 Children

The epidemiology of sepsis in children has received comparatively little attention. Only one large administrative database study has been published [70], using 1995 hospital discharge data from seven US states. The incidence of sepsis is highest in infants and falls dramatically in older children. Incidence was 15% higher in boys than girls (0.60 versus 0.52 per 1000, $p < 0.001$). Severe sepsis in neonates is dominated by perinatal events, in particular prematurity and low birthweight. Overall, the majority of infections causing severe sepsis were respiratory (37.2%) or primary bacteremia (25.0%), which was particularly common in neonates (41.8% of all causes of severe sepsis in this age group). Sepsis accounted for 7% of deaths in children in 1995, more than cancer.

2.8 Resource Utilization

Treatment of patients with sepsis is expensive. The financial costs of treating patients with sepsis are around six times that for patients with other diagnoses in the ICU [71], reflecting both a higher daily cost and a longer ICU stay. In the United Kingdom between 1996 and 2003, 27% of admissions to intensive care units had severe sepsis, but these patients used 46% of ICU bed days as their ICU length of stay (median 4.6 days for survivors) was longer than that of the average ICU patient [34]. Of note is the preference to quote median rather than mean figures for length of stay, as the data are markedly skewed. A minority of patients with sepsis do not recover quickly, and in suffering complications of both the septic episode and underlying comorbidities require a prolonged ICU stay. For example, while the median length of stay in United States teaching hospital ICUs is 7 days, the mean is 13.8 ± 20.0 days.

In 1993–94, median total charges for patients with severe sepsis ranged from US$42 802 to US$181 758 in eight different hospitals [72]. Costs varied mainly due to differences in length of stay, and remained significantly different even after adjustment for illness severity. The use of various costly therapeutic modalities also varied considerably. After adjustment for illness severity, therapeutic modalities and total resources used did not correlate with outcome [72]. This may reflect the increasingly marginal benefit obtained by use of new and expensive technologies, and highlights the importance of applying simple treatments (such as early administration of antibiotics) in the most effective manner possible.

Analysis of the age-specific financial cost of sepsis reveals an interesting phenomenon: in patients aged >85 years, hospital costs, rate of admission to ICU and ICU length of stay are all less than in younger patients (Figure 2.4) [18]. This almost certainly reflects less aggressive care being afforded to the elderly. While the elderly will always suffer in cost-per-life-year saved calculations for any therapy, many elderly patients are able to return to their previously acceptable standard of living after intensive care [73] leading some to suggest that age alone should not determine the aggressiveness of care offered.

The cost of treating patients with sepsis can be quantified in terms of staff workload as well as financial cost. A French cohort study found the Omega score (a weighted sum of ICU procedures) higher in ICU patients with severe sepsis than in those without [9]. Treatment cost (calculated based on the Omega score) was twice as high in ICU patients with severe sepsis. Patients with peritonitis (compared to infection at other sites) required the greatest intensity of therapeutic intervention, as did older patients, those who had undergone emergency surgery, those with septic shock, those with a higher APACHE II score and those with hospital acquired infection.

An economic analysis of sepsis limited to direct acute care hospital costs is incomplete. Unfortunately there is very little data available regarding costs of healthcare following discharge from an acute hospital. Impairment of functional status persists long after discharge in many ICU patients [6, 74], and over time more people are requiring post-hospital inpatient care [17]. Sepsis affects patients who are still of an economically productive age, and productivity loss due to premature death was found to account for 72% of the total cost of severe sepsis [75].

The estimates of the financial costs of sepsis quoted above unfortunately make no distinction between community-acquired and nosocomial sepsis. The costs incurred by a patient admitted with the primary diagnosis of community-acquired sepsis are all attributable to sepsis. However, separating the costs for a surgical patient who subsequently develops nosocomial infection into sepsis and non-sepsis components is difficult, and has not been attempted in large epidemiological studies.

2.9 Prevention, Early Intervention, and Systems of Care

ICU management is only a small part of the impact of a health care system on the incidence and severity of sepsis. For example, early,

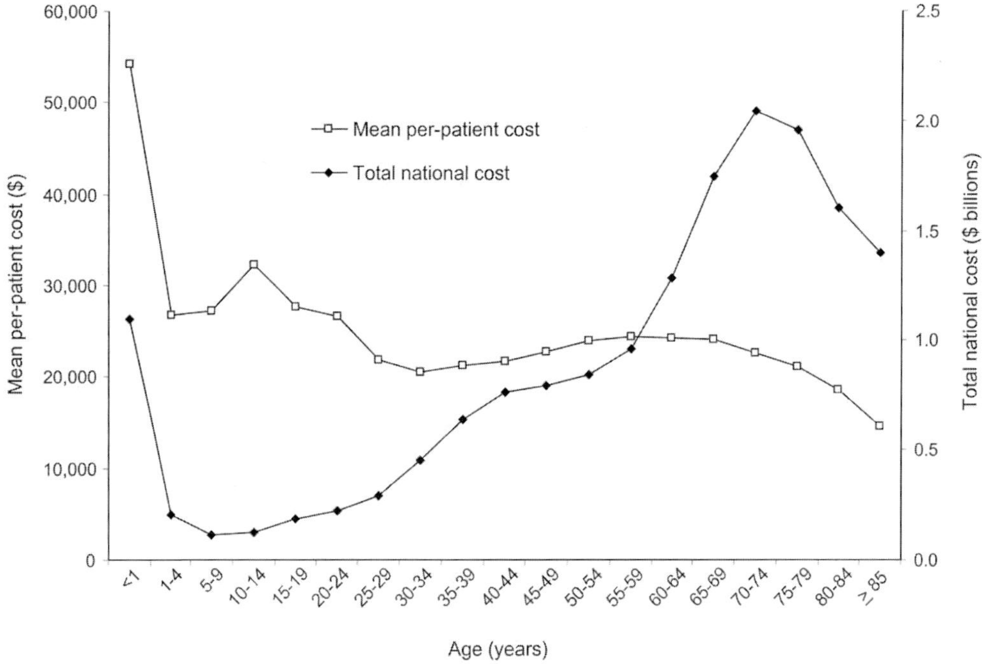

Figure 2.4 Unites States national age-specific average and total hospital costs for severe sepsis. Costs are calculated by multiplying total hospital charges by the hospital-specific cost-to-charge ratio derived from the Health Care Financing Administration Provider Specific File. All costs are expressed as 1995 U.S. dollars. National estimates are generated from the seven-state cohort using state and national age- and gender-specific population estimates from the National Center for Health Statistics and the U.S. Census [18].

physiological-goal-directed aggressive care in the emergency department during the first 6 hours after presentation with severe sepsis or septic shock reduced hospital mortality in Detroit, USA from 46.5 to 30.5% [76]. The impact of such interventions appears to be highly dependent on the existing primary and hospital care systems. Only 1% of patients presenting to a major Australian teaching hospital between 2000 and 2003 with an infectious diagnosis had severe sepsis or septic shock [77], compared to 3.4% in the United Kingdom [78] and anecdotally an even greater number in Detroit, USA [79]. This may reflect the impact of free and rapid access to early primary care in Australia [77]. The hospital mortality of patients with severe sepsis in the Australian study was considerably lower (30.2%) than that in the United Kingdom (43%) and in the control arm of the Early Goal Directed Therapy trial (46.5%) [76]. This might reflect a persistent benefit of earlier pre-hospital care in Australia, or perhaps some the unique features of ICU care there [77, 80].

2.10
Outcome other than Mortality

Morbidity due to sepsis is difficult to quantify. Morbidity during ICU and hospital stay is measured using indices such as organ dysfunction scores and duration of organ system support. The relevance of such measures to the patient is questionable, as most will have been unconscious or heavily sedated whilst undergoing such therapy [81]. While important in quantifying costs, measures of morbidity have much more meaning to patients after their discharge from the intensive care unit and hospital. Only recently have long term outcomes other than mortality been studied. As the cost of aggressive treatment of sepsis becomes increasingly unsustainable, analysis of quality of life in survivors will need to better inform health care policy.

Patients who are treated in intensive care for sepsis are increasingly frail. In the United States in 1979, 78.5% of patients who survived their admission for septicemia were discharged home. This decreased to 56.4% in 2000, while the proportion discharged to other health care facilities (such as rehabilitation and long-term care facilities) increased from 16.8 to 31.8% [17]. A number of factors probably contributed to this trend, including increased availability of such long-term care facilities, increasing age and comorbidity of the patients, and decreasing hospital mortality leading to the greater survival of more frail patients. A follow-up study of patients initially identified in a clinical trial found that survivors of sepsis had significant ongoing physical limitations, including problems with work and activities of daily living, pain, and perceived general health. Interestingly, mental health and vitality were reported to be higher than the general population [6], highlighting the need to incorporate subjective as well as objective measures into studies of morbidity. An observational follow-up study of 30 sepsis survivors in Canada found sepsis associated with poorer reported physical functioning, general health, vitality and social functioning than average population values [74]. The impact of morbidity in sepsis survivors is not only felt by the patient. Depression, lifestyle disruption, and employment reduction were common among informal caregivers of critical illness survivors [82].

A criticism of expensive new sepsis therapies with marginal mortality benefit is their potential to only increase the number of patients who survive with an unacceptable quality of life. The incorporation of quality of life measures as outcomes in clinical trials is therefore an encouraging development. Patients enrolled in a clinical trial of antithrombin III were evaluated using a quality of life questionnaire up to 90 days after enrolment. One subgroup of patients who received the trial drug (those who did not also receive heparin) recorded better outcomes in social and psychologic functioning [83]. This paper is not so much interesting for the conclusion of its subgroup analysis but rather for the fact that it represents the first attempt to incorporate quality of life indices as an outcome in a pharmaceutical trial.

2.11
Implications for Planning of Health Services

The changing epidemiology of sepsis will require changes in the health care systems of developed countries. The incidence is sepsis is increasing [32], partly but not exclusively due to the increasing mean age of the population. Patients are older, with more comorbidities, and consume greater hospital resources. They are more likely to survive their acute illness episode, but also more likely to require long-term care, and to suffer long-term disability [17]. The considerable variation in the resources used in treating sepsis without a demonstrable effect on outcome [72] suggests it may be possible to contain costs without affecting mortality. However in the United States in particular this will require a different paradigm of care, where intensive care resources are reserved for patients who have a reasonable chance of returning to a quality of life that they would deem acceptable.

References

1. Vincent, J.L. (**1997**) Dear SIRS, I'm sorry to say that I don't like you. *Crit. Care Med.*, **25**, 372–374.
2. Brun-Buisson, C. (**2000**) The epidemiology of the systemic inflammatory response. *Intensive Care Med.*, **26**(Suppl 1), S64–S74.
3. Salvo, I., de Cian, W., Musicco, M., Langer, M., Piadena, R., Wolfler, A., Montani, C. and Magni, E. (**1995**) The Italian SEPSIS study: preliminary results on the incidence and evolution of SIRS, sepsis, severe sepsis and septic shock. *Intensive Care Med.*, **21**(Suppl 2), S244–S249.
4. Rangel-Frausto, M.S., Pittet, D., Costigan, M., Hwang, T., Davis, C.S. and Wenzel, R.P. (**1995**) The natural history of the systemic inflammatory response syndrome (SIRS). A prospective study. *JAMA*, **273**, 117–123.
5. Cohen, J., Guyatt, G., Bernard, G.R., Calandra, T., Cook, D., Elbourne, D., Marshall, J., Nunn, A., *et al.* (**2001**) New strategies for clinical trials in patients with sepsis and septic shock. *Crit. Care Med.*, **29**, 880–886.
6. Perl, T.M., Dvorak, L., Hwang, T. and Wenzel, R.P. (**1995**) Long-term survival and function after suspected gram-negative sepsis. *JAMA*, **274**, 338–345.
7. Sasse, K.C., Nauenberg, E., Long, A., Anton, B., Tucker, H.J. and Hu, T.W. (**1995**) Long-term survival after intensive care unit admission with sepsis. *Crit. Care Med.*, **23**, 1040–1047.
8. Laupland, K.B., Gregson, D.B., Zygun, D.A., Doig, C.J., Mortis, G. and Church, D.L. (**2004**) Severe bloodstream infections: a population-based assessment. *Crit. Care Med.*, **32**, 992–997.
9. Adrie, C., Alberti, C., Chaix-Couturier, C., Azoulay, E., De Lassence, A., Cohen, Y., Meshaka, P., Cheval, C., *et al.* (**2005**) Epidemiology and economic evaluation of severe sepsis in France: age, severity, infection site, and place of acquisition (community, hospital, or intensive care unit) as determinants of workload and cost. *J. Crit. Care*, **20**, 46–58.
10. Bone, R.C., Balk, R.A., Cerra, F.B., Dellinger, R.P., Fein, A.M., Knaus, W.A., Schein, R.M. and Sibbald, W.J. (**1992**) Definitions for sepsis and organ failure and guidelines for the use of innovative therapies in sepsis. The ACCP/SCCM Consensus Conference Committee.

American College of Chest Physicians/Society of Critical Care Medicine. *Chest*, **101**, 1644–1655.
11. Pittet, D., Thievent, B., Wenzel, R.P., Li, N., Auckenthaler, R. and Suter, P.M. (**1996**) Bedside prediction of mortality from bacteremic sepsis. A dynamic analysis of ICU patients. *Am. J. Respir. Crit. Care Med.*, **153**, 684–693.
12. Kieft, H., Hoepelman, A.I., Zhou, W., Rozenberg-Arska, M., Struyvenberg, A. and Verhoef, J. (**1993**) The sepsis syndrome in a Dutch university hospital. Clinical observations. *Arch. Intern. Med.*, **153**, 2241–2247.
13. Finfer, S., Bellomo, R., Lipman, J., French, C., Dobb, G. and Myburgh, J. (**2004**) Adult-population incidence of severe sepsis in Australian and New Zealand intensive care units. *Intensive Care Med.*, **30**, 589–596.
14. Vincent, J.L., Sakr, Y., Sprung, C.L., Ranieri, V.M., Reinhart, K., Gerlach, H., Moreno, R., Carlet, J., *et al.* (**2006**) Sepsis in European intensive care units: results of the SOAP study. *Crit. Care Med.*, **34**, 344–353.
15. Levy, M.M., Fink, M.P., Marshall, J.C., Abraham, E., Angus, D., Cook, D., Cohen, J., Opal, S.M., *et al.* (**2003**) 2001 SCCM/ESICM/ACCP/ATS/SIS International Sepsis Definitions Conference. *Crit. Care Med.*, **31**, 1250–1256.
16. Ollendorf, D.A., Fendrick, A.M., Massey, K., Williams, G.R. and Oster, G. (**2002**) Is sepsis accurately coded on hospital bills? *Value Health*, **5**, 79–81.
17. Martin, G.S., Mannino, D.M., Eaton, S. and Moss, M. (**2003**) The epidemiology of sepsis in the United States from 1979 through 2000. *N. Engl. J. Med.*, **348**, 1546–1554.
18. Angus, D.C., Linde-Zwirble, W.T., Lidicker, J., Clermont, G., Carcillo, J. and Pinsky, M.R. (**2001**) Epidemiology of severe sepsis in the United States: analysis of incidence, outcome, and associated costs of care. *Crit. Care Med.*, **29**, 1303–1310.
19. Linde-Zwirble, W.T., Cooper, L.M., Corp, C., Sample, G.A. and Angus, D.C. (**2006**) How have hospitals adapted to the new severe sepsis ICD-9-CM codes? Analysis of US Medicare data. *Crit. Care Med.*, **34**, A135.
20. Ponce de Leon-Rosales S.P., Molinar-Ramos, F., Dominguez-Cherit, G., Rangel-Frausto, M.S. and Vazquez-Ramos, V.G. (**2000**) Prevalence of infections in intensive care units in Mexico: a multicenter study. *Crit. Care Med.*, **28**, 1316–1321.
21. van Gestel, A., Bakker, J., Veraart, C.P. and van Hout, B.A. (**2004**) Prevalence and incidence of severe sepsis in Dutch intensive care units. *Crit. Care*, **8**, R153–R162.
22. Sands, K.E., Bates, D.W., Lanken, P.N., Graman, P.S., Hibberd, P.L., Kahn, K.L., Parsonnet, J., Panzer, R., *et al.* (**1997**) Epidemiology of sepsis syndrome in 8 academic medical centers. *JAMA*, **278**, 234–240.
23. Brun-Buisson, C., Doyon, F. and Carlet, J. (**1996**) Bacteremia and severe sepsis in adults: a multicenter prospective survey in ICUs and wards of 24 hospitals. French Bacteremia–Sepsis Study Group. *Am. J. Respir. Crit Care Med.*, **154**, 617–624.
24. Linde-Zwirble, W.T. and Angus, D.C. (**2004**) Severe sepsis epidemiology: sampling, selection, and society. *Crit. Care*, **8**, 222–226.
25. Angus, D.C., Pereira, C.A. and Silva, E. (**2006**) Epidemiology of severe sepsis around the world. *Endocr. Metab. Immune. Disord. Drug Targets*, **6**, 207–212.
26. Esteban, A., Frutos-Vivar, F., Ferguson, N.D., Penuelas, O., Lorente, J.A., Gordo, F., Honrubia, T., Algora, A., *et al.* (**2007**) Sepsis incidence and outcome: Contrasting the intensive care unit with the hospital ward. *Crit. Care Med.*, **35**, 1284–1289.
27. Angus, D.C., Sirio, C.A., Clermont, G. and Bion, J. (**1997**) International

comparisons of critical care outcome and resource consumption. *Crit. Care Clin.*, **13**, 389–407.
28. Traub, M., Bradt, D.A. and Joseph, A.P. (**2007**) The Surge Capacity for People in Emergencies (SCOPE) study in Australasian hospitals. *Med. J. Aust.*, **186**, 394–398.
29. Jaimes, F. (**2005**) A literature review of the epidemiology of sepsis in Latin America. *Rev. Panam. Salud. Publica.*, **18**, 163–171.
30. Silva, E., Pedro, M.A., Sogayar, A.C., Mohovic, T., Silva, C.L., Janiszewski, M., Cal, R.G., de Sousa, E.F., *et al.* (**2004**) Brazilian Sepsis Epidemiological Study (BASES study). *Crit. Care*, **8**, R251–R260.
31. Lopez, A.D., Mathers, C.D., Ezzati, M., Jamison, D.T. and Murray, C.J. (**2006**) Global and regional burden of disease and risk factors, 2001: systematic analysis of population health data. *Lancet*, **367**, 1747–1757.
32. Dombrovskiy, V.Y., Martin, A.A., Sunderram, J. and Paz, H.L. (**2007**) Rapid increase in hospitalization and mortality rates for severe sepsis in the United States: A trend analysis from 1993 to 2003. *Crit. Care Med.*, **35**, 1244–1250.
33. Friedman, G., Silva, E. and Vincent, J.L. (**1998**) Has the mortality of septic shock changed with time. *Crit. Care Med.*, **26**, 2078–2086.
34. Harrison, D.A., Welch, C.A. and Eddleston, J.M. (**2006**) The epidemiology of severe sepsis in England, Wales and Northern Ireland, 1996 to 2004: secondary analysis of a high quality clinical database, the ICNARC Case Mix Programme Database. *Crit. Care*, **10**, R42.
35. Brun-Buisson, C., Meshaka, P., Pinton, P. and Vallet, B. (**2004**) EPISEPSIS: a reappraisal of the epidemiology and outcome of severe sepsis in French intensive care units. *Intensive Care Med.*, **30**, 580–588.
36. Fry, A.M., Shay, D.K., Holman, R.C., Curns, A.T. and Anderson, L.J. (**2005**) Trends in hospitalizations for pneumonia among persons aged 65 years or older in the United States, 1988–2002. *JAMA*, **294**, 2712–2719.
37. Karon, J.M., Fleming, P.L., Steketee, R.W. and De Cock, K.M. (**2001**) HIV in the United States at the turn of the century: an epidemic in transition. *Am. J. Public Health*, **91**, 1060–1068.
38. Earle, C.C., Neville, B.A., Landrum, M.B., Ayanian, J.Z., Block, S.D. and Weeks, J.C. (**2004**) Trends in the aggressiveness of cancer care near the end of life. *J. Clin. Oncol.*, **22**, 315–321.
39. Croce, M.A., Fabian, T.C., Malhotra, A.K., Bee, T.K. and Miller, P.R. (**2002**) Does gender difference influence outcome? *J. Trauma*, **53**, 889–894.
40. Crabtree, T.D., Pelletier, S.J., Gleason, T.G., Pruett, T.L. and Sawyer, R.G. (**1999**) Gender-dependent differences in outcome after the treatment of infection in hospitalized patients. *JAMA*, **282**, 2143–2148.
41. Wichmann, M.W., Inthorn, D., Andress, H.J. and Schildberg, F.W. (**2000**) Incidence and mortality of severe sepsis in surgical intensive care patients: the influence of patient gender on disease process and outcome. *Intensive Care Med.*, **26**, 167–172.
42. Esper, A.M., Moss, M., Lewis, C.A., Nisbet, R., Mannino, D.M. and Martin, G.S. (**2006**) The role of infection and comorbidity: Factors that influence disparities in sepsis. *Crit. Care Med.*, **34**, 2576–2582.
43. Asai, K., Hiki, N., Mimura, Y., Ogawa, T., Unou, K. and Kaminishi, M. (**2001**) Gender differences in cytokine secretion by human peripheral blood mononuclear cells: role of estrogen in modulating LPS-induced cytokine secretion in an ex vivo septic model. *Shock*, **16**, 340–343.
44. van Eijk, L.T., Dorresteijn, M.J., Smits, P., van der Hoeven, J.G., Netea, M.G. and Pickkers, P. (**2007**) Gender differences in the innate immune response and vascular

reactivity following the administration of endotoxin to human volunteers. *Crit. Care Med.*, **35**, 1464–1469.

45. Valentin, A., Jordan, B., Lang, T., Hiesmayr, M. and Metnitz, P.G. (**2003**) Gender-related differences in intensive care: a multiple-center cohort study of therapeutic interventions and outcome in critically ill patients. *Crit. Care Med.*, **31**, 1901–1907.
46. Eachempati, S.R., Hydo, L., Shou, J. and Barie, P.S. (**2006**) Sex differences in creation of do-not-resuscitate orders for critically ill elderly patients following emergency surgery. *J. Trauma*, **60**, 193–197.
47. Centers for Disease Control and Prevention (**2005**) Health disparities experienced by black or African Americans – United States. *MMWR Morb. Mortal Wkly. Rep.*, **54**, 1–3.
48. Ho, K.M., Finn, J., Dobb, G.J. and Webb, S.A. (**2006**) The outcome of critically ill Indigenous patients. *Med. J. Aust.*, **184**, 496–499.
49. Dombrovskiy, V.Y., Martin, A.A., Sunderram, J. and Paz, H.L. (**2007**) Occurrence and outcomes of sepsis: influence of race. *Crit. Care Med.*, **35**, 763–768.
50. Asch, S.M., Kerr, E.A., Keesey, J., Adams, J.L., Setodji, C.M., Malik, S. and McGlynn, E.A. (**2006**) Who is at greatest risk for receiving poor-quality health care? *N. Engl. J. Med.*, **354**, 1147–1156.
51. Saleh, M., Vaillancourt, J.P., Graham, R.K., Huyck, M., Srinivasula, S.M., Alnemri, E.S., Steinberg, M.H., Nolan, V., *et al.* (**2004**) Differential modulation of endotoxin responsiveness by human caspase-12 polymorphisms. *Nature*, **429**, 75–79.
52. Moss, M. (**2005**) Epidemiology of sepsis: race, sex, and chronic alcohol abuse. *Clin. Infect. Dis.*, **41**(Suppl 7), S490–S497.
53. Rangel-Frausto, M.S. (**1999**) The epidemiology of bacterial sepsis. *Infect. Dis. Clin. North Am.*, **13**, 299–312, vii.
54. Cohen, J., Cristofaro, P., Carlet, J. and Opal, S. (**2004**) New method of classifying infections in critically ill patients. *Crit. Care Med.*, **32**, 1510–1526.
55. Reacher, M.H., Shah, A., Livermore, D.M., Wale, M.C., Graham, C., Johnson, A.P., Heine, H., Monnickendam, M.A., *et al.* (**2000**) Bacteraemia and antibiotic resistance of its pathogens reported in England and Wales between 1990 and 1998: trend analysis. *BMJ*, **320**, 213–216.
56. Carlet, J., Ben Ali, A. and Chalfine, A. (**2004**) Epidemiology and control of antibiotic resistance in the intensive care unit. *Curr. Opin. Infect. Dis.*, **17**, 309–316.
57. Peres-Bota, D., Rodriguez, H., Dimopoulos, G., DaRos, A., Melot, C., Struelens, M.J. and Vincent, J.L. (**2003**) Are infections due to resistant pathogens associated with a worse outcome in critically ill patients? *J. Infect.*, **47**, 307–316.
58. Gleason, T.G., Crabtree, T.D., Pelletier, S.J., Raymond, D.P., Karchmer, T.B., Pruett, T.L. and Sawyer, R.G. (**1999**) Prediction of poorer prognosis by infection with antibiotic-resistant gram-positive cocci than by infection with antibiotic-sensitive strains. *Arch. Surg.*, **134**, 1033–1040.
59. Garcia-Garmendia, J.L., Ortiz-Leyba, C., Garnacho-Montero, J., Jimenez-Jimenez, F.J., Monterrubio-Villar, J. and Gili-Miner, M. (**1999**) Mortality and the increase in length of stay attributable to the acquisition of Acinetobacter in critically ill patients. *Crit. Care Med.*, **27**, 1794–1799.
60. Padkin, A., Goldfrad, C., Brady, A.R., Young, D., Black, N. and Rowan, K. (**2003**) Epidemiology of severe sepsis occurring in the first 24 hrs in intensive care units in England, Wales, and Northern Ireland. *Crit. Care Med.*, **31**, 2332–2338.
61. Danai, P.A., Sinha, S., Moss, M., Haber, M.J. and Martin, G.S. (**2007**) Seasonal variation in the epidemiology of sepsis. *Crit. Care Med.*, **35**, 410–415.
62. Alberti, C., Brun-Buisson, C., Burchardi, H., Martin, C., Goodman, S., Artigas, A.,

Sicignano, A., Palazzo, M., *et al.* (**2002**) Epidemiology of sepsis and infection in ICU patients from an international multicentre cohort study. *Intensive Care Med.*, **28**, 108–121.
63. Klastersky, J. (**1995**) Intensive care support in oncology: a contradiction? *Support. Care Cancer*, **3**, 88.
64. Benoit, D.D., Vandewoude, K.H., Decruyenaere, J.M., Hoste, E.A. and Colardyn, F.A. (**2003**) Outcome and early prognostic indicators in patients with a hematologic malignancy admitted to the intensive care unit for a life-threatening complication. *Crit. Care Med.*, **31**, 104–112.
65. Williams, M.D., Braun, L.A., Cooper, L.M., Johnston, J., Weiss, R.V., Qualy, R.L. and Linde-Zwirble, W. (**2004**) Hospitalized cancer patients with severe sepsis: analysis of incidence, mortality, and associated costs of care. *Crit. Care*, **8**, R291–R298.
66. Danai, P.A., Moss, M., Mannino, D.M. and Martin, G.S. (**2006**) The epidemiology of sepsis in patients with malignancy. *Chest.*, **129**, 1432–1440.
67. Rosenberg, A.L., Seneff, M.G., Atiyeh, L., Wagner, R., Bojanowski, L. and Zimmerman, J.E. (**2001**) The importance of bacterial sepsis in intensive care unit patients with acquired immunodeficiency syndrome: implications for future care in the age of increasing antiretroviral resistance. *Crit. Care Med.*, **29**, 548–556.
68. Proctor, R.A. (**2001**) Bacterial sepsis in patients with acquired immunodeficiency syndrome. *Crit. Care Med.*, **29**, 683–684.
69. Timsit, J.F. (**2005**) Open the intensive care unit doors to HIV-infected patients with sepsis. *Crit. Care*, **9**, 629–630.
70. Watson, R.S., Carcillo, J.A., Linde-Zwirble, W.T., Clermont, G., Lidicker, J. and Angus, D.C. (**2003**) The epidemiology of severe sepsis in children in the United States. *Am. J. Respir. Crit. Care Med.*, **167**, 695–701.
71. Edbrooke, D.L., Hibbert, C.L., Kingsley, J.M., Smith, S., Bright, N.M. and Quinn, J.M. (**1999**) The patient-related costs of care for sepsis patients in a United Kingdom adult general intensive care unit. *Crit. Care Med.*, **27**, 1760–1767.
72. Yu, D.T., Black, E., Sands, K.E., Schwartz, J.S., Hibberd, P.L., Graman, P.S., Lanken, P.N., Kahn, K.L., *et al.* (**2003**) Severe sepsis: variation in resource and therapeutic modality use among academic centers. *Crit. Care*, **7**, R24–R34.
73. Marik, P.E. (**2007**) Should age limit admission to the intensive care unit? *Am. J. Hosp. Palliat. Care*, **24**, 63–66.
74. Heyland, D.K., Hopman, W., Coo, H., Tranmer, J. and McColl, M.A. (**2000**) Long-term health-related quality of life in survivors of sepsis. Short Form 36: a valid and reliable measure of health-related quality of life. *Crit. Care Med.*, **28**, 3599–3605.
75. Schmid, A., Schneider, H., Adlof, A., Smolle, K.H., Edelmann, G., Sporn, P., Frass, M., Sumann, G., *et al.* (**2002**) Economic burden of illness imposed by severe sepsis in Austria. *Wien. Klin. Wochenschr.*, **114**, 697–701.
76. Rivers, E., Nguyen, B., Havstad, S., Ressler, J., Muzzin, A., Knoblich, B., Peterson, E. and Tomlanovich, M. (**2001**) Early goal-directed therapy in the treatment of severe sepsis and septic shock. *N. Engl. J. Med.*, **345**, 1368–1377.
77. Ho, B.C., Bellomo, R., McGain, F., Jones, D., Naka, T., Wan, L. and Braitberg, G. (**2006**) The incidence and outcome of septic shock patients in the absence of early-goal directed therapy. *Crit. Care*, **10**, R80.
78. Reuben, A.D., Appelboam, A.V. and Higginson, I. (**2006**) The outcomes of severe sepsis and septic shock in the UK. *Crit. Care*, **10**, 417.
79. Rivers, E. (**2006**) The outcome of patients presenting to the emergency department with severe sepsis or septic shock. *Crit. Care*, **10**, 154.
80. Bellomo, R., Stow, P.J. and Hart, G.K. (**2007**) Why is there such a difference in outcome between Australian

intensive care units and others? *Curr. Opin. Anaesthesiol.*, **20**, 100–105.

81. Vincent, J.L. (**2004**) Endpoints in sepsis trials: more than just 28-day mortality? *Cri. Care Med.*, **32**, S209–S213.
82. Van Pelt, D.C., Milbrandt, E.B., Qin, L., Weissfeld, L.A., Rotondi, A.J., Schulz, R., Chelluri, L., Angus, D.C., *et al.* (**2007**) Informal caregiver burden among survivors of prolonged mechanical ventilation. *Am. J. Respir. Crit. Care Med.*, **175**, 167–173.
83. Rublee, D., Opal, S.M., Schramm, W., Keinecke, H.O. and Knaub, S. (**2002**) Quality of life effects of antithrombin III in sepsis survivors: results from the KyberSept trial [ISRCTN22931023]. *Crit. Care*, **6**, 349–356.
84. Pittet, D., Rangel-Frausto, S., Li, N., Tarara, D., Costigan, M., Rempe, L., Jebson, P. and Wenzel, R.P. (**1995**) Systemic inflammatory response syndrome, sepsis, severe sepsis and septic shock: incidence, morbidities and outcomes in surgical ICU patients. *Intensive Care Med.*, **21**, 302–309.
85. Brun-Buisson, C., Doyon, F., Carlet, J., Dellamonica, P., Gouin, F., Lepoutre, A., Mercier, J.C., Offenstadt, G., *et al.* (**1995**) Incidence, risk factors, and outcome of severe sepsis and septic shock in adults. A multicenter prospective study in intensive care units. French ICU Group for Severe Sepsis. *JAMA*, **274**, 968–974.
86. Teres, D., Rapoport, J., Lemeshow, S., Kim, S. and Akhras, K. (**2002**) Effects of severity of illness on resource use by survivors and nonsurvivors of severe sepsis at intensive care unit admission. *Crit. Care Med.*, **30**, 2413–2419.
87. Annane, D., Aegerter, P., Jars-Guincestre, M.C. and Guidet, B. (**2003**) Current epidemiology of septic shock: the CUB-Rea Network. *Am. J. Respir. Crit. Care Med.*, **168**, 165–172.
88. Zahorec, R., Firment, J., Strakova, J., Mikula, J., Malik, P., Novak, I., Zeman, J. and Chlebo, P. (**2005**) Epidemiology of severe sepsis in intensive care units in the Slovak Republic. *Infection*, **33**, 122–128.
89. Flaatten, H. (**2004**) Epidemiology of sepsis in Norway in 1999. *Crit. Care*, **8**, R180–R184.
90. Sundararajan, V., Macisaac, C.M., Presneill, J.J., Cade, J.F. and Visvanathan, K. (**2005**) Epidemiology of sepsis in Victoria, Australia. *Crit. Care Med.*, **33**, 71–80.

3
Similarities of the Clinical Aspects of Sepsis and Non-infectious SIRS

Anne Claire Lukaszewicz and Didier Payen

3.1
Introduction

The systemic inflammation syndrome groups clinical and biological symptoms that are sensitive but non-specific: abnormal temperature (>38 °C or <36 °C), white blood cell count (>12000 or <4000 cells/mm^3), tachycardia (HR > 90/min) and tachypnea (RR > 20/min) [1]. The sepsis syndrome describes the systemic inflammation related to an infection, with a wide range of severity from sepsis only to the most severe pattern, septic shock. The later clinical entity is mainly defined on hemodynamic items such as systemic hypotension persisting despite an adequate fluid resuscitation, which requires vasopressors. Despite intense hemodynamic resuscitation, septic shock still has a poor prognosis since the mortality rate ranges from 35 and 50% [2]. Examination of the cause of death in septic shock patients shows a mortality that varies both in relation to the number and the severity of induced organ failures [2, 3], and to any pre-existing co-morbidity. As a consequence, a moderate septic injury may precipitate the occurrence of the shock when cardio-vascular reserve is limited by a chronic disease. Organ function fails not in direct relation to pathogen invasion, but in relation to the intensity of inflammatory response. Despite intense research, the link between systemic inflammation and the remote organ failures remains unclear. Several hypotheses may account for such a failure. First, the organ hypo-perfusion procedure that lasts for a sufficient time to induce organ ischemia leading to cellular energetic failure. Second, a relative "immune toxicity" of the immune cells infiltrating the tissues, with observed re-programming according to a permissive genetic background [4, 5]. In addition, the prognosis for failure of each organ seems to differ largely. As an example, for a similar severity the occurrence of an acute renal failure considerably worsen the outcome [6, 7].

The clinical symptoms in sepsis mainly result from the intensity of host response to pathogen-associated molecular patterns (PAMPs), which use different and complex signaling pathways. The host response depends on the

Sepsis and Non-infectious Systemic Inflammation. Edited by J.-M. Cavaillon and C. Adrie

ISBN: 978-3-527-31935-0

type of stimulation, ability of the immune system to respond, and the nidus of infection. The persistence of infection and then of pathogen stimulation was shown to be an unfavorable condition corresponding to higher rates of death. This has been well documented by studies proving the benefit of using an early adequate antibiotic strategy adapted to the causal bacteria, or to rapidly perform the surgical removal of the infection nidus, as recommended by the Surviving Sepsis Campaign [8]. The persistence of systemic inflammation would promote organ failures by complex mechanisms, especially if the tolerance of the organ is limited. Since the immune system has a phenotype moving rapidly towards an anti-inflammatory profile, such "immune depression" may in turn increase the risk of secondary or nosocomial infection [9].

In ICU patients, systemic inflammation may also result from non-infection etiologies that can mimic sepsis. Among these etiologies, ischemia–reperfusion after hemorrhagic shock, the use of extra-corporeal circulation for cardiac surgery, severe pancreatitis, mesenteric ischemia, severe trauma or major surgery are the most common [10–12]. The pathways by which the systemic inflammatory response is triggered differ largely. They may be triggered by the release of reactive oxygen species (ROS), proteolytic enzymes, products of tissue or cellular damage, etc. The clinical presentation may then mimic the systemic inflammation observed during severe sepsis or septic shock, but mostly of a shorter duration if associated with a procedure (cardiac surgery for example). As a consequence, the prognosis appears better than that for severe sepsis [7]. The clinical imbroglio between infectious and non-infectious causes of systemic inflammation also arises from the possible association of a non-infectious cause of inflammation complicated by an infection.

Ascribing inflammation to an infection or another cause could be then rather difficult. That is the reason why the context, as stated in the definition of sepsis, is of crucial importance for the etiological diagnosis of inflammation. Diagnosis is often simple when infection is obvious as in the case of acute pneumonia or peritonitis, however, it may be more difficult if the infection is only suspected or if infection is a complication of the initial injury. If the clinical symptoms of systemic inflammation are easy to recognize, the task might be more complex when these symptoms occur in a complex context such as surgery, trauma, ischemia–reperfusion after cardiac arrest. This task is of importance not only for resuscitation strategy, but also for clinical trials, to enroll suitable patients. It might be assumed that the initial mistake would not be a major problem, since over-treatment (i.e. treatment for severe sepsis or septic shock even when this is not the case) has limited consequences. This of course depends on the type of resuscitation undertaken, and on the type of treatment given. Since gluco-corticosteroids or activated protein C may have their side-effects, these drugs cannot be given to non-septic shock patients [13, 14]. The *a posteriori* proof of the infectious origin of the systemic inflammation is then of particular importance. It should also be remembered that if bacteria can be found, only 30% of blood cultures prove to be positive, and pulmonary samples may not provide a definite conclusion, thus the diagnosis of sepsis remains a difficult challenge for clinicians, especially when sepsis is not the primary injury motivating the hospitalization.

3.2 Case Report

The following case illustrates the difficulty facing the clinician in diagnosing severe sepsis or septic shock at the early phase, during the so-called "golden hours" [1]. A 61-year-old man was admitted to the emergency room after 3 days of fever (39–40 °C) and abdominal pain. On arrival, the patient had an episode of confusion, mild hypotension (BP 100/52 mmHg), tachycardia (HR 110/min), tachypnea (30/min), with a pulse oximetry saturation (SpO^2) of 97% at room air. The physical examination found a depressible painful abdomen (right hypocondrium), jaundice and dark oliguria. The routine laboratory tests showed hepatic cytolysis, thrombocytopenia, elevated lactate (9 mmol/L) and a WBC count of 12 000/ml. As the patient presented with SIRS criteria, the medical team started to search for an infectious cause of SIRS. After sampling blood, sputum, and urine for bacteriological screening, and also for the most common viral infections, an empirical antimicrobial therapy was initiated. The abdominal CT scan and echography were negative for a diagnosis of angiocholitis or peritonitis, but showed hepato-splenomegaly with coelio-mesenteric and ileocecal nodes. Since the patient became severely hypotensive, a laparotomy was performed, which eliminated peritonitis, angiocholitis, or mesenteric ischemia. Although the liver appeared normal, a liver biopsy was performed. A few hours later, because of a severe thrombocytopenia, a bone smear was taken which showed hemophagocytosis syndrome and the presence of large T cells, leading to the suspicion of lymphoma. The patient rapidly developed acute refractory shock and died from acute liver failure. The liver biopsy confirmed the infiltration of the liver by the abnormal T-cell lymphoma.

This case illustrates the fact that SIRS may exist even in the absence of infection, and may mimic severe sepsis or septic shock. In the absence of a definitive diagnosis, the early phase of treatment is adapted for severe sepsis.

3.3 Epidemiology of SIRS

Several clinical studies or surveys have reported a very high prevalence of SIRS (50 to 80%) in patients admitted to hospital, particularly to the ICU [7, 15, 16]. The recognition of SIRS criteria does not imply that it is related to an infection. If only 30–40% of ICU patients with SIRS criteria have a documented infection, the rate of infection will be higher in patients with more than two SIRS criteria. The number of SIRS criteria provides information regarding severity, since the number of organ failures and mortality increases with number of SIRS criteria. This allows the severity to be graded from SIRS to sepsis, severe sepsis and septic shock with 28-day mortality of 10 to 20%, 20–40% and 40–60% respectively [16].

However, this outcome classification must to be interpreted in the light of quality of care, co-morbidity, genetic susceptibility, and progresses for care. In this respect, it is remarkable to note that the death rate in the control groups of major clinical trials [17, 18] is relatively lower than those previously reported.

This may reflect the selection of patients and the quality of care.

3.4 SIRS Criteria reflect Systemic Inflammation

In this section, the molecular signaling events related to the clinical manifestation of SIRS will be analyzed, with particular emphasis on the differences between infection- and non-infection-induced SIRS.

3.4.1 Fever

Temperature elevation has been shown to enhance several parameters of immune function, including antibody production, T-cell activation, production of cytokines, and enhanced neutrophil and macrophage functions [19]. It is known that increasing body temperature is associated with improved outcome from infectious diseases [19]. Fever is representative of the neuroendocrine changes that characterize the acute-phase response. Several circulating cytokines may induce fever such as TNFα, IL-1β and IL-6 [20]. Several distinct pathways may be implicated in thermogenesis. Possible mechanisms include entry across the organum vasculosum laminae terminalis (OVLT) or induction of central mediators in this area, saturable transport across the blood–brain barrier and activation of neuronal afferent pathways. Blood vessels in the brain contain specific receptors for circulating cytokines in proximity to the pre-optic region of the anterior hypothalamus. Some of these mediators such as TNFα and IL-1α elicit second messengers such as prostanoids (PGE2) in the brain while others such as IL-1β, IL-8 and IL-6 seem to partially induce fever via the release of corticotrophin releasing hormone (CRH). Secondly, IL-1β produced in the brain stem is required for the final steps leading to fever (PGE2) and the central response to inflammation [21, 22]. However, circulating cytokines are not the unique inducers of fever, a neural route exists between inflamed tissue and brain. The recent observation that sub-diaphragmatic vagotomy blocks fever after intraperitoneal (but not intramuscular) injection of lipopolysaccharide (LPS) implicates this neural transmission in the febrile response [23, 24]. Nerve associated lymphoid cells may release pro-inflammatory cytokines and trigger vagal endings [25]. No clear differences have been reported regarding the differences between the infectious and non-infectious activation of these pathways.

3.4.2 Tachycardia

Heart rate (HR) variability is under the control of the autonomic nervous system with a low frequency component (0.4–1 Hz) regulated by the sympathetic tone and a high frequency component (1–4 Hz) regulated by the vagal tone. In systemic inflammation, tachycardia has been related to an imbalance in the autonomic nervous system with an increased sympathetic branch outflow. Tachycardia in progressing inflammation has been associated with a reduction in heart rate variability and sympathetic branch modulation [26, 27]. Circulating mediators and cytokines may influence cardiovascular reflexes leading to the impairment of autonomic balance, the exact mechanism of which is unknown [26]. Most experimental studies using septic models have described an autonomic imbalance with an increase in sympathetic activity following endotoxin injection [28]. In an endotoxinemic animal model, pharmacological reinforcement of the cholinergic vagal central control lowered the magnitude of the inflammatory response and increased the high frequency component of the HR without changing the sympathetic tone [29]. In severe cases as in human septic shock, tachycardia and the high concentration of catecholamine contrast with the hyporeactivity of the vessels and heart [30, 31]. Circulating TNFα, IL-1 and IL-6 may increase the heart rate and reduce the vessel response to β-adrenergic agonists [32–34]. TNFα has been shown to induce tachycardia to a greater extent than hypotension in septic and inflammatory animal models [35] and together with the administration of IL-1β was found to worsen the circulatory changes. In humans, administration of immuno-modulating drugs such as IL-2 which is used in the treatment of neoplasia (100 000 U/kg every 8 h for 4 days), was associated with tachycardia (around 50 beats/min increase), hypotension (it is necessary to use an infusion of vasopressor in four over five patients) and slight changes in cardiac function [36]. These hemodynamic alterations were totally reversible when treatment was interrupted.

3.4.3 Leucocytosis

Leucocytosis in systemic inflammation is characterized by an increase in circulating monocytes and granulocytes. These cells originate essentially from the bone marrow, which is the target for many signaling molecules. Among these, the growth factors (granulocyte colony-stimulating factor (G-CSF), granulocyte-macrophage colony-stimulating factor (GM-CSF), stem cell factor (SCF), erythropoietin) and cytokines (IL-3, IL-6, IL-1, IL-8, macrophage inflammatory protein-1α, and Flt3 ligand) stimulate the mobilization of stem/progenitor cells into circulating forms [37]. The neural control of myelopoiesis has also been described as a result of sympathetic innervation caused by alpha adrenergic agonists such as norepinephrine [38]. Maestroni *et al.* [38] showed the presence of adrenergic receptors in bone marrow immune cells and also reported that adrenergic agonists stimulate lymphopoiesis while attenuating myelopoiesis in the absence of injury. Tang *et al.* [39] looked at the impact of sepsis and burn compared to burn-only or sham animals. These authors demonstrated an increase of norepinephrine content in bone marrow associated with an increase in colony forming unit cells, an effect that was reduced when the release of norepinephrine is attenuated. The major result was that inhibition of an increase in norepinephrine levels dramatically improved the survival rate (62% versus 0% in norepinephrine-intact animals). This was largely different between burn-only versus burn and sepsis (*Pseudomonas aeruginosa*), suggesting that sepsis induced specific pathways and stimulation. These findings at the bone marrow level fit well with the similar concept developed at the circulating level [40]. As a consequence, it can be concluded that sympathetic stimulation participates in availability and activation of monocytes/macrophages originating from bone marrow, and promotes systemic inflammation. The functional phenotypes of bone marrow progenitor-derived macrophages, such as clonogenic potential and LPS-induced cytokine release, are greatly influenced by norepinephrine in burn or burn + sepsis [41]. Based on these observations, correlations between clinical symptoms related to sympathetic activation and inflammation would be expected. There are no clear data to demonstrate such a relationship in human SIRS, whatever the cause. As counter proof, the impact of the type of anesthesia on immune consequences may support this hypothesis. Additional epidural or spinal anesthesia in major surgery results in lower circulating epinephrine and norepinephrine levels during the surgery, which in the case of norepinephrine persist during the postoperative period as compared to general anesthesia alone [42]. Immunological testing of circulating cells reported preservation of lymphocyte functions (CD4/CD8 ratio [42]), increased TH1/TH2 cytokine ratio [43], increased B-cell and T-helper count [44], and proliferation capacity [44] and no difference in monocyte phenotype or innate immunity [44].

With regard to cells, stimulation with norepinephrine of peritoneal macrophages [45] and Kupffer cells [46] increased production of tumor necrosis factor (TNF). These observations strongly suggest that catecholamines have the potential to stimulate macrophage cytokine release. Little is known about the consequences on anti-inflammatory mediator release. The adrenergic pro-inflammatory response of immune cells seems to be mediated by α-adrenoceptors, an effect balanced by the anti-inflammatory effect of β-adrenergic receptors [47]. Again, the net effect would be determined by adrenoceptor density and type, which may differ within cells and tissues.

3.4.4
Tachypnea

The pathophysiology of achypnea observed during SIRS has not received a great deal of attention. Frequently, tachypnea is associated with hyperventilation due to increased respiratory rate and minute ventilation. The usual respiratory drive stimulator, hypoxia or acidosis, are not necessary for inflammation-induced tachypnea [48]. Such hyperventilation can induce hypocapnia (SIRS criteria) with respiratory alkalosis. The peripheral stimulation of the respiratory drive is routed by vagal afferents, especially via the non-myelinated fibers of the vagal nerve. The stimulation is triggered by lung sensory receptors, arterial chemo-receptors or skeletal metabolo-receptors [48, 49].

Several inflammatory mediators have been proposed to interfere with respiratory drive by direct or indirect mechanisms such as bradykinin [50], or thromboxane. Cytokines such as TNF are also mentioned because of their action on capillary permeability and lung parenchymal edema would result from stimulation of the vagal pathway. Another stimulus of respiratory drive could be the relative hypovolemia secondary to the distributive shock in severe systemic inflammation. Because the thoracic pump phenomenon constitutes a compensatory response to reduced blood pressure in order to maintain the cardiac pump filling and systemic blood flow, this links tachypnea to the compensatory phenomena of the cardiovascular system [51, 52]. Some similarities in mechanisms implicated in hyperventilation in SIRS or cardiac failure should be mentioned. As in heart failure, hyperventilation is due to several causes including, alterations in lung mechanics, reduced lung diffusion, increased ventilatory needs due to increased CO_2 production, increased dead space ventilation, and overactive reflexes from metabolo-receptors, baro-receptors and chemo-receptors. All of these reflexes interfere with the derangement of cardiovascular reflex control in heart failure [53].

3.5
Phenomenon Triggering Inflammation

3.5.1
Infectious Etiology

Infection is defined by an abnormal presence of pathogen in a tissue or fluid. Innate immunity cells are present in tissue and epithelia as sentries which detect, isolate, and clear pathogens and initiate the inflammatory network. Many studies have reported on the chronology of mediator and hormone release after an initial injury, such as sepsis. The first line of defense essentially involves pro-inflammatory mediators, followed simultaneously by the release of anti-inflammatory mediators or re-programmation of immune cells. Mediators with a long lasting effect are then released during the recovery phase and healing. Based on this chronology, clinicians have proposed an adapted therapeutic strategy [1].

3.5.2
Non-infectious Etiology

Large tissue injury such as occurs in major surgery, trauma, severe heart failure [54] or pancreatitis [11] may induce alterations in the surrounding cells which lead to an inflammatory phenotype. The most frequently described non-infectious systemic inflammation occurs in cardiac surgery with or without cardiopulmonary bypass (CPB) [12]. In this clinical situation postoperative SIRS leads to multiple organ failure (MOF) in 11% of patients, giving a death rate of 41% [12]. The severity of postoperative SIRS has been shown to be associated with the occurrence of multiple organ dysfunction syndrome and with the prognosis [55]. This postoperative systemic inflammation syndrome relates to imbricate mechanisms such as, ischemia–reperfusion, tissue damage related to surgery, and contact with extra-corporeal devices. The main characteristics are whole body activation, but for a limited period of time. This may partially explain why systemic mediators or cell patterns move rapidly from the pro- to anti-inflammatory profile. As a consequence, organs may fail for a variety of reasons, not all being related to systemic inflammation [56] such as cardio-vascular failure, coagulation activation [57], thermic abnormalities, allergic reactions, or LPS or bacterial translocation. Neutrophils are a major group of cells implicated in inflammation after cardiopulmonary bypass. The degranulation of activated neutrophils may induce tissue damage, edema, tissue factor expression and clotting. As a consequence, major efforts have been made to limit blood-circuit interactions, which mostly implies the activation of neutrophil functions. This includes elastase release [56], release of cytokines such as IL-8 and IL-1ra [58–60], and expression of adhesion molecules such as CD11b and L-selectin. The off-pump coronary artery bypass grafting (CABG) technique avoiding extracorporeal circulation (ECC) has been developed with the aim of reducing the consequent inflammation and coagulation. Stress hormones (such as cortisol [61]) and acute phase inflammatory response molecules such as plasma CRP and IL-6 [62, 63] seem to be only slightly modified by surgery without ECC, with a reduced elevation of proinflammatory cytokines (TNFα and IL-1 [55, 63]). Transcriptomic data indicate that circulating leucocytes overexpressed adhesion molecules (ICAM, L-selectin) and signaling factors (cytokines and TLR receptors) after contact with CPB that may facilitate their trapping in tissue [64]. If neutrophil activation can be prevented in off-pump surgery [65], then monocytes are down-regulated similarly with or without bypass.

In other situations, systemic inflammatory syndrome results from tissue injury (trauma) or enzyme release (pancreatitis), or from gut translocation of bacteria or bacterial products (gut ischemia, major surgery). Even the triggers of inflammation differ although a common pathway is finally activated such as TLR-4 and Pamp or Damp. The perfusion deficiency in these critical clinical conditions might increase gut permeability allowing bacterial antigens or other fragments from bacteria (LPS or other PAMPs) to reach the systemic circulation. Then, LPS or other antigens may trigger the TLR-4 pathway, which in turn is followed by a common inflammatory cascade similar to that described in sepsis-induced inflammation. Others studies have suggested a possible activation of the TLR-4 pathway, independent of the presence of LPS or PAMPs. As an example, TLR-4 activation has been reproduced in liver in a murine model of ischemia/reperfusion [66], without implicating LPS. Different triggering molecules have been proposed such as proteases, the S100s family [67] or extracellular matrix alteration of endothelial cells (release of heparin sulfate, proteoglycan) [68], that may induce the TLR-4 pathway in the same manner as LPS [69, 70]. Injection of heparin sulfate in mouse mimicked the septic syndrome. Conversely, mice deficient for TLR-4 or CD14 (TLR-4 co-receptor) were protected [71]. The proposed hypothesis in sepsis or after tissue injury, is that mediators such as heparin sulfate released from endothelial cell surfaces (through the action of elastases) or proteases would be sufficient to activate the immune cells locally. The effect of activation of the TLR-4 pathway on the immune cell would be the key step in the initiation of the response to infection or other types of tissue lesion. This step would be necessary for initiation of SIRS and propagation of inflammation at sites distant from the initial lesion [72]. Following the activation of TLR-4, signaling would take common pathways and lead to similarities in clinical events and compensatory anti-inflammation mechanisms.

These triggering molecules are released by different types of cells, or from granules (defensins from polymorphonuclear cells) or after necrosis of the cell (HMGB1). They act through receptors with chemotactic and activating effect on host cells for enhancing the local inflammatory network. These molecules are grouped under the heading of DAMPs (damage-associated molecular-pattern molecules) or alarmins [68].

As a consequence, there are few SIRS clinical differences among the various origins, the clinical differences resulted more from the etiology of SIRS than from the SIRS itself.

3.6 Circulating Markers for Systemic Inflammation Characterization

In practice, if the clinic cannot differentiate the infectious versus non-infectious origin of SIRS, blood markers could be used to diagnose SIRS originating from an infection. To fulfill effectiveness criteria, such markers need to have the following characteristics: rapid assay method; adequate specificity and sensitivity; coherent with the evolution (outcome and treatment); and a satisfactory cost/benefit ratio. In the recent past, the most debated marker for sepsis diagnosis was the value of the plasma level of procalcitonin (PCT). The recent publication from Tang *et al.* [73] reviewed 18 studies and showed that PCT measurement gave a low diagnostic performance. The mean value for both sensitivity and specificity was 71% and it was concluded that PCT cannot reliably differentiate sepsis from other non-infectious causes of systemic inflammatory response syndrome in critically ill adult patients. Since no marker fulfills the criteria to diagnose SIRS related to infection, other techniques need to be developed. Among these, the potential use of molecular technology methods may help in future for this purpose. It has been proposed to use 16SRNA gene detection in blood as a marker of bacterial presence at any time [74]. As an example, this technique was used for acute pancreatitis to diagnose complications caused by infection. The authors found only one positive blood culture (*E. coli*) in 22 patients. Using amplification technique for 16SRNA, they found eight other samples to be positive. This allowed the infection and therefore the complications to be treated. We also published the use of random PCR to diagnose systemic candidosis in patients with negative blood culture. This PCR technique was positive in 33% of patients with a negative yeast culture, allowing them to be treated with adapted anti-fungi therapy [75].

3.7 Conclusion

SIRS is a syndrome with high sensitivity for systemic inflammation but with a low specificity for infection etiology. Symptoms quoted to define SIRS are simple to collect clinically, but do not help in distinguishing between an infectious or non-infectious origin. No clear information can be gained from inspection of the mechanisms responsible for these symptoms regarding the different etiologies. As a consequence, it is mainly the suspicion of infection-related SIRS that leads to a tentative diagnosis of sepsis in a patient, while waiting for secondary confirmation from bacteriological screening. Despite this, some situations remain difficult to classify, since symptoms are intimately associated with the underlying disease and co-morbidity. Since treatment needs to be delivered rapidly using adapted anti-microbial therapy, new diagnostic tools are warranted. Among the future perspectives, the use of molecular biology techniques might be promising, such as bacterial 16SRNA amplification, micro-chips for genes expression or proteomic patterns.

Acknowledgments

This work was supported in part by the Ministry of Research, Plan Quadriennal 2005–2008, the PHRC 2002–2007, Direction de la Recherche Clinique, Ile de France, France.

References

1. Bone, R.C., Sibbald, W.J., and Sprung, C.L. (**1992**) The ACCP-SCCM consensus conference on sepsis and organ failure. *Chest*, **101**, 1481–3.
2. Vincent, J.L., Sakr, Y., Sprung, C.L., Ranieri, V.M., Reinhart, K., Gerlach, H., Moreno, R., Carlet, J., Le Gall, J.R. and Payen, D. (**2006**) Sepsis in European intensive care units: results of the SOAP study. *Crit. Care Med.*, **34**, 344–53.
3. Levy, M.M., Fink, M.P., Marshall, J.C., Abraham, E., Angus, D., Cook, D., Cohen, J., Opal, S.M., Vincent, J.L. and Ramsay, G. (**2003**) 2001 SCCM/ESICM/ACCP/ATS/SIS International Sepsis Definitions Conference. *Crit. Care Med.*, **31**, 1250–6.
4. Singer, M., De Santis, V., Vitale, D. and Jeffcoate, W. (**2004**) Multi-organ failure is an adaptive, endocrine-mediated, metabolic response to overwhelming systemic inflammation. *Lancet*, **364**, 545–8.
5. Arcaroli, J., Fessler, M.B. and Abraham, E. (**2005**) Genetic polymorphisms and sepsis. *Shock*, **24**, 300–12.
6. Alberti, C., Brun-Buisson, C., Goodman, S.V., Guidici, D., Granton, J., Moreno, R., Smithies, M., Thomas, O., Artigas, A. and Le Gall, J.R. (**2003**) Influence of systemic inflammatory response syndrome and sepsis on outcome of critically ill infected patients. *Am. J. Respir. Crit. Care Med.*, **168**, 77–84.
7. Sprung, C.L., Sakr, Y., Vincent, J.L., Le Gall, J.R., Reinhart, K., Ranieri, V.M., Gerlach, H., Fielden, J., Groba, C.B. and Payen, D. (**2006**) An evaluation of systemic inflammatory response syndrome signs in the Sepsis Occurrence In Acutely Ill Patients (SOAP) study. *Intensive Care Med.*, **32**, 421–7.
8. Surviving Sepsis Campaign: International guidelines for management of severe sepsis and septic shock. (**2008**) *Crit. Care Med.*, **36**, 296–327.
9. Venet, F., Tissot, S., Debard, A.L., Faudot, C., Crampe, C., Pachot, A., Ayala, A. and Monneret, G. (**2007**) Decreased monocyte human leukocyte antigen-DR expression after severe burn injury: Correlation with severity and secondary septic shock. *Crit. Care Med.*, **35**, 1910–7.
10. Arumugam, T.V., Shiels, I.A., Woodruff, T.M., Granger, D.N. and Taylor, S.M. (**2004**) The role of the complement system in ischemia-reperfusion injury. *Shock*, **21**, 401–9.
11. Riche, F.C., Cholley, B.P., Laisne, M.J., Vicaut, E., Panis, Y.H., Lajeunie, E.J., Boudiaf, M. and Valleur, P.D. (**2003**) Inflammatory cytokines, C reactive protein, and procalcitonin as early predictors of necrosis infection in acute necrotizing pancreatitis. *Surgery*, **133**, 257–62.
12. Laffey, J.G., Boylan, J.F. and Cheng, D.C. (**2002**) The systemic inflammatory response to cardiac surgery: implications for the anesthesiologist. *Anesthesiology*, **97**, 215–52.
13. Payen, D., Sablotzki, A., Barie, P.S., Ramsay, G., Lowry, S., Williams, M., Sarwat, S., Northrup, J., Toland, P. and Booth, F.V. (**2007**) International integrated database for the evaluation of severe sepsis and drotrecogin alfa (activated) therapy: analysis of efficacy and safety data in a large surgical cohort. *Surgery*, **141**, 548–61.
14. Sprung, C.L., Annane, D., Keh, D., Moreno, R., Singer, M., Freivogel, K., Weiss, Y.G., Benbenishty, J., Kalenka, A., Forst, H., Laterre, P.F., Reinhart, K., Cuthbertson, B.H., Payen, D. and Briegel, J. (**2008**) Hydrocortisone therapy for patients with septic shock. *New Engl. J. Med.*, **358**, 111–24.
15. Brun-Buisson, C. (**2000**) The epidemiology of the systemic inflammatory response. *Intensive Care Med.*, **26**(Suppl 1), S64–74.
16. Rangel-Frausto, M.S., Pittet, D., Costigan, M., Hwang, T., Davis, C.S.

and Wenzel, R.P. (**1995**) The natural history of the systemic inflammatory response syndrome (SIRS). A prospective study. *JAMA*, **273**, 117–23.
17. Bernard, G.R., *et al.* (**2001**) PROWESS Study Group. *New Engl. J. Med.*, **344**, 699–709.
18. Sprung, C.L., *et al.* (**2008**) CORTICUS Study Group. *New Engl. J. Med.*, **358**, 111–24.
19. Marik, P.E. (**2000**) Fever in the ICU. *Chest*, **117**, 855–69.
20. Rothwell, N.J. and Hopkins, S.J. (**1995**) Cytokines and the nervous system II: Actions and mechanisms of action. *Trends Neurosci.*, **18**, 130–6.
21. Licinio, J. and Wong, M.L. (**1997**) Pathways and mechanisms for cytokine signaling of the central nervous system. *J. Clin. Invest.*, **100**, 2941–7.
22. Watkins, L.R. and Maier, S.F. (**1999**) Implications of immune-to-brain communication for sickness and pain. *Proc. Natl. Acad. Sci. USA*, **96**, 7710–3.
23. Goldbach, J.M., Roth, J. and Zeisberger, E. (**1997**) Fever suppression by subdiaphragmatic vagotomy in guinea pigs depends on the route of pyrogen administration. *Am. J. Physiol.*, **272**, R675–81.
24. Fleshner, M., Goehler, L.E., Schwartz, B.A., McGorry, M., Martin, D., Maier, S.F. and Watkins, L.R. (**1998**) Thermogenic and corticosterone responses to intravenous cytokines (IL-1beta and TNF-alpha) are attenuated by subdiaphragmatic vagotomy. *J. Neuroimmunol.*, **86**, 134–41.
25. Goehler, L.E., Gaykema, R.P., Nguyen, K.T., Lee, J.E., Tilders, F.J., Maier, S.F. and Watkins, L.R. (**1999**) Interleukin-1beta in immune cells of the abdominal vagus nerve: a link between the immune and nervous systems? *J. Neurosci.*, **19**, 2799–806.
26. Griffin, M.P. and Moorman, J.R. (**2001**) Toward the early diagnosis of neonatal sepsis and sepsis-like illness using novel heart rate analysis. *Pediatrics*, **107**, 97–104.
27. Pontet, J., Contreras, P., Curbelo, A., Medina, J., Noveri, S., Bentancourt, S. and Migliaro, E.R. (**2003**) Heart rate variability as early marker of multiple organ dysfunction syndrome in septic patients. *J. Crit. Care*, **18**, 156–63.
28. Palsson, J., Ricksten, S.E., Delle, M. and Lundin, S. (**1988**) Changes in renal sympathetic nerve activity during experimental septic and endotoxin shock in conscious rats. *Circ. Shock*, **24**, 133–41.
29. Pavlov, V.A., Ochani, M., Gallowitsch-Puerta, M., Ochani, K., Huston, J.M., Czura, C.J., Al-Abed, Y. and Tracey, K.J. (**2006**) Central muscarinic cholinergic regulation of the systemic inflammatory response during endotoxemia. *Proc. Natl. Acad. Sci. USA*, **103**, 5219–23.
30. Annane, D., Bellissant, E., Sebille, V., Lesieur, O., Mathieu, B., Raphael, J.C. and Gajdos, P. (**1998**) Impaired pressor sensitivity to noradrenaline in septic shock patients with and without impaired adrenal function reserve. *Br. J. Clin. Pharmacol.*, **46**, 589–97.
31. Annane, D., Trabold, F., Sharshar, T., Jarrin, I., Blanc, A.S., Raphael, J.C. and Gajdos, P. (**1999**) Inappropriate sympathetic activation at onset of septic shock: a spectral analysis approach. *Am. J. Respir. Crit. Care Med.*, **160**, 458–65.
32. Oddis, C.V., Simmons, R.L., Hattler, B.G. and Finkel, M.S. (**1994**) Chronotropic effects of cytokines and the nitric oxide synthase inhibitor, L-NMMA, on cardiac myocytes. *Biochem. Biophys. Res. Commun.*, **205**, 992–7.
33. Kuster, H., Weiss, M., Willeitner, A.E., Detlefsen, S., Jeremias, I., Zbojan, J., Geiger, R., Lipowsky, G. and Simbruner, G. (**1998**) Interleukin-1 receptor antagonist and interleukin-6 for early diagnosis of neonatal sepsis 2 days before clinical manifestation. *Lancet*, **352**, 1271–7.
34. Aronson, D., Mittleman, M.A. and Burger, A.J. (**2001**) Interleukin-6 levels are inversely correlated with heart rate variability in patients with decompensated heart failure. *J. Cardiovasc. Electrophysiol.*, **12**, 294–300.

35. Gardiner, S.M., Kemp, P.A., March, J.E., Woolley, J. and Bennett, T. (**1998**) The influence of antibodies to TNF-alpha and IL-1beta on haemodynamic responses to the cytokines, and to lipopolysaccharide, in conscious rats. *Br. J. Pharmacol.*, **125**, 1543–50.
36. Ognibene, F.P., Rosenberg, S.A., Lotze, M., Skibber, J., Parker, M.M., Shelhamer, J.H. and Parrillo, J.E. (**1988**) Interleukin-2 administration causes reversible hemodynamic changes and left ventricular dysfunction similar to those seen in septic shock. *Chest*, **94**, 750–4.
37. Mohle, R. and Kanz, L. (**2007**) Hematopoietic growth factors for hematopoietic stem cell mobilization and expansion. *Semin. Hematol.*, **44**, 193–202.
38. Maestroni, G.J. (**2006**) Sympathetic nervous system influence on the innate immune response. *Ann. NY Acad. Sci.*, **1069**, 195–207.
39. Tang, Y., Shankar, R., Gamboa, M., Desai, S., Gamelli, R.L. and Jones, S.B. (**2001**) Norepinephrine modulates myelopoiesis after experimental thermal injury with sepsis. *Ann. Surg.*, **233**, 266–75.
40. Munford, R.S. and Pugin, J. (**2001**) Normal responses to injury prevent systemic inflammation and can be immunosuppressive. *Am. J. Respir. Crit. Care Med.*, **163**, 316–21.
41. Cohen, M.J., Shankar, R., Stevenson, J., Fernandez, R., Gamelli, R.L. and Jones, S.B. (**2004**) Bone marrow norepinephrine mediates development of functionally different macrophages after thermal injury and sepsis. *Ann. Surg.*, **240**, 132–41.
42. Yokoyama, M., Itano, Y., Katayama, H., Morimatsu, H., Takeda, Y., Takahashi, T., Nagano, O. and Morita, K. (**2005**) The effects of continuous epidural anesthesia and analgesia on stress response and immune function in patients undergoing radical esophagectomy. *Anesth. Analg.*, **101**, 1521–7.
43. Le Cras, A.E., Galley, H.F. and Webster, N.R. (**1998**) Spinal but not general anesthesia increases the ratio of T helper 1 to T helper 2 cell subsets in patients undergoing transurethral resection of the prostate. *Anesth. Analg.*, **87**, 1421–5.
44. Volk, T., Schenk, M., Voigt, K., Tohtz, S., Putzier, M. and Kox, W.J. (**2004**) Postoperative epidural anesthesia preserves lymphocyte, but not monocyte, immune function after major spine surgery. *Anesth. Analg.*, **98**, 1086–92, table of contents.
45. Spengler, R.N., Chensue, S.W., Giacherio, D.A., Blenk, N. and Kunkel, S.L. (**1994**) Endogenous norepinephrine regulates tumor necrosis factor-alpha production from macrophages *in vitro*. *J. Immunol.*, **152**, 3024–31.
46. Zhou, M., Yang, S., Koo, D.J., Ornan, D.A., Chaudry, I.H. and Wang, P. (**2001**) The role of Kupffer cell alpha(2)-adrenoceptors in norepinephrine-induced TNF-alpha production. *Biochim. Biophys. Acta*, **1537**, 49–57.
47. Elenkov, I.J., Wilder, R.L., Chrousos, G.P. and Vizi, E.S. (**2000**) The sympathetic nerve–an integrative interface between two supersystems: the brain and the immune system. *Pharmacol. Rev.*, **52**, 595–638.
48. Tang, G.J., Kou, Y.R. and Lin, Y.S. (**1998**) Peripheral neural modulation of endotoxin-induced hyperventilation. *Crit. Care Med.*, **26**, 1558–63.
49. Fregosi, R.F. and Seals, D.R. (**1993**) Hypoxic potentiation of the ventilatory response to dynamic forearm exercise. *J. Appl. Physiol.*, **74**, 2365–72.
50. Kaufman, M.P., Coleridge, H.M., Coleridge, J.C. and Baker, D.G. (**1980**) Bradykinin stimulates afferent vagal C-fibers in intrapulmonary airways of dogs. *J. Appl. Physiol.*, **48**, 511–7.
51. Plaisance, P., Adnet, F., Vicaut, E., Hennequin, B., Magne, P., Prudhomme, C., Lambert, Y., Cantineau, J.P., Leopold, C., Ferracci, C., Gizzi, M. and Payen, D. (**1997**) Benefit of active compression-decompression

cardiopulmonary resuscitation as a prehospital advanced cardiac life support. A randomized multicenter study. *Circulation*, **95**, 955–61.
52. Plaisance, P., Lurie, K.G., Vicaut, E., Adnet, F., Petit, J.L., Epain, D., Ecollan, P., Gruat, R., Cavagna, P., Biens, J. and Payen, D. (**1999**) A comparison of standard cardiopulmonary resuscitation and active compression-decompression resuscitation for out-of-hospital cardiac arrest. French Active Compression–Decompression Cardiopulmonary Resuscitation Study Group. *New Engl. J. Med.*, **341**, 569–75.
53. Agostoni, P., Cattadori, G., Bussotti, M. and Apostolo, A. (**2007**) Cardiopulmonary interaction in heart failure. *Pulm. Pharmacol. Ther.*, **20**, 130–4.
54. Hasper, D., Hummel, M., Kleber, F.X., Reindl, I. and Volk, H.D. (**1998**) Systemic inflammation in patients with heart failure. *Eur. Heart J.*, **19**, 761–5.
55. Mei, Y.Q., Ji, Q., Liu, H., Wang, X., Feng, J., Long, C., Cheng, B., Xing, Y., Li, J. and Hu, D. (**2007**) Study on the relationship of APACHE III and levels of cytokines in patients with systemic inflammatory response syndrome after coronary artery bypass grafting. *Biol. Pharm. Bull.*, **30**, 410–4.
56. Takayama, H., Soltow, L.O. and Aldea, G.S. (**2007**) Differential expression in markers for thrombin, platelet activation, and inflammation in cell saver versus systemic blood in patients undergoing on-pump coronary artery bypass graft surgery. *J. Cardiothorac. Vasc. Anesth.*, **21**, 519–23.
57. Bruins, P., te Velthuis, H., Yazdanbakhsh, A.P., Jansen, P.G., van Hardevelt, F.W., de Beaumont, E.M., Wildevuur, C.R., Eijsman, L., Trouwborst, A. and Hack, C.E. (**1997**) Activation of the complement system during and after cardiopulmonary bypass surgery: postsurgery activation involves C-reactive protein and is associated with postoperative arrhythmia. *Circulation*, **96**, 3542–8.
58. Marie, C., Fitting, C., Cheval, C., Losser, M.R., Carlet, J., Payen, D., Foster, K. and Cavaillon, J.M. (**1997**) Presence of high levels of leukocyte-associated interleukin-8 upon cell activation and in patients with sepsis syndrome. *Infect. Immun.*, **65**, 865–71.
59. Marie, C., Muret, J., Fitting, C., Losser, M.R., Payen, D. and Cavaillon, J.M. (**1998**) Reduced *ex vivo* interleukin-8 production by neutrophils in septic and nonseptic systemic inflammatory response syndrome. *Blood*, **91**, 3439–46.
60. Marie, C., Muret, J., Fitting, C., Payen, D. and Cavaillon, J.M. (**2000**) Interleukin-1 receptor antagonist production during infectious and noninfectious systemic inflammatory response syndrome. *Crit. Care Med.*, **28**, 2277–82.
61. Hoda, M.R., El-Achkar, H., Schmitz, E., Scheffold, T., Vetter, H.O. and De Simone, R. (**2006**) Systemic stress hormone response in patients undergoing open heart surgery with or without cardiopulmonary bypass. *Ann. Thorac. Surg.*, **82**, 2179–86.
62. Harmoinen, A., Kaukinen, L., Porkkala, T., Tarkka, M. and Kaukinen, S. (**2006**) Off-pump surgery does not eliminate microalbuminuria or other markers of systemic inflammatory response to coronary artery bypass surgery. *Scand. Cardiovasc. J.*, **40**, 110–6.
63. Orhan, G., Sargin, M., Senay, S., Yuksel, M., Kurc, E., Tasdemir, M., Ozay, B. and Aka, S.A. (**2007**) Systemic and myocardial inflammation in traditional and off-pump cardiac surgery. *Tex. Heart Inst. J.*, **34**, 160–5.
64. Tomic, V., Russwurm, S., Moller, E., Claus, R.A., Blaess, M., Brunkhorst, F., Bruegel, M., Bode, K., Bloos, F., Wippermann, J., Wahlers, T., Deigner, H.P., Thiery, J., Reinhart, K. and Bauer, M. (**2005**) Transcriptomic and proteomic patterns of systemic inflammation in on-pump and off-pump coronary artery bypass grafting. *Circulation*, **112**, 2912–20.

65. Matata, B.M., Sosnowski, A.W. and Galinanes, M. (**2000**) Off-pump bypass graft operation significantly reduces oxidative stress and inflammation. *Ann. Thorac. Surg.*, **69**, 785–91.
66. Zhai, Y., Shen, X.D., O'Connell, R., Gao, F., Lassman, C., Busuttil, R.W., Cheng, G. and Kupiec-Weglinski, J.W. (**2004**) Cutting edge: TLR4 activation mediates liver ischemia/reperfusion inflammatory response via IFN regulatory factor 3-dependent MyD88-independent pathway. *J. Immunol.*, **173**, 7115–9.
67. Vogl, T., Tenbrock, K., Ludwig, S., Leukert, N., Ehrhardt, C., van Zoelen, M.A., Nacken, W., Foell, D., van der Poll, T., Sorg, C. and Roth, J. (**2007**) Mrp8 and Mrp14 are endogenous activators of Toll-like receptor 4, promoting lethal, endotoxin-induced shock. *Nat. Med.*, **13**, 1042–9.
68. Rubartelli, A. and Lotze, M.T. (**2007**) Inside, outside, upside down: damage-associated molecular-pattern molecules (DAMPs) and redox. *Trends Immunol.*, **28**, 429–36.
69. Kodaira, Y. and Platt, J.L. (**2000**) Modification of antigen-presenting cell functions by heparan sulfate. *Transplant. Proc.*, **32**, 947.
70. Johnson, G.B., Brunn, G.J., Kodaira, Y. and Platt, J.L. (**2002**) Receptor-mediated monitoring of tissue well-being via detection of soluble heparan sulfate by Toll-like receptor 4. *J. Immunol.*, **168**, 5233–9.
71. Johnson, G.B., Brunn, G.J. and Platt, J.L. (**2004**) Cutting edge: an endogenous pathway to systemic inflammatory response syndrome (SIRS)-like reactions through Toll-like receptor 4. *J. Immunol.*, **172**, 20–4.
72. Brunn, G.J. and Platt, J.L. (**2006**) The etiology of sepsis: turned inside out. *Trends Mol. Med.*, **12**, 10–6.
73. Tang, B.M., Eslick, G.D., Craig, J.C. and McLean, A.S. (**2007**) Accuracy of procalcitonin for sepsis diagnosis in critically ill patients: systematic review and meta-analysis. *Lancet Infect. Dis.*, **7**, 210–7.
74. Zhang, W.Z., Han, T.Q., Tang, Y.Q. and Zhang, S.D. (**2001**) Rapid detection of sepsis complicating acute necrotizing pancreatitis using polymerase chain reaction. *World J. Gastroenterol.*, **7**, 289–92.
75. Bougnoux, M., Dupont, C., Mateo, J., Saulnier, P., Faivre, V., Payen, D. and Nicolas-Chanoine, M. (**1999**) Serum is more suitable than whole blood for diagnosis of systemic candidiasis by nested PCR. *J. Clin. Microbiol.*, **37**, 925–30.

4
Organ Dysfunctions during Severe Sepsis and Septic-like Syndromes: Epidemiology, Classification, and Mechanisms

Olfa Hamzaoui and Jean Carlet

4.1
Definitions

MODS is defined as the presence of altered organ function in acutely ill patient such that homeostasis cannot be maintained without intervention [1] and is characterized by the failure of at least two organs occurring either simultaneously or in a sequential manner. Since the consensus expert meeting in 1980, Bone and colleagues [2], when describing patients with a "sepsis syndrome" proposed that organ failure is evidence of inadequate perfusion of the given organ leading to its dysfunction.

The severity of the multiple organ failure has been shown to be correlated with prognosis in septic patients [3, 4], and this concept has been used in the most recent clinical trials [5, 6]. Indeed, in order to quantify the degree of the organ failure many modern scores have been developed [5] and became an integral part of critical care practice by classifying patients according to an objective evaluation of the severity of their illness.

4.2
Classification of Organ Dysfunctions

The ICU mortality rate in septic patients has been correlated with the number of failing organs and the degree of failure [4, 5]. Quantification of organs using scoring systems is useful for describing different groups of patients with regard to the severity of their illness for purposes of ICU management, clinical trials and quality control. The multiple organ dysfunction score (MODS) (Table 4.1) was developed by choosing variables which defined organ failure from publications between 1969 and 1993. These variable were then assessed for their ability to predict ICU mortality [3]. The other most commonly used organ dysfunction score is sequential organ failure assessment (SOFA) (Table 4.2) originally called the "sepsis-related" organ failure assessment

Sepsis and Non-infectious Systemic Inflammation. Edited by J.-M. Cavaillon and C. Adrie

ISBN: 978-3-527-31935-0

Table 4.1 The **M**ultiple **O**rgan **D**ysfunction **S**core (**MODS**).

	Score				
Organ system	**0**	**1**	**2**	**3**	**4**
Respiratory: PaO_2/FiO_2 (mmHg)	>300	226–300	151–225	76–150	≤75
Renal: serum creatinine (mg/dl)	≤1.1	1.2–2.2	2.3–3.9	4–5.6	≥5.6
Hepatic: serum Bilirubin (mg/dl)	≤1.2	1.3–3.5	3.6–7	7–14	>14
Cardiovascular: PAR	≤10	10.1–15	15.1–20	20.1–30	>30
Hematologic: platelet count (10^3)	>120	81–120	51–80	21–50	≤20
Neurologic: Glasgow Coma Scale	15	13–14	10–12	7–9	≤6

PAR: Pressure adjusted heart rate calculated as (heart rate × (right atrial pressure/mean arterial systemic pressure)

Table 4.2 The Sequential Organ Failure Assessment (SOFA).

	0	**1**	**2**	**3**	**4**
Respiratory: PaO_2/FiO_2 ratio (mmHg)	>400	≤400	≤300	≤200	≤100
Renal: creatinine (mg/dl) or urine ouput	≤1.2	1.2–1.9	2.0–3.4	3.5–4.9 or <500 ml/d	>5.0 or <200 ml/d
Hepatic: Bilirubin (mg/dl)	≤1.2	1.2–1.9	2.0–5.9	6.0–11.9	>12
Cardiovascular: Hypotension	No Hypotension	MAP <70 mmHg	Dopamine ≤5 or dobutamine (any dose)	Dopamine >5 or epinephrine ≤0.1 or norepi nephrine ≤0.1	Dopamine >15 or epinephrine >0.1 or norepi nephrine >0.1
Hematologic: platelet count ($10.3/mm^3$)	>150	≤150	≤100	≤50	≤20
Neurologic: Glasgow Coma Score	15	13–14	10–12	6–9	≤6

PAR: Pressure adjusted heart rate calculated as the product of the heart rate multiplied by the ratio of the right atrial pressure to the mean arterial pressure

which can be applied to all ICU patients [4]. The main difference between the two scores is the evaluation of cardiovascular function. For the MODS this evaluation is based on the "pressure-adjusted heart rate", defined as the product of the heart rate multiplied by the ratio of the right atrial pressure to the mean arterial pressure. The SOFA score takes into account the mean arterial pressure and therapeutic interventions with vasopressors.

A model based on the presence or absence of organ dysfunctions and/or infection (the ODIN model) which was developed in a general intensive care

Table 4.3 The PIRO system [7].

Predisposition	Premorbid illness Cultural or religious beliefs Age, sex Genetic polymorphism: in the inflammatory response
Infection	Site Type Extent
Response	SIRS Sepsis Septic shock Biological markers: CRP, PCT
Organ dysfunction	Number of failing organs: MOF, MODS, SOFA

SIRS: systemic inflammatory response; CRP: C-reactive protein; **PCT**: procalcitonine, MOF: multiple organ failure, MODS: multiple organ dysfunction syndrome, SOFA: sepsis-related organ failure assessement

unit, offered a reliable method for characterizing ICU patients. Indeed, the ODIN score was shown to correlate with the death rate [7].

In addition, a staging system which stratifies patients according to their baseline risk of an adverse outcome and their potential to respond to therapy, was recently developed by analogy with the Tumor Nodule Metastases (TNM) classification. This classification, known as PIRO [8] is not a new severity score and classifies patients on the basis of their Predisposing conditions, the nature and extent of the Injury (in the case of sepsis, Infection), the nature and magnitude of the host Response, and the degree of concomitant Organ dysfunction (Table 4.3). This score is not aimed at predicting mortality but rather at describing the patient in a more precise manner.

4.3 Epidemiology

Severe sepsis and MOF remain the leading cause of death in critically-ill patients. It is estimated that 150 000 deaths each year in Europe and more than 200 000 in USA can be attributed to these cause [1]. Many reports of recent epidemiological data are available [8, 9] in particular focusing on, the associated mortality which remains high (30–45%), the consumption of bed days in ICU and in the hospital, 45% and 33% respectively, and finally the cost (16.5 billion dollars in the US) [8, 10].

The incidence of severe sepsis was recently reported to be 95/100 000 inhabitants in a national prospective multicenter trial in French ICUs (EPISEPSIS) with a mortality rate at 30 days of 35% [11]. Similar results were found by Martin *et al.* [12] who reported an incidence of 81 cases per 100 000 inhabitants in a selected population in the USA [2]. Moreover, the number of cases of severe sepsis rose from 82.7 to 240.4 per 100 000 population in the USA and to 51 per 100 000 population in the UK (8, 10). Multiple organ failure as a consequence of tissue hypoperfusion, cellular hypoxia and metabolic dysfunction as commonly observed in severe sepsis and septic shock, was first described in 1977. Eiseman *et al.* [9] reported in their study that bacterial sepsis was the main cause of multiple organ failure in 69% of cases.

The most common manifestation of the multiple organ dysfunction is acute lung injury which is defined as refractory hypoxemia related to high permeability pulmonary edema [13] and its extreme manifestation is acute respiratory distress syndrome (ARDS) which occurs in more than 40% of patients with sepsis and severe sepsis [8, 13].

In addition to sepsis, other situations are known to be associated with MOF such as trauma, pancreatitis, severe hemorraghic shock, extra-corporeal circulation and burn. Trauma is known to activate the systemic release of inflammatory mediators leading to the initiation of MOF [14]. The incidence of MOF varies from 7% [15] to 17% [16] and the mortality rate related to the MOF has decreased during the last years from 85 to 33% [16].

4.4
Pathophysiology of Organ Dysfunctions in Sepsis

Organ dysfunctions and the multiple organ dysfunction syndrome (MODS) or failure (MOF) occur frequently in patients with sepsis. However it is not clear why sepsis may progress to MODS in certain patients, even after restoration of adequate systemic blood flow. The most commonly accepted pathophysiology is that the inflammatory process can become dysregulated in response to the infectious pathogens and this can lead to the liberation of an excessive concentration of inflammatory mediators over a prolonged period or to the depletion or downregulation of other substances, leading to microvascular and/or mitochondrial dysfunctions. These processes can induce widespread tissue hypoperfusion and MODS.

4.4.1
Microvascular Dysfunction

The vascular endothelium plays a central role in microvascular flow which is essential for the efficient delivery of oxygen to the tissues. The vascular endothelium is composed of endothelium cells (EC) with specific properties that are essential for the appropriate functioning of the vascular endothelium. The EC count varies from 1 to 6×10^{13} cells, and the cells cover a total of 1–7 m^2 of the body surface area and account for 1 kg/70 kg body weight [17]. They have multiple functions in the body, in particular they form the lining of blood vessels, mediate vasomotor tone, promote antithrombosis and profibrinolysis and inhibit platelet and leukocyte adhesion [18, 19].

Microvascular dysfunction is recognized as a major factor contributing to organ dysfunction and death in septic patients. Endothelial insult which is produced by stimulating the aggregation of leukocytes and of platelets and by inducing abnormal coagulation, leads to impaired tissue perfusion and subsequent organ dysfunction.

4.4.2
Endothelial Injury and Coagulopathy

Endothelial injury is defined by microscopic changes in the shape of endothelial cells as well as defects in the endothelial lining or elevated soluble markers of endothelial injury [20]. These anatomical changes have been reported in several studies [21–22], and lipopolysaccharide (LPS) has long been identified as being responsible for the detachment of the endothelium from its underlying layer and for other cellular injuries as early as 15 min after the administration of LPS [23]. On the other hand, proinflammatory cytokines are also responsible for EC injuries by increasing the permeability of the cells and allowing the inflammatory fluid to move from the blood to the interstitial space within 6 h of the onset of inflammation, producing a maximal effect over a 12–24-h period [24, 25]. Indeed, in order to assess the EC injuries many authors have suggested the measurement of plasma levels of inflammatory markers: high levels of thrombomodulin (TM), E-selectin and von Willebrand factor (vWF) have been reported in different inflammatory diseases (sepsis, acute lung injury), situations in which endothelium damage was observed [26–28].

Moreover, the outer membrane of the EC expresses molecules with anticoagulant properties and these molecules accelerate the inactivation of coagulation proteases and activate the tissue factor pathway inhibitor (TFPi). This latter factor controls the extrinsic pathway in the clotting cascade [19]. Under the influence of inflammatory and septic stimuli the EC rapidly acquire procoagulant behavior by different mechanisms: (1) increasing the transcriptional upregulation of tissue factor expression in EC, monocytes and other cells; (2) decreasing the levels of endothelial anticoagulant components by internalizing the TM and the release of inactive TM into the bloodstream [29, 30]. Under *in vitro* conditions the addition of LPS and/or cytokines to ECs has been shown to decrease the tissue-type plasminogen activator, to increase the expression of tissue factor and the plasminogen activator inhibitor 1 (PAI-1) and to generate procoagulant microparticles [19, 30, 31]. This interaction between the inflammatory mediators and ECs induces a strong procoagulant phenotype.

The activation of coagulation and the altered fibrinolytic mechanisms might then induce tissue ischemia and tissue necrosis and consequently MODS [32].

4.4.3 Endothelial Activation

Endothelial activation refers to the increased expression or release of endothelial adhesion molecules; these adhesion molecules include the ICAMs, the E-selectins, and the platelet EC adhesion molecules. The activated endothelium induces the migration of activated leukocytes and their adhesion to the ECs. A large number of experimental studies have shown that the inhibition of adherence to the ECs has a beneficial effect [33, 34].

4.4.4 Endothelial Dysfunction

ECs produce vasoactive molecules that regulate vasomotor tone and contribute to the control of blood pressure; these molecules include vasodilators such as nitric oxide (NO) and prostacyclin, vasoconstrictors such as endothelin, thromboxane A2 and platelet activating factor (PAF). The activated endothelium is associated with an imbalance between its vasonstrictor and vasodilatory properties [35].

During sepsis, a decreased endothelial-dependent vascular relaxation has been shown in several studies that have reported attenuated acetylcholine-induced relaxation in vascular rings isolated from large arteries [35]. Several mechanisms induce these abnormalities such as the alteration in EC surface receptors or the altered expression or activity of the constitutive endothelial NO synthase (ec NOS) [36]. The combined effect of altered vascular relaxation, altered blood flow distribution, increased leukocyte adhesion and increased coagulation contribute to the loss of equilibrium between the supply and demand for oxygen, the heterogeneity of the microcirculation in sepsis and hence to hypoperfusion. In summary, evidence suggests that the endothelial function is altered during sepsis in a heterogeneous manner from one site of the vascular tree to another; these changes initiate and promote repetitive cycles of inflammation, coagulation and cellular interactions leading to microvascular occlusion and hypoxia and finally to organ dysfunction. Accordingly, studies using recent advances in technology which allow the bedside monitoring of the microcirculation in septic patients, have led us to a better understanding of circulatory failure by visualizing and quantifying the perfusion of the microcirculation [37, 38].

4.4.5 Cytopathic Hypoxia

Cytopathic hypoxia is defined in sepsis as an intrinsic defect in cellular respiration leading to a decrease in oxygen consumption [39]. It is an important element in the pathophysiology of established sepsis and MOF. A number of biochemical mechanisms have been reported to explain the development of cytopathic hypoxia under pathological conditions. These mechanisms include the following.

4.4.5.1 Inhibition of Pyruvate Dehydrogenase

Inhibition of pyruvate dehyrogenase leads to a diminished delivery of pyruvate to the mitochondrial tricarboxylic acid (TCA) cycle. The accumulation of excess pyruvate in cells also leads to the increased production of lactate [40, 41].

4.4.5.2 Inhibition of Cytochrome a_1a_3

The endogenous production of NO is capable of causing reversible inhibition of the terminal element in the mitochondrial respiratory chain (cytochrome a_1a_3). This phenomenon is more pronounced when the arterial oxygen pressure is low [42, 43] leading to reduced cellular respiration [44].

4.4.5.3 Inhibition of Mitochondrial Enzyme under the Influence of Peroxynitrite ($ONOO^-$)

Peroxynitrite is the product of the reaction between NO with O_2^-. Under certain conditions which are present during various acute inflammatory conditions, including ischemia/reperfusion injury and sepsis, large quantities of peroxynitrite are produced by mitochondria leading to impaired mitochondrial respiration [39].

4.4.5.4 The Poly (ADP-ribose) Polymerase (PARP-1) Hypothesis

Recent *in vitro* studies [45] have reported the importance of (PARP-1)-dependent NAD^+/NADH depletion as the mechanism responsible for cytopathic hypoxia caused by inflammatory mediators.

4.4.6 Inflammatory Cytokines

It is widely accepted that the inflammatory response plays an important role in mediating sepsis. Pathogens activate the complement cascade and induce a rapid release of inflammatory mediators from a number of different cell types (e.g. monocytes, EC). This latter reaction contributes to the eradication of invading microorganisms. Simultaneously, anti-inflammatory pathways are activated to dampen and control this response [46, 47].

Proinflammatory mediators such as Tumor Necrosis Factor (TNF), interleukine 1 and interferon γ induce the synthesis of eicosanoides (prostaglandins and leucotrienes) which promote inflammation, alter vascular tone and increase vascular permeability [48]. In contrast, anti-inflammatory mediators (interleukin 10, interleukin 4 and 6) are produced in order to downregulate the host response and initiate leukocyte reprogramming and changes in the immune response [49]. These two phases are ideally coordinated to defend the host against the pathogen invasion. Nevertheless, an excessive inflammatory response, an inadequate anti-inflammatory response or possibly an imbalance between these two phases may contribute to tissue damage [50].

4.5 Main Organ Dysfunctions

4.5.1 Cardiocirculatory Dysfunction

Cardiocirculatory dysfunction results from the failure of the peripheral circulation and cardiac dysfunction. Circulatory failure is due to a dilation of capacitance vessels combined with microvascular hyperpermeability, both lead to an acute decrease in the left ventricular preload and can thereby compromise cardiac output. In addition, left ventricular performance can be further compromised by an intrinsic decrease in myocardial contractility which is mainly observed in septic patients. The hemodynamic pattern in sepsis commonly includes a high cardiac output, arterial hypotension, an increased consumption or transport of oxygen to counter the reduced utilization of systemic O_2 and in more severe cases a low cardiac output is also observed.

4.5.1.1 Cardiac Dysfunction

According to experimental and clinical studies, myocardial depression is observed early after the induction of sepsis [51] and is reversible after a few days. Myocardial depression is due to both intrinsic (calcium homeostasis, NO and apoptosis) and extrinsic factors (circulating factors, cytokines).

Intrinsic Factors

Calcium homeostasis: during septic myocardial dysfunction, two main mechanisms have been described: the decreased number of calcium "L" channels inducing a decrease in calcium flow [52] and reduced Ca^{2+} responsiveness in myofilaments [53]. This perturbation in calcium homeostasis is mainly due to a change in the molecular composition of the β-adrenergic receptor. Alterations in myofilament sensitivity are frequently associated with an increase in the length of cardiomyocytes and sarcomeres which explains the fluid responsiveness observed during the early phase of sepsis and the acute ventricular dilation observed after fluid infusion [54].

The role of the nitric oxide (NO): during sepsis an over-production of NO is associated chronologically with myocardial dysfunction accompanied by an equivalent expression of the inducible form of the NO synthase (iNOS) in the four cardiac cavities [55]. NO contributes to cardiac dysfunction either directly under the influence of LPS [56] or indirectly via peroxynitrite which is produced after stimulation by inflammatory cytokines (IL1-β, TNF-α, IFN-γ) [57]. Indeed the alteration in cardiac performance is rather linked to the increase of the expression of the iNOS than to the NO itself. These perturbations are reversible and cardiac function recovers when the altered cardiac proteins regain their functionality [55].

Other factors: cardiac dysfunction and alteration in cardiac contractility can result from other direct mechanisms such as apoptosis [58], structural abnormalities (myocyte necrosis, interstitial edema, deposition of fibrin in vessels) [59] and alterations in the respiratory cycle of mitochondria [60].

Extrinsic Factors

Examples of extrinsic factors include circulating and paracrine substances. The most important circulating factor is the myocardial depressant factor (MDF) which induces a negative inotropic effect under the influence of cytokines in particular TNF-α and the IL1-β [61]. Moreover, the ECs which surround the myocytes secrete numerous mediators (prostaglandins, endothelins) which modify the function of the myocytes and consequently worsen cardiac contractility [62].

4.5.1.2 Vascular Dysfunction

In cardiogenic shock or hypovolemic shock, vasoconstriction is the most common mechanism used to avoid hypotension. However, the vasodilatation which occurs during sepsis is related to an excessive production of NO by iNOS stimulated by inflammatory cytokines [63]. NO induces relaxation of smooth muscle cells by stimulating the transformation of GTP to GMPc leading to arteriolar vasodilatation. In addition, during sepsis NO production is responsible for the increased vascular permeability of many organs including the heart, liver, kidney and splanchnic sector [64].

Moreover, in septic patients adrenal failure is commonly observed [65–67] resulting in a low level of mineralocorticoids and more importantly in glucocorticoids which are involved in vascular reactivity. Indeed, glucocorticoids modulate vascular reactivity to angiotensine II and to cathecolamines [68], they also modulate vascular permeability and decrease production of NO as well as of other vasodilators. Therefore adrenal failure can contribute to the hemodynamic instability observed in these patients and their high dependency on cathecolamines despite the control of the infection source. The favorable effect of low doses of glucocorticoids was confirmed in a phase III trial [69].

In addition, impairment of microvascular blood flow during sepsis is observed and is due to the increase in the number of capillaries where no blood flow is observed, in particular in the skeletal muscles, intestinal villosities [70, 71], diaphragm [72] and hepatic sinusoids [73]. This alteration in the microvascular flow results from an impaired distribution of the flow of red blood cells in the microcirculation and the imbalance between the demand and supply of oxygen to the tissues. These latter effects are also enhanced by the excessive production of NO. Moreover, despite the over-compensation of the oxygen supply by the functional capillaries, this heterogeneity in oxygen flow still alter oxygen extraction [74].

Finally, a combination between the effect of the inflammatory mediators and the activated coagulation contributes to a reduced micro-vascular perfusion pressure leading to the formation of micro-thrombi and to the obstruction of the vessels [75].

4.5.2
Renal Dysfunction

Acute renal failure (ARF) is common among patients with sepsis.

It has been observed in 19% of patients with sepsis, 23% of patients with severe sepsis and 51% of patients presenting with septic shock [76, 77]. Despite the new supporting therapies, ARF remains an independent risk factor for mortality in critically ill patients which varies from 45 to 70% when associated with sepsis [78, 79].

The underlying pathogenesis of ARF during sepsis is complex. Current knowledge is mainly based on animal models or *in vitro* experiments. Sepsis is characterized by arterial vasodilatation related mainly to the excessive production of NO leading to hypotension and a fall in renal blood flow associated with selective renal vasoconstriction. The main mechanism presumed in ARF is a change in hemodynamics [80]; this includes preglomerular afferent arteriolar vasoconstriction [81], tubular hypoperfusion [82] and the redistribution of the cortical to medullary blood flow [83]. Indeed ARF is mainly due to two categories of factors:

1. Constitutive mechanisms which usually compensate for hemodynamic disorders are altered in sepsis. This results in abnormal activation of the sympathetic system, the renin–angiotensin–aldosterone system, the kinin–kallikrein system and the release of α-atrial natriuretic peptide [84].
2. Sepsis-induced mechanisms among which are the involvement of several inflammatory factors in the modulation and deterioration of renal hemodynamics, the effect of cytokines (TNF-α, IL-1β, IL-8) which play a central role in the occurrence of various renal lesions observed during sepsis by decreasing the ultrafiltration coefficient, altering vascular tone and inducing tubulo-interstitial lesions [85, 86].

4.5.3
Respiratory Dysfunction

Sepsis is commonly complicated by respiratory dysfunction and increased risk of respiratory failure. Nearly 50% of patients with sepsis will develop acute lung

Table 4.4 The criteria for acute lung injury (ALI) and acute respiratory distress syndrome (ARDS) (see [13]).

	Criteria: of oxygenation, chest radiograph, pulmonary artery wedge pressure (PWP)
ALI	Acute onset $PaO_2/FiO_2 = 300$ mmHg (regardless of the PEEP level) Bilateral infiltrat on chest radiograph PWP = 18 mmHg or no evidence of left atrial hypertension
ARDS	Acute onset $PaO_2/FiO_2 = 200$ mmHg (regardless of the PEEP level) Bilateral infiltrat on chest radiograph PWP = 18 mmHg or no evidence of left atrial hypertension

injury (ALI) or acute respiratory distress syndrome (ARDS) (Table 4.4) [87]. Several reasons may explain this frequent association: pneumonia is a frequent cause of sepsis [10], infection can lead to distinct alterations in morphology and function of the lung parenchyma and cardiovascular changes induced by sepsis can influence pulmonary hemodynamics and gas exchange. However, other circumstances characterized by a systemic inflammatory response can induce respiratory failure. Among them, pancreatitis and hemorrhagic shock can commonly trigger an inflammatory response and induce respiratory failure [88, 89].

The inflammatory response observed in many clinical conditions such as sepsis, severe burns, pancreatitis, hemorrhagic shock and trauma is responsible for pulmonary dysfunction. After initiation, multiple inflammatory cascades induce diffuse endothelial cell injury and increased capillary permeability via inflammatory mediators such as TNF-α, interleukin-1 beta, interleukin-6 and Platelet Activation Factor (PAF) [90]. This causes interstitial pulmonary edema which may progress to alveolar flooding and collapse [91]. Endothelial cell injury also leads to microvascular thrombosis and hemorrhage which reduce the capillary surface available for gas exchange. In addition, alveolar epithelial cells are also injured leading to a loss in the surfactant essential for reducing the alveolar surface tension and maintaining alveolar oxygen exchange. These abnormalities are responsible for a diffuse loss of functional alveolar volume and a reduction in pulmonary compliance leading to profound hypoxemia. Hypoxemia is the result of both ventilation–perfusion inequalities and physiologic shunt [91].

4.5.4 Brain Dysfunction

Sepsis is commonly associated with a wide variety of cerebral damage and dysfunction with an incidence varying between 9 and 71% (depending on the definition) [92, 93] and the related mortality varies from 16 to 63% according to the severity of the brain dysfunction, confirming its prognostic value [94]. The term "delirium" has recently been replaced with "encephalopathy" and sepsis-associated delirium (SAD) is often used to describe brain dysfunction occurring during sepsis or systemic inflammation [95], although it is only a symptom.

The diagnostic approach to SAD relies mainly on clinical and electrophysiological criteria. Clinical examination shows an impaired cognitive function varying from inattention, disorientation to stupor and coma. The diagnosis also requires the exclusion of direct infection of the central nervous system, head trauma, fat embolism and the side-effects of drugs [95]. Electroncephalography represents a sensitive diagnostic tool for brain dysfunction, in particular in critically ill patients where clinical assessment is difficult [96]. Whether brain dysfunction occurs as a result of infectious or non-infectious conditions, the main cause seems to be systemic inflammation [97]. Inflammatory mediators have profound effects on endothelial cells and astrocytes by disrupting the blood–brain barrier which results in impaired neuronal function. This latter phenomenon leads to peri-microvessel edema and disruption of astrocyte endfeet. Moreover, free radicals produced by leukocytes damage erythrocytes and limit oxygen delivery to the brain allowing aromatic amino acids to enter the brain and disturb neurotransmission. Other mechanisms implicated in the development of brain dysfunction include inadequate perfusion pressure and disturbances of the cerebral microcirculation [98, 99] and dysfunction of mitochondrial respiration leading to apoptosis of brain cells [100]. All these abnormalities seem to cause extensive neuronal injury. Indeed, in a recent magnetic resonance imaging (MRI) study which included nine septic shock patients with clinical symptoms of neurological impairment, it was shown that SAD can be documented by MRI. The lesions were mainly situated in the white matter and were associated with a poor outcome [101].

4.5.5 Hematologic Dysfunction

Exposure of the endothelium to inflammatory and/or septic stimuli can rapidly lead to procoagulant activity [102]. These abnormalities can range from thrombocytopenia to disseminated intravascular coagulation (DIC).

Thrombocytopenia has been established as an important independent risk factor for the development of multiple organ failure [103]. Indeed, laboratory and clinical studies have now confirmed that thrombocytopenia-associated multiple organ failure is a thrombotic microangiopathy syndrome which varies from thrombocytopenic purpura to DIC [103]. Although DIC occurs in fewer than 20% of patients with severe sepsis, a sepsis related coagulopathy characterized by high levels of D-dimers and abnormalities in the plasma proteins is always present [104, 105]. DIC is a consumptive syndrome represented in the most severe form by purpura fulminans and in its less severe form by abnormalities in platelet count and prothrombin time (PT) or activated partial thromboplastin (PTT). It has been described clinically as the constellation of thrombocytopenia, decreased factors V and X, decreased fibrinogen and increased D-dimers. Increased anti-fibrinolytic activity also contributes to sustained thrombosis in patients with DIC [103]. DIC is a primary determinant of outcome in critically-ill patients. Accordingly, despite the reversal of shock, there are still patients in whom DIC and coagulopathy is a predictor of mortality [103].

4.5.6 Hepatic Dysfunction

Hepatic dysfunction is an important but poorly understood condition. Damage to the hepatic microcirculation is the main cause of liver failure in sepsis [106]. The observed damage to the hepatocytes is related to several mechanisms: the release of cytokines from activated Küpffer cells has cytotoxic effects on hepatocytes [107–109], the early reduction in the portal blood flow under the effect of nitric oxide released by the inflammatory intestinal cells [110] and later a decrease in the perfusion of liver sinusoids. In addition the adherence of platelets initiates leukocyte–endothelial interactions leading to impaired hepatic microperfusion [106]. The increase in hepatocellular enzymes commonly observed during septic shock is the result of hypoxia due to decreased liver perfusion and the liberation of cytotoxic mediators. All these microcirulatory disturbances result in hepatocellular damage as a result of hypoxia and cytotoxic cellular damage.

During sepsis structural abnormalities (cholestasis and steatosis), an increase in hepatocellular injury and loss of hepatocyte regeneration were observed. Different mechanisms were advanced in order to explain hepatic dysfunction during sepsis. Among them the hypothesis reported by Deutshman *et al.* in their study that the decreased hepatic response to the cytokine interleukine 6 is responsible for the development of structural abnormalities [111].

4.6 Critique of the Organ Dysfunction Concept and of the Scores used in ICU Patients

According to the 1992 consensus conference, sepsis is characterized by the presence of SIRS (Systemic Inflammatory Response Syndrome) and is associated with symptoms of infection. Severe sepsis is defined by the failure of at least one organ in addition to the presence of SIRS and septic shock is characterized by the occurrence of profound hypotension which persists after volemic expansion and requires the administration of vasoactive drugs. According to the definition used during various clinical trials, the failure of another organ associated with hypotension may or may not be required to meet the criteria of severe sepsis. The main problem is that most organ dysfunctions or failures are a heterogeneous mix of the symptoms of septic shock (oliguria, hypotension), consequences of septic shock, cause of sepsis or septic shock (i.e. hypoxemia due to severe pneumonia which nowadays represent 50% of the syndrome), specific and direct effects of infection on organs and the underlying abnormalities of this given organ (lung, brain and liver in particular).

The cardiovascular system is considered as an organ, although its dysfunction is characterized by either non-specific symptoms (hypotension) or arbitrary therapeutic intervention (vasoactive drugs in the SOFA score).

In the PIRO concept [8], septic shock is considered to be a "response", although it might rather be considered as an organ dysfunction or as a consequence of the severity of infection (which may in part be due to the intensity of the response). In most cases during clinical management, as well as clinical trials, the role of hypotension is over-emphasized, although, as clearly mentioned in the recent international consensus statement [112] it is a symptom which is poorly predictive of shock and can dramatically underestimate the frequency of septic shock. Lactate levels [112] or possibly oxygen saturation of the tissues (STO_2) seem to provide more accurate information regarding the condition of the tissues.

Most of the scores currently used have been "constructed" empirically, after a consensus of experts, although scores assessing severity of illness (APS part of Apache, or of SAPS scores) derived from very large and international data bases were already available, as were scores for specific organ dysfunction which had also been extracted from these same data bases, as exemplified by LOD [113]. There are specific and dramatic problems associated with the use of the Glasgow Coma Score to define neurological dysfunction, due to the extensive variability in the methods used to calculate the score in sedated patients. Finally, the prognostic "power" of the dysfunction of each organ has been defined arbitrarily, and has been used to describe dysfunction or failure in all the organs, despite the fact that this can differ quite dramatically from one organ to another. The very poor prognosis of septic patients with acute renal failure is possibly the best example [78, 79].

4.7
Organ Failure: Severity Marker or Protective Phenomenon?

As emphasized in the review by Hotchkiss *et al.* [114] the discordance between the degree of organ dysfunction and the histological findings in patients dying from sepsis is very intriguing. In most patients cell death and tissue necrosis is limited, and we know that complete recovery of organ function can occur rapidly in surviving patients as is well known for acute renal tubulopathy. Some authors [114, 115] have speculated that organ dysfunction might be a process of "hibernation" or cell "stunning", aimed at protecting the organs from hypoxia. The role of mitochondria is considered to be important in both the pathophysiology of severe sepsis and septic shock, and in this so-called cellular hibernation [115].

4.8 Conclusion

The mechanisms which lead to organ dysfunction and multiple organ failure are extremely complex, and still not completely understood. They may vary from one organ to another, although it seems that there are some common mechanisms. We have made obvious progress in the methods used to grade these organ dysfunctions, both on a daily basis, and during clinical trials. There are often overlaps between a certain dysfunction and the symptoms of septic shock, and further research is needed to clarify this issue. Whether organ dysfunction represents "protection" of the organs has still to be demonstrated, and investigated.

References

1. Members of the American College of Chest physicians/Society of Crit Care Consensus Conference Committee: American College of Chest physicians/Society of Crit Care Consensus Conference Committee American College of Chest physicians/Society of Crit Care Consensus Conference Committee (**1992**) Definitions for sepsis and organ Failure and Guidelines for the use of innovative therapies in sepsis. *Crit. Care Med.*, **20**, 864–874.
2. Bone, R.C., Fisher, C.J., Clemmer, T.P., *et al.* (**1989**) Sepsis syndrome: a valid clinical entity. *Crit. Care Med.*, **17**, 389–393.
3. Marshall, J.C., Cook, D.J., Christou, N.V., *et al.* (**1995**) Multiple organ dysfunction score: A reliable descriptor of a complex clinical outcome. *Crit. Care Med.*, **23**, 1638–1652.
4. Vincent, J.L., Moreno, R., Takala, J., *et al.* (**1996**) The SOFA (sepsis-related Organ Failure Assessement) score to describe organ dysfunction/failure. *Intensive. Care Med.*, **22**, 707–710.
5. Cook, R., Cook, D.J., Tilley, J., *et al.* (**2001**) Multiple organ dysfunctions: baseline and serial component scores. *Crit. Care Med.*, **29**, 2046–2050.
6. Bernard, G.R., Vincent, J.L., Laterre, P.F., *et al.* (**2001**) Efficacy and safety of recombinant human activated protein C for severe sepsis. *N. Engl. J. Med.*, **344**, 699–709.
7. Fagon, J.Y., Chastre, J., Novara, A., *et al.* (**1993**) Characterisation of intensive care unit patients using a model based on the presence or absence of organ dysfunctions and/or infection: the ODIN model. *Intensive Care Med.*, **19**, 137–144.
8. Levy, M., Fink, M.P., Marshall, J.C., *et al.* for the International Sepsis Definitions Conference (**2003**) 2001 SCCM/ESICM/ACCP/ATS/SIS International Sepsis Definitions Conference. *Crit. Care Med.*, **31**, 1250–1256.
9. Eiseman, B., Beart, R. and Norton, L. (**1977**) Multiple organ failure. *Surg. Gynecol. Obstet.*, **144**, 323–326.
10. Padkin, A., Goldfrad, C., Brady, A.R., *et al.* (**2003**) Epidemiology of severe sepsis occurring in the first 24 hours in intensive care units in England,Wales, and Northern Ireland. *Crit. Care Med.*, **31**, 2332–2338.
11. The EPISEPSIS group (**2004**) EPISEPSIS: a reappraisal of the epidemiology and outcome of severe sepsis in French intensive care units. *Intensive Care Med.*, 2004, **30**, 580–588.
12. Martin, G.S., Mannino, D.M., Eaton, S., *et al.* (**2003**) The

epidemiology of sepsis in the United States from 1979 to 2000. *N. Engl. J. Med.*, **348**, 1546–1554.
13. Bernard, G.R., Artigas, A., Brigham, K.L., *et al.* (**1994**) The American-European Consensus Conference on ARDS. Definitions, mechanisms, relevant outcomes, and clinical trial coordination. *Am. J. Respir. Crit. Care Med.*, **149**, 818–824.
14. DeLong Jr, W.G. and Born, C.T. (**2004**) Cytokines in patients with polytrauma. *Clin. Orthop.*, **422**, 57–65.
15. Durham, R.M., Moran, J.J., Mazusk, J.E., *et al.* (**2003**) Multiple Organ Failure in trauma patients. *J. Trauma*, **55**, 608–616.
16. Nast-Kolb, D., Aufkolk, M., Rucholtz, S., *et al.* (**2001**) Multiple organ failure still a major cause of morbidity but not mortality in blunt multiple trauma. *J. Trauma*, **51**, 835–842.
17. Cines, D.B., Pollak, E.S., Buck, C.A., *et al.* (**1998**) Endothelial cells in physiology and in the pathophysiology of vascular disorders. *Blood*, **91**, 3527–3561.
18. Gross, P.L. and Aird, W.C. (**2000**) The endothelium and thrombosis. *Semin. Thromb. Hemost.*, **26**, 463–478.
19. Bombeli, T., Mueller, M., Haeberli, A., *et al.* (**1997**) Anticoagulant properties of the vascular endothelium. *Thromb. Hemost.*, **77**, 408–423.
20. Stefanec, T. (**2000**) Endothelial apoptosis: could it have a role in the pathogenesis and treatment of disease. *Chest*, **117**, 841–854.
21. Reidy, M.A. and Schwartz, S.M. (**1983**) Endothelial injury and regeneration. IV. Endotoxin: a non denuding injury to aortic endothelium. *Lab. Invest.*, **48**, 25–34.
22. Reidy, M.A. and Bowyer, D.E. (**1977**) Scanning electron microscopy: morphology of aortic endothelium following injury by endotoxin and during subsequent repair. *Atherosclerosis*, **26**, 319–328.
23. Leclerc, J., PU, Q., Corseaux, D., *et al.* (**2000**) A single endotoxin injection in the rabbit causes prolonged blood vessel dysfunction and a procoagulant state. *Crit. Care Med.*, **28**, 3672–3678.
24. Lee, M. and Schuessler, G. and Chien, S. (**1998**) Time dependant effects of endotoxin on the ultrastructure of the aortic endothelium. *Artery*, **15**, 71–89.
25. Wang, P., Wood, T.J., Zhou, M., *et al.* (**1996**) Inhibition of the biological activity of tumor necrosis factor maintains vascular endothelial cell function during hyperdynamic sepsis. *J. Trauma*, **40**, 694–701.
26. Bauer, P.R. (**2002**) Microvascular responses to sepsis: clinical significance. *Pathophysiology*, **8**, 141–148.
27. Kayal, S., Jais, J.P., Anuini, N., *et al.* (**1998**) Elevated circulating E-selectin, intercellular adhesion molecule 1, and von Willebrand factor in patients with severe infection. *Am. J. Respir. Crit. Care Med.*, **157**, 776–784.
28. Rubin, D.B., Wiener-Kronish, J.P., Murray, J.F., *et al.* (**1990**) Elevated von Willebrand factor antigen is an early plasma predictor of acute lung injury in nonpulmonary sepsis syndrome. *J. Clin. Invest.*, **86**, 474–480.
29. Rapaport, S. and Rao, L. (**1992**) Initiation and regulation of tissue factor-dependent blood coagulation. *Arterioscler. Thromb.*, **12**, 1111–1121.
30. Moore, K.L., Andreoli, S.P., Esmon, N.L., *et al.* (**1987**) Endotoxin enhances tissue factor and suppresses thrombomodulin expression of human vascular endpthelium *in vitro*. *J. Clin. Invest.*, **79**, 124–130.
31. Combes, V., Simon, A.C., Grau, G.E., *et al.* (**1999**) In vitrogeneration of endothelial microparticles and possible prothrombotic activity in patients with lupus anticoagulant. *J. Clin. Invest.*, **104**, 93–102.
32. Regoeczi, E. and Brain, M.C. (**1969**) Organ distribution of disseminated intravascular coagulation. *Br. J. Haematol.*, **17**, 73–81.

33. Munoz, F.M., Hawkins, E.P., Bullard, D.C., *et al.* (**1997**) Host defense against systemic infection with *Streptococcus pneumoniae* is impaired in the E-, P- and E-/P-selectin-deficient mice. *J. Clin. Invest.*, **100**, 2099–2106.
34. Steeber, D.A., Tang, M.L., Green, N.E., *et al.* (**1999**) Leukocyte entry intosites of inflammation requires overlapping interactions between the L-selectin and ICAM-1 pathways. *J. Immunol.*, **163**, 2176–2186.
35. Wiel, E. and Vallet, B. (**2001**) Vascular endothelial cell dysfunction in septic shock. *Crit. Care Med.*, **29**, S36–S41.
36. Vallet, B. (**2003**) Bench-to-bedside: endothelial cell dysfunction in severe sepsis: a role in organ dysfunction? *Crit. Care*, **7**, 130–138.
37. De Backer, D., Creteur, J., Preisser, J.C., *et al.* (**2002**) Microvascular blood flow is altered in patients with sepsis. *Am. J. Respir. Crit. Care Med.*, **166**, 98–104.
38. Sakr, Y., Dubois, M.J., De Backer, D., *et al.* (**2004**) Persistant microcirculatory alterations are associated with organ failure and death in patients with septic shock. *Crit. Care Med.*, **32**, 1825–1831.
39. Fink, M.P. (**2002**) Bench-to-bedside review: cytopathic hypoxia. *Crit. Care*, **6**, 491–499.
40. Vary, T.C. (**1996**) Sepsis-induced alterations in pyruvate dehydrogenase complex activity in rat skeletal muscle: effects on plasma lactate. *Shock*, **6**, 89–94.
41. Vary, T.C. and Hazen, S. (**1999**) Sepsis alters pyruvate dehydrogenase kinase activity in skeletal muscle. *Mol. Cell. Biochem.*, **198**, 113–118.
42. Cassina, A. and Radi, R. (**1996**) Differential inhibitory action of nitric oxide and peroxynitrite on mitochondrial electron transport. *Arch. Biochem. Biophys.*, **328**, 309–316.
43. Takehara, Y., Kanno, T. and Yoshioka, T., *et al.* (**1995**) Oxygen-dependant regulation of mitochondrial energy metabolism by nitric oxide. *Arch. Biochem. Biophys.*, **323**, 27–32.
44. Brown, G.C., Bolanos, J.P., Heal, S.J., *et al.* (**1995**) Nitric oxide produced by activated astrocyte rapidly and reversibly inhibits cellular respiration. *Neursci. Lett.*, **193**, 201–204.
45. Khan, A.U., Delude, R.L., Han, Y.H., *et al.* (**2002**) Liposomal NAD+ prevents diminished O_2 consumption by immunostimulated Caco-2 cells. *Am. J. Physiol.*, **282**, L1082–L1091.
46. Van der Poll, T. and Van Deventer, S.J. (**1999**) Cytokines and anti-cytokines in the pathogenesis of sepsis. *Infect. Dis. Clin. North. Am.*, **13**, 413–426.
47. Pinsky, M. (**2000**) Immune balance in critically ill patients. *Arch. Immunol. Ther. Exp.*, **48**, 439–442.
48. Hanasaki, K., Yokota, Y., Ishizaki, J., *et al.* (**1997**) Resistance to endotoxic shock in phospholipase A2 receptor-deficient mice. *J. Biol. Chem.*, **272**, 32792–32797.
49. Annane, D., Bellissant, E. and Cavaillon, J.M. (**2005**) Septic shock. *Lancet*, **365**, 63–78.
50. Aird, W.C. (**2003**) The role of the endothelium in severe sepsis and multiple organ dysfunction syndrome. *Blood*, **101**, 3765–3777.
51. Parrillo, J. (**1990**) Myocardial depression during septic shock in humans. *Crit. Care Med.*, **18**, 1183–1184.
52. Abi Gerges, N., Tavernier, B., Mebazza, A., *et al.* (**1999**) Sequential changes in autonomic regulation of cardiac myocyte after *in vivo* endotoxin injection in rat. *Am. J. Resp. Crit. Care Med.*, **160**, 1196–1204.
53. Tavernier, B., Mebazza, A., Mateo, P., *et al.* (**2001**) Phosphorylation-dependent alteration in myofilaments calcium sensitivity but normal mitochondrial function in septic heart. *Am. J. Resp. Crit. Care Med.*, **163**, 362–367.

54. Komukai, K. and Kurihara, S. (**1997**) Length-dependence of calcium-tension relationship in aequorin-injected ferret papillary muscles. *Am. J. Physiol.*, **273**, H1068–H1074.
55. Rabuel, C., Renaud, E., Ratajzak, P., *et al.* (**2003**) Les proteines cardiaques sont nitrées par le peroxynitrite puis dégradés par l'ubiquitine chez les patients septiques. *Ann. Fr. Anesth. Réanim.*, **22**, S348.
56. Brady, A., Pool-Wilson, P., Harding, S., *et al.* (**1992**) Nitric oxide production within cardiac myocytes reduces their contractility in endotoxemia. *Am. J. Physiol.*, **263**, H1963–H1966.
57. Kelly, R., Balligand, J.L. and Smith, T.W. (**1996**) Nitric oxide and cardiac function. *Circ. Res.*, **79**, 368–380.
58. Fauvel, H., Marchetti, P., Chopin, C., *et al.* (**2001**) Differential effects of caspase inhibitors on endotoxin-induced myocardial dysfunction and heart apoptosis. *Am. J. Physiol.*, **280**, H1608–H1614.
59. Fernandes Jr, C., Iervolino, M., *et al.* (**1994**) Interstitial myocarditis in sepsis. *Am. J. Cardiol.*, **74**, 958.
60. Brealey, D., Brand, M., Hargreaves, I., *et al.* (**2002**) Association between mitochondrial dysfunction and severity and outcome of septic shock. *Lancet*, **360**, 219–223.
61. Kumar, A., Thota, V., Dee, L., *et al.* (**1996**) Tumor necrosis factor alpha and interleukin 1 beta are responsible for *in vitro* myocardial cell depression induced by human septic shock serum. *J. Exp. Med.*, **183**, 949–958.
62. Mebazza, A., De Keulenaer, G., Paqueron, X., *et al.* (**2001**) Activation of cardiac endotheliumas a compensatory component in endotoxin-induced cardiomyopathy: role of endothelin, prostaglandin, prostaglandins and nitric oxide. *Circulation*, **104**, 3137–3144.
63. Beasley, D. and Eldridge, M. (**1994**) Interleukin-1 beta and tumor necrosis factor alpha synergistically induce NO synthase in rat vascular smooth muscle cells. *Am. J. Physiol.*, **266**, R1197–1203.
64. Laszlo, F., Whittle, B.J., Evans, S.M., *et al.* (**1995**) Association of microvascular leakage with induction of nitric oxide synthase: effects of nitric oxide synthas inhibitors in various organs. *Eur. J. Pharmacol.* **283**, 47–53.
65. Annane, D., Sebille, V., Troche, G., *et al.* (**2000**) A 3-Level prognostic classification in septic shock based on cortisol levels and cortisol response to corticotrophin. *JAMA*, **283**, 1038–1045.
66. Marik, P.E. and Zaloga, G.P. (**2003**) Adrenal insufficiency during septic shock. *Crit. Care Med.*, **31**, 141–145.
67. Briegel, J., Scelling, G., Haller, M., *et al.* (**1996**) A Comparison of the adrenocortical response during septic shock and after complete recovery. *Intensive Care Med.*, **22**, 894–899.
68. Annane, D. and Bellissant, E. (**2002**) Impact of corticosteroids on the vascular reponse to cathecolamines in septic shock. *Reanimation*, **11**, 111–116.
69. Annane, D., Sebille, V., Charpentier, C., *et al.* (**2002**) Effect of treatment with low doses of hydrocortisone and fluorocortisone on mortality in patients with septic shock. *JAMA*, **288**, 862–871.
70. Lam, C., Tyml, K., Martin, C., *et al.* (**1994**) Microvascular perfusion impaired in a rat model of normotensive sepsis. *J. Clin. Invest.*, **94**, 2077–2083.
71. Piper, R.D., Pitt Hyde, M., Li, F., *et al.* (**1996**) Miccrocirculatory changes in rat skeletal muscle in sepsis. *Am. J. Respir. Crit. Care Med.*, **154**, 931–937.
72. Boczkowski, J., Vicaut, E. and Aubier, M. (**1992**) *In vivo* effects of *Escherichia coli* endotoxemia on diaphragmatic microcirculation in rats. *J. Appl. Physiol.*, **72**, 2219–2224.
73. Unger, L.S., Cryer, H.M., Garrison, R.N., *et al.* (**1989**) Differential response of the microvasculature in the liver during bacteremia. *Circ. Shock*, **29**, 335–344.

74. Hinshaw, L.B. (**1996**) Sepsis/septic shock: Participation of the microcirculation: an abbreviated review. *Crit. Care Med.*, **24**, 1072–1078.
75. Diaz, N.L., Finol, H.J., Torres, S.H., *et al.* (**1998**) Histochemical and ultrastructural study of skeletal muscle in patients with sepsis and multiple organ failure syndrome (MOFS). *Histol. Histopathol.*, **13**, 121–128.
76. Riedmann, N.C., Guo, R.F. and Ward, P.A. (**2003**) The enigma of sepsis. *J. Clin. Invest.*, **112**, 460–467.
77. Rangel-Frausto, M.S., Pittet, D., Costigan, M., *et al.* (**1995**) The natural history of systemic inflammatory response syndrome (SIRS): a prospective study. *JAMA*, **273**, 117–123.
78. Levy, E.M., Viscoli, C.M., Horowitz, R.I., *et al.* (**1996**) The effect of acute renal failure on mortality: a cohort analysis. *JAMA*, **275**, 1489–1494.
79. Angus, D.C., Linde-Zwirble, W.T., Lidicker, J., *et al.* (**2001**) Epidemiology of severe sepsis in the United States: analysis of incidence, outcome, and associated costs of care. *Crit. Care Med.*, **29**, 1303–1310.
80. Groenveld, A.B.J., Tran, D.D., Van Der Meulen, J., *et al.* (**1991**) Acute renal failure in the medical intensive care unit: predisposing, complicating factors and outcome. *Nephron*, **59**, 602–610.
81. Lugon, J.R., Boim, M.A., Ramos, O.L., *et al.* (**1989**) Renal function and glomular hemodynamics in male endotoxic rats. *Kidney Int.*, **36**, 570–575.
82. Millar, C.G.M. and Thiemermann, C. (**1997**) Interenal hemodynamics and renal dysfunction in endotoxaemia: effects of nitric oxide synthase inhibition. *Br. J. Pharmacol.*, **121**, 1824–1830.
83. Van Lambalgen, A.A., Bouriquet, N. and Cassellas, D. (**1996**) Effects of endotoxin on tone and pressure-responsiveness of preglomerular juxtamedullary vessels. *Eur. J. Physiol.*, **432**, 574–577.
84. Vervloet, M.G., Piepot, H.A. and Groeneveld, A.B.J. (**2002**) The kidney in sepsis. The sepsis text.
85. Kohan, D.E. (**1994**) Role of endothelin and tumor necrosis factor in the renal response to sepsis. *Nephrol. Dial. Transplant.*, **9**(suppl 4), 73–77.
86. Baud, L. and Ardaillou, R. (**1995**) Tumor necrosis factor in renal injury. *Miner. Electrolyte Metab.*, **21**, 336–341.
87. Bernard, G.R., Artigas, A., Brigham, K.L., *et al.* (**1994**) The American European consensus conference on ARDS. Definitions, mechanisms, relevant outcomes, and clinical trial coordinations. *Am. J. Respir. Crit. Care Med.*, **149**, 818–824.
88. Zhu, A.J., Shi, J.S. and Sun, X.J. (**2003**) Organ failure associated with severe acute pancreatitis. *World. J. Gastroenterol.*, **9**, 2570–2573.
89. Crimi, E., Zhang, H., Han, R.N.N., *et al.* (**2006**) Ischemia and reperfusion increases susceptibility to ventilator-induced lung injury in rats. *Am. J. Respir. Crit. Care Med.*, **174**, 178–186.
90. Bahtia, M. and Moochhala, S. (**2004**) Role of inflammatory mediators in the pathophysiology of acute respiratory distress syndrome. *J. Pathol.*, **202**, 145–156.
91. Awad, S.S. (**2003**) State of the art-therapy for severe sepsis and multisystem organ dysfunction. *Am. J. Surg.*, **28**, 23S–30S.
92. Papadopoulos, M.C., Davies, D.C., Moss, R.F., *et al.* (**2000**) Pathophysiology of septic encephalopathy a review. *Crit. Care Med.*, **28**, 3019–3024.
93. Eidelman, L.A., Putterman, D., Putterman, C., *et al.* (**1996**) The spectrum mortality of septic encephalopathy. Definitions, etiologies and mortalities. *JAMA*, **275**, 470–473.
94. Eidelman, L.A., Putterman, D., Putterman, C., *et al.* The spectrum of septic encephalopathy. Definitions,

etiologies, and mortalities. *JAMA*, **275**, 470–473.
95. Ebersoldt, M., Sharshar, T. and Annane, D. (**2007**) Sepsis-associated delirium. *Intensive Care Med.*, **33**, 941–950.
96. Young, C.B., Bolton, C.F. and Archibald, Y.M. (**1992**) The electroencephalogram in sepsis-associated encephalopathy. *J. Clin. Neurophysiol.*, **9**, 145–152.
97. Goris, R.J.A. (**1990**) Mediators of multiple organ failure. *Intensive Care Med.*, **16**, S192–S196.
98. Terborg, C., Schummer, W., Albrecht, M., *et al.* (**2001**) Dysfunction of vasomotor reactivity in severe sepsis and septic shock. *Intensive Care Med.*, **27**, 1231–1234.
99. Pollard, V., Prough, D.S., Deyo, D.J., *et al.* (**1997**) Cerebral blood flow during experimental endotoxemia in volunteers. *Crit. Care Med.*, **25**, 1700–1706.
100. Chuang, Y.C., Tsai, J.L., Chang, A.Y., *et al.* (**2002**) Dysfunction of the mitochondrial respiratory chain in the rostral ventro-lateral medulla during experimental endotoxemia in the rat. *J. Biomed. Sci.*, **9**, 542–548.
101. Sharshar, T., Carlier, R., Bernard, F., *et al.* (**2007**) Brain lesions in septic shock: a magnetic resonance imaging study. *Intensive Care Med.*, **33**, 798–806.
102. Vallet, B. and Wiel, E. (**2001**) Endothelial cell dysfunction and coagulation. *Crit. Care Med.*, **29**, S36–S41.
103. Ngyuen, T.C. and Carcillo, J.A. (**2006**) Bench-to-bedside review: thrombocytopenia-associated multiple organ failure – a newly appreciated syndrome in the critically ill. *Crit. Care*, **10**, 235–242.
104. Sadler, J.E. (**1998**) Biochemistry and genetics of von Willebrand factor. *Annu. Rev. Biochem.*, **67**, 395–424.
105. Levy, G.G., Nichols, W.C., Lian, E.C., *et al.* (**2001**) Mutations in a member of the ADAMTS gene family cause thrombotic thrombocytopenic purpura. *Nature*, **413**, 488–494.
106. Corner, R.S., Hoerer, E., Kulu, Y., *et al.* (**2006**) Hepatic platelet and leukocyte adherence during endotoxemia. *Crit. Care*, **10**, R15.
107. Koo, D.J., Chaudry, I.H. and Wang, P. (**1999**) Kupffer cells are responsible for producing inflammatory cytokines and hepatocellular dysfunction during early sepsis. *J. Surg. Res.*, **83**, 151–157.
108. Wang, P., Ba, Z.F. and Chaudry, I.H. (**1995**) Hepatocellular dysfunction occurs earlier than the onset of hyperdynamic circulation during sepsis. *Shock*, **3**, 21–26.
109. Wang, P., Ba, Z.F. and Chaudry, I.H. (**1997**) Mechanism of hepatocellular dysfunction during early sepsis. Key role of increased gene expression and release of proinflammatory cytokines tumor necrosis factor and interleukin-6. *Arch. Surg.*, **132**, 364–370.
110. Smets, D., Spapen, H., Diltoer, M., *et al.* (**1999**) Liver perfusion and hepatocellular inflammatory response in sepsis. *Acta. Clin. Belg.*, **54**, 201–206.
111. Deutschman, C.S., Cereda, M., Ochroch, A., *et al.* (**2006**) Sepsis-induced cholestasis, steatosis, hepatocellular injury, and impaired hepatocellular regeneration are enhanced in interleukin-6 – /–mice. *Crit. Care Med.*, **34**, 2613–2620.
112. Antonelli, M., Levy, M., Andrews, P.J.D., *et al.* (**2007**) Hemodynamic monitoring in shock and implications for management. *Intensive Care Med.*, **33**, 575–590.
113. Le Gall, J.R., Klar, J., Lemeshow, S., *et al.* (**1996**) The Logistic Organ Dysfunction system. A new way to assess organ dysfunction in the intensive care unit. ICU Scoring Group. *JAMA*, **276**, 802–810.
114. Hotchkiss, R.S. and Karl, I.E. (**2003**) The pathophysiology and treatment of sepsis. *N. Engl. J. Med.*, **348**, 138–150.
115. Abraham, E. and Singer, M. (**2007**) Mechanisms of sepsis-induced organ dysfunction. *Crit. Care Med.*, **35**, 1–9.

PART II
Pathogens

Sepsis and Non-infectious Systemic Inflammation. Edited by J.-M. Cavaillon and C. Adrie

ISBN: 978-3-527-31935-0

5
Pathogens in Sepsis: Gram-negative Bacterial PAMPs and PRRs

Andra Schromm, Christian Alexander, Thomas Gutsmann, Jörg Andrä and Cordula Stamme

5.1
Introduction

In the US, the rate of sepsis increased from 83 cases per 100 000 population in 1979 to 240 cases per 100 000 population in 2000 corresponding to an annual increase of 8.7% [1]. In that study, mortality rates were up to 70%, depending on severity. Gram-positive bacteria (GPB) were the predominant causative organism after 1987, accumulating in 52% of cases in 2000 compared to 38% Gram-negative bacteria (GNB). The recently published Sepsis Occurrence in Acutely Ill Patients study across Europe showed a sepsis incidence of 35% in ICU patients, with a mortality rate of 27% and GPB in 40% of cases, and GNB in 38% of cases [2]. However, a re-emergence of Gram-negative health care-associated bloodstream infections between 1996 and 2003 was reported most recently, reflecting the inherent unanticipated evolution of microbial epidemiology [3]. The primary source of infection switched from intra-abdominal to pulmonary sites with pneumonia being the leading cause (~40%) of sepsis, followed by the abdomen (~20%), primary infections of the blood (15%), and the urinary tract (10%) [4]. Pulmonary dysfunction was the most common organ failure; its odds ratio for predicting ICU death was the highest [4]. Transmission of *Pseudomonas aeruginosa* in either community-acquired pneumonia or nosocomial pulmonary infection may have a mortality rate of up to 60%. The most recent National Nosocomial Infection Surveillance indicates that *P. aeruginosa* is the second most common cause of nosocomial pneumonia after *Staphylococcus aureus* [5]. Organisms that were formerly classified as primarily 'commensal', namely enterococci, *Escherichia coli*, as well as environmental organisms such as *P. aeruginosa* and *Acinetobacter baumannii*, have become emerging pathogens. Antimicrobial resistance is reported in ~30% of ICU infections involving GNB [6]. Infections due to antimicrobial-resistant bacteria, particularly GNB, are associated with higher mortality, prolonged hospitalization, and increased health-care costs and are now a serious threat to patients and clinicians world-wide [7, 8].

Sepsis and Non-infectious Systemic Inflammation. Edited by J.-M. Cavaillon and C. Adrie

ISBN: 978-3-527-31935-0

In this chapter we review the current knowledge on GNB-derived pathogen-associated molecular patterns (PAMPs), constituting highly conserved motifs of pathogens, and their cognate eukaryotic sensor proteins, known collectively as pattern recognition receptors (PRRs) [9]. Since endotoxin, (lipopolysaccharide, LPS), is the most potent GNB-derived mediator implicated in the pathogenesis of sepsis, this chapter focuses in detail on the endotoxic principles of LPS activity, the LPS signaling pathways and their negative regulators as well as the modulation of LPS biological activity.

5.2
Pathogen-associated Molecular Pattern (PAMP) Molecules of GNB

During the past decade, enormous progress has been made in the assignment of major PAMP molecules and the corresponding PRRs of innate immunity. Bacteria-specific structural motifs within several major bacterial PAMPs have been identified that activate individual members of PRRs (Table 5.1). Due to their sole presence in bacteria, these PAMP substructures apparently represent optimal molecular markers for the detection of infections. In mammalian innate immunity, a pivotal class of signaling PRRs designated as Toll-like receptors (TLRs) representing type1-transmembrane receptors localized on the cell surface or in endosomal compartments has been intensively characterized. In addition, two classes of cytoplasmic signaling PRRs termed nucleotide-binding oligomerization domain (Nod)-like receptors (NLRs) and retinoic acid-inducible gene (RIG)-I-like receptors (RLRs) have been identified [10–12]. Accessory PRRs such as LPS-binding protein (LBP) and CD14 assist particular types of signaling PRRs like the TLR4/MD-2- or the TLR2-dependent systems by promoting sensitivity in the recognition of particular PAMP molecules. Table 5.1 summarizes the currently assigned major classes of GNB-derived PAMPs, their particular immuno-activating substructures and the corresponding PRRs of mammalian innate immunity.

Aside from partial structures of bacterial nucleic acids such as non-methylated CpG oligonucleotides of DNA and as yet undefined bacterial RNA substructures, the majority of GNB PAMPs originates from the bacterial cell wall (Figure 5.1). The cell wall of GNB is characterized by the presence of two lipid bilayers, the outer membrane (OM) and the inner membrane (IM), that are separated by the periplasmic space containing a three-dimensional network of peptidoglycan (PG), also known as murein [13, 14]. The OM of common GNB such as *E. coli* is extremely asymmetrical. While the inner leaflet is composed of phospholipids, the outward directed leaflet is formed predominantly by lipid A, constituting about 75% of the total membrane surface. In addition to LPS, the polysaccharide coat on the OM surface of enterobacteria contains enterobacterial common antigen and, in some species and strains, capsular polysaccharides [15]. The inner layer of the OM contains the N-terminal acyl moieties of two major bacterial lipoproteins, murein lipoprotein (MLP) and

Table 5.1 PAMPs of GNB and PRRs of mammalian innate immunity.

GNB		Mammalian host	
PAMP	PAMP-specific substructure	Signaling PRR(s)	Accessory PRR(s)
LPS	Lipid A domain	TLR4/MD-2	LBP, CD14
Lipoproteins	S-Diacylglyceryl-N-acyl-oligopeptide	TLR2 *plus* TLR1	LBP, CD14
Flagella	D1-domain	TLR5 Ipaf or Naip5 (cytoplasmic)	n.a.
Bacterial DNA	Non-methylated CpG-oligonucleotides	TLR9	n.a.
mDAP-type peptidoglycan	Muramyldipeptide (MDP)	Nod2 (cytoplasmic)	n.a.
	iso-Glu-mDAP-dipeptide	Nod1 (cytoplasmic)	n.a.
Bacterial RNA	?	NALP3 (cytoplasmic)	n.a.

n.a., not assigned.

PG-associated lipoprotein (PAL) anchoring the OM to the periplasmic PG network in GNB [16]. In addition, many wt-strains of *E. coli* and *S. enterica* sv Typhimurium and several other species of GNB express flagellar filaments on their cell surface [17] representing a polymeric assembly of proteins of flagellin (FliC). Highly conserved evolutionary and bacteria-specific partial structures of LPS, lipoproteins, flagellin and PG (Table 5.1, Figures 5.2, 5.3) represent primary PAMP motifs that mediate the rapid activation of innate immunity at minute concentrations. The following sections describe the structural features of these major PAMPs of GNB in more detail.

5.2.1 Lipopolysaccharides

Lipopolysaccharides (LPS) are prototypic PAMPs of GNB. According to early observations on the induction of lethal septic shock in animal models, LPS have been synonymously termed as endotoxins [18, 19]. LPS represent a class of heat-stable amphiphilic molecules composed of a predominantly lipophilic and evolutionarily highly conserved membrane-anchoring region, lipid A, and a covalently linked hydrophilic poly- or oligosaccharide region. Based on the variability of the monosaccharide composition, the polysaccharide region in wild type (wt)-GNB is subdivided into the terminal highly variable O-specific chain and the phylogenetically more conserved core region most proximal to

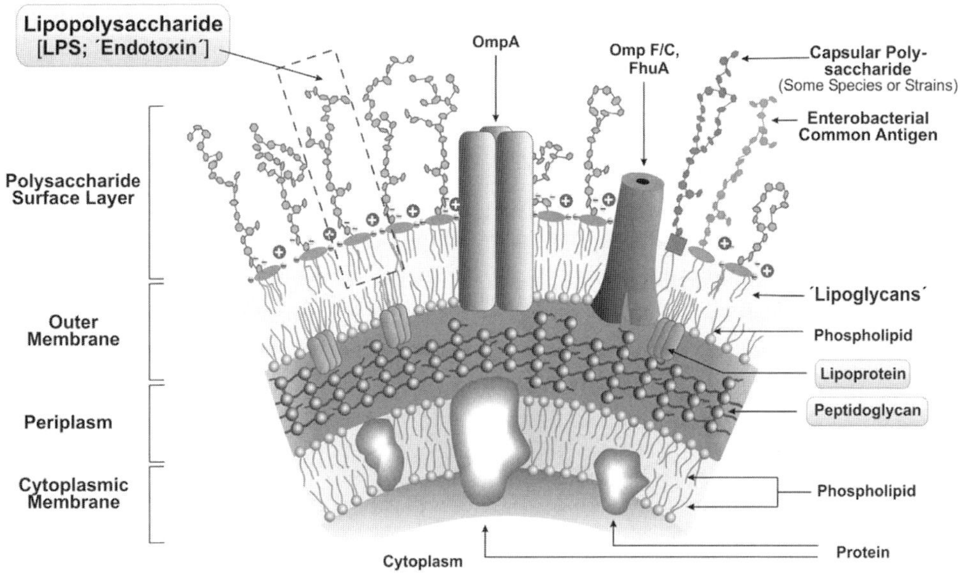

Figure 5.1 Cell wall-associated PAMPs of GNB. Scheme of the cell wall composition of Gram-negative enterobacteria such as *E. coli* and *S. enterica* sv. Typhimurium. LPS, BLP, and PG (highlighted) are major PAMPs of GNB.

lipid A (Figure 5.2A). LPS of these wt-strains containing a full core oligosaccharide and an O-specific chain are known as smooth (S)-forms of LPS. Many wt-species of pathogenic GNB colonizing the surface of the respiratory and urogenital tract, such as *N. meningitidis* or *H. influenzae*, lack the O-specific chain. LPS from these bacterial species are termed low molecular weight LPS (LMW-LPS) or lipooligosaccharides (LOS) representing natural equivalents of the rough (R)-forms of LPS found in certain mutant laboratory strains of enterobacteria [20–22].

The lipid A region is the minimal motif of LPS recognition by cellular and humoral PRRs of innate immunity [23]. Lipid A consists of a dimeric $\beta(1 \rightarrow 6)$-linked D-glucosamine unit carrying phosphate groups at position 1 of the proximal, reducing glucosamine (GlcNI) and at position 4′ of the distal, non-reducing glucosamine residue (GlcNII) (Figure 5.2B). This 1,4′-bisphosphorylated disaccharide backbone is asymmetrically acylated by amide- or esterbound (R)-3-hydroxymyristoyl residues at positions 2 and 3 of the GlcNI unit as well as by two characteristic acyloxyacyl groups residues linked to positions 2′ and 3′ of the GlcNII unit. The poly- or oligosaccharide region is linked to the primary hydroxyl function at position 6′ of the lipid A domain. In mammalian assay systems, maximal endotoxicity is induced by the complete hexaacylated and bisphosphorylated enterobacterial lipid A structure. Lipid A structures of LPS derived from various GNB species have been characterized

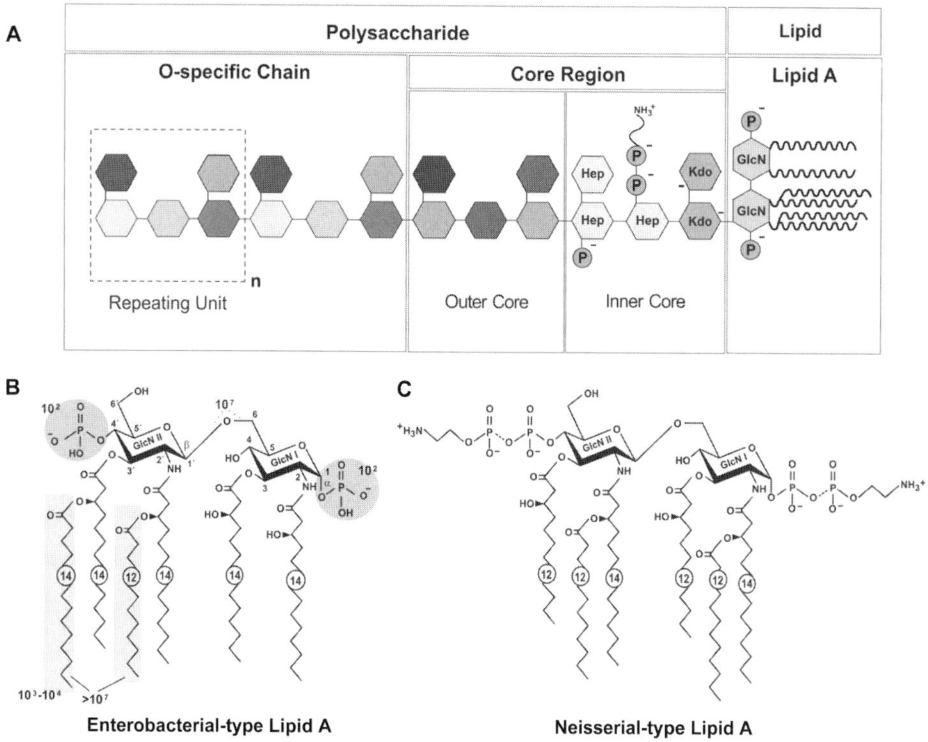

Figure 5.2 Scheme of LPS from GNB and chemical structures of highly endotoxic forms of lipid A. (A) General architecture of LPS from S-type enterobacteria. Hexaacylated lipid A structures of the major enterobacterial (B) and Neisserial (C) chemotypes display maximal endotoxic activities. Circled numbers represent the chain lengths of the corresponding acyl residues. The core oligosaccharide region (A) is bound to the position C6′ of the distal glucosamine (GlcNII). The approximate values of reduction in endotoxic activities due to removal or lack of individual structural elements of enterobacterial-type lipid A are indicated. In the symmetrically acylated form of lipid A from *N. meningitidis* (C) both backbone phosphate residues are 'capped' by phosphoethanolamine groups. GlcN, glucosamine; Kdo, '2-keto-3-deoxyoctulosonic acid' (3-deoxy-D-*manno*-octulosonic acid); Hep, D-glycero-D-*manno*-heptose.

in terms of biological activities. The majority of LPS is characterized by the predominant presence of asymmetrically acylated forms identical or rather similar to the (4 + 2) acylation pattern of enterobacterial-type lipid A (Figure 5.2B). Another major form of hexaacylated lipid A, characterized by a symmetrical distribution of the acyl residues along the backbone structure, was identified in the highly endotoxic LMW-LPS from *N. meningitidis* (Figure 5.2C). Several low or non-activating LPS chemotypes in certain human pathogenic bacteria such as *Legionella pneumophila* contain more exceptional lipid A structures as compared to the enterobacterial- or Neisserial-type of lipid A.

LPS belong to the most powerful classes of immunostimulators known, functioning physiologically as specific indicators of infection by GNB. The complex formed by TLR4 and the accessory protein MD-2 constitutes the central cellular signaling system that mediates the recognition of lipid A [24, 25]. The TLR4/MD-2 system is strongly enhanced by the combined actions of LBP and the membrane or soluble forms of CD14 that provide the opsonization of LPS aggregates or intact bacteria and the transfer of LPS to the receptor complex [26]. At acute-phase-like concentrations, LBP and CD14, in combination with plasma lipoproteins, are implicated in the neutralization and clearance of LPS [27–30]. Although increased plasma levels of endotoxin have been observed in some clinical setting of severe sepsis [31], the prognostic value of systemic endotoxin levels has remained relatively vague [32].

5.2.2 Lipoproteins of GNB

The lipoproteins of GNB generally consist of a phylogenetically conserved N-terminal lipid anchor linked to a variable polypeptide region that is responsible for the particular cellular function(s) of each individual lipoprotein species [33]. Human enterobacteria may express over 100 different lipoprotein genes. For instance, in the genome of *E. coli* K12 a set of 81 proven and an additional 44 predicted lipoproteins have been assigned including MLP and PAL, both constituting prototypic enterobacterial lipoproteins [34]. As initially revealed by the biochemical analysis of MLP from *E. coli*, the lipoproteins of Gram-negative enterobacteria share a common triacylated N-terminal lipid modification consisting of a characteristic N-acyl-S-diacylglyceryl-cysteinyl group (Figure 5.3A), representing the central immunologically active substructure of GNB-derived lipoproteins [35]. This triacylated lipopeptide activates mammalian cells via the specific heteromeric combination of TLR2 and TLR1 [36–38] with accessory functions of LBP and CD14 in TLR2-dependent immune cell activation [39]. Synergistic effects and reciprocal cross-tolerance induction by LPS and lipoproteins are documented in immune cell activation and corresponding effects of these major GNB-derived PAMPs may also occur during the course of septic disorders [40, 41].

5.2.3 Flagellins

Many wt-strains of *E. coli* and *S. enterica* sv Typhimurium and several other species of GNB express flagellar filaments of variable length (typically 5 to 10 μm) on their cell surface, known as flagellin [17]. Flagellin monomers constitute another class of potent activators of innate immune responses displaying a half-maximal cytokine induction in human myeloid cells at

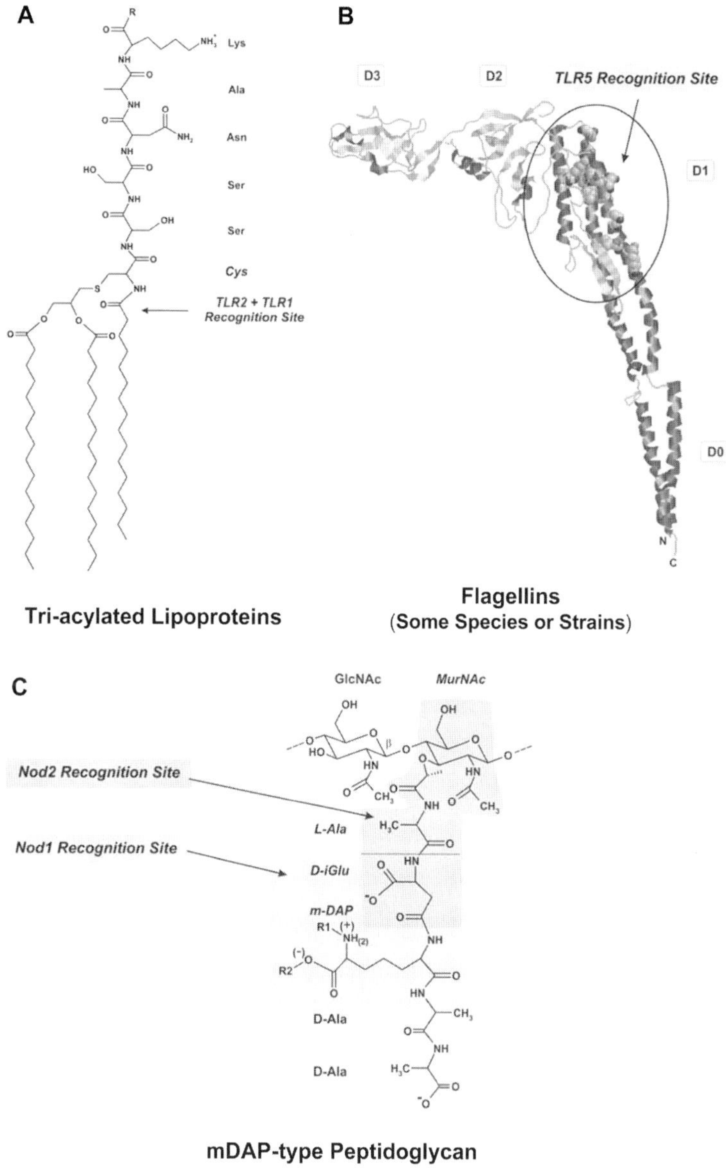

Figure 5.3 Activating domains in lipoproteins, flagellins and PG of GNB. As shown by the substructure of MLP from *E. coli* (A) the N-terminal lipopeptide domain including the highly conserved N-acyl-S-diacylglyceryl-cysteine moiety is recognized by the heteromeric assembly of TLR 2 and TLR1. The D1 domain of the flagellin monomer from *S. enterica* sv. Typhimurium (B) represents the particular recognition structure of TLR5 [45, 46]. In mDAP-type PGs (C), the common bacterial muramyldipeptide (MDP) unit [MurNAc-(L-Ala)-(D-iso-Glu)] and the (D-iso-Glu)-(mDAP) substructure of the stem peptide region (highlighted) activate Nod2 and Nod1, respectively. Assignment of groups R1 and R2 at the m-DAP residue in enterobacterial PG. (C) R1 = H or amide-linkage to D-Ala′ in a neigboring stem peptide, R2 = H or amide linkage to the side chain ε-amino group of the C-terminal Lys-residue in MLP.

picomolar concentrations [42–43]. TLR5 is the central surface receptor of bacterial flagellins in mammalian cells [44]. A highly conserved region in the D1 domain of flagellins [45, 46], essential for the multimeric formation of protofilaments, is the substructure of flagellin recognition by TLR5 (Figure 5.3B) [46]. Flagellins of *S. enterica* sv. Typhimurium and *Legionella pneumophila* have been identified to activate the cytoplasmic Nod-like receptors Ipaf and Naip5, respectively [47–49]. Since intravenous flagellin challenge of mice can induce acute lung inflammation even more pronounced than that induced by LPS, flagellin is considered to be another key pathogenicity factor in Gram-negative human sepsis [50]. Comparable to the cross-regulatory activities of LPS and lipoproteins, reciprocal synergistic and tolerance-inducing effects between flagellins and LPS or BLP have been observed in mammalian myeloid cells [51, 52].

5.2.4
m-DAP-type PG

Between the IM and the OM of GNB, the peptidoglycan (PG) macropolymer forms the cell wall skeleton. Bacterial PGs consist of linear glycan strands composed of β(1–4)-linked alternating units of D-N-acetylglucosamine (GlcNAc) and D-N-acetylmuramic acid (MurNAc) that are crosslinked by oligopeptide structures. These short oligopeptides connect MurNAc residues of single glycan strands in the three-dimensional murein network and essentially contain an evolutionarily highly conserved stem peptide substructure that may be expanded in a species-dependent fashion by a far more phylogenetically variable linker peptide [14, 53, 54]. The stem peptides in PG and corresponding biosynthetic precursor molecules are covalently attached to the lactyl groups of muraminic acid residues and share a common sequence motif of five amino acid residues including D-alanine (D-Ala), D-iso-glutamate (D-iGlu) or D-iso-glutamine (D-iGln) and an essential diamino acid, either meso-diaminopimelic acid (m-DAP) or L-lysine (L-Lys), that represents the central crosslinking residue. Accordingly, bacterial PGs have been classified into m-DAP- and L-Lys-type structures [55]. The mureins of enterobacteria and a variety of other GNB have been assigned to the m-DAP-type class of PGs [54, 55]. The chemical structure of the disaccharide-stem peptide unit present in PG of *E. coli* is shown in Figure 5.3. In PG, neighboring glycan chains are predominantly interlinked via a peptide bond formed in a transpeptidase reaction between the side-chain amino group of the m-DAP residue of one stem peptide unit and the carboxy group of the D-Ala residue at position 4 of another unit. About one half of the disaccharide–peptide repeating units of murein are involved in corresponding cross bridges [14, 53, 54]. Despite its apparent high overall stability, the murein network is constantly remodelled by *de novo* biosynthesis and recycling of murein [53, 54].

In mDAP-type PGs two partially overlapping substructures – the common muramyldipeptide (MDP) unit [MurNAc-(L-Ala)-(D-iso-Glu)] and the (D-iGlu)-(m-DAP)-dipeptide region (iE-DAP-motif) of the stem peptide activate Nod2 and Nod1, respectively [56–59]. Nod1 has been implicated in the immune recognition of invasive and intracellular pathogens such as *Shigella flexneri* [60] and *P. aeruginosa* [61]. Nod1-induced immune response appears to be relatively moderate compared to the strong response to TLR ligands. Acylated derivatives of the iE-DAP ligand have a strongly increased immune stimulation compared to the parent iE-DAP molecule [62]. One of these optimized ligands induces sepsis in mice in a Nod1-dependent manner, indicating that Nod1 might also be involved in acute inflammatory reactions to bacterial ligands [63]. Synergistic enhancing or priming effects of Nod1 and Nod2 stimulation on LPS-induced immunoactivation have been documented *in vitro* and *in vivo* [64–68]. Moreover, the outlined attachment of MLP and PAL to murein represents a particular combination of these major PAMP molecules that may make a considerable contribution to the pathogenesis of human septic diseases.

5.3
The LPS-activated TLR4/MD-2-signaling system: Central PRR Signaling Pathways and Negative Regulatory Factors

The TLR4/MD-2 complex is the central mammalian transmembrane signaling PRR mediating the recognition of the LPS lipid A domain (Figure 5.2). Loss of function [69, 70] and gene knockout studies have established the role of TLR4 and MD-2 in LPS recognition [71, 72]. TLR4/MD-2-activation by LPS/lipid A is assisted by an extracellular enhancement system provided by the combined actions of LBP and the soluble form of myeloid cell surface antigen CD14 (sCD14). The membrane form of CD14 (mCD14) serves as the main cellular acceptor of LPS delivered by LBP or sCD14 [73, 74]. Throughout the last decade the complex network of TLR4/MD-2-activated intracellular signaling pathways as well as several negative regulatory mechanisms, have been elaborated in detail (Figure 5.4).

Intriguing data recently published by Beutler and coworkers revealed that mCD14 is essential for activation of the TLR4/MD-2 complex by S-form LPS from wt-strains but not by R-form LPS of rough-mutant strains [75, 76]. Thus, TLR4/MD-2 distinguishes between LPS chemotypes, requiring CD14 for activation of MyD88-dependent signaling by S-LPS but not by R-LPS. For half a century, S-LPS has been considered to be the 'classical' form of endotoxically active LPS. However, these data inferred a paradigm shift.

In addition to mCD14, TLR4, and MD-2 other proteins such as heat shock proteins 70 and 90, CXCR5 [77], the potassium channel MaxiK [78–80], as well as a membrane-associated form of LBP [81, 82] are engaged in building

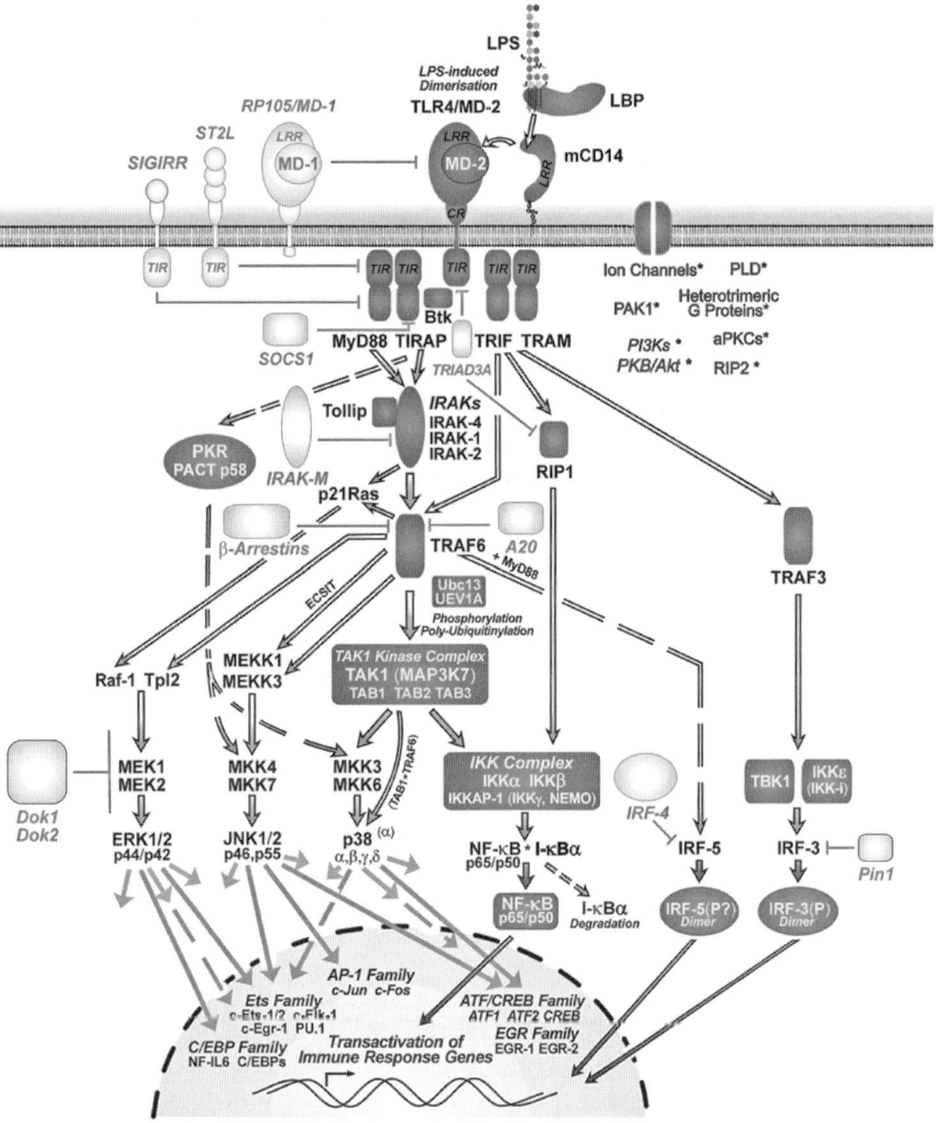

a complex receptor cluster contributing to LPS-induced signaling. Moreover, several studies suggest that recruitment of these additional receptor molecules into distinct membrane domains is necessary for the formation of an active signaling complex [83, 84] finally leading to LPS-induced homodimerization of TLR4 and subsequent transmembrane signaling [85, 86].

Following delivery of LPS to the cell surface, TLR4/MD-2-dependent transmembrane signaling is triggered by as yet not fully elucidated mechanisms leading to the rapid intracellular recruitment of four Toll/IL-1R homology (TIR) domain-containing adapter proteins: MyD88, TIRAP (Mal),

◀ **Figure 5.4** LPS-activated TLR4/MD-2-signal transduction pathway and negative regulators. Current scheme showing mechanisms of immune cell activation via the TLR4/MD-2-dependent system of LPS-induced signaling including negative-regulatory factors (light gray and dashed lines). Not yet fully assigned to the outlined network (asterisk): PI3Ks, phosphatidylinositol 3-kinases; PKB/Akt, protein kinase B/Akt kinase; PAK1, p21-activated kinase 1; aPKC, atypical protein kinase C; PLD, phospholipase D; RIP2, receptor interacting protein-2; and ion channels like the high-conductance Ca^{2+}-activated potassium (MaxiK) channel. LBP, LPS binding protein; mCD14, membrane-bound CD14; TLR4, Toll-like receptor 4. MD-2, myeloid differentiation protein-2; MyD88, myeloid differentiation factor 88; TIRAP, Toll-interleukin 1 receptor domain-containing adapter protein; TRIF, TIR domain-containing adapter inducing IFN-β; TRAM, Trif-related adaptor molecule; IRAK-1, IRAK-2, IRAK-4, interleukin-1 receptor-associated kinase-1, -2, -4; TRAF6, TRAF3, TNF receptor-associated factor 6 and 3; TAK1, TGFβ-activated kinase 1; NF-κB, nuclear factor κB; I-κB, inhibitor of NF-κB; IKKα, IKKβ, IKK-ε, I-κB kinases α, β, ε. IKKγ, I-κB kinase γ subunit; NEMO, NF-κB essential modulator; MAP kinases, mitogen-activated protein kinases; ECSIT, evolutionarily conserved signaling intermediate in Toll pathways; TAB1, TAB2, TAB3, TAK1-binding protein 1, 2, 3; MAP3Ks, mitogen-activated protein kinase kinase kinases. MEKK1, MEKK2, MEKK3, MAPK/ERK kinase kinase 1, 2, 3; MAP2Ks, MAP kinase kinases; ERK1, ERK2, Extracellular-signal-regulated kinase 1 and 2; JNK1, JNK2, c-Jun N-terminal kinase 1 and 2; Tpl2, tumor progression locus 2; RIP1, receptor interacting protein-1; TBK1, TANK-binding kinase-1; IRF, interferon regulatory factor. RP105, radioprotective 105-kDa protein; MD-1, myeloid differentiation protein-1; ST2L, protein of cDNA library clone ST2, membrane-bound form; SIGIRR, single immunoglobulin IL-1 receptor-related molecule; SOCS-1, suppressor of cytokine signalling-1; IRAK-M, interleukin-1 receptor-associated kinase-M; Dok1, Dok2, downstream of tyrosine kinase 1 and 2; Pin1, Peptidyl-prolyl cis-trans isomerase NIMA-interacting 1. LRR, leucine-rich repeat domain; CR, cysteine-rich region; TIR, Toll/IL-1 receptor homology domain; DD, death domain.

TRIF (TICAM-1), and TRAF (TICAM-2). Subsequently, MyD88- and TRIF-dependent activation of multiple intracellular signaling pathways is induced, including a set of NF-κB-activating pathways, the triad of major MAP kinase cascades and the activation of the interferon-regulatory factors IRF-3 and IRF-5 [10, 87]. The 'commmon core' subset of MyD88-dependent signaling pathways include the interleukin-1 receptor-associated kinase (IRAK) family members IRAK-1 and IRAK-4, the central adapter protein TRAF6 and the mitogen-activated kinase kinase kinase (MAP3K) TAK1. Moreover, the TLR4 signaling system has been found to be unique in the recruitment of the complete panel of known TIR domain-containing adapter proteins leading in turn to the downstream regulation of a considerably larger number of immunoresponsive genes as compared to other TLR family members [88, 89].

Several factors negatively regulate TLR4/MD-2 signaling and contribute to the fine-tuning or termination of LPS signaling (Figure 5.4). RP105, a TLR4 homolog lacking a signaling domain extracellularly associated with the MD-2 homolog MD-1, interferes with the formation of the activating LPS-TLR4/MD-2-receptor complex [90]. The TIR domain-containing transmembrane proteins ST2L, a key regulator in the induction of endotoxin tolerance, and SIGIRR inhibit the assembly of TLR4 signaling complexes by sequestering MyD88 and TIRAP, and by negative regulatory interactions with TLR4, IRAK-1 and TRAF6, respectively [91, 92]. Intracellularly, a suppressor of cytokine signaling 1 turns off the initial phase of TLR4 signaling by promoting the ubiquitination and degradation of the phosphorylated form of TIRAP (Mal) [93]. Moreover, the IRAK family member, IRAK-M is an inducible inhibitor of TLR4 signaling at the level of IRAK-4 and IRAK-1 activation [94]. TRIAD3A and A20 modulate the ubiquitination state of TLR4 and TRAF6, respectively. TRIAD3A is an E3 ubiquitin-protein ligase that mediates the ubiquitination and subsequent degradation of TLR4 [95], whereas A20 is a deubiquitinating enzyme removing ubiquitin chains from the activated form of TRAF6 [96]. In addition, binding of β-arrestins-1 and -2 to TRAF6 prevents its autoubiquitination and activation [97, 98]. The adapter proteins Dok1 and Dok2 exert the specific inhibition of MEK1/MEK2-mediated ERK1/ERK2 activation [99], whereas the peptidyl-prolyl isomerase Pin1 binds to the activated form of IRF-3 promoting the ubiquitination and degradation of this pivotal factor in the induction of IFN-β expression [100]. In addition, IRF-4 interferes with the IRF-5 pathway of TLR4 signaling by preventing IRF-5 binding to MyD88 [101, 102].

Central questions in TLR4/MD-2 signaling concern the initial molecular events in the formation of the LPS–receptor complex assembly and subsequent proximal signaling steps. Further, the ongoing analysis of 'PRR cross-talk' between individual TLRs and that between the TLR, NLR, and RLR families will provide a more detailed understanding of the integrated multisensory recognition of microbial pathogens.

5.4 Endotoxic Activity of Enterobacterial LPS is determined by Physico-chemical Properties

The recognition of LPS by the host immune system and the initiation of an inflammatory response, are the basic immunological processes promoting either a successful immune-defense or, at an unrestricted stage, the development of Gram-negative sepsis. Investigations into the structural prerequisites for GNB-derived LPS biological activity revealed that the immune response critically depends on a defined chemical structure of the lipid A moiety.

The physical state of biologically active LPS has long been obscure. Because of its amphiphilic structure, LPS forms large aggregates in an aqueous environment above a certain threshold concentration, i.e. the critical micellar concentration (CMC) [103]. At concentrations above the CMC, aggregates are in a dynamic equilibrium with monomers which remain at a constant concentration corresponding to the CMC. Information on the CMC value for LPS or lipid A is sparse. Measurements with partial structures indicate very low CMC values for lipid A ($<10^{-8}$ M, [104, 105]. Data reported by Takayama *et al.* [106, 107] indicate for deep rough mutant LPS Re values between 10^{-7} and 10^{-8} M.

The aggregates are very stable even after dilution to concentrations below the CMC. These data suggest that aggregates predominate in the concentration range relevant for biological responses. The role of aggregates or monomers in the process of cell activation has been controversially discussed [108–110]. Monomers dissociated from aggregate suspensions of LPS or lipid A in a dialysis system do not confer activation of human mononuclear cells (MNC) which sensitively respond with cytokine production to equal concentrations of the corresponding aggregate suspensions, indicating that monomers themselves are not able to induce cell activation, whereas aggregates are biologically active [111]. Thus, in the initial phase of cell activation the presence of aggregates seems to be decisive.

Serum proteins have a peculiar role in the recognition and transport of LPS. In particular, LBP and sCD14 interact with LPS aggregates, mediate the transport of LPS to the cell surface and enhance cell activation in response to minute amount of LPS. However, several pathways for these transport processes have been suggested which might operate alternatively or in parallel, depending on the conditions. LBP has initially been described to extract LPS monomers from aggregates and to subsequently transfer these to mCD14 on the host cell [112]. Gioannini *et al.* [113] recently suggested a vectorial transport along soluble proteins, with LBP binding to the LPS aggregate, opsonizing it for sCD14 which then monomerizes the aggregate. This sCD14–LPS monomer complex binds to sMD-2, and concomitantly to TLR4 [114, 115]. The model proposes that this protein-bound monomer first interacts with the host cell. A third transport pathway has recently been suggested operating via a membrane-bound form of LBP (mLBP). This cell-associated LBP is present on MNC and is important for cell activation by LPS [30]. Employing model membranes, a transport function was assigned to mLBP leading to an intercalation of LPS into the phospholipid membrane [82, 116, 117]. Intercalation of LPS into the host cell membrane changes the physical functionality of the cytoplasmic membrane by altering the membrane fluidity and provides an alternative pathway of bringing LPS into close proximity with the proteins of the signaling complex. Elucidation of the participation of the different LPS transport pathways and proteins involved, which might *in vivo* operate in parallel, will help to provide a better understanding of the initial phase of cell activation.

The three-dimensional structure of the aggregate strongly depends on the chemical structure of the aggregate-forming molecules. Depending on the ratio of the cross-sections of the hydrophilic and hydrophobic moieties of the lipid A moiety, different types of aggregates can be formed: micellar in the case of a significantly greater cross-section of the hydrophilic than the hydrophobic moiety, lamellar for similar cross-sections of both moieties, and nonlamellar inverted (cubic, inverted hexagonal II) for lipid A molecules with a greater cross-section in the hydrophobic than the hydrophilic moiety. The aggregate structure can be investigated by small-angle X-ray diffraction (SAXS) measurements on LPS suspensions under near physiological conditions. From the aggregate structure determined, the molecular shape of the individual lipid A molecules forming the aggregate can be deduced to be cylindrical for lamellar aggregates and conical for nonlamellar-inverted aggregates. Investigations of a variety of LPS and lipid A samples demonstrate that only conical molecules, which form cubic inverted aggregate structures, exhibit high endotoxic activity, whereas cylindrical molecules, forming lamellar aggregate structures, exhibit low or no endotoxic activity [118]. Importantly, enterobacterial lipid A prefers to adopt nonlamellar cubic structures under near physiological conditions. Thus, the shape of the aggregate-forming molecules correlates with the ability to activate the innate immune response of the host.

This principle of an 'endotoxic conformation' has been extended to bacterial virulence factors of the lipopeptide family. The synthetic lipopeptides Pam_2CSK_4 and Pam_3CSK_4 like LPS form cubic-inverted aggregates structures, whereas the lipolanthionine peptide lipolan associates with lamellar aggregate structures. Consequently, the former two structures express high biological activity whereas the latter is inactive but expresses antagonistic activity [119], suggesting that the mechanisms of cell activation by amphiphilic molecules are governed by a general principle.

5.5 Modulation of Biological Activity of LPS

The immune cell-stimulating activity of LPS may be modulated by direct interaction of host proteins and peptides with LPS, indirectly by molecules interfering with LPS-recognition proteins, or with components of the intracellular LPS-signaling cascade. The direct binding of cationic peptides to lipid A results in neutralization of the biological activity of LPS both *in vitro* and *in vivo*. Prominent representatives of this group of proteins are bactericidal/permeability-increasing protein (BPI) [120] and lactoferrin [121]. Plasma lipoproteins, namely chylomicrons, very low density lipoproteins (VLDL), low density lipoproteins (LDL), and high density lipoproteins (HDL) have a significant impact on LPS-induced cell activation both *in vitro* and *in vivo* [122, 123]. *In vivo*, circulating lipoproteins significantly contribute to the clearance of LPS [124, 125]. However, the mechanisms involved are only partially understood [126–128]. Collectins inhibit LPS activity by a diversity of mechanisms [129]. In addition, acyloxyacyl hydrolase, an endogenous lipase, detoxifies LPS by partial enzymatic cleavage of the lipid A acyl chains [130], and liver alkaline phosphatase generates inactive monophosphoryl lipid A [131].

5.5.1 Antimicrobial Peptides and Proteins

Antimicrobial peptides (AMPs), also referred to as host defense peptides, are part of the innate immunity of almost all species [132]. In mammals, they provide an effective first line of defense against invading pathogens on epithelial surfaces (e.g. β-defensins [133] and cathelicidins [134]), in body fluids (e.g. lactoferrin [135]), or as part of the antimicrobial armament of T-cells and NK-cells (e.g. granulysin [136] and NK-lysin [137]). Despite great heterogeneity in origin and amino acid sequence, AMPs have in common a positive net charge in combination with an amphipathic secondary structure. These properties are prerequisites for their interaction with negatively-charged bacterial membranes and membrane components such as LPS and anionic phospholipids. In addition to a direct lytic action on bacterial membranes resulting in instant cell death, they bind to and partially neutralize LPS activity. This anti-endotoxic activity has been observed *in vitro* and *in vivo* [138], and has inspired great interest in these compounds as promising structures for drug development [139].

The LPS-neutralizing peptides that have been reviewed are fragments of larger proteins, liberated under physiological conditions (lactoferricin and LL-37), or represent the active part of a parent molecule (NK-2). *Lactoferrin* is an iron-binding protein with bacteriostatic and bactericidal activity against GNB. It modulates the inflammatory process by preventing the release of cytokines by monocytes and by regulating the proliferation and differentiation of immune cells. Lactoferrin has been isolated from breast milk, tear fluid, vaginal secretions, gut-lining fluid, exocrine and respiratory secretions, and is released from polymorphonuclear neutrophils (PMN) in inflamed areas [140]. Lactoferrin bears a low and a high affinity LPS-binding site. One binding site is located in the N-terminal fragment of lactoferrin, and is known as *lactoferricin*, which is released *in vivo* after pepsin cleavage [135, 141].

Cathelicidins are another important group of proteins with antimicrobial and LPS-neutralizing properties identified in various mammals including humans. Cathelicidins contain a highly conserved N-terminal domain called cathelin and a C-terminal domain that comprises an antimicrobial peptide. Human (h) and rabbit (r) cathelicidins are termed 18-kDa cationic antibacterial protein hCAP18 and rCAP18, respectively. Cathelicidin or its processed C-terminal active 37-residue, α-helical peptide *LL-37*, is present in granules of PMN [134], alveolar macrophages, bronchial epithelial cells, and bronchial glands [142]. It has been isolated from seminal plasma, blood, sweat, saliva, surface airway liquid, and the urinary tract [143]. The LPS-neutralizing activity was first described by Hirata *et al.* showing that rCAP18 and its synthetic N-terminal fragment inhibits LPS-induced tissue factor generation by human blood monocytes [144]. The same authors demonstrated that synthetic peptides significantly protect C57BL/6 mice from lethal LPS challenge. Lung epithelial injury and edema formation caused by the instillation of the cytotoxic strain *P. aeruginousa* PA103 was reduced by the concomitant instillation of the peptide and, moreover, it decreased the release of TNFα, IL-6, and nitric oxide, induced by the application of the antibiotic aztreonam.

NK-lysin is a representative of the lytic saposin-like proteins which are structurally characterized by a compact α-helical fold, stabilized by disulfide bonds [145]. Mammalian homologs have been found in human (granulysin) [146], cattle [147], and horse [148]. NK-lysin binds to S-form LPS from *E. coli, P. aeruginosa,* and *S. minnesota,* and inhibits the LPS-induced cell activation of murine bone marrow cells and the binding of LPS to granulocytes [149]. Most strikingly, NK-lysin is an effective LPS neutralizer in the galactosamine-sensitized sepsis mouse model [149]. A 27-amino acid residue peptide derivative of NK-lysin, termed *NK-2,* was found to inhibit LPS-induced release of TNFα by human monouclear cells [150].

5.5.2
Mechanisms of LPS Neutralization by Cationic Peptides

Lactoferrin binds to lipid A and reduces LPS-induced cytokine release by MNC [141], while synthesized lactoferricin-derived peptides based on the bovine sequence bind to charges in the core oligosaccharide rather than to lipid A [151]. Binding sites for NK-lysin were proposed in the lipid A and oligosaccharide moieties of LPS [149]. NK-2, the active fragment of NK-lysin, prevents LPS-induced pro-inflammatory cytokine release by human MNC and macrophages in part by competing with LBP for LPS binding [150]. Since the NK-2 concentrations necessary to completely neutralize LPS progressively increase from Re-, R- to S-form LPS, the oligosaccharide moiety of LPS seems to protect bacteria by serving as a trap for this cationic peptide [150]. A similar effect has been observed for LF11 and lauryl-LF11, peptides derived from human lactoferricin [152]. To be an effective inhibitor of LPS activity, peptides must bind to LPS with sufficient affinity to compete with LBP and CD14. In fact, LL-37 and derived peptides inhibit the binding of LPS to LBP and, furthermore, bind to mCD14 on human immune cells thereby inhibiting the binding of LPS to these cells [153]. LL-37 directly competes with CD14 for LPS binding and partially displaces FITC-LPS bound to CD14 on RAW264.7 cells [153]. The structural origin of the neutralization of ReLPS by LF11 was analyzed by NMR spectroscopy and molecular modeling. LF11 folds into a T-shaped arrangement of a hydrophobic core and two clusters of basic residues that match exactly the distance between the two phosphate groups in the lipid A region of LPS [154]. Hence demonstrating the importance of molecular fit combined with electrostatic and hydrophobic interactions.

By comparing an 11-amino acid residue fragment of LF11 with its acylated counterpart (lauryl-LF11) it was found that lauryl-LF11 exhibits much stronger antimicrobial activity and inhibition of LPS-induced cytokine release. Higher bioactivity of lauryl-LF11 is associated with a higher ability to (over) compensate the surface charges of LPS aggregates. LPS aggregate structures were changed from a unilamellar/inverted cubic into a multilamellar form [152]. Charge overcompensation and reaggregation of LPS aggregates have also been observed with the NK-lysin-derived peptide NK-2. Multilamellar stacks of LPS bilayers, induced by a granulysin-derived peptide, were visualized by freeze-fracture electron microscopy [155]. According to the conformational concept of endotoxicity, LPS in this supramolecular arrangement no longer functions as an activator of immune cells. An interesting study by Shai *et al.* also demonstrated an effect of LL-37 on the aggregation state of LPS [156]. Hancock *et al.* demonstrated that LL-37 inhibits LPS-mediated immune cell activation and protects mice from lethal endotoxemia [157, 158]. Gene expression profiling on macrophages revealed that LL-37 directly up-regulates 29 genes encoding chemokines and chemokine receptors. LL-37 consistently up-regulated the expression of MCP-1 and IL-8 without stimulating the expression of TNFα and in part, altered gene expression by acting on the TLR-NFκB pathway [157, 158].

5.5.3 Pulmonary Collectins

Pulmonary surfactant is a lipoprotein complex that is synthesized by type II pneumocytes and by airway Clara cells and is secreted into the alveolar liquid layer of the lung. The unique composition of surfactant facilitates the functional combination of different biological effects such as preventing alveolar collapse at expiration and immunomodulating innate and adaptive pulmonary host defense responses [129]. Immunomodulation is primarily mediated by surfactant proteins (SP)-A and SP-D both belonging together with the serum mannose binding protein (MBP), to a family of mammalian C-type lectins that have in common an N-terminal collagen-like region and a C-terminal carbohydrate recognition domain (CRD) [159]. MBP, mainly produced in the liver and secreted into the bloodstream, can neutralize LPS by preventing its binding to TLR4/MD-2 or by activating complement (C3) [160, 161]. SP-A and SP-D bind and aggregate a variety of GNB and GPB and enhance their phagocytosis and killing both *in vivo* and *in vitro* [129]. SP-A-deficient mice exhibit delayed microbial clearance and higher levels of bronchoalveolar inflammatory mediators after intratracheal inoculation with clinically relevant pathogens [162], and isolated LPS [163]. SP-A and SP-D directly interact with alveolar macrophages (AM), the major effector cells of the pulmonary innate immune system, through binding to cell surface receptors, resulting in the modulation of chemotaxis, phagocytosis, and modified pro- or anti-inflammatory immune responses.

Airway inflammation associated with local or systemic LPS release from GNB is still a major cause of life-threatening pulmonary diseases [164]. Efficient negative signaling cascades to prevent autotoxic mediator release by AM in response to LPS and, in particular, the modulation of NF-κB activation threshold by pulmonary collectins, have been studied intensively in recent years. SP-A has been shown to inhibit LPS-induced TNF-α production [165–167], inducible nitric oxide synthase protein expression [168] and NF-κB activity [166, 169, 170] in immunocompetent cells. Inhibition of LPS-induced NF-κB activity by SP-A and SP-D has been suggested to occur via direct interaction of the collectins with components of the LPS receptor complex, including LBP, CD14, TLR-4, and MD-2, but also independently of LPS-specific signaling pathways [166–172]. Intracellulary, SP-A can exert its anti-inflammatory effects on LPS-challenged AM via a mechanism involving an SP-A-mediated direct modulation of the basal and LPS-coupled IκB-α (inhibitor of NF-κB) turnover in these cells [170]. In addition, the collectins selectively permeabilize GNB [173]. Future experiments will explore the temporal regulation of direct and indirect antimicrobial effects of SP-A and SP-D on the pulmonary immune balance.

5.6 Therapeutic Approaches

Evidently, factors other than sepsis itself have a significant impact on patients' survival. These include the severity of the underlying disease, the presence of comorbidities, the severity of hemodynamic disturbances, and the degree of resultant coagulopathy, all of which are treated in parallel in an established and multifaceted clinical setting. Importantly, the role of antibiotics in the treatment of Gram-negative infection is essential for the foreseeable future [8]. Three broad therapeutic strategies aimed at blocking TLR function have been suggested recently [174] and include the use of soluble TLRs, peptides or antibodies that interfere with extracellular domains of TLRs, and interference with intracellular events such as the recruitment of pivotal adapter proteins. Among several potent LPS antagonistic compounds identified and characterized, the lipid A mimetic antagonist E5564 is currently the most advanced and its efficacy has been tested in phase II clinical trials [175, 176]. The applicability of synthetic antimicrobial peptides derived from naturally-occurring templates as adjunctive therapy is under debate. These peptides can combine antimicrobial activity, a low cytotoxic activity and the capability to neutralize PAMPs, e.g. LPS. However, problems concerning potential development of resistances, enzymatic degradation in serum, and relatively high cost of the peptides have to be solved. The studies on pulmonary collectins described in this chapter raise the intriguing possibility that therapy with SP-A, SP-D, or both might be efficacious in treating Gram-negative infectious diseases. As recently proposed by Hopkins and Cohen [177], these strategies, due to a number of inherent limitations when considered as mono-interventions, will help to improve established therapies. The optimal targeting of established and/or novel therapeutic approaches towards specific Gram-negative-infected patient subgroups comprising the identification and valuation of anatomic sites of infection is a pivotal challenge for both clinicians and basic scientists aiming to reduce sepsis mortality.

Acknowledgments

This chapter is dedicated to Professor Ernst Theodor Rietschel in grateful recognition of his encouragement, sound scientific advice and inspiring enthusiasm, all of which we strive to perpetuate through our students and fellows.

Grant Support: DFG STA 609/1-4, Emmy Noether Grant SCHR 621/2-2, SFB 617 (project A17), Clinique La Prairie Research.

References

1. Martin, G.S., Mannino, D.M., Eaton, S. and Moss, M. (**2003**) The epidemiology of sepsis in the United States from 1979 through 2000. *N. Engl. J. Med.*, **348**, 1546–1554.
2. Vincent, J.L., Sakr, Y., Sprung, C.L., Ranieri, V.M., Reinhart, K., Gerlach, H., Moreno, R., Carlet, J., Le, G., Jr. and Payen, D. (**2006**) Sepsis in European intensive care units: results of the SOAP study. *Crit. Care Med.*, **34**, 344–353.
3. Albrecht, S.J., Fishman, N.O., Kitchen, J., Nachamkin, I., Bilker, W.B., Hoegg, C., Samel, C., Barbagallo, S., Arentzen, J. and Lautenbach, E. (**2006**) Reemergence of gram-negative health care-associated bloodstream infections. *Arch. Intern. Med.*, **166**, 1289–1294.
4. Guidet, B., Aegerter, P., Gauzit, R., Meshaka, P. and Dreyfuss, D. (**2005**) Incidence and impact of organ dysfunctions associated with sepsis. *Chest*, **127**, 942–951.
5. NNIS System (**2003**) National Nosocomial Infections Surveillance (NNIS) System Report, data summary from January 1992 through June 2003. *Am. J. Infect. Control*, **31**, 481–498.
6. Merz, L.R., Warren, D.K., Kollef, M.H., Fridkin, S.K. and Fraser, V.J. (**2006**) The impact of an antibiotic cycling program on empirical therapy for gram-negative infections. *Chest*, **130**, 1672–1678.
7. Neuhauser, M.M., Weinstein, R.A., Rydman, R., Danziger, L.H., Karam, G. and Quinn, J.P. (**2003**) Antibiotic resistance among gram-negative bacilli in US intensive care units: implications for fluoroquinolone use. *JAMA*, **289**, 885–888.
8. Levy, S.B. and Marshall, B. (**2004**) Antibacterial resistance worldwide: causes, challenges and responses. *Nat. Med.*, **10**, S122–S129.
9. Janeway, C.A., Jr. and Medzhitov, R. (**2002**) Innate immune recognition. *Annu. Rev. Immunol.*, **20**, 197–216.
10. Akira, S., Uematsu, S. and Takeuchi, O. (**2006**) Pathogen recognition and innate immunity. *Cell*, **124**, 783–801.
11. Creagh, E.M. and O'Neill, L.A. (**2006**) TLRs, NLRs and RLRs: a trinity of pathogen sensors that co-operate in innate immunity. *Trends Immunol.*, **27**, 352–357.
12. Trinchieri, G. and Sher, A. (**2007**) Cooperation of Toll-like receptor signals in innate immune defence. *Nat. Rev. Immunol.*, **7**, 179–190.
13. Nikaido, H. (**1996**) Outer Membrane. In *Escherichia coli and Salmonella: Cellular and Molecular Biology*, Neidhardt, F.C., *et al.* (eds.), 2nd ed., pp 29–47, American Society for Microbiology.
14. Park, J.T. (**1996**) The Murein Sacculus. In *Escherichia coli and Salmonella: Cellular and Molecular Biology*, Neidhardt, F.C., *et al.* (eds.), 2nd ed., pp 48–57, American Society for Microbiology.
15. Rick, P.D. and Silver, R.P. (**1996**) Enterobacterial Common Antigen and Capsular Polysaccharides. In *Escherichia coli and Salmonella: Cellular and Molecular Biology*, Neidhardt, F.C., *et al.* (eds.), 2nd ed., pp 104–122, American Society for Microbiology.
16. Wu, H.C. (**1996**) Biosynthesis of Lipoproteins. In *Escherichia coli and Salmonella: Cellular and Molecular Biology*, Neidhardt, F.C. *et al.* (eds.), 2nd ed., pp 1005–1014, American Society for Microbiology.

17. Mac Nab, R.M. (**1996**) Flagella and Motility. In *Escherichia coli and Salmonella: Cellular and Molecular Biology*, Neidhardt, F.C., *et al.* (eds.), 2nd ed., pp 123–145, American Society for Microbiology.
18. Rietschel, E.T. and Westphal, O. (**1999**) Endotoxin: Historical Perspectives. In *Endotoxin in Health and Disease*, Brade, H., Opal, S.M., Vogel, S.N. and Morrison, D.C. (eds.), pp 1–30, Marcel Dekker, Inc., New York, Basel.
19. Beutler, B. and Rietschel, E.T. (**2003**) Innate immune sensing and its roots: the story of endotoxin. *Nat. Rev. Immunol.*, **3**, 169–176.
20. Rietschel, E.T., Brade, H., Holst, O., Brade, L., Muller-Loennies, S., Mamat, U., Zähringer, U., Beckmann, F., Seydel, U., Brandenburg, K., Ulmer, A.J., Mattern, T., Heine, H., Schletter, J., Loppnow, H., Schonbeck, U., Flad, H.D., Hauschildt, S., Schade, U.F., Di, P.F., Kusumoto, S. and Schumann, R.R. (**1996**) Bacterial endotoxin: Chemical constitution, biological recognition, host response, and immunological detoxification. *Curr. Top. Microbiol. Immunol.*, **216**, 39–81.
21. Holst, O., Ulmer, A.J., Brade, H., Flad, H.D. and Rietschel, E.T. (**1996**) Biochemistry and cell biology of bacterial endotoxins. *FEMS Immunol. Med. Microbiol.*, **16**, 83–104.
22. Alexander, C. and Rietschel, E.T. (**2001**) Bacterial lipopolysaccharides and innate immunity. *J. Endotoxin Res.*, **7**, 167–202.
23. Zähringer, U., Lindner, B. and Rietschel, E.T. (**1999**) Chemical Structure of Lipid A: recent advances in structural analysis of biologically active molecule. In *Endotoxin in Health and Disease*, Brade, H., Opal, S.M., Vogel, S.N. and Morrison, D.C. (eds.), pp 93–114, Marcel Dekker, Inc., New York, Basel.
24. Beutler, B. (**2002**) TLR4 as the mammalian endotoxin sensor. *Curr. Top. Microbiol. Immunol.*, **270**, 109–120.
25. Akira, S. and Takeda, K. (**2004**) Toll-like receptor signalling. *Nat. Rev. Immunol.*, **4**, 499–511.
26. Schumann, R.R. and Latz, E. (**2000**) Lipopolysaccharide-binding protein. *Chem. Immunol.*, **74**, 42–60.
27. Zweigner, J., Gramm, H.J., Singer, O.C., Wegscheider, K. and Schumann, R.R. (**2001**) High concentrations of lipopolysaccharide-binding protein in serum of patients with severe sepsis or septic shock inhibit the lipopolysaccharide response in human monocytes. *Blood*, **98**, 3800–3808.
28. Kitchens, R.L. and Thompson, P.A. (**2003**) Impact of sepsis-induced changes in plasma on LPS interactions with monocytes and plasma lipoproteins: roles of soluble CD14, LBP, and acute phase lipoproteins. *J. Endotoxin Res.*, **9**, 113–118.
29. Hamann, L., Alexander, C., Stamme, C., Zähringer, U. and Schumann, R.R. (**2005**) Acute-phase concentrations of lipopolysaccharide (LPS)-binding protein inhibit innate immune cell activation by different LPS chemotypes via different mechanisms. *Infect. Immun.*, **73**, 193–200.
30. Gutsmann, T., Müller, M., Carroll, S.F., MacKenzie, R.C., Wiese, A. and Seydel, U. (**2001**) Dual role of lipopolysaccharide (LPS)-binding protein in neutralization of LPS and enhancement of LPS-induced activation of mononuclear cells. *Infect. Immun.*, **69**, 6942–6950.
31. Marshall, J.C., Foster, D., Vincent, J.L., Cook, D.J., Cohen, J., Dellinger, R.P., Opal, S., Abraham, E., Brett, S.J., Smith, T., Mehta, S., Derzko, A. and Romaschin, A. (**2004**) Diagnostic and prognostic implications of endotoxemia in critical illness: results of the MEDIC study. *J Infect. Dis.*, **190**, 527–534.
32. Opal S.M., Scannon P.J., Vincent J.L., White M., Carroll S.F., Palardy J.E., Parejo N.A., Pribble J.P. and Lemke J.H. (**1999**) Relationship between plasma levels of

lipopolysaccharide (LPS) and LPS-binding protein in patients with severe sepsis and septic shock. *J. Infect. Dis.*, **180**, 1584–1589.

33. Hayashi, S. and Wu, H.C. (**1990**) Lipoproteins in bacteria. *J. Bioenerg. Biomembr.*, **22**, 451–471.
34. Gonnet, P., Rudd, K.E. and Lisacek, F. (**2004**) Fine-tuning the prediction of sequences cleaved by signal peptidase II: a curated set of proven and predicted lipoproteins of Escherichia coli K-12. *Proteomics.*, **4**, 1597–1613.
35. Bessler, W.G., Cox, M., Lex, A., Suhr, B., Wiesmüller, K.H. and Jung, G. (**1985**) Synthetic lipopeptide analogs of bacterial lipoprotein are potent polyclonal activators for murine B lymphocytes. *J. Immunol.*, **135**, 1900–1905.
36. Aliprantis, A.O., Yang, R.B., Mark, M.R., Suggett, S., Devaux, B., Radolf, J.D., Klimpel, G.R., Godowski, P. and Zychlinsky, A. (**1999**) Cell activation and apoptosis by bacterial lipoproteins through toll-like receptor-2. *Science*, **285**, 736–739.
37. Takeuchi, O., Sato, S., Horiuchi, T., Hoshino, K., Takeda, K., Dong, Z., Modlin, R.L. and Akira, S. (**2002**) Cutting edge: role of Toll-like receptor 1 in mediating immune response to microbial lipoproteins. *J. Immunol.*, **169**, 10–14.
38. Nakata, T., Yasuda, M., Fujita, M., Kataoka, H., Kiura, K., Sano, H. and Shibata, K. (**2006**) CD14 directly binds to triacylated lipopeptides and facilitates recognition of the lipopeptides by the receptor complex of Toll-like receptors 2 and 1 without binding to the complex. *Cell Microbiol.*, **8**, 1899–1909.
39. Schröder, N.W., Heine, H., Alexander, C., Manukyan, M., Eckert, J., Hamann, L., Gobel, U.B. and Schumann, R.R. (**2004**) Lipopolysaccharide binding protein binds to triacylated and diacylated lipopeptides and mediates innate immune responses. *J. Immunol.*, **173**, 2683–2691.
40. Sato, S., Nomura, F., Kawai, T., Takeuchi, O., Muhlradt, P.F., Takeda, K. and Akira, S. (**2000**) Synergy and cross-tolerance between toll-like receptor (TLR) 2- and TLR4-mediated signaling pathways. *J. Immunol.*, **165**, 7096–7101.
41. Liang, M.D., Bagchi, A., Warren, H.S., Tehan, M.M., Trigilio, J.A., Beasley-Topliffe, L.K., Tesini, B.L., Lazzaroni, J.C., Fenton, M.J. and Hellman, J. (**2005**) Bacterial peptidoglycan-associated lipoprotein: a naturally occurring toll-like receptor 2 agonist that is shed into serum and has synergy with lipopolysaccharide. *J. Infect. Dis.* **191**, 939–948.
42. Wyant, T.L., Tanner, M.K. and Sztein, M.B. (**1999**) Salmonella typhi flagella are potent inducers of proinflammatory cytokine secretion by human monocytes. *Infect. Immun.*, **67**, 3619–3624.
43. McDermott, P.F., Ciacci-Woolwine, F., Snipes, J.A. and Mizel, S.B. (**2000**) High-affinity interaction between gram-negative flagellin and a cell surface polypeptide results in human monocyte activation. *Infect. Immun.*, **68**, 5525–5529.
44. Hayashi, F., Smith, K.D., Ozinsky, A., Hawn, T.R., Yi, E.C., Goodlett, D.R., Eng, J.K., Akira, S., Underhill, D.M. and Aderem, A. (**2001**) The innate immune response to bacterial flagellin is mediated by Toll-like receptor 5. *Nature*, **410**, 1099–1103.
45. Samatey, F.A., Imada, K., Nagashima, S., Vonderviszt, F., Kumasaka, T., Yamamoto, M. and Namba, K. (**2001**) Structure of the bacterial flagellar protofilament and implications for a switch for supercoiling. *Nature*, **410**, 331–337.
46. Smith, K.D., Andersen-Nissen, E., Hayashi, F., Strobe, K., Bergman, M.A., Barrett, S.L., Cookson, B.T. and Aderem, A. (**2003**) Toll-like receptor 5 recognizes a conserved site on flagellin required for protofilament formation and bacterial motility. *Nat. Immunol.*, **4**, 1247–1253.
47. Miao, E.A., Alpuche-Aranda, C.M., Dors, M., Clark, A.E., Bader, M.W., Miller, S.I. and Aderem, A. (**2006**) Cytoplasmic flagellin activates

caspase-1 and secretion of interleukin 1beta via Ipaf. *Nat. Immunol.*, **7**, 569–575.
48. Franchi, L., Amer, A., Body-Malapel, M., Kanneganti, T.D., Ozoren, N., Jagirdar, R., Inohara, N., Vandenabeele, P., Bertin, J., Coyle, A., Grant, E.P. and Nunez, G. (**2006**) Cytosolic flagellin requires Ipaf for activation of caspase-1 and interleukin 1beta in salmonella-infected macrophages. *Nat. Immunol.*, **7**, 576–582.
49. Molofsky, A.B., Byrne, B.G., Whitfield, N.N., Madigan, C.A., Fuse, E.T., Tateda, K. and Swanson, M.S. (**2006**) Cytosolic recognition of flagellin by mouse macrophages restricts Legionella pneumophila infection. *J. Exp. Med.*, **203**, 1093–1104.
50. Liaudet, L., Szabo, C., Evgenov, O.V., Murthy, K.G., Pacher, P., Virag, L., Mabley, J.G., Marton, A., Soriano, F.G., Kirov, M.Y., Bjertnaes, L.J. and Salzman, A.L. (**2003**) Flagellin from gram-negative bacteria is a potent mediator of acute pulmonary inflammation in sepsis. *Shock*, **19**, 131–137.
51. Liaudet, L., Murthy, K.G., Mabley, J.G., Pacher, P., Soriano, F.G., Salzman, A.L. and Szabo, C. (**2002**) Comparison of inflammation, organ damage, and oxidant stress induced by Salmonella enterica serovar Muenchen flagellin and serovar Enteritidis lipopolysaccharide. *Infect. Immun.*, **70**, 192–198.
52. Bagchi, A., Herrup, E.A., Warren, H.S., Trigilio, J., Shin, H.S., Valentine, C. and Hellman, J. (**2007**) MyD88-dependent and MyD88-independent pathways in synergy, priming, and tolerance between TLR agonists. *J. Immunol.*, **178**, 1164–1171.
53. Van Heijenoort, J. (**1996**) Murein synthesis. In *Escherichia coli and Salmonella: Cellular and Molecular Biology*, Neidhardt, F.C., *et al.* (eds.), 2nd ed., pp 1015–1024, American Society for Microbiology.
54. Holtje, J.V. (**1998**) Growth of the stress-bearing and shape-maintaining murein sacculus of Escherichia coli. *Microbiol. Mol. Biol. Rev.*, **62**, 181–203.
55. Schleifer, K.H. and Kandler, O. (**1972**) Peptidoglycan types of bacterial cell walls and their taxonomic implications. *Bacteriol. Rev.*, **36**, 407–477.
56. Inohara, N., Ogura, Y., Fontalba, A., Gutierrez, O., Pons, F., Crespo, J., Fukase, K., Inamura, S., Kusumoto, S., Hashimoto, M., Foster, S.J., Moran, A.P., Fernandez-Luna, J.L. and Nunez, G. (**2003**) Host recognition of bacterial muramyl dipeptide mediated through NOD2. Implications for Crohn's disease. *J. Biol. Chem.*, **278**, 5509–5512.
57. Girardin, S.E., Boneca, I.G., Carneiro, L.A., Antignac, A., Jehanno, M., Viala, J., Tedin, K., Taha, M.K., Labigne, A., Zähringer, U., Coyle, A.J., DiStefano, P.S., Bertin, J., Sansonetti, P.J. and Philpott, D.J. (**2003**) Nod1 detects a unique muropeptide from gram-negative bacterial peptidoglycan. *Science*, **300**, 1584–1587.
58. Girardin, S.E., Boneca, I.G., Viala, J., Chamaillard, M., Labigne, A., Thomas, G., Philpott, D.J. and Sansonetti, P.J. (**2003**) Nod2 is a general sensor of peptidoglycan through muramyl dipeptide (MDP) detection. *J. Biol. Chem.*, **278**, 8869–8872.
59. Chamaillard, M., Hashimoto, M., Horie, Y., Masumoto, J., Qiu, S., Saab, L., Ogura, Y., Kawasaki, A., Fukase, K., Kusumoto, S., Valvano, M.A., Foster, S.J., Mak, T.W., Nunez, G. and Inohara, N. (**2003**) An essential role for NOD1 in host recognition of bacterial peptidoglycan containing diaminopimelic acid. *Nat. Immunol.*, **4**, 702–707.
60. Girardin, S.E., Tournebize, R., Mavris, M., Page, A.L., Li, X., Stark, G.R., Bertin, J., DiStefano, P.S., Yaniv, M., Sansonetti, P.J. and Philpott, D.J. (**2001**) CARD4/Nod1

mediates NF-kappaB and JNK activation by invasive Shigella flexneri. *EMBO Rep.*, **2**, 736–742.

61. Travassos, L.H., Carneiro, L.A., Girardin, S.E., Boneca, I.G., Lemos, R., Bozza, M.T., Domingues, R.C., Coyle, A.J., Bertin, J., Philpott, D.J. and Plotkowski, M.C. (**2005**) Nod1 participates in the innate immune response to *Pseudomonas aeruginosa*. *J. Biol. Chem.*, **280**, 36714–36718.
62. Hasegawa, M., Kawasaki, A., Yang, K., Fujimoto, Y., Masumoto, J., Breukink, E., Nunez, G., Fukase, K. and Inohara, N. (**2007**) A role of lipophilic peptidoglycan-related molecules in induction of Nod1-mediated immune responses. *J. Biol. Chem.*, **282**, 11757–11764.
63. Cartwright, N., Murch, O., McMaster, S.K., Paul-Clark, M.J., van Heel, D.A., Ryffel, B., Quesniaux, V.F., Evans, T.W., Thiemermann, C. and Mitchell, J.A. (**2007**) Selective NOD1 agonists cause shock and organ injury/dysfunction *in vivo*. *Am J Respir. Crit. Care Med.*, **175**, 595–603.
64. Uehara, A., Sugawara, Y., Kurata, S., Fujimoto, Y., Fukase, K., Kusumoto, S., Satta, Y., Sasano, T., Sugawara, S. and Takada, H. (**2005**) Chemically synthesized pathogen-associated molecular patterns increase the expression of peptidoglycan recognition proteins via toll-like receptors, NOD1 and NOD2 in human oral epithelial cells. *Cell Microbiol.*, **7**, 675–686.
65. van Heel, D.A., Ghosh, S., Butler, M., Hunt, K., Foxwell, B.M., Mengin-Lecreulx, D. and Playford, R.J. (**2005**) Synergistic enhancement of Toll-like receptor responses by NOD1 activation. *Eur. J. Immunol.*, **35**, 2471–2476.
66. Yang, S., Tamai, R., Akashi, S., Takeuchi, O., Akira, S., Sugawara, S. and Takada, H. (**2001**) Synergistic effect of muramyldipeptide with lipopolysaccharide or lipoteichoic acid to induce inflammatory cytokines in human monocytic cells in culture. *Infect. Immun.*, **69**, 2045–2053.
67. Wolfert, M.A., Murray, T.F., Boons, G.J. and Moore, J.N. (**2002**) The origin of the synergistic effect of muramyl dipeptide with endotoxin and peptidoglycan. *J. Biol. Chem.*, **277**, 39179–39186.
68. Ribi, E.E., Cantrell, J.L., Von Eschen, K.B. and Schwartzman, S.M. (**1979**) Enhancement of endotoxic shock by N-acetylmuramyl-L-alanyl-(L-seryl) -D-isoglutamine (muramyl dipeptide). *Cancer Res.*, **39**, 4756–4759.
69. Poltorak, A., He, X., Smirnova, I., Liu, M.Y., Van, H.C., Du, X., Birdwell, D., Alejos, E., Silva, M., Galanos, C., Freudenberg, M., Ricciardi-Castagnoli, P., Layton, B. and Beutler, B. (**1998**) Defective LPS signaling in C3H/HeJ and C57BL/10ScCr mice: mutations in Tlr4 gene. *Science*, **282**, 2085–2088.
70. Schromm, A.B., Lien, E., Henneke, P., Chow, J.C., Yoshimura, A., Heine, H., Latz, E., Monks, B.G., Schwartz, D.A., Miyake, K. and Golenbock, D.T. (**2001**) Molecular genetic analysis of an endotoxin nonresponder mutant cell line: a point mutation in a conserved region of MD-2 abolishes endotoxin-induced signaling. *J Exp. Med.*, **194**, 79–88.
71. Hoshino, K., Takeuchi, O., Kawai, T., Sanjo, H., Ogawa, T., Takeda, Y., Takeda, K. and Akira, S. (**1999**) Cutting edge: Toll-like receptor 4 (TLR4)-deficient mice are hyporesponsive to lipopolysaccharide: evidence for TLR4 as the LPS gene product. *J. Immunol.*, **162**, 3749–3752.
72. Nagai, Y., Akashi, S., Nagafuku, M., Ogata, M., Iwakura, Y., Akira, S., Kitamura, T., Kosugi, A., Kimoto, M. and Miyake, K. (**2002**) Essential role of MD-2 in LPS responsiveness and TLR4 distribution. *Nat. Immunol.*, **3**, 667–672.
73. Wright, S.D., Ramos, R.A., Tobias, P.S., Ulevitch, R.J. and Mathison, J.C. (**1990**) CD14, a receptor for complexes of

lipopolysaccharide (LPS) and LPS binding protein. *Science*, **249**, 1431–1433.

74. Schumann, R.R., Leong, S.R., Flaggs, G.W., Gray, P.W., Wright, S.D., Mathison, J.C., Tobias, P.S. and Ulevitch, R.J. (**1990**) Structure and function of lipopolysaccharide binding protein. *Science*, **249**, 1429–1431.
75. Jiang, Z., Georgel, P., Du, X., Shamel, L., Sovath, S., Mudd, S., Huber, M., Kalis, C., Keck, S., Galanos, C., Freudenberg, M. and Beutler, B. (**2005**) CD14 is required for MyD88-independent LPS signaling. *Nat. Immunol.*, **6**, 565–570.
76. Huber, M., Kalis, C., Keck, S., Jiang, Z., Georgel, P., Du, X., Shamel, L., Sovath, S., Mudd, S., Beutler, B., Galanos, C. and Freudenberg, M.A. (**2006**) R-form LPS, the master key to the activation of TLR4/MD-2-positive cells. *Eur. J. Immunol.*, **36**, 701–711.
77. Triantafilou, K., Triantafilou, M. and Dedrick, R.L. (**2001**) A CD14-independent LPS receptor cluster. *Nat. Immunol.*, **2**, 338–345.
78. Blunck, R., Scheel, O., Müller, M., Brandenburg, K., Seitzer, U. and Seydel, U. (**2001**) New insights into endotoxin-induced activation of macrophages: involvement of a K+ channel in transmembrane signaling. *J. Immunol.*, **166**, 1009–1015.
79. Scheel, O., Papavlassopoulos, M., Blunck, R., Gebert, A., Hartung, T., Zahringer, U., Seydel, U. and Schromm, A.B. (**2006**) Cell activation by ligands of the toll-like receptor and interleukin-1 receptor family depends on the function of the large-conductance potassium channel MaxiK in human macrophages. *Infect. Immun.*, **74**, 4354–4356.
80. Papavlassopoulos, M., Stamme, C., Thon, L., Adam, D., Hillemann, D., Seydel, U. and Schromm, A.B. (**2006**) MaxiK blockade selectively inhibits the lipopolysaccharide-induced I kappa B-alpha/NF-kappa B signaling pathway in macrophages. *J. Immunol.*, **177**, 4086–4093.
81. Gutsmann, T., Haberer, N., Carroll, S.F., Seydel, U. and Wiese, A. (**2001**) Interaction between lipopolysaccharide (LPS), LPS-binding protein (LBP), and planar membranes. *Biol. Chem.*, **382**, 425–434.
82. Roes, S., Mumm, F., Seydel, U. and Gutsmann, T. (**2006**) Localization of the lipopolysaccharide-binding protein in phospholipid membranes by atomic force microscopy. *J. Biol. Chem.*, **281**, 2757–2763.
83. Triantafilou, M., Morath, S., Mackie, A., Hartung, T. and Triantafilou, K. (**2004**) Lateral diffusion of Toll-like receptors reveals that they are transiently confined within lipid rafts on the plasma membrane. *J. Cell Sci.*, **117**, 4007–4014.
84. Triantafilou, M., Brandenburg, K., Kusumoto, S., Fukase, K., Mackie, A., Seydel, U. and Triantafilou, K. (**2004**) Combinational clustering of receptors following stimulation by bacterial products determines lipopolysaccharide responses 7. *Biochem. J.*, **381**, 527–536.
85. Saitoh, S., Akashi, S., Yamada, T., Tanimura, N., Kobayashi, M., Konno, K., Matsumoto, F., Fukase, K., Kusumoto, S., Nagai, Y., Kusumoto, Y., Kosugi, A. and Miyake, K. (**2004**) Lipid A antagonist, lipid IVa, is distinct from lipid A in interaction with Toll-like receptor 4 (TLR4)-MD-2 and ligand-induced TLR4 oligomerization. *Int. Immunol.*, **16**, 961–969.
86. Saitoh, S., Akashi, S., Yamada, T., Tanimura, N., Matsumoto, F., Fukase, K., Kusumoto, S., Kosugi, A. and Miyake, K. (**2004**) Ligand-dependent Toll-like receptor 4 (TLR4)-oligomerization is directly linked with TLR4-signaling. *J. Endotoxin Res.*, **10**, 257–260.
87. Liew, F.Y., Xu, D., Brint, E.K. and O'Neill, L.A. (**2005**) Negative regulation of toll-like receptor-mediated immune responses. *Nat. Rev. Immunol.*, **5**, 446–458.

88. Weighardt, H., Jusek, G., Mages, J., Lang, R., Hoebe, K., Beutler, B. and Holzmann, B. (**2004**) Identification of a TLR4- and TRIF-dependent activation program of dendritic cells. *Eur. J. Immunol.*, **34**, 558–564.
89. Schmitz, F., Mages, J., Heit, A., Lang, R. and Wagner, H. (**2004**) Transcriptional activation induced in macrophages by Toll-like receptor (TLR) ligands: from expression profiling to a model of TLR signaling. *Eur. J. Immunol.*, **34**, 2863–2873.
90. Divanovic, S., Trompette, A., Atabani, S.F., Madan, R., Golenbock, D.T., Visintin, A., Finberg, R.W., Tarakhovsky, A., Vogel, S.N., Belkaid, Y., Kurt-Jones, E.A. and Karp, C.L. (**2005**) Negative regulation of Toll-like receptor 4 signaling by the Toll-like receptor homolog RP105. *Nat. Immunol.*, **6**, 571–578.
91. Brint, E.K., Xu, D., Liu, H., Dunne, A., McKenzie, A.N., O'Neill, L.A. and Liew, F.Y. (**2004**) ST2 is an inhibitor of interleukin 1 receptor and Toll-like receptor 4 signaling and maintains endotoxin tolerance. *Nat. Immunol.*, **5**, 373–379.
92. Wald, D., Qin, J., Zhao, Z., Qian, Y., Naramura, M., Tian, L., Towne, J., Sims, J.E., Stark, G.R. and Li, X. (**2003**) SIGIRR, a negative regulator of Toll-like receptor-interleukin 1 receptor signaling. *Nat. Immunol.*, **4**, 920–927.
93. Mansell, A., Smith, R., Doyle, S.L., Gray, P., Fenner, J.E., Crack, P.J., Nicholson, S.E., Hilton, D.J., O'Neill, L.A. and Hertzog, P.J. (**2006**) Suppressor of cytokine signaling 1 negatively regulates Toll-like receptor signaling by mediating Mal degradation. *Nat. Immunol.*, **7**, 148–155.
94. Kobayashi, K., Hernandez, L.D., Galan, J.E., Janeway, C.A., Jr., Medzhitov, R. and Flavell, R.A. (**2002**) IRAK-M is a negative regulator of Toll-like receptor signaling. *Cell*, **110**, 191–202.
95. Chuang, T.H. and Ulevitch, R.J. (**2004**) Triad3A, an E3 ubiquitin-protein ligase regulating Toll-like receptors. *Nat. Immunol.*, **5**, 495–502.
96. Boone, D.L., Turer, E.E., Lee, E.G., Ahmad, R.C., Wheeler, M.T., Tsui, C., Hurley, P., Chien, M., Chai, S., Hitotsumatsu, O., McNally, E., Pickart, C. and Ma, A. (**2004**) The ubiquitin-modifying enzyme A20 is required for termination of Toll-like receptor responses. *Nat. Immunol.*, **5**, 1052–1060.
97. Wang, Y., Tang, Y., Teng, L., Wu, Y., Zhao, X. and Pei, G. (**2006**) Association of beta-arrestin and TRAF6 negatively regulates Toll-like receptor-interleukin 1 receptor signaling. *Nat. Immunol.*, **7**, 139–147.
98. Parameswaran, N., Pao, C.S., Leonhard, K.S., Kang, D.S., Kratz, M., Ley, S.C. and Benovic, J.L. (**2006**) Arrestin-2 and G protein-coupled receptor kinase 5 interact with NFkappaB1 p105 and negatively regulate lipopolysaccharide-stimulated ERK1/2 activation in macrophages. *J. Biol. Chem.*, **281**, 34159–34170.
99. Shinohara, H., Inoue, A., Toyama-Sorimachi, N., Nagai, Y., Yasuda, T., Suzuki, H., Horai, R., Iwakura, Y., Yamamoto, T., Karasuyama, H., Miyake, K. and Yamanashi, Y. (**2005**) Dok-1 and Dok-2 are negative regulators of lipopolysaccharide-induced signaling. *J. Exp. Med.*, **201**, 333–339.
100. Saitoh, T., Tun-Kyi, A., Ryo, A., Yamamoto, M., Finn, G., Fujita, T., Akira, S., Yamamoto, N., Lu, K.P. and Yamaoka, S. (**2006**) Negative regulation of interferon-regulatory factor 3-dependent innate antiviral response by the prolyl isomerase Pin1. *Nat. Immunol.*, **7**, 598–605.
101. Negishi, H., Ohba, Y., Yanai, H., Takaoka, A., Honma, K., Yui, K., Matsuyama, T., Taniguchi, T. and Honda, K. (**2005**) Negative regulation of Toll-like-receptor signaling by IRF-4. *Proc. Natl. Acad. Sci. USA*, **102**, 15989–15994.

102. Honma, K., Udono, H., Kohno, T., Yamamoto, K., Ogawa, A., Takemori, T., Kumatori, A., Suzuki, S., Matsuyama, T. and Yui, K. (**2005**) Interferon regulatory factor 4 negatively regulates the production of proinflammatory cytokines by macrophages in response to LPS. *Proc. Natl. Acad. Sci. USA*, **102**, 16001–16006.
103. Israelachvili, J.N. (**1991**) *Intermolecular and Surface Forces*. Academic Press, London.
104. Maurer, N. (**1991**) Determination of size and structure of lipid IV a vesicles by quasi-elastic light scattering and small-angle X-ray scattering. *J. Appl. Cryst.*, **24**, 832–835.
105. Seydel, U. (**1999**) A biophysical view on the function and activity of endotoxins. In *Endotoxin in Health and Disease*, Brade, H., Opal, S., Vogel, S.N. and Morrison, D.C. (eds.), pp 195–220, Marcel Dekker, New York.
106. Takayama, K., Din, Z.Z., Mukerjee, P., Cooke, P.H. and Kirkland, T.N. (**1990**) Physicochemical properties of the lipopolysaccharide unit that activates B lymphocytes. *J. Biol. Chem.*, **265**, 14023–14029.
107. Takayama, K., Mitchell, D.H., Din, Z.Z., Mukerjee, P., Li, C. and Coleman, D.L. (**1994**) Monomeric Re lipopolysaccharide from Escherichia coli is more active than the aggregated form in the Limulus amebocyte lysate assay and in inducing Egr-1 mRNA in murine peritoneal macrophages. *J. Biol. Chem.*, **269**, 2241–2244.
108. Yu, B. and Wright, S.D. (**1996**) Catalytic properties of lipopolysaccharide (LPS) binding protein. Transfer of LPS to soluble CD14. *J. Biol. Chem.*, **271**, 4100–4105.
109. Shnyra, A., Hultenby, K. and Lindberg, A.A. (**1993**) Role of the physical state of Salmonella lipopolysaccharide in expression of biological and endotoxic properties. *Infect. Immun.*, **61**, 5351–5360.
110. Kitchens, R.L. and Munford, R.S. (**1998**) CD14-dependent internalization of bacterial lipopolysaccharide (LPS) is strongly influenced by LPS aggregation but not by cellular responses to LPS. *J. Immunol.*, **160**, 1920–1928.
111. Müller, M., Lindner, B., Kusumoto, S., Fukase, K., Schromm, A.B. and Seydel, U. (**2004**) Aggregates are the biologically active units of endotoxin. *J. Biol. Chem.*, **279**, 26307–26313.
112. Hailman, E., Lichenstein, H.S., Wurfel, M.M., Miller, D.S., Johnson, D.A., Kelley, M., Busse, L.A., Zukowski, M.M. and Wright, S.D. (**1994**) Lipopolysaccharide (LPS)-binding protein accelerates the binding of LPS to CD14. *J. Exp. Med.*, **179**, 269–277.
113. Gioannini, T.L., Teghanemt, A., Zhang, D., Coussens, N.P., Dockstader, W., Ramaswamy, S. and Weiss, J.P. (**2004**) Isolation of an endotoxin-MD-2 complex that produces Toll-like receptor 4-dependent cell activation at picomolar concentrations. *Proc. Natl. Acad. Sci. USA*, **101**, 4186–4191.
114. Akashi, S., Saitoh, S., Wakabayashi, Y., Kikuchi, T., Takamura, N., Nagai, Y., Kusumoto, Y., Fukase, K., Kusumoto, S., Adachi, Y., Kosugi, A. and Miyake, K. (**2003**) Lipopolysaccharide interaction with cell surface Toll-like receptor 4-MD-2: higher affinity than that with MD-2 or CD14. *J. Exp. Med.*, **198**, 1035–1042.
115. Prohinar, P., Re, F., Widstrom, R., Zhang, D., Teghanemt, A., Weiss, J.P. and Gioannini, T.L. (**2007**) Specific high affinity interactions of monomeric endotoxin.protein complexes with Toll-like receptor 4 ectodomain. *J. Biol. Chem.*, **282**, 1010–1017.
116. Schromm, A.B., Brandenburg, K., Rietschel, E.T., Flad, H.D., Carroll, S.F. and Seydel, U. (**1996**) Lipopolysaccharide-binding protein mediates CD14-independent intercalation of lipopolysaccharide into phospholipid membranes. *FEBS Lett.*, **399**, 267–271.

117. Gutsmann, T., Fix, M., Larrick, J.W. and Wiese, A. (**2000**) Mechanisms of action of rabbit CAP18 on monolayers and liposomes made from endotoxins or phospholipids. *J. Membr. Biol*, **176**, 223–236.
118. Brandenburg, K., Mayer, H., Koch, M.H., Weckesser, J., Rietschel, E.T. and Seydel, U. (**1993**) Influence of the supramolecular structure of free lipid A on its biological activity. *Eur. J. Biochem.*, **218**, 555–563.
119. Schromm, A.B., Howe, J., Ulmer, A.J., Wiesmüller, K.H., Seyberth, T., Jung, G., Rössle, M., Koch, M.H., Gutsmann, T. and Brandenburg, K. (**2007**) Physicochemical and biological analysis of synthetic bacterial lipopeptides: validity of the concept of endotoxic conformation. *J. Biol. Chem.*, **282**, 11030–11037.
120. Elsbach, P., Weiss, J., Doerfler, M., Shu, C., Kohn, F., Ammons, W.S., Kung, A.H., Meszaros, K.K. and Parent, J.B. (**1994**) The bactericidal/permeability increasing protein of neutrophils is a potent antibacterial and anti-endotoxin agent *in vitro* and *in vivo*. *Prog. Clin. Biol. Res.*, **388**, 41–51.
121. Zhang, G.H., Mann, D.M. and Tsai, C.M. (**1999**) Neutralization of endotoxin *in vitro* and *in vivo* by a human lactoferrin-derived peptide. *Infect. Immun.*, **67**, 1353–1358.
122. Emancipator, K., Csako, G. and Elin, R.J. (**1992**) *In vitro* inactivation of bacterial endotoxin by human lipoproteins and apolipoproteins. *Infect. Immun.*, **60**, 596–601.
123. Feingold, K.R., Funk, J.L., Moser, A.H., Shigenaga, J.K., Rapp, J.H. and Grunfeld, C. (**1995**) Role for circulating lipoproteins in protection from endotoxin toxicity. *Infect. Immun.*, **63**, 2041–2046.
124. Munford, R.S., Andersen, J.M. and Dietschy, J.M. (**1981**) Sites of tissue binding and uptake *in vivo* of bacterial lipopolysaccharide-high density lipoprotein complexes: studies in the rat and squirrel monkey. *J. Clin. Invest*, **68**, 1503–1513.
125. Read, T.E., Harris, H.W., Grunfeld, C., Feingold, K.R., Calhoun, M.C., Kane, J.P. and Rapp, J.H. (**1993**) Chylomicrons enhance endotoxin excretion in bile. *Infect. Immun.*, **61**, 3496–3502.
126. Sprong, T., Netea, M.G., van der, L.P., Verver-Jansen, T.J., Jacobs, L.E., Stalenhoef, A., van der Meer, J.W. and van, D.M. (**2004**) Human lipoproteins have divergent neutralizing effects on *E. coli* LPS, *N. meningitidis* LPS, and complete Gram- negative bacteria. *J. Lipid Res.*, **45**, 742–749.
127. Parker, T.S., Levine, D.M., Chang, J.C., Laxer, J., Coffin, C.C. and Rubin, A.L. (**1995**) Reconstituted high-density lipoprotein neutralizes gram-negative bacterial lipopolysaccharides in human whole blood. *Infect. Immun.*, **63**, 253–258.
128. Brandenburg, K., Jürgens, G., Andrä, J., Lindner, B., Koch, M.H., Blume, A. and Garidel, P. (**2002**) Biophysical characterization of the interaction of high-density lipoprotein (HDL) with endotoxins. *Eur. J. Biochem.*, **269**, 5972–5981.
129. Wright, J.R. (**2005**) Immunoregulatory functions of surfactant proteins. *Nat. Rev. Immunol.*, **5**, 58–68.
130. Gioannini, T.L., Teghanemt, A., Zhang, D., Prohinar, P., Levis, E.N., Munford, R.S. and Weiss, J.P. (**2007**) Endotoxin-binding proteins modulate the susceptibility of bacterial endotoxin to deacylation by acyloxyacyl hydrolase. *J. Biol. Chem.*, **282**, 7877–7884.
131. Bentala, H., Verweij, W.R., Huizinga-Van, der Vlag A., Loenen-Weemaes, A.M., Meijer, D.K. and Poelstra, K. (**2002**) Removal of phosphate from lipid A as a strategy to detoxify lipopolysaccharide. *Shock*, **18**, 561–566.
132. Zasloff, M. (**2002**) Antimicrobial peptides of multicellular organisms. *Nature*, **415**, 389–395.
133. Harder, J., Bartels, J., Christophers, E. and Schroder, J.M.

(**1997**) A peptide antibiotic from human skin. *Nature*, **387**, 861.

134. Cowland, J.B., Johnsen, A.H. and Borregaard, N. (**1995**) hCAP-18, a cathelin/pro-bactenecin-like protein of human neutrophil specific granules. *FEBS Lett.*, **368**, 173–176.
135. Vogel, H.J., Schibli, D.J., Jing, W., Lohmeier-Vogel, E.M., Epand, R.F. and Epand, R.M. (**2002**) Towards a structure-function analysis of bovine lactoferricin and related tryptophan- and arginine-containing peptides. *Biochem. Cell Biol.*, **80**, 49–63.
136. Pena, S.V., Hanson, D.A., Carr, B.A., Goralski, T.J. and Krensky, A.M. (**1997**) Processing, subcellular localization, and function of 519 (granulysin), a human late T cell activation molecule with homology to small, lytic, granule proteins. *J. Immunol.*, **158**, 2680–2688.
137. Andersson, M., Gunne, H., Agerberth, B., Boman, A., Bergman, T., Sillard, R., Jornvall, H., Mutt, V., Olsson, B., Wigzell, H., Dagerlind, A., Boman, H. G. and Gudmundsson, G. H. (**1995**) NK-lysin, a novel effector peptide of cytotoxic T and NK cells. Structure and cDNA cloning of the porcine form, induction by interleukin 2, antibacterial and antitumour activity. *EMBO J*, **14**, 1615–1625.
138. Brogden, K.A., Ackermann, M., Zabner, J. and Welsh, M.J. (**2004**) Antimicrobial peptides suppress microbial infections and sepsis in animal models. In *Mammalian host defence peptides*, Devine, D.A. and Hancock, R.E.W. (eds.), pp 189–228, Cambridge University Press, Cambridge, UK.
139. Falla, T.J. and Zhang, L. (**2004**) Therapeutic potential and applications of innate immunity peptides. In *Mammalian host defence peptides*, Devine, D.A. and Hancock, R.E.W. (eds.), pp 69–110, Cambridge University Press, Cambridge, UK.
140. Baynes, R.D. and Bezwoda, W.R. (**1994**) Lactoferrin and the inflammatory response. In *Lactoferrin: Structure and Function*, Hutchens, T.W., Lönnerdal, B. and Rumball, S. (eds.), Plenum Press, New York.
141. Appelmelk, B.J., An, Y.Q., Geerts, M., Thijs, B.G., de Boer, H.A., MacLaren, D.M., de, G.J. and Nuijens, J.H. (**1994**) Lactoferrin is a lipid A-binding protein. *Infect. Immun.*, **62**, 2628–2632.
142. Agerberth, B., Grunewald, J., Castanos-Velez, E., Olsson, B., Jornvall, H., Wigzell, H., Eklund, A. and Gudmundsson, G.H. (**1999**) Antibacterial components in bronchoalveolar lavage fluid from healthy individuals and sarcoidosis patients. *Am. J. Respir. Crit. Care Med.*, **160**, 283–290.
143. Chromek, M., Slamova, Z., Bergman, P., Kovacs, L., Podracka, L., Ehren, I., Hokfelt, T., Gudmundsson, G.H., Gallo, R.L., Agerberth, B. and Brauner, A. (**2006**) The antimicrobial peptide cathelicidin protects the urinary tract against invasive bacterial infection. *Nat. Med.*, **12**, 636–641.
144. Hirata, M., Zhong, J., Wright, S.C. and Larrick, J.W. (**1995**) Structure and functions of endotoxin-binding peptides derived from CAP18. *Prog. Clin. Biol. Res.*, **392**, 317–326.
145. Liepinsh, E., Andersson, M., Ruysschaert, J.M. and Otting, G. (**1997**) Saposin fold revealed by the NMR structure of NK-lysin. *Nat. Struct. Biol.*, **4**, 793–795.
146. Pena, S.V. and Krensky, A.M. (**1997**) Granulysin, a new human cytolytic granule-associated protein with possible involvement in cell-mediated cytotoxicity. *Semin. Immunol.*, **9**, 117–125.
147. Endsley, J.J., Furrer, J.L., Endsley, M.A., McIntosh, M.A., Maue, A.C., Waters, W.R., Lee, D.R. and Estes, D.M. (**2004**) Characterization of bovine homologues of granulysin and NK-lysin. *J. Immunol.*, **173**, 2607–2614.
148. Davis, E.G., Sang, Y., Rush, B., Zhang, G. and Blecha, F. (**2005**) Molecular cloning and characterization of equine NK-lysin.

Vet. Immunol. Immunopathol., **105**, 163–169.
149. Andersson, M., Girard, R. and Cazenave, P. (**1999**) Interaction of NK lysin, a peptide produced by cytolytic lymphocytes, with endotoxin. *Infect. Immun.*, **67**, 201–205.
150. Andrä, J., Koch, M.H., Bartels, R. and Brandenburg, K. (**2004**) Biophysical characterization of endotoxin inactivation by NK-2, an antimicrobial peptide derived from mammalian NK-lysin. *Antimicrob. Agents Chemother.*, **48**, 1593–1599.
151. Farnaud, S., Spiller, C., Moriarty, L.C., Patel, A., Gant, V., Odell, E.W. and Evans, R.W. (**2004**) Interactions of lactoferricin-derived peptides with LPS and antimicrobial activity. *FEMS Microbiol. Lett.*, **233**, 193–199.
152. Andrä, J., Lohner, K., Blondelle, S.E., Jerala, R., Moriyon, I., Koch, M.H., Garidel, P. and Brandenburg, K. (**2005**) Enhancement of endotoxin neutralization by coupling of a C12-alkyl chain to a lactoferricin-derived peptide. *Biochem. J.*, **385**, 135–143.
153. Nagaoka, I., Hirota, S., Niyonsaba, F., Hirata, M., Adachi, Y., Tamura, H. and Heumann, D. (**2001**) Cathelicidin family of antibacterial peptides CAP18 and CAP11 inhibit the expression of TNF-alpha by blocking the binding of LPS to CD14(+) cells. *J. Immunol.*, **167**, 3329–3338.
154. Japelj, B., Pristovsek, P., Majerle, A. and Jerala, R. (**2005**) Structural origin of endotoxin neutralization and antimicrobial activity of a lactoferrin-based peptide. *J. Biol. Chem.*, **280**, 16955–16961.
155. Chen, X., Howe, J., Andrä, J., Rössle, M., Richter, W., da Silva, A.P., Krensky, A.M., Clayberger, C. and Brandenburg, K. (**2007**) Biophysical analysis of the interaction of granulysin-derived peptides with enterobacterial endotoxins. *Biochim. Biophys. Acta*, (in press).
156. Rosenfeld, Y., Papo, N. and Shai, Y. (**2006**) Endotoxin (lipopolysaccharide) neutralization by innate immunity host-defense peptides. Peptide properties and plausible modes of action. *J. Biol. Chem.*, **281**, 1636–1643.
157. Mookherjee, N., Brown, K.L., Bowdish, D.M., Doria, S., Falsafi, R., Hokamp, K., Roche, F.M., Mu, R., Doho, G.H., Pistolic, J., Powers, J.P., Bryan, J., Brinkman, F.S. and Hancock, R.E. (**2006**) Modulation of the TLR-mediated inflammatory response by the endogenous human host defense peptide LL-37. *J. Immunol.*, **176**, 2455–2464.
158. Scott, M.G., Davidson, D.J., Gold, M.R., Bowdish, D. and Hancock, R.E. (**2002**) The human antimicrobial peptide LL-37 is a multifunctional modulator of innate immune responses. *J. Immunol.*, **169**, 3883–3891.
159. Drickamer, K., Dordal, M.S. and Reynolds, L. (**1986**) Mannose-binding proteins isolated from rat liver contain carbohydrate-recognition domains linked to collagenous tails. Complete primary structures and homology with pulmonary surfactant apoprotein. *J. Biol. Chem.*, **261**, 6878–6887.
160. Van de Wetering, J.K., van Golde, L.M. and Batenburg, J.J. (**2004**) Collectins: players of the innate immune system. *Eur. J. Biochem.*, **271**, 1229–1249.
161. Zhao, L., Ohtaki, Y., Yamaguchi, K., Matsushita, M., Fujita, T., Yokochi, T., Takada, H. and Endo, Y. (**2002**) LPS-induced platelet response and rapid shock in mice: contribution of O-antigen region of LPS and involvement of the lectin pathway of the complement system. *Blood*, **100**, 3233–3239.
162. Levine, A.M. and Whitsett, J.A. (**2001**) Pulmonary collectins and innate host defense of the lung. *Microbes Infect.*, **3**, 161–166.
163. Borron, P., McIntosh, J.C., Korfhagen, T.R., Whitsett, J.A., Taylor, J. and Wright, J.R. (**2000**) Surfactant-associated protein A inhibits LPS-induced cytokine and nitric oxide production *in vivo*. *Am. J. Physiol. Lung Cell Mol. Physiol.*, **278**, L840–L847.

164. Piantadosi, C.A. and Schwartz, D.A. (**2004**) The acute respiratory distress syndrome. *Ann. Intern. Med.*, **141**, 460–470.
165. McIntosh, J.C., Mervin-Blake, S., Conner, E. and Wright, J.R. (**1996**) Surfactant protein A protects growing cells and reduces TNF-alpha activity from LPS-stimulated macrophages. *Am. J. Physiol.*, **271**, L310–L319.
166. Gardai, S.J., Xiao, Y.Q., Dickinson, M., Nick, J.A., Voelker, D.R., Greene, K.E. and Henson, P.M. (**2003**) By binding SIRPalpha or calreticulin/CD91, lung collectins act as dual function surveillance molecules to suppress or enhance inflammation. *Cell*, **115**, 13–23.
167. Garcia-Verdugo, I., Sanchez-Barbero, F., Soldau, K., Tobias, P.S. and Casals, C. (**2005**) Interaction of SP-A (surfactant protein A) with bacterial rough lipopolysaccharide (Re-LPS), and effects of SP-A on the binding of Re-LPS to CD14 and LPS-binding protein. *Biochem. J.*, **391**, 115–124.
168. Stamme, C., Walsh, E. and Wright, J.R. (**2000**) Surfactant protein A differentially regulates IFN-gamma- and LPS-induced nitrite production by rat alveolar macrophages. *Am. J. Respir. Cell Mol. Biol.*, **23**, 772–779.
169. Alcorn, J.F. and Wright, J.R. (**2004**) Surfactant protein A inhibits alveolar macrophage cytokine production by CD14-independent pathway. *Am. J. Physiol. Lung Cell Mol. Physiol.*, **286**, L129–L136.
170. Wu, Y., Adam, S., Hamann, L., Heine, H., Ulmer, A.J., Buwitt-Beckmann, U. and Stamme, C. (**2004**) Accumulation of inhibitory kappaB-alpha as a mechanism contributing to the anti-inflammatory effects of surfactant protein-A. *Am. J. Respir. Cell Mol. Biol.*, **31**, 587–594.
171. Sano, H., Chiba, H., Iwaki, D., Sohma, H., Voelker, D.R. and Kuroki, Y. (**2000**) Surfactant proteins A and D bind CD14 by different mechanisms. *J. Biol. Chem.*, **275**, 22442–22451.
172. Stamme, C., Müller, M., Hamann, L., Gutsmann, T. and Seydel, U. (**2002**) Surfactant protein A inhibits lipopolysaccharide-induced immune cell activation by preventing the interaction of lipopolysaccharide with lipopolysaccharide-binding protein. *Am. J. Respir. Cell Mol. Biol.*, **27**, 353–360.
173. Wu, H., Kuzmenko, A., Wan, S., Schaffer, L., Weiss, A., Fisher, J.H., Kim, K.S. and McCormack, F.X. (**2003**) Surfactant proteins A and D inhibit the growth of Gram-negative bacteria by increasing membrane permeability. *J. Clin. Invest.*, **111**, 1589–1602.
174. Opal, S.M. and Huber, C.E. (**2002**) Bench-to-bedside review: Toll-like receptors and their role in septic shock. *Crit. Care*, **6**, 125–136.
175. Lynn, M., Rossignol, D.P., Wheeler, J.L., Kao, R.J., Perdomo, C.A., Noveck, R., Vargas, R., D'Angelo, T., Gotzkowsky, S. and McMahon, F.G. (**2003**) Blocking of responses to endotoxin by E5564 in healthy volunteers with experimental endotoxemia. *J. Infect. Dis.*, **187**, 631–639.
176. Bennett-Guerrero, E., Grocott, H.P., Levy, J.H., Stierer, K.A., Hogue, C.W., Cheung, A.T., Newman, M.F., Carter, A.A., Rossignol, D.P. and Collard, C.D. (**2007**) A phase II, double-blind, placebo-controlled, ascending-dose study of Eritoran (E5564), a lipid A antagonist, in patients undergoing cardiac surgery with cardiopulmonary bypass. *Anesth. Analg.*, **104**, 378–383.
177. Hopkins, P. and Cohen, J. (**2002**) Toll-like receptors: the key to the stable door?. *Crit. Care*, **6**, 99–101.

6
Pathogens in Sepsis: Gram-positive Bacterial PAMPs, PRRs and Superantigens

Christian Alexander, Romney Haylett, Andra Schromm, Ulrich Zähringer and Lothar Rink

6.1
Introduction

Prototypic members of the low G + C DNA group of Gram-positive bacteria phylogenetically assigned to the group of firmicutes have been increasingly documented to evoke several severe human diseases such as septic disorders, otitis media or meningitis and also represent the predominant class of pathogens in the formation of bacterial biofilms on the surface of medical devices [1–4]. Serious concerns have been raised due to the increasing frequency of multiply antibiotic-resistant strains of Gram-positive pathogens in nosocomial infections with particularly high incidences in intensive care units (ICUs) [5–8]. In a majority of clinics in Northern and Latin America as well as in European hospitals, *S. aureus* and coagulase-negative staphylococci (CNS) with primary incidence of *S. epidermidis* have become the predominant pathogens isolated in septic diseases [9–13]. Due to these epidemiological data, we will largely restrict the following sections of our review to the pathogenic properties of *S. aureus* that represents a paradigmatic Gram-positive species in many aspects. Analogous to other bacteria found in septic diseases, *S. aureus* is a normal human commensal that preferentially colonizes the nasal passages but also other anatomical surfaces such as the skin or the gastrointestinal mucosa and may become an opportunistic pathogen in the immunocompromised host [14–16]. Almost from the beginning of the antibiotic era in the early 1940s *S. aureus* has successively gained resistance to the vast majority of clinically applied antibiotics including the sequential emergence and rapidly increasing incidence of penicillin G-resistant, methicillin/oxacillin resistant (MRSA/ORSA), vancomycin intermediate-resistant (VISA) and complete vancomycin-resistant strains (VRSA) that were initially detected in hospital settings, but were also consistently found in the community in subsequent years [7, 8]. Individual clinical isolates and community-derived strains of *S. aureus* display a considerable degree of genetic and phenotypic heterogeneity [17–19] and

Sepsis and Non-infectious Systemic Inflammation. Edited by J.-M. Cavaillon and C. Adrie

ISBN: 978-3-527-31935-0

may express a wide variety of pathogenicity factors including multiple antibiotic resistance proteins, invasins and adhesins such as hyaluronidase or fibronectin/fibrinogen-binding proteins, immune escape factors like protein A or leucocidins, capsular polysaccharides (CPS), biofilm-constituting components and the 'aureal' antioxidant cell wall carotinoids of the staphyloxanthin group as well as bacterial superantigens [1, 18–22]. Several members of the latter class of Gram-positive pathogenicity factors have been identified to induce severe septic shock-like symptoms in humans and will thus be referred to more in detail in a separate section of this chapter. More recently, the unique pathogenic versatility of *S. aureus* has been further documented by a series of genome sequencing projects providing complete genomic data of 12 strains to date [23].

6.2 Pathogen-associated Molecular Pattern (PAMP) Molecules of Gram-positive Bacteria

The classical definition of pathogen-associated molecular patterns (PAMPs) and pattern recognition receptors (PRRs) introduced by Charles Janeway [24, 25] has been refined during the past decade concerning the rather specific interactions of signaling PRRs such as Toll-like receptors (TLRs), nucleotide-binding oligomerization domain (Nod)-like receptors (NLRs) and retinoic acid-inducible gene (RIG)-I-like receptors (RLRs) with comparatively small, but highly conserved substructures of bacterial PAMP molecules. TLR2 is the central mammalian surface receptor for the immunorecognition of PAMP-structures of Gram-positive bacteria. Sensitive recognition of major Gram-positive bacterial PAMPs via TLR2 is aided by a group of so-called accessory PRRs including lipopolysaccharide-binding protein (LBP), CD14 and CD36 that apparently function as extracellular ligand delivery factors for these particular PAMP-PRR signaling systems [26–28]. Moreover, recent progress in revealing the activation of mammalian PRRs by PAMP structures of Gram-positive bacteria has also provided several new molecular aspects in our understanding of the pathogenesis of Gram-positive septic infections and may correspondingly lead to novel adjunctive therapeutic approaches. In Table 6.1 the currently assigned major classes of immuno-activating PAMPs found in Gram-positive bacteria, the particular immunorecognized substructures of these bacterial macromolecules and the corresponding PRRs of mammalian innate immunity are summarized.

Two major PAMP structures of Gram-positive bacteria – peptidoglycan (PG) and lipoproteins (LP) – are localized in the bacterial cell wall that is schematically shown in Figure 6.1. Whereas the central roles of TLR2 and the cytoplasmic Nod proteins in the immunorecognition of these Gram-positive cell wall components have been rather well characterized, the contributions of the intracellular PAMP structures – non-methylated CpG-DNA [29] and

Table 6.1 PAMPs of Gram-positive bacteria and corresponding cognate PRRs of mammalian innate immunity.

Gram-positive bacteria		Mammalian host	
PAMP	**PAMP-specific substructure**	**Signaling PRR(s)**	**Accessory PRR(s)**
Lipoproteins[a]	S-Diacylglyceryl oligopeptide[a]	TLR2 *plus* TLR6[a]	CD36, LBP, CD14
LTA (?)[b]	Gentiobiosyl-sn-diacylglycerol domain and D-Ala residues	TLR2 *plus* TLR6	CD36, LBP, CD14
Bacterial DNA	Non-methylated CpG-Oligonucleotides	TLR9	n.a.
L-Lys-type peptidoglycan[c]	Muramyldipeptide (MDP)	Nod2 (cytoplasmic)	n.a.
m-DAP-type peptidoglycan[c]	Muramyldipeptide (MDP)	Nod2 (cytoplasmic)	n.a.
	iso-Glx-*m*DAP dipeptide	Nod1 (cytoplasmic)	n.a.
Bacterial RNA	?	NALP3 (cytoplasmic)	n.a.

n.a., not assigned.
[a] Predicted diacylated structure and TLR2-TLR6-dependent immunorecognition of lipoproteins from *S. aureus* and other classical Gram-positive bacteria of the firmicute phylum.
[b] LTA preparations derived from *S. aureus* and other Gram-positive bacteria have been reported to activate mammalian innate immunity via the heteromeric combination of TLR2 with TLR6, but according to more recent studies residual amounts of lipoproteins or lipopeptides appear to represent the TLR2-activating component(s) in these LTA preparations.
[c] While most Gram-positive bacteria including *S. aureus* express L-Lys-type peptidoglycan, some Gram-positive species of the low G + C DNA content group such as *B. subtilis* and *L. monocytogenes* have been identified to contain a *m*-DAP-type form of murein.

bacterial RNA [30] – still need to be defined. Distinct from the cell envelopes of Gram-negative bacteria, the cell walls of Gram-positive 'firmicute' species such as *S. aureus* are characterized by the presence of a single (cytoplasmic) membrane and a rather thick (usually about 60 to 80 nm) multilayered and highly cross-linked peptidoglycan network. This PG matrix additionally contains covalently linked wall teichoic acid (WTA) polymers and LPXTG-type surface proteins such as protein A (spA) or fibronectin binding proteins in *S. aureus*, whereas bacterial lipoproteins and lipoteichoic acid (LTA) polymers are attached by lipid-anchoring to the outward directed leaflet of the cytoplasmic membrane [1, 31–33]. Moreover, many clinically relevant strains of *S. aureus* also express an outer shell of capsular polysaccharides (CPS) that are commonly

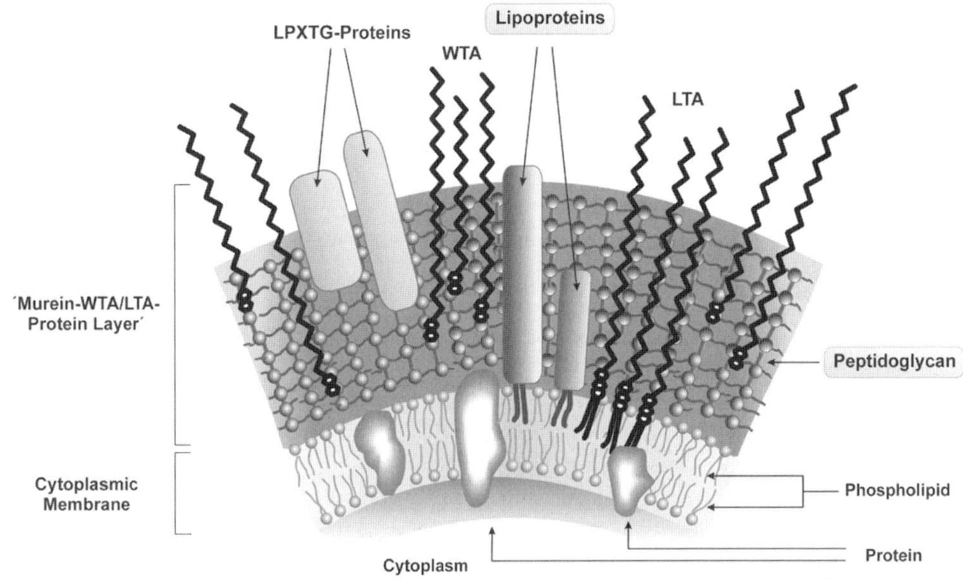

Figure 6.1 Cell wall and major wall-associated PAMPs of Gram-positive bacteria. The general cell wall composition of *S. aureus* representing the prototypic Gram-positive species in human septic diseases is schematically shown. The highlighted cell wall components, bacterial lipoproteins (LP) and peptidoglycan (PG) have been shown to be major PAMPs of Gram-positive bacteria. As indicated, lipoteichoic acids (LTA) and wall teichoic acids (WTA) are anchored to the outer surface of the cytoplasmic membrane and to the peptidoglycan network, respectively. The rigid cell walls of Gram-positive bacteria are additionally characterized by a specific series of LPXTG-proteins including virulence factors such as protein A and fibronectin-binding proteins in *S. aureus* that are also covalently bound to the PG matrix. Moreover, many clinical isolates of *S. aureus* have been shown to express capsular polysaccharides (CPS) although the method of attachment to the cell wall has yet to be determined.

assigned to a set of 11 serotypes to date [19]. During the past decade the increasing capability of many staphylococcal strains to form biofilms i.e. matrix-embedded multilayered cell clusters on plastic or glass surfaces of medical devices, has been intensively studied. This clustered embedding of staphylococci as well as other Gram-positive bacteria in biofilms represents an inducible feature of the Gram-positive cell wall surface providing rather effective protection from antibiotics and antibacterial defense peptides [3, 4, 34–37]. Specific linear surface glycans consisting of modified β-1,6-linked N-acetyl-D-glucosamine (GlcNAc) residues designated as polysaccharide intercellular adhesin (PIA) or polymeric N-acetyl-glucosamine (PNAG) have been identified to play a central role in biofilm formation by corresponding strains of *S. aureus* and *S. epidermidis* [35] and the surface net charge provided predominantly by LTA and WTA molecules in the cell wall of *S. aureus* has been also found to contribute to this particular pathogenic process [38].

As the structural and immunological features of peptidoglyan, lipoproteins and the teichoic acid polymers in the cell wall of Gram-positive bacteria have been a particular focus of many investigations, these major components of the Gram-positive cell wall will be described in more detail in the following sections.

6.2.1 Lipoproteins of Gram-positive Bacteria

As already revealed in primary studies of TLR assignment, TLR2 is a central mammalian signaling PRR in the immunorecognition of Gram-positive bacteria [39–42], but the identification of the corresponding TLR2-activating PAMP molecules has been a matter of considerable controversial investigation [43–48]. Bacterial lipoproteins are characterized by a specific N-terminal lipid anchor structure consisting of a common and evolutionarily invariant S-diacylglyceryl-cysteine moiety that can be modified by N-acylation. The N-terminal lipopeptide region containing this bacteria-specific membrane anchoring motif and a short far more variable oligopeptide sequence has been identified to be the pivotal activator of TLR2-centered signaling systems in mammalian innate immunity [49]. Although the existence of lipoproteins in *S. aureus* and other Gram-positive bacteria had already been well documented more than a decade ago [50], the PRR-directed activities of Gram-positive lipoproteins have only very recently attracted attention. According to several more recent studies employing staphylococcal strains deficient in enzymes of the terminal (pro)lipoprotein processing pathway, the membrane-anchored lipoproteins of *S. aureus* and other Gram-positive bacteria appear to be the predominant activators of TLR2-dependent immunorecognition in the mammalian host [46, 51–53], whereas the prior assignments of TLR2-activating properties to staphylococcal peptidoglycan and lipoteichoic acids have been suggested to be mediated by residual and highly immuno-active lipoproteins in the corresponding PG and LTA preparations [44, 46].

The first strong evidence concerning the presence of lipoproteins in *S. aureus* and in *Bacillus* species was reported in the early 1980s, revealing the existence of an N-terminal lipopeptide structure in the membrane-bound form of the classical penicillin-resistance factor β-lactamase (BlaZ) [54, 55]. In a more recent study, metabolic labeling with [^{14}C] palmitate revealed the expression of at least 22 lipoproteins in exponentially-growing *S. aureus* cells [51]. According to a well-elaborated bioinformatical analysis, a total of 47 to 55 lipoprotein candidate genes have been assigned in the complete genomes of six *S. aureus* strains, to date [56]. In addition, orthologs of the essential first enzymes of the (pro)lipoprotein modification pathway in *E. coli*, phosphatidylglycerol-prolipoprotein diacylglyceryltransferase (Lgt) and prolipoprotein signalpeptidase (Lsp), have been identified to be present in complete genome sequences of *S. aureus*, whereas corresponding database searches for an ortholog of the third enterobacterial enzyme, apolipoprotein N-acyltransferase (Lnt), catalyzing the attachment of a third fatty acid to the N-terminus of lipoproteins in *E. coli*, have retrieved no positive results [51, 53]. These *in silico* data have led to the proposal, that lipoproteins of *S. aureus* and other Gram-positive bacteria may contain a diacylated N-terminal lipid anchor structure [51, 53], analogous to the prototypic mycoplasma lipopeptide MALP-2 purified from *M. fermentans* [57]. In Figure 6.2A a corresponding diacylated N-terminal lipopeptide structure is indicated for peptidyl-prolyl cis/trans isomerase (PrsA) representing a primary lipoprotein candidate in *S. aureus* [51]. As MALP-2 has been shown to be an extremely strong and stereospecific activator of TLR2 in the particular combination with TLR6 displaying half maximal activities at picomolar concentrations [57–59], the lipoproteins or lipopeptides of Gram-positive bacteria have additionally been predicted to also require the combination of TLR2 and TLR6 for immuno-activation of mammalian cells [51, 53]. However, certain diacylated synthetic lipopeptides have been more recently found to activate murine macrophages in a TLR6-independent fashion [60, 61]. Therefore, the chemical structure of the N-terminal membrane anchoring motif as well as the assignment of the particular co-receptor profile in the predicted TLR2-dependent immuno-activation of lipoproteins or lipopeptides of *S. aureus* and other Gram-positive bacteria remain to be investigated to date. With respect to the rather strong LPS-like 'endotoxic' activities of MALP-2 in murine models of septic shock [62] the lipoproteins anticipated in *S. aureus* and other Gram-positive bacteria may also display profound pathogenic effects in the course of corresponding septic diseases.

6.2.2
Peptidoglycans of Gram-positive Bacteria

As compared to the murein structure of Gram-negative bacteria, the cell walls of Gram-positive 'firmicute' species contain a rather thick multilayered peptidoglycan matrix. In accordance with the general architecture of murein found throughout the bacterial kingdom, the PG network of Gram-positive bacteria consists of linear glycan strands composed of β(1–4)-linked alternating units of β-N-acetyl-D-glucosamine (GlcNAc) and β-N-acetyl-D-muramic acid (MurNAc) that are crosslinked by characteristic oligopeptide structures. These peptide substructures are covalently bound to MurNAc residues in the glycan chains and comprise of an evolutionarily highly conserved stem peptide region and a far more phylogenetically variable linker segment. In Figure 6.2B the chemical structure of the elementary disaccharide-stem peptide unit present in peptidoglycan of *S. aureus* is shown. According to the essential crosslinking diamino acid residue in position 3 of the stem peptide region, PG of *S. aureus* has been assigned to the L-lysine (L-Lys)-type family of peptidoglycans [1, 31, 63]. Though many other Gram-positive bacteria of the firmicute group have

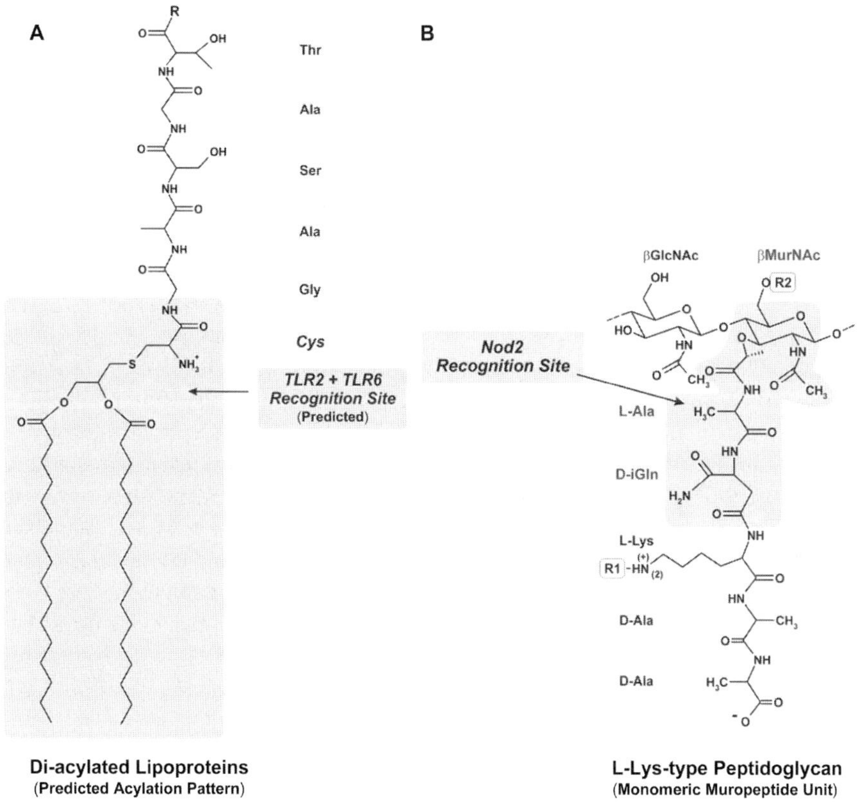

Figure 6.2 Immuno-activating domains in lipoproteins and peptidoglycan of Gram-positive bacteria. As indicated for the membrane-anchoring substructure of peptidyl-prolyl cis/trans isomerase (PrsA) from *S. aureus* (A) lipoproteins of Gram-positive bacteria have been predicted to contain a 'MALP-2-like' diacylated N-terminal lipopeptide domain including the highly conserved S-diacylglyceryl-cysteine moiety that has been correspondingly proposed to be recognized by the heteromeric combination of Toll-like receptor 2 (TLR2) and TLR6 in the mammalian host. In L-Lys-type PG found in *S. aureus* and many other Gram-positive bacteria (B) the common bacterial muramyldipeptide (MDP) unit that has been identified to activate the mammalian cytoplasmic sensorprotein Nod2 is highlighted. Assignment of groups R1 and R2 in PG of *S. aureus*, respectively: R1 = H or a penta-glycine-[(Gly)5] unit that may either mediate crosslinking to a D-Ala′ residue at position 4 in a neighboring stem peptide or provide covalent coupling to the C-terminal threonyl residue in LPXTG proteins. R2 = H or a phospho-diester linkage to wall teichoic acid (WTA) polymers (see also Figure 6.3B).

also been shown to express L-Lys-type PG structures, some Gram-positive species such as *B. subtilis* and *L. monocytogenes* have been identified to contain an 'enterobacteria-like' *meso*-diaminopimelic acid (*m*-DAP)-type form of murein [64, 65]. The PG structure of *S. aureus* has been characterized and shown to contain penta-glycyl [$(Gly)_5$] linkage units that are amide-bound to the ε-amino groups of the stem peptide Lys-residues and provide either the formation of cross-bridges to neighboring glycan strands or the covalent coupling of LPXTG-proteins within the rather compact outer layer of the staphylococcal cell wall. Both types of coupling reactions have been identified to proceed *via* a transpeptidase mechanism. Crosslinking of PG is catalyzed by members of the penicillin-binding protein (PBP) superfamily [66], whereas the covalent anchoring of LPXTG-type proteins to the N-terminal residue of the penta-glycyl linkage units is mediated by the more recently identified enzyme sortase (SrtA) [67]. In addition, wall teichoic acids have been shown to be covalently attached to the PG matrix of *S. aureus* and other Gram-positive bacteria via a phosphodiester-linkage to the C6 position of MurNAc residues [32]. Concerning the overall structure of murein from *S. aureus* a novel model characterized by a perpendicular orientation of the glycan strands to the plasma membrane has been proposed more recently [68, 69] that has been controversially discussed by other authors [70]. Two major classes of clinically administered antibacterial agents – the group of β-lactam antibiotics including classical members such as penicillin G and the stabilized second generation agents methicillin and oxacillin as well as the group of glycopeptide antibiotics such as vancomycin and teicoplanin – have been well characterized and shown to interfere with the transpeptidase reaction of PG crosslinking. As summarized in several excellent reviews on this topic, *S. aureus* has successively evolved very effective resistance mechanisms to each of these antibiotics including the early plasmid-encoded expression of β-lactamases, the development of the strongly penicillin-insensitive PBP2 variant MecA (PBP2a, PBP2′) in MRSA/ORSA strains, the production of thickened hypo-crosslinked murein functioning as a type of vancomyin trap in VISA strains and most recently the achievement of complete vancomycin resistance due to the expression of the *vanA* gene in VRSA strains. The latter enzyme has been shown to provide the incorporation of a novel terminal D-Ala-D-Lac-substructure into the stem-peptide region of PG precursors, thus replacing the common vancomycin-targeted D-Ala-D-Ala-unit [5–8].

The common muramyldipeptide (MDP) unit [MurNAc-(L-Ala)-(D-*iso*-Glu/Gln)] present in both the L-Lys-type and the *m*-DAP-type families of peptidoglycan has been identified to activate the cytoplasmic sensorprotein Nod2 in mammalian innate immunity thus contributing also to the immunorecognition of major Gram-positive pathogens including *S. aureus* [71, 72]. In addition to this broad spectrum sensing of the MDP motif in partial structures of PG via Nod2, *m*-DAP-type forms of murein as found in some Gram-positive species such as *B. subtilis* and *L. monocytogenes* [70, 71] have also been shown to activate the intracellular sensor Nod1 which specifically recognizes the (D-*i*Glu/Gln)-(*m*-DAP)-dipeptide [*i*E/Q-*m*-DAP] motif of the stem peptide region in corresponding PG biosynthesis intermediates or recycling products [73, 74]. In general, the combined immunorecognition of phylogenetically conserved partial structures in lipoproteins and peptidoglycans of

Gram-positive bacteria via TLR2- and Nod protein-centered mechanisms is currently considered to be a central issue in the pathogenesis of septic diseases related to *S. aureus* and other Gram-positive species.

6.2.3
Lipoteichoic Acids and Wall Teichoic Acids

According to distinct cell wall anchoring modes two general types of teichoic acid polymers, wall teichoic acids (WTA) and lipoteichoic acids (LTA), have been distinguished in *S. aureus* and other Gram-positive bacteria (Figure 6.3). In general, LTA and WTA share common building blocks of a short oligosaccharide linkage unit and a terminal polymeric region consisting of substituted alditol-phosphate (glycerol-phosphate or ribitol-phosphate) repeating units. In both, WTA and LTA of *S. aureus* D-alanine (D-Ala) and N-acetyl-D-glucosamine (GlcNAc) residues have been shown to be esterified to the respective poly-glycerol-phosphate and poly-ribitol-phosphate main chains. LTA is anchored to the outer leaflet of the cytoplasmic membrane by a common diacylglycerol unit [75, 76], while the WTA polymers are covalently attached to the C6 positions of MurNAc residues in the PG network via a phosphodiester linkage [32]. Though WTA contribute up to 60% of the dry mass of purified cell walls from Gram-positive bacteria, these teichoic acid polymers have been found to be dispensable for the immunostimulatory activities of cell wall preparations of *S. aureus* and *S. pneumoniae* in human myeloid cells [77]. Concerning the substitution pattern of the terminal repeating units, however, in a series of investigations analyzing a *dltA* deficient strain of *S. aureus* significant contributions of the D-alanine residues in WTA and LTA to major mechanisms of pathogenicity including the resistance to various cationic antibacterial peptides and to vancomycin have been revealed [78–80]. Moreover, this *S. aureus* mutant bearing a stronger negative surface charge due to the lack of D-alanine esters in its teichoic acids was found to display a strong reduction in the colonization of polystyrene or glass indicating a key role for the net charge of LTA and WTA in the pathogenic process of biofilm formation in corresponding wild-type strains [38].

The immuno-activation potential of purified LTA preparations has been intensely characterized throughout the past decade. In an initial set of studies employing commercially available preparations, LTA have been reported to activate mammalian cells via Toll-like receptor 4 (TLR4) and the TLR4-specific accessory protein MD-2 [40, 81, 82] but these TLR4/MD-2-directed activities have subsequently been shown to be mediated by contaminating lipopolysaccharides (LPS) of Gram-negative bacteria [83, 84]. In another series of investigations, more intensively purified preparations of LTA from *S. aureus* and other Gram-positive bacteria have then been assigned to the activation of mammalian cells via a TLR2-dependent mechanism, and additional evidence was obtained in some of these studies for the additional requirement of

A

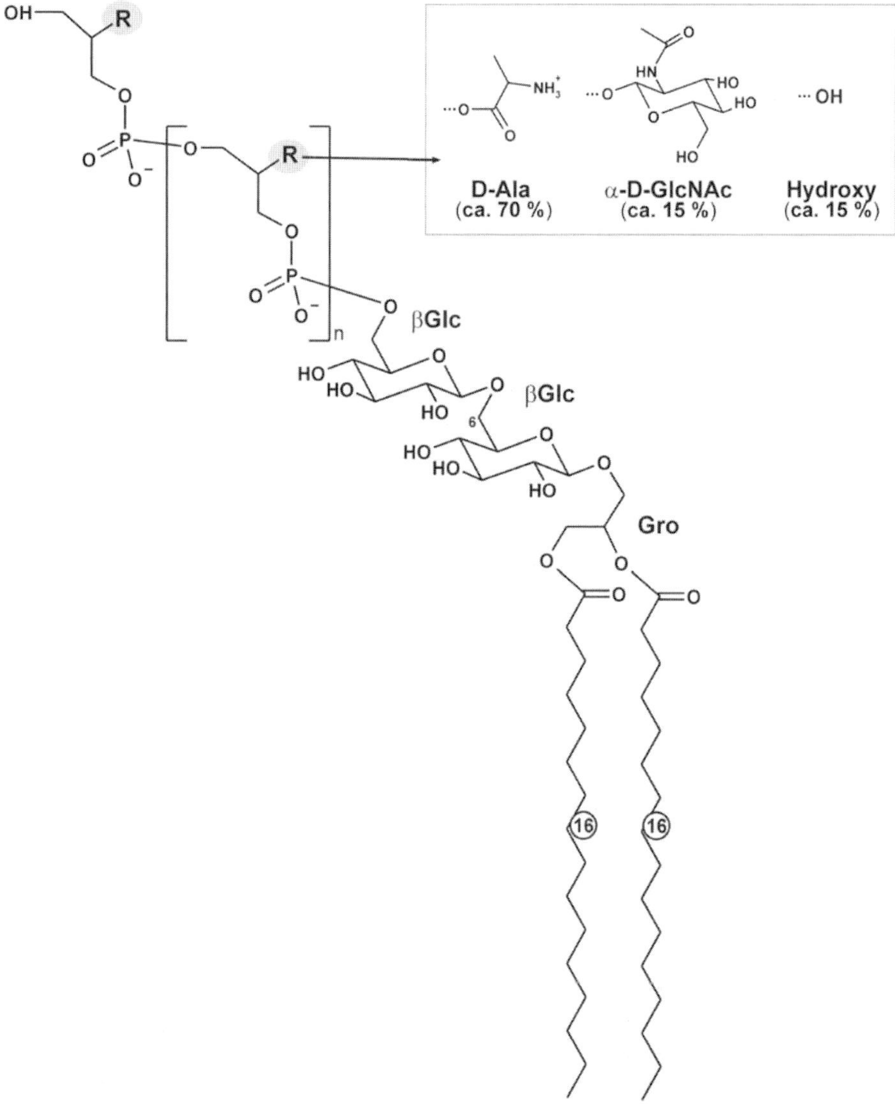

Figure 6.3A Chemical structures of lipoteichoic acids and wall teichoic acids of *S. aureus*. Lipoteichoic acids (LTA) and wall teichoic acids (WTA) represent major cell wall components of Gram-positive bacteria. The chemical structures of LTA and WTA of *S. aureus* are shown in panels A and B. On a per weight basis WTA make up about 50 to 60% of the overall composition of the cell wall in *S. aureus* and other Gram-positive bacteria. As indicated in the inserts the group R in the glycerol-phosphate repeating units in LTA of *S. aureus* has been identified as D-alanine (D-Ala), N-acetyl-glucosamine (GlcNAc) or a free hydroxy group present at the given percentage ratios and these side chain structures have also been found in the ribitol-phosphate repeating units in WTA of *S. aureus*.

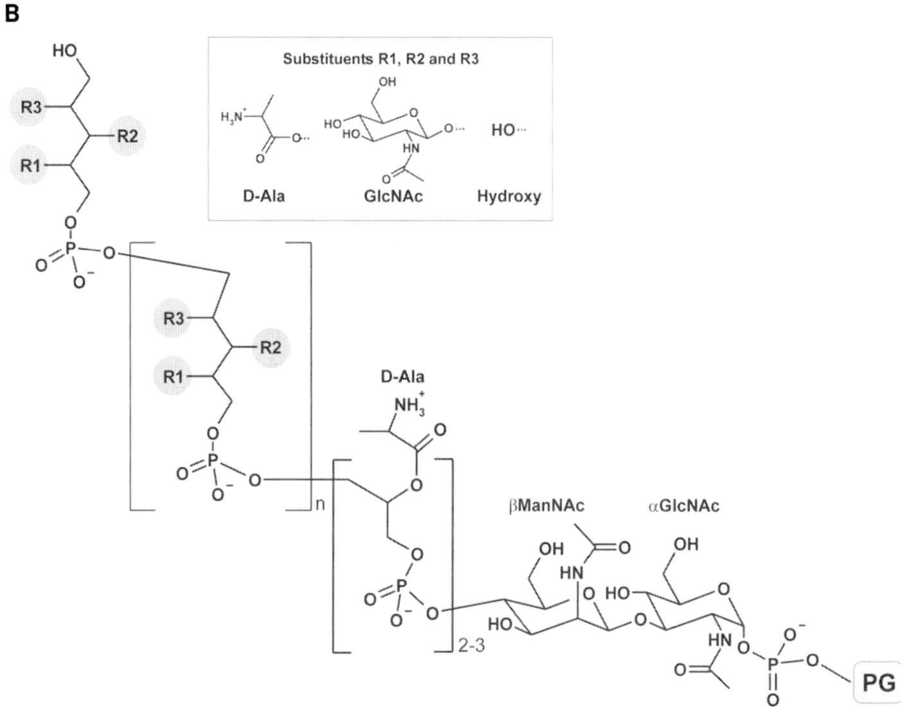

Figure 6.3B *(continued).*

TLR6 as well as for modulating functions of lipopolysaccharide-binding protein (LBP) and CD14 in these TLR2-centered activities of lipoteichoic acids [43, 85–88]. However, the origin of TLR2-activity in cell wall-derived preparations of Gram-positive bacteria has recently been challenged by studies employing mutant strains of *S. aureus* completely lacking N-terminally lipid-modified mature lipoproteins due to deficiency in the primary enzyme of the prolipoprotein processing pathway, phosphatidylglycerol-prolipoprotein diacylglyceryltransferase (Lgt) [51, 53]. Cellular lysates as well as structurally complete LTA preparations from these *lgt*-deficient strains of *S. aureus* have been shown to display drastically reduced or virtually no activation of myeloid and TLR2-transfected mammalian cells even at very high concentrations and thus, the authors concluded that lipoproteins but not LTA represent the predominant TLR2-activating component of *S. aureus* [46, 51, 89], a conclusion that is currently under debate [47, 48]. Though several lines of evidence have indicated a role for LTA in immuno-escape mechanisms of *S. aureus* and other Gram-positive bacteria, these membrane-anchored teichoic acid polymers apparently do not activate TLR2 or any other member of the mammalian TLR family. The important role of LTA in the viability and pathogenicity of *S. aureus* has additionally been substantiated by the finding that the bactericidal actions of the novel antibiotic daptomycin are due to the interference of this agent with the early stages of LTA biosynthesis [90–92].

6.3 Activation of TLR2-dependent Signaling by Lipoproteins of Gram-positive Bacteria: Central Signaling Pathways and Major Negative Regulatory Factors

To date, the chemical structure of the N-terminal lipid anchor region in lipoproteins or lipopeptides of firmicute Gram-positive bacteria as well as the activation mode of TLR2-dependent signaling by purified members of this subfamily of bacterial lipoproteins/peptides remain to be determined. According to the analysis of genome data, *S. aureus* and other Gram-positive bacteria have been proposed to express a 'MALP-2-like' diacylated subclass of lipoproteins that may in turn activate mammalian cells via the combination of TLR2 with TLR6 [51, 53]. Concerning this current working hypothesis, the extracellular proteins LBP, CD14 and CD36 have been reported to enhance the TLR2/TLR6-dependent immunorecognition of diacylated bacterial lipoproteins or synthetic lipopeptides [93, 94] and thus may have analogous accessory functions in TLR2-centered signaling activated by lipoproteins or lipopeptides of Gram-positive bacteria. Moreover, physico-chemical analysis of synthetic lipopeptides has recently revealed, that analogously to LPS of Gram-negative bacteria, the molecular conformation and the supramolecular

Figure 6.4 Activation of TLR2-dependent signaling by lipoproteins of Gram-positive bacteria: central signaling pathways and major negative regulatory factors. A current model of major signaling pathways in TLR2-TLR6-dependent activation of mammalian myeloid cells by the predicted diacylated 'MALP-2-like' type of lipoproteins (LP) of *S. aureus* and other Gram-positive bacteria is schematically shown. In addition, putative accessory functions of the extracellular PRRs LBP, mCD14 and CD36 in TLR2-TLR6-activation by diacylated lipoproteins or lipopeptides are indicated and major negative-regulatory factors of TLR2-dependent signaling systems (indicated in light grey and by broken lines) are also assigned. LBP, LPS binding protein; mCD14, membrane-associated CD14; CD36, CD36 antigen; TLR4, Toll-like receptor 4; MD-2, myeloid differentiation protein-2; MyD88, myeloid differentiation factor 88; TIRAP, Toll-interleukin 1 receptor (TIR) domain-containing adapter protein; Btk, Bruton's tyrosine kinase; IRAK-1, IRAK-2, IRAK-4, interleukin-1 receptor-associated kinase-1, -2, and -4. TRAF6, TNF receptor-associated factor 6; TAK1, TGFβ-activated kinase 1; NF-κB, nuclear factor κB; I-κB, inhibitor of NF-κB; IKKα, IKKβ, I-κB kinases α and β; IKKγ, I-κB kinase γ subunit; NEMO, NF-κB essential modulator; TAB1, TAB2, TAB3, TAK1-binding protein 1, 2 and 3; MAP3K7, mitogen-activated protein kinase kinase kinase 7; MEKK1, MEKK2, MAPK/ERK kinase kinase 1 and 2; MKKs, MAP kinase kinases; MEK1, MEK2, MAPK/ERK kinase 1 and 2; MAPK, mitogen-activated protein kinase; ERK1, ERK2, extracellular-signal-regulated kinase 1 and 2; JNK1, JNK2, c-Jun N-terminal kinase 1 and 2. Negative regulatory factors: SOCS-1, suppressor of cytokine signaling-1; IRAK-M, interleukin-1 receptor-associated kinase-M. Major types of signaling domains: LRR, leucine-rich repeat domain; CR, cysteine-rich region; TIR, Toll/IL-1 receptor homology domain. ▶

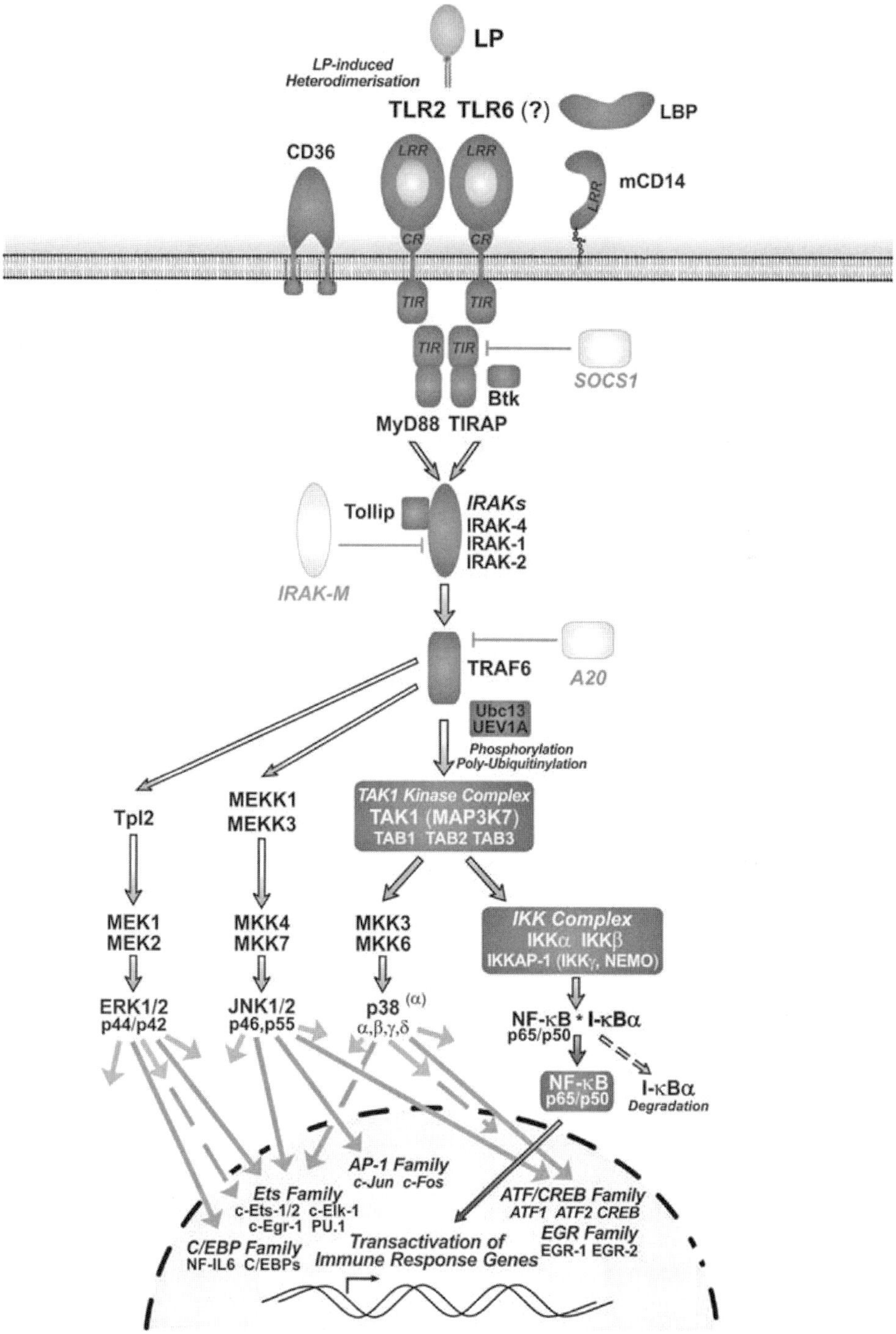
LP
LP-induced Heterodimerisation
TLR2 TLR6 (?)
LBP
CD36
LRR
LRR
mCD14
LRR
CR
CR
TIR
TIR
TIR
TIR
SOCS1
Btk
MyD88 TIRAP
Tollip
IRAKs
IRAK-4
IRAK-1
IRAK-2
IRAK-M
TRAF6
A20
Ubc13
UEV1A
Phosphorylation
Poly-Ubiquitinylation
TAK1 Kinase Complex
TAK1 (MAP3K7)
TAB1 TAB2 TAB3
Tpl2
MEKK1
MEKK3
MEK1
MEK2
MKK4
MKK7
MKK3
MKK6
IKK Complex
IKKα IKKβ
IKKAP-1 (IKKγ, NEMO)
ERK1/2
p44/p42
JNK1/2
p46,p55
p38 (α)
α,β,γ,δ
NF-κB * I-κBα
p65/p50
NF-κB
p65/p50
I-κBα
Degradation
AP-1 Family
c-Jun c-Fos
Ets Family
c-Ets-1/2 c-Elk-1
c-Egr-1 PU.1
ATF/CREB Family
ATF1 ATF2 CREB
EGR Family
EGR-1 EGR-2
C/EBP Family
NF-IL6 C/EBPs
Transactivation of
Immune Response Genes

aggregate structures formed by lipopeptides are important determinants of their biological activity [95]. A MALP-2-based current working model for the major signaling pathways of TLR2/TLR6-dependent activation of mammalian myeloid cells by these predicted diacyl-type of Gram-positive lipoproteins or lipopeptides, is shown schematically in Figure 6.4. By analogy to MALP-2 triggering of TLR2- and TLR6-dependent signaling by diacylated lipoproteins of *S. aureus* and other Gram-positive firmicute species this would lead to the rapid intracellular recruitment of two Toll/IL-1R homology (TIR) domain-containing adapter proteins MyD88 and TIRAP (Mal) and the subsequent activation of major MyD88-dependent intracellular signaling pathways including NF-κB-inducing pathways and the triad of major MAP kinase cascades [26, 96–98]. The adapter protein TIRAP(Mal) has been revealed to function as a specific membrane recruitment factor for MyD88 in the TLR2- and TLR4-dependent signaling systems and thus represents a particular regulatory element in the recognition of major bacterial cell wall structures by these TLRs [99–101]. As summarized in more detail in recent review articles [97, 98] and also indicated in Figure 6.4, a set of intracellular factors – suppressor of cytokine signaling-1 (SOCS-1), interleukin-1 receptor-associated kinase-M (IRAK-M) and the TRAF6-deubiquitinating enzyme A20 – has been identified that negatively regulate distinct steps of TLR2-dependent signaling systems and may thus contribute to the promotion of a lipoprotein-tolerant cellular state during septic infections caused by Gram-positive bacteria. Most of these negative regulatory factors have been found to be shared by all mammalian TLR signaling systems with the single exception of TLR3 and may thus also provide a state of cellular cross-tolerance for the corresponding panel of bacterial or viral PAMP structures. Similar to current pathogenesis models of Gram-negative septic infections, a stage- and probably also organ-specific sequence of imbalanced and inadequately regulated modes of TLR2 activation by lipoproteins of Gram-positive bacteria ranging from a hyposensitive to a hyperinflammatory state, may at least partially contribute to the development of Gram-positive septic complications.

6.4 Superantigens: A Specific Class of Gram-positive Pathogenicity Factors

Superantigens are toxins of microbial or viral origin that can simultaneously bind to major histocompatibility complex (MHC) class II molecules found on antigen presenting cells (APCs) and to the variable β chain of the T cell receptor (TCRVβ). The most well-studied superantigens are those produced by the Gram-positive bacteria, *Staphylococcus aureus* and *Streptococcus pyogenes, Streptococcus dysgalactiae* and *Streptococcus equi*, but *Mycoplasma* and *Yersinia* are other microbes that can produce superantigens. With the help of genome sequencing, the number of members in the microbial superantigen family has grown and now consists of approximately 40 toxins and additional allelic variants including the classical staphylococcal enterotoxin (SE) and streptococcal pyrogenic exotoxin (SPE) subfamilies as well as toxic shock syndrome toxin-1 (TSST-1) [102–105] (Table 6.2).

The term 'superantigens' was first coined by Kappler and Marrack to describe a group of antigens that behaved differently from the classical antigens normally found on MHC molecules [106]. By binding outside the peptide groove and crosslinking MHC-II and TCR, superantigens are able to activate T cells oligoclonally leading to the production of large amounts of cytokines from either T cells (IL-2, TNF-α, and IFN-γ) or from antigen presenting cells (TNF-α and IL-1) [107–110] (Figure 6.5). The extensive cytokine production overstimulates the immune system leading to hypertension, fever, and shock [111, 112]. Serious consequences of superantigen release include toxic shock, food poisoning, and scarlet fever [113, 114].

Much progress has been made on the characterization of binding modes to TCRVβ and MHC-II due to the growing availability of crystal structures. The vast array of sequenced microbial genomes has also contributed greatly to the localization of superantigens on various genetic elements including phages, plasmids, and pathogenicity islands. The aim of this section is to give insight into the superantigens and their mode of binding to and activating APCs and T cells and deciphering their role in human diseases with a particular emphasis on sepsis.

6.4.1 Superantigen Structure

Most superantigens share a characteristic two-domain architecture with a pseudo β barrel resembling the oligosaccharide-/oligonucleotide-binding fold (OB fold) at the amino-terminal domain and a β grasp motif at the carboxy-terminal domain. Detailed structural information was first provided in 1992 with the successful crystallization of SEB [115]. Since then, various crystal structures have been published either as superantigens alone, superantigens complexed with MHC-II molecules or with TCRVβ, and recently, a ternary crystal structure has been reported with MHC-II/peptide-superantigen-TCR [116].

Although it has been speculated that superantigens also bind to MHC-I [117] and CD1 molecules [118], all available crystal structures and most experimental evidence show specific superantigen binding to MHC-II structures. The first MHC-II molecule was crystallized in 1993 revealing an α/β heterodimeric quaternary structure [119]. The interactions between superantigens and MHC-II molecules can be classified into three groups:

1. contact with the MHC-IIα chain peripheral to MHC-bound peptide (e.g. SEB)
2. contact with the MHC-IIα over the MHC-bound peptide (e.g. TSST-1)
3. zinc-mediated contact with the MHC-IIβ chain (e.g. SPE-C)

Table 6.2 Superantigens and their properties.

Superantigens[a]	Molecular weight (kD)	Crystal structure	Human TCR (Vβ)	MHC α/β chain	Diseases[b]	Gene localization[c]
SEA	27.1	+	1, 5.3, 6.3–6.4, 6.9, 7.3–7.4, 9, 16, 21.3, 22, 23	+/+	Food poisoning	Phage DNA
SEB	28.4	+	1, 3.2, 6.4, 15	+/−	Food poisoning	Plasmid-borne Pathogenicity island
SEC (alleles 1–3)	27.5	−	3.2, 6.4, 6.9, 12, 15	+/−	Food poisoning	Pathogenicity island
SED	26.9	+	1, 5.3, 6.9, 7.4, 8, 12	+/+	Food poisoning	Plasmid-borne
SEE	26.8	−	5, 6.3, 6.4, 6.9, 8	+/+	Food poisoning	Phage DNA
SEG	27.0	−	3, 12, 13.1–13.2, 14, 15	+/−	Food poisoning	Pathogenicity island, egc
SEH	25.2	+	Vα?	−/+	Toxic shock	
SEI	24.9	+	1, 5.1, 5.3, 23	−/+	Food poisoning	Pathogenicity island, egc
SEJ	28.5	−	?		Food poisoning	Plasmid-borne
SElK	25.3	−	5.1, 5.2, 6.7			Pathogenicity island
SElL	24.7	−	5.1, 5.2, 6.7, 16, 22			Pathogenicity island
SElM	24.8	−	6, 8, 9, 18, 21.3			Pathogenicity island, egc
SElN	26.1	−				egc
SElO	26.7	−				Pathogenicity island, egc

SElP	26.4	−	5.1, 6, 8, 16, 18, 21.3			Pathogenicity island
SElQ	26.0	−	2, 5, 21.3			Pathogenicity island
SElR		−	3, 11, 12, 13.2, 14			Plasmid-borne
SElU		−				egc
SElU2		−	13.2, 14			egc
SElV		−	6,18, 21			egc
TSST-1	21.9	+	2	+/−	Toxic shock syndrome	Pathogenicity island
SPE-A (alleles 1–6)	26.0	+	2, 12.2, 14, 15	+/−	Scarlet fever, STSS	Phage DNA
SPE-C	24.4	+	2, 3.2, 12.5, 15	−/+	STSS, KD?	Phage DNA
SPE-G	24.6	−	2, 4, 6.9, 9, 12.3	−/+		
SPE-H	23.6	+	2, 7.3, 9, 23	−/+		Mobile element
SPE-I	26.0	+	6.9, 9, 18, 22	−/+		in *S.p.* M1 genome
SPE-J	24.6	+	2	−/+		In *S.p.* M1 genome
SPE-L/K	27.4	−	1, 5, 23	−/+	Rheumatic fever?	Phage DNA
SPE-M	26.2	−	1, 5, 23	−/+	Rheumatic fever?	Mobile element

(*continued overleaf*)

Table 6.2 *(continued)*.

Superantigens[a]	Molecular weight (kD)	Crystal structure	Human TCR (Vβ)	MHC α/β chain	Diseases[b]	Gene localization[c]
SSA	26.9	+	1, 3, 15, 17, 19	+/−	STSS	
SMEZ (alleles 1–34)	24.3	+	2, 4, 7.3, 8	−/+	KD?	SMEZ gene
SDM	25.0	−	1, 23	−/+		
Spe-G^{dys}	24.4	−	1, 4, 10 (bovine)			
SePE-H	23.6	−			ES?	
SePE-I	25.7	−			ES?	
SePE-L	27.4	−			ES?	
SePE-M	26.2				ES?	
MAM	25.2	+	6, 8	+/+	Arthritis?	
YPM (a-c)	14.5	+	3, 9, 13.1–13.2		KD?	Unstable chromosomal locus

[a] *Staphylococcus aureus* produces the staphylococcal enterotoxin (SE-) and toxic shock syndrome toxin-1 (TSST-1) superantigens, streptococcal pyrogenic exotoxins (SPE-), streptococcal superantigen (SSA) and the streptococcal mitogenic exotoxins (SMEZ) are produced by *Streptococcus pyogenes, Streptococcus dysgalactiae*-derived mitogen (SDM) and a superantigen similar to SPE-G (Spe-G^{dys}) by *Streptococcus dysgalactiae, Streptococcus equi* pyrogenic exotoxin (SePE-) by *Streptococcus equi*, and the superantigens originally described as mitogens are produced by *Mycoplasma arthritidis* (MAM) and *Yersinia pseudotuberculosis* (YPM)

[b] KD, Kawasaki Disease; STSS, streptococcal toxic shock syndrome; ES, equine strangles.

[c] egc, enterotoxin gene cluster

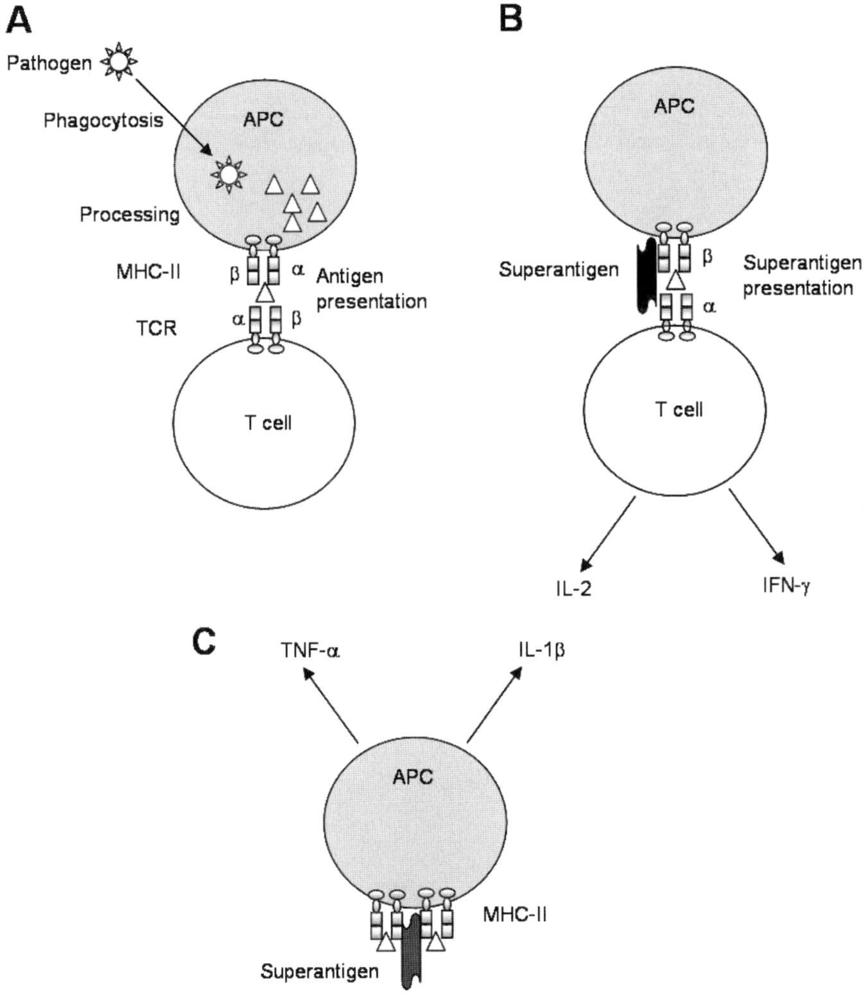

Figure 6.5 Antigen presentation and pathogenic effects of bacterial superantigens. Classical antigen presentation ordinarily involves phagocytosis of a pathogen, processing within the cell, and presentation on MHC-II molecules, which can then be recognized by the T cell receptor (TCR) on T cells (A). Superantigen presentation bypasses the phagocytosis and processing steps and cross-links the T cell receptor with a MHC-II molecule (B). Some superantigens can also ligate MHC-II molecules and increase the activity of antigen presenting cells by triggering cytokine production (C).

Some superantigens are even able to crosslink MHC-II molecules by either simultaneous binding to MHC-IIα and MHC-IIβ chains (SEA, SEE), two MHC-IIα chains (SED), or two MHC-IIβ chains (SPE-C).

6.4.2
Gene Localization, Superantigen Groups, and Allelic Variation

Superantigens are often found encoded on mobile genetic elements contributing to the fact that no streptococcal or staphylococcal isolate has been identified with a complete repertoire of known superantigens [105]. The mobile genetic elements transporting superantigen genes include prophages/bacteriophages, plasmids, and a major source appears to be located within pathogenicity islands (Table 6.2). The presence of most superantigens on pathogenicity islands may help decipher the origins and mode of transfer of superantigens from one bacterial strain to another.

Although a bounty of information is now available aiding in the identification of new superantigens, this has also led to quite a bit of confusion. There are many proteins which are structurally and functionally related to superantigens. The nearest cousin would be the staphylococcal enterotoxin-like proteins (SEls). This name has been given to proteins with proven superantigenic activity but without emetic properties. SEls are almost identical in biological nature to superantigens since they require MHC-II molecules for activation of T cells bearing specific TCRVβ domains [120]. However, these molecules have not yet been shown or do not induce emesis following oral administration in a primate model. Due to this novel panel of pathological properties, certain superantigens had to be reclassified into this new group of SEls (SElK-SElV) (Table 6.2).

Another group of superantigen-like proteins found in Gram-positive bacteria are the staphylococcal superantigen-like (SSL) proteins formerly known as the staphylococcal enterotoxin-like toxins (SET). SSL (SET) proteins are structurally related, but do not bear a functional resemblance to superantigens. A SET gene cluster was first described by Williams and colleagues as a group of related genes that can stimulate cytokine production in human peripheral blood mononuclear cells (PBMCs) [121]. Despite the homology in structure, recombinant SETs do not bind to MHC-II, nor do they stimulate T cells. The exact function of the SET family is yet to be determined, however, it was recently found that SSL5 abrogates neutrophil extravasation towards the site of infection by interfering with the interaction between PSGL-1 and P-selectin [122].

Allelic variation has also enhanced the number of superantigens. At least six allelic variants have been described for the speA gene [123] and the different variants appear to show varying mitogenic activity [124]. Shortly after the discovery of SMEZ [125], an allelic variant of SMEZ was identified and given the name SMEZ-2 [126]. Since then, 34 SMEZ alleles have been reported [127]. SMEZ is different from other superantigens due to its highly polymorphic nature stemming from genetic mosaicism at a single genetic locus [126]. Occasionally, the allelic variation can be dramatic enough to permit the superantigens to bind to different TCRVβ domains, which is seen in the case of SEC1 and SEC2 [128].

6.4.3 Superantigen Involvement in Sepsis

Superantigens have been implicated in various diseases such as toxic shock syndrome (TSS) [129], food poisoning [114], streptococcal toxic shock syndrome [130], septic shock [131], Kawasaki disease [132], and acute rheumatic fever [133]. This section is designed to provide an overview of the role of the superantigen in causing sepsis-related diseases including toxic shock syndrome and streptococcal toxic shock syndrome.

Superantigens express enormous immuno-activating potency that may contribute to septic shock by initiating a proinflammatory/Th1 cytokine storm through T cell activation and TNF-α release. A biphasic model has been suggested where there is an early cytokine burst comprised of IL-2, TNF-α, and IL-6 followed by a rise in the Th1 cytokines, IL-12 and IFN-γ, whereby the high IFN-γ levels coincide with the time of death [134]. Although superantigens can induce Th1 cytokines, these cytokines may not play the central role in septic shock. Treating animals with antibodies against Th1 cytokines such as IL-2, IL-12, and IFN-γ has not been proven effective against toxic shock [134–136], whereas anti-TNF-α pretreatment protected the animals from the early cytokine burst, which is thus more likely to be responsible for lethality than the later rise in Th1 cytokines [134].

Until recently, the prevalence and distribution of superantigens in septic shock patients was unclear. In one study, 11 *S. aureus* superantigen genes were screened in sepsis patients with or without septic shock, and it was determined that septic shock was more likely to be associated with the *sea* gene, but less associated with the *egc* gene [137]. SEA appeared more capable of inducing a strong Th1 response by triggering MIP-1α release and an increased proinflammatory response with elevated levels of TNF-α production [138]. SEA appears to be more effective at inducing a large cytokine burst that can lead to septic shock compared to other related superantigens.

6.4.3.1 Toxic Shock Syndrome (TSS)

From a clinical perspective, toxic shock syndrome is a special entity of sepsis. Toxic shock syndrome was first identified as a serious illness in young children infected with *Staphylococcus aureus* [139]. The actual importance of and widespread concern about TSS arose in the 1980s during an epidemic of TSS involving young females using tampons during menstruation. It was later discovered that this menstrual disease was caused by intravaginal colonization of strains of *S. aureus* that produce the superantigen TSST-1 [129, 140, 141]. The disease is characterized by symptoms such as rash, fever, severe hypotension, desquamation, and could eventually lead to fatal shock [104, 142]. In addition to menstruation-associated TSS, there are also non-menstrual TSS cases. TSST-1 appears to be the primary superantigen involved in menstrual TSS, whereby this list expands to include SEB and SEC for non-menstrual

TSS [143]. Many forms of this disease have been reported such as postsurgical TSS [144] and influenza-associated TSS [145]. A significant decline has been described in the incidence of menstruation-associated TSS cases, but not in the frequency of non-menstrual TSS cases. The superantigenic contribution leading to the pathogenesis of TSS includes an increase in Vβ-specific T cell proliferation and release of excessive amounts of cytokines such as TNF-α and IL-1 that can induce fever and shock.

The involvement of TSST-1 in TSS has been shown with rabbit and rodent models [108, 146], a harmless strain of *S. aureus* was made pathogenic through bacteriophage-mediated transfer of TSST-1 DNA [147], and by an increase in anti-TSST-1 antibody titer, albeit that immunity against TSST-1 is very weak in a few seroconverted patients [129]. Normal mice have served as the primary model for studying the lethal effects of superantigens even though this model may not be the most appropriate since mice are highly resistant to the lethal effects of these toxins [148] partly due to the lower binding affinity of superantigens to murine MHC-II compared to human HLA class II [149]. The development of HLA class II transgenic mice (HLA-tg) has become a steadily more popular option, which displays a similar response to superantigens as humans [150, 151]. Rabbits are also applicable due to the manifestation of a disease similar to TSS after continuous infusion with low doses of superantigens [152].

An important risk factor in the development of either STSS (streptococcal toxic shock syndrome) or TSS seems to be a lack of antibody against superantigens [153–156]. Less than half of menstrual TSS patients have sero-positive titers to TSST-1 after 2 months of their illness [155] and some were unable to develop significant TSST-1 titers to prevent reinfections with TSS [129]. The inability to produce antibodies against TSST-1 was also observed in approximately 50% of rabbits [157]. Of particular interest is the mechanism(s) leading to an inadequate antibody response to TSS and STSS. It has been suggested that TSST-1 may suppress immunoglobulin (Ig)-synthesizing cells through two possible mechanisms:

1. Superantigens suppress Ig synthesis indirectly by activating primarily $CD4^+$ T cells, which often triggers a wave of Th1 cytokines thereby overpowering the Th2 response. A lack of Th2 cytokines corresponds to a decline in IL-4 and IL-5 levels necessary for B cell proliferation and differentiation [158].
2. Using high concentrations of TSST-1 can directly mediate T cell cytotoxicity towards B cells, which would decrease the number of Ig-producing cells.

A correlation between a lack of neutralizing antibodies to superantigens and TSS/STSS should be a key consideration for the treatment of these diseases. An option could be the administration of intravenous immunoglobulin (IVIG), which in the majority of cases contains neutralizing antibodies against superantigens [159].

An interesting aspect of TSST-1 is its ability to cross endothelial cell barriers, which can induce systemic toxicity [160]. Not only may passing the endothelial barrier be important for toxic shock syndrome, it may also be significant in the pathogenesis of food poisoning.

6.4.3.2 Streptococcal Toxic Shock Syndrome (STSS)

The first cases of STSS were reported in the late 1980s and were attributed to group A Streptococcus (GAS) [109]. STSS may also be caused by Group C and Group G streptococcal (GCS and GGS) strains [161], although superantigens have not been identified in every STSS case since the discovery of superantigens in these strains is still underway.

The major differences between STSS and TSS are that STSS is often associated with bacteremia, myositis, or necrotizing fascitiitis, STSS is correlated with a higher mortality rate, and STSS is caused by streptococcal superantigens. Streptococcal superantigens are linked to STSS since there is a strong correlation between *spea* and *spec* genes in certain patients [130, 162] and some patients have circulating superantigens during the time course of infection [163]. Superantigens may be responsible for the rapid progression of STSS into sepsis.

The immunogenetic and molecular background for the severity of GAS infections are currently under investigation. As mentioned previously, low levels of protective antibodies elevate the risk of developing a GAS infection, but once the bacteria invade a normally sterile site, other factors are involved in determining the severity of the disease. The patient's response to the superantigen by means of proliferative capacity and cytokine response becomes of primary importance. Particular haplotypes of HLA-II may play a significant role since allelic variation can influence cytokine responses to superantigen [164]. One protective haplotype, DRB1*1501/DQB1*0602, has already been identified, which confers resistance to developing severe systemic disease as a result of GAS infection [165]. The HLA-DQ6 allele was confirmed for its protective effect in HLA transgenic mice demonstrated by lower levels of TNF-α and IFN-γ and thus a decrease in severity of infection [166].

There is a growing need to directly target the causative bacteria-derived molecular agents of sepsis. The rapid transition from sepsis into septic shock could be attributed to the presence of superantigens, making the identification of superantigens to be of extreme importance. A number of molecular biological approaches are currently available including multiplex PCRs designed to detect superantigens produced by *S. aureus* [167, 168] and *S. pyogenes* [169]. After identifying the superantigens, intravenous immunoglobulin (IVIG) therapy might be a plausible resort for limiting the toxic effect of superantigens assuming that the number of neutralizing anti-superantigen antibodies is adequate [170]. Determining the extent of superantigen involvement in sepsis will help in generating specific therapeutic treatments designed to combat this potentially lethal disease.

6.5 Novel Therapeutic Concepts in Gram-positive Septic Diseases

In contrast to Gram-negative bacteria where endotoxin represents the prominent inducer of inflammation, no corresponding prototypic activator of innate immunity has been defined for Gram-positive bacteria. According to current understanding, several immunologically-active molecules may act synergistically in orchestrating the immune response to Gram-positive infections. With regard to the prevalence and still increasing incidence of highly antibiotic-resistant strains of *S. aureus* and coagulase-negative staphylococci (CNS) in nosocomial septic infections [9–13] there is an ongoing demand for novel approaches in therapeutic intervention and in protective prevention of Gram-positive septic disorders. In the short term, the clinical implementation of new types of antibiotics such as daptomycin [91, 92] or platensimycin [171] is expected to be an effective conventional therapeutic option, but the extensive administration of these novel antibacterial agents in clinical settings would most probably in turn favor the rapid emergence of corresponding resistant strains of *S. aureus* or other Gram-positive pathogens. The protective vaccination or passive immunization of patients at risk of developing septic complications represents another promising option and in particular the inclusion of surface-exposed pathogenicity factors of Gram-positive bacteria such as capsular polysaccharides, LPXTG-proteins and lipoproteins as well as toxoid derivatives of superantigens in corresponding vaccine formulations may provide significant protective effects [172]. As also discussed in the previous sections of this chapter, adjunctive immunotherapeutic approaches including more precise immunomonitoring of patients and the early initiation of stage-specific immunomodulatory interventions are currently considered to be a favorite option for the treatment or even prevention of Gram-positive as well as of Gram-negative septic complications [173–175]. Furthermore, in co-infections superantigens enhance endotoxin-induced activity and thus the presence of both Gram-positive and Gram-negative bacteria should be addressed in the treatment of sepsis patients [176, 177]. As compared to the rather detailed physico-chemical and immunological characterization of LPS from Gram-negative bacteria, the corresponding analysis of major PAMP structures of Gram-positive bacteria including in particular Gram-positive members of the bacterial lipoprotein family, is much less advanced to date. The actual efforts to define the physico-chemical and immunological properties of major PAMP molecules from Gram-positive bacteria in more detail may hopefully also contribute to the development of novel molecular approaches for the therapy of Gram-positive septic disorders.

Acknowledgments

The financial support of Clinique La Prairie Research and the Deutsche Forschungsgemeinschaft (Emmy Noether Grant SCHR 621/2-2 and Priority Programm 1110) is gratefully acknowledged.

References

1. Götz, F., Bannerman, T. and Schleifer K.H. (**2006**) The genera *Staphylococcus* and *Macrococcus*. in: *The Prokaryotes – A Handbook on the Biology of Bacteria*: Vol. 4: Bacteria: Firmicutes, Cyanobacteria, 3rd ed. Falkow, S., *et al.* (eds.). Springer Verlag, New York, 5–75.
2. Koontz, F.P. (**2000**) Trends in post-operative infections by Gram-positive bacteria. *Int. J. Antimicrob. Agents.*, **16**(Suppl 1), S35–S37.
3. Costerton, J.W., Stewart, P.S. and Greenberg, E.P. (**1999**) Bacterial biofilms: a common cause of persistent infections. *Science*, **284**, 1318–1322.
4. Stewart, P.S. and Costerton, J.W. (**2001**) Antibiotic resistance of bacteria in biofilms. *Lancet*, **358**, 135–138.
5. Berger-Bächi, B. (**2002**) Resistance mechanisms of gram-positive bacteria. *Int. J. Med. Microbiol.*, **292**, 27–35.
6. Tenover, F.C. (**2006**) Mechanisms of antimicrobial resistance in bacteria. *Am. J. Med.*, **119**(6 Suppl 1), S3–S10.
7. Chambers, H.F. (**2001**) The changing epidemiology of *Staphylococcus aureus*? *Emerg. Infect. Dis.*, **7**, 178–182.
8. Lowy, F.D. (**2003**) Antimicrobial resistance: the example of *Staphylococcus aureus*. *J. Clin. Invest.*, **111**, 1265–1273.
9. Weinstein, M.P., Towns, M.L., Quartey, S.M., Mirrett, S., Reimer, L.G., Parmigiani, G. and Reller, L.B. (**1997**) The clinical significance of positive blood cultures in the 1990s: a prospective comprehensive evaluation of the microbiology, epidemiology, and outcome of bacteremia and fungemia in adults. *Clin. Infect. Dis.*, **24**, 584–602.
10. Crowe, M., Ispahani, P., Humphreys, H., Kelley, T. and Winter, R. (**1998**) Bacteraemia in the adult intensive care unit of a teaching hospital in Nottingham, UK, 1985–1996. *Eur. J. Clin. Microbiol. Infect. Dis.*, **17**, 377–384.
11. Fluit, A.C., Schmitz, F.J. and Verhoef, J. (**2001**) Frequency of isolation of pathogens from bloodstream, nosocomial pneumonia, skin and soft tissue, and urinary tract infections occurring in European patients. *Eur. J. Clin. Microbiol. Infect. Dis.*, **20**, 188–191.
12. Diekema, D.J., Pfaller, M.A., Schmitz, F.J., Smayevsky, J., Bell, J., Jones, R.N. and Beach, M. (**2001**) Survey of infections due to Staphylococcus species: frequency of occurrence and antimicrobial susceptibility of isolates collected in the United States, Canada, Latin America, Europe, and the Western Pacific region for the SENTRY Antimicrobial Surveillance Program, 1997–1999. *Clin. Infect. Dis.*, **32**(Suppl 2), S114–S132.
13. Martin, G.S., Mannino, D.M., Eaton, S. and Moss, M. (**2003**) The epidemiology of sepsis in the United States from 1979 through 2000. *N. Engl. J. Med.*, **348**, 1546–1554.
14. Kluytmans, J., van Belkum, A. and Verbrugh, H. (**1997**) Nasal carriage of *Staphylococcus aureus*: epidemiology, underlying mechanisms, and associated risks. *Clin. Microbiol. Rev.*, **10**, 505–520.

15. Wertheim, H.F., Vos, M.C., Ott, A., van Belkum, A., Voss, A., Kluytmans, J.A., van Keulen, P.H., Vandenbroucke-Grauls, C.M., Meester, M.H. and Verbrugh, H.A. (**2004**) Risk and outcome of nosocomial *Staphylococcus aureus* bacteraemia in nasal carriers versus non-carriers. *Lancet*, **364**, 703–705.
16. Noble, W.C. (**1998**) Skin bacteriology and the role of *Staphylococcus aureus* in infection. *Br. J. Dermatol.*, **139**(Suppl 53), 9–12.
17. Fitzgerald, J.R., Sturdevant, D.E., Mackie, S.M., Gill, S.R. and Musser, J.M. (**2001**) Evolutionary genomics of *Staphylococcus aureus*: insights into the origin of methicillin-resistant strains and the toxic shock syndrome epidemic. *Proc. Natl. Acad. Sci. USA*, **98**, 8821–8826.
18. Lindsay, J.A. and Holden, M.T. (**2004**) *Staphylococcus aureus*: superbug, super genome? *Trends Microbiol.*, **12**, 378–385.
19. O'Riordan, K. and Lee, J.C. (**2004**) *Staphylococcus aureus* capsular polysaccharides. *Clin. Microbiol. Rev.*, **17**, 218–234.
20. Liu G.Y., Essex, A., Buchanan, J.T., Datta, V., Hoffman, H.M., Bastian, J.F., Fierer, J. and Nizet, V. (**2005**) *Staphylococcus aureus* golden pigment impairs neutrophil killing and promotes virulence through its antioxidant activity. *J. Exp. Med.*, **202**, 209–215.
21. Dinges. M.M., Orwin, P.M. and Schlievert, P.M. (**2000**) Exotoxins of *Staphylococcus aureus. Clin. Microbiol. Rev.*, **13**, 16–34.
22. Alouf, J.E. and Müller-Alouf, H. (**2003**) Staphylococcal and streptococcal superantigens: molecular, biological and clinical aspects. *Int. J. Med. Microbiol.*, **292**, 429–440.
23. Entrez Genome Project, National Center for Biotechnology Information: http://www.ncbi.nlm.nih.gov/genomes/lproks.cgi.
24. Janeway, C.A. Jr. (**1989**) Approaching the asymptote? Evolution and revolution in immunology. *Cold Spring Harbour Symp. Quant. Biol.*, **54**, 1–13.
25. Janeway, C.A. Jr. and Medzhitov, R. (**2002**) Innate immune recognition. *Annu. Rev. Immunol.*, **20**, 197–216.
26. Akira, S., Uematsu, S. and Takeuchi, O. (**2006**) Pathogen recognition and innate immunity. *Cell*, **124**, 783–801.
27. Creagh, E.M. and O'Neill, L.A. (**2006**) TLRs, NLRs and RLRs: a trinity of pathogen sensors that co-operate in innate immunity. *Trends Immunol.*, **27**, 352–357.
28. Trinchieri, G. and Sher, A. (**2007**) Cooperation of Toll-like receptor signals in innate immune defence. *Nat. Rev. Immunol.*, **7**, 179–190.
29. Hemmi, H., Takeuchi, O., Kawai T., Kaisho, T., Sato, S., Sanjo, H., Matsumoto M., Hoshino, K., Wagner, H., Takeda, K. and Akira, S. (**2000**) A Toll-like receptor recognizes bacterial DNA. *Nature*, **408**, 740–745.
30. Kanneganti, T.D., Ozoren, N., Body-Malapel, M., Amer, A., Park, J.H., Franchi L., Whitfield, J., Barchet, W., *et al.* (**2006**) Bacterial RNA and small antiviral compounds activate caspase-1 through cryopyrin/Nalp3. *Nature*, **440**, 233–236.
31. Giesbrecht, P., Kersten, T., Maidhof, H. and Wecke, J. (**1998**) Staphylococcal cell wall: morphogenesis and fatal variations in the presence of penicillin. *Microbiol. Mol. Biol. Rev.*, **62**, 1371–1414.
32. Coley, J., Tarelli, E., Archibald, A.R. and Baddiley, J. (**1978**) The linkage between teichoic acid and peptidoglycan in bacterial cell walls. *FEBS Lett.*, **88**, 1–9.
33. Mazmanian, S.K., Ton-That, H. and Schneewind, O. (**2001**) Sortase-catalysed anchoring of surface proteins to the cell wall of *Staphylococcus aureus. Mol. Microbiol.*, **40**, 1049–1057.
34. Götz, F. (**2002**) *Staphylococcus* and biofilms. *Mol. Microbiol.*, **43**, 1367–1378.
35. Mack, D., Becker, P., Chatterjee, I., Dobinsky, S., Knobloch, J.K.,

Peters G., Rohde H. and Herrmann M. (**2004**) Mechanisms of biofilm formation in *Staphylococcus epidermidis* and *Staphylococcus aureus*: functional molecules, regulatory circuits, and adaptive responses. *Int. J. Med. Microbiol.*, **294**, 203–212.

36. Fux, C.A., Costerton, J.W., Stewart, P.S. and Stoodley, P. (**2005**) Survival strategies of infectious biofilms. *Trends Microbiol.*, **13**, 34–40.
37. O'Gara, J.P. (**2007**) ica and beyond: biofilm mechanisms and regulation in *Staphylococcus epidermidis* and *Staphylococcus aureus*. *FEMS Microbiol. Lett.*, **270**, 179–188.
38. Gross, M., Cramton, S.E., Götz, F. and Peschel, A. (**2001**) Key role of teichoic acid net charge in Staphylococcus aureus colonization of artificial surfaces. *Infect. Immun.*, **69**, 3423–3426.
39. Yoshimura, A., Lien, E., Ingalls, R.R., Tuomanen, E., Dziarski, R. and Golenbock, D. (**1999**) Cutting edge: recognition of Gram-positive bacterial cell wall components by the innate immune system occurs via Toll-like receptor 2. *J. Immunol.*, **163**, 1–5.
40. Takeuchi, O., Hoshino, K., Kawai, T., Sanjo, H., Takada, H., Ogawa, T., Takeda, K. and Akira, S. (**1999**) Differential roles of TLR2 and TLR4 in recognition of gram-negative and gram-positive bacterial cell wall components. *Immunity*, **11**, 443–451.
41. Underhill, D.M., Ozinsky, A., Hajjar, A.M., Stevens, A., Wilson, C.B., Bassetti, M. and Aderem, A. (**1999**) The Toll-like receptor 2 is recruited to macrophage phagosomes and discriminates between pathogens. *Nature*, **401**, 811–815.
42. Takeuchi, O., Hoshino, K. and Akira, S. (**2000**) Cutting edge: TLR2-deficient and MyD88-deficient mice are highly susceptible to *Staphylococcus aureus* infection. *J. Immunol.*, **165**, 5392–5396.
43. Schwandner, R., Dziarski, R., Wesche, H., Rothe, M. and Kirschning, C.J. (**1999**) Peptidoglycan- and lipoteichoic acid-induced cell activation is mediated by Toll-like receptor 2. *J. Biol. Chem.*, **274**, 17406–17409.
44. Travassos, L.H., Girardin, S.E., Philpott, D.J., Blanot, D., Nahori, M.A., Werts, C. and Boneca, I.G. (**2004**) Toll-like receptor 2-dependent bacterial sensing does not occur via peptidoglycan recognition. *EMBO Rep.*, **5**, 1000–1006.
45. Dziarski, R. and Gupta, D. (**2005**) *Staphylococcus aureus* peptidoglycan is a Toll-like receptor 2 activator: a reevaluation. *Infect. Immun.*, **73**, 5212–5216.
46. Hashimoto, M., Tawaratsumida, K., Kariya, H., Kiyohara, A., Suda, Y., Krikae, F., Kirikae, T. and Götz, F. (**2006**) Not lipoteichoic acid but lipoproteins appear to be the dominant immunobiologically active compounds in *Staphylococcus aureus*. *J. Immunol.*, **177**, 3162–3169.
47. von Aulock, S., Hartung, T. and Hermann, C. (**2007**) Comment on "Not lipoteichoic acid but lipoproteins appear to be the dominant immunobiologically active compounds in *Staphylococcus aureus*". *J. Immunol.*, **178**, 2610.
48. Hashimoto, M., Furuyashiki, M. and Suda, Y. (**2007**) Response to comment on "Not lipoteichoic acid but lipoproteins appear to be the dominant immunobiologically active compounds in *Staphylococcus aureus*". *J. Immunol.*, **178**, 2610–2611.
49. Takeda, K., Takeuchi, O. and Akira, S. (**2002**) Recognition of lipopeptides by Toll-like receptors. *J. Endotoxin. Res.*, **8**, 459–463.
50. Sutcliffe, I.C. and Russell, R.R. (**1995**) Lipoproteins of Gram-positive bacteria. *J. Bacteriol.*, **177**, 1123–1128.
51. Stoll, H., Dengjel, J., Nerz, C. and Götz, F. (**2005**) *Staphylococcus aureus* deficient in lipidation of prelipoproteins is attenuated in growth and immune activation. *Infect. Immun.*, **73**, 2411–2423.

52. Hashimoto, M., Tawaratsumida, K., Kariya, H., Aoyama, K., Tamura, T. and Suda, Y. (**2006**) Lipoprotein is a predominant Toll-like receptor 2 ligand in *Staphylococcus aureus* cell wall components. *Int. Immunol.*, **18**, 355–362.
53. Bubeck Wardenburg, J., Williams, W.A. and Missiakas, D. (**2006**) Host defenses against *Staphylococcus aureus* infection require recognition of bacterial lipoproteins. *Proc. Natl. Acad. Sci. USA*, **103**, 13831–13836.
54. Nielsen, J.B., Caulfield, M.P. and Lampen, J.O. (**1981**) Lipoprotein nature of *Bacillus licheniformis* membrane penicillinase. *Proc. Natl. Acad. Sci. USA*, **78**, 3511–3515.
55. Nielsen, J.B. and Lampen, J.O. (**1982**) Membrane-bound penicillinases in Gram-positive bacteria. *J. Biol. Chem.*, **257**, 4490–4495.
56. Babu, M.M., Priya, M.L., Selvan, A.T., Madera, M., Gough, J., Aravind, L. and Sankaran, K. (**2006**) A database of bacterial lipoproteins (DOLOP) with functional assignments to predicted lipoproteins. *J. Bacteriol.*, **188**, 2761–2773.
57. Mühlradt, P.F., Kiess, M., Meyer, H., Süssmuth, R. and Jung, G. (**1997**) Isolation, structure elucidation, and synthesis of a macrophage stimulatory lipopeptide from *Mycoplasma fermentans* acting at picomolar concentration. *J. Exp. Med.*, **85**, 1951 1958.
58. Takeuchi, O., Kaufmann, A., Grote, K., Kawai, T., Hoshino, K., Morr, M., Mühlradt, P.F. and Akira, S. (**2000**) Cutting edge: preferentially the R-stereoisomer of the mycoplasmal lipopeptide macrophage-activating lipopeptide-2 activates immune cells through a Toll-like receptor 2- and MyD88-dependent signaling pathway. *J. Immunol.*, **164**, 554–557.
59. Takeuchi, O., Kawai, T., Mühlradt, P.F., Morr, M., Radolf, J.D., Zychlinsky, A., Takeda, K. and Akira S. (**2001**) Discrimination of bacterial lipoproteins by Toll-like receptor 6. *Int. Immunol.*, **13**, 933–940.
60. Buwitt-Beckmann, U., Heine, H., Wiesmüller, K.H., Jung, G., Brock, R., Akira, S. and Ulmer, A.J. (**2005**) Toll-like receptor 6-independent signaling by diacylated lipopeptides. *Eur. J. Immunol.*, **35**, 282–289.
61. Buwitt-Beckmann, U., Heine H., Wiesmüller, K.H., Jung, G., Brock, R., Akira, S. and Ulmer, A.J. (**2006**) TLR1- and TLR6-independent recognition of bacterial lipopeptides. *J. Biol. Chem.*, **281**, 9049–9057.
62. Galanos, C., Gumenscheimer, M., Mühlradt, P.F, Jirillo, E. and Freudenberg, M. (**2000**) MALP-2, a mycoplasma lipopeptide with classical endotoxic properties: end of an era of LPS monopoly? *J. Endotoxin. Res.*, **6**, 471–476.
63. Schleifer, K.H. and Kandler, O. (**1972**) Peptidoglycan types of bacterial cell walls and their taxonomic implications. *Bacteriol. Rev.*, **36**, 407–477.
64. Warth, A.D. and Strominger, J.L. (**1971**) Structure of the peptidoglycan from vegetative cell walls of *Bacillus subtilis*. *Biochemistry*, **10**, 4349–4358.
65. Kamisango, K., Saiki, I., Tanio, Y., Okumura, H., Araki, Y., Sekikawa, I., Azuma, I. and Yamamura, Y. (**1982**) Structures and biological activities of peptidoglycans of *Listeria monocytogenes* and *Propionibacterium acnes*. *J. Biochem. (Tokyo)*, **92**, 23–33.
66. Goffin, C. and Ghuysen, J.M. (**1998**) Multimodular penicillin-binding proteins: an enigmatic family of orthologs and paralogs. *Microbiol. Mol. Biol. Rev.*, **62**, 1079–1093.
67. Mazmanian, S.K., Liu, G., Jensen, E.R., Lenoy, E. and Schneewind, O. (**2000**) *Staphylococcus aureus* sortase mutants defective in the display of surface proteins and in the pathogenesis of animal infections. *Proc. Natl. Acad. Sci. USA*, **97**, 5510–5515.
68. Dmitriev, B.A., Ehlers, S. and Rietschel, E.T. (**1999**) Layered

murein revisited: a fundamentally new concept of bacterial cell wall structure, biogenesis and function. *Med. Microbiol. Immunol. (Berl.)*, **187**, 173–181.

69. Dmitriev, B.A., Toukach, F.V., Holst, O., Rietschel, E.T. and Ehlers, S. (**2004**) Tertiary structure of *Staphylococcus aureus* cell wall murein. *J. Bacteriol.*, **186**, 7141–7148.
70. Vollmer, W. and Höltje, J.V. (**2004**) The architecture of the murein (peptidoglycan) in gram-negative bacteria: vertical scaffold or horizontal layer(s)? *J. Bacteriol.*, **186**, 5978–5987.
71. Inohara, N., Ogura, Y., Fontalba, A., Gutierrez, O., Pons F., Crespo, J., Fukase, K., Inamura, S., *et al.* (**2003**) Host recognition of bacterial muramyl dipeptide mediated through NOD2: implications for Crohn's disease. *J. Biol. Chem.*, **278**, 5509–5512.
72. Girardin, S.E., Boneca, I.G., Viala, J., Chamaillard, M., Labigne, A., Thomas, G., Philpott, D.J. and Sansonetti, P.J. (**2003**) Nod2 is a general sensor of peptidoglycan through muramyl dipeptide (MDP) detection. *J. Biol. Chem.*, **278**, 8869–8872.
73. Girardin, S.E., Boneca, I.G., Carneiro, L.A., Antignac, A., Jehanno, M., Viala, J., Tedin, K., Taha, M.K., *et al.* (**2003**) Nod1 detects a unique muropeptide from gram-negative bacterial peptidoglycan. *Science*, **300**, 1584–1587.
74. Chamaillard, M., Hashimoto, M., Horie, Y., Masumoto, J., Su, Q., Saab, L., Ogura, Y., Kawasaki, A., *et al.* (**2003**) An essential role for NOD1 in host recognition of bacterial peptidoglycan containing diaminopimelic acid. *Nature Immunol.*, **4**, 702–707.
75. Fischer, W., Mannsfeld, T. and Hagen, G. (**1990**) On the basic structure of poly(glycerophosphate) lipoteichoic acids. *Biochem. Cell. Biol.*, **68**, 33–43.
76. Morath, S., Geyer, A. and Hartung, T. (**2001**) Structure-Function relationship of cytokine induction by lipoteichoic acid from *Staphylococcus aureus*. *J. Exp. Med.*, **193**, 393–398.
77. Majcherczyk, P.A., Rubli, E., Heumann, D., Glauser, M.P. and Moreillon, P. (**2003**) Teichoic acids are not required for *Streptococcus pneumoniae* and *Staphylococcus aureus* cell walls to trigger the release of tumor necrosis factor by peripheral blood monocytes. *Infect. Immun.*, **71**, 3707–3713.
78. Peschel, A., Otto, M., Jack, R.W., Kalbacher, H., Jung, G. and Götz, F. (**1999**) Inactivation of the *dlt* operon in *Staphylococcus aureus* confers sensitivity to defensins, protegrins, and other antimicrobial peptides. *J. Biol. Chem.*, **274**, 8405–8410.
79. Peschel, A., Vuong, C., Otto, M. and Götz, F. (**2000**) The D-alanine residues of *Staphylococcus aureus* teichoic acids alter the susceptibility to vancomycin and the activity of autolytic enzymes. *Antimicrob. Agents Chemother.*, **44**, 2845–2847.
80. Collins, L.V., Kristian, S.A., Weidenmaier, C., Faigle, M., Van Kessel, K.P., Van Strijp, J.A., Götz F., Neumeister, B. and Peschel, A. (**2002**) *Staphylococcus aureus* strains lacking D-alanine modifications of teichoic acids are highly susceptible to human neutrophil killing and are virulence attenuated in mice. *J. Infect. Dis.*, **186**, 214–219.
81. Dziarski, R., Wang, Q., Miyake, K., Kirschning, C.J. and Gupta, D. (**2001**) MD-2 enables Toll-like receptor 2 (TLR2)-mediated responses to lipopolysaccharide and enhances TLR2-mediated responses to Gram-positive and Gram-negative Bacteria and their cell wall components. *J. Immunol.*, **166**, 1938–1944.
82. Nagai, Y., Akashi, S., Nagafuku, M., Ogata, M., Iwakura, Y., Akira, S., Kitamura, T., Kosugi, A., Kimoto, M. and Miyake, K. (**2002**) Essential role of MD-2 in LPS responsiveness and

TLR4 distribution. *Nat. Immunol.*, **3**, 667–672.

83. Gao, J.J., Xue, Q., Zuvanich, E.G., Haghi, K.R. and Morrison, D.C. (**2001**) Commercial preparations of lipoteichoic acid contain endotoxin that contributes to activation of mouse macrophages *in vitro*. *Infect. Immun.*, **69**, 751–757.
84. Morath, S., Geyer, A., Spreitzer, I., Hermann, C. and Hartung, T. (**2002**) Structural decomposition and heterogeneity of commercial lipoteichoic acid preparations. *Infect. Immun.*, **70**, 938–944.
85. Michelsen, K.S., Aicher, A., Mohaupt, M., Hartung, T., Dimmeler, S., Kirschning, C.J. and Schumann, R.R. (**2001**) The role of Toll-like receptors (TLRs) in bacteria-induced maturation of murine dendritic cells – PGN and LTA are inducers of DC maturation and require TLR2. *J. Biol. Chem.*, **276**, 25680–25686.
86. Morr, M., Takeuchi, O., Akira, S., Simon, M.M. and Mühlradt, P.F. (**2002**) Differential recognition of structural details of bacterial lipopeptides by Toll-like receptors. *Eur. J. Immunol.*, **32**, 3337–3347.
87. Schröder, N.W., Morath, S., Alexander, C., Hamann, L., Hartung, T., Zähringer, U., Göbel, U.B., Weber, J.R. and Schumann, R.R. (**2003**) Lipoteichoic acid (LTA) of *S. pneumoniae* and *S. aureus* activates immune cells *via* Toll-like receptor (TLR)-2, LPS binding protein (LBP) and CD14 while TLR-4 and MD-2 are not involved. *J. Biol. Chem.*, **278**, 15587–15594.
88. Lotz, S., Aga E., Wilde, I., Van Zandbergen, G., Hartung, T., Solbach, W. and Laskay, T. (**2004**) Highly purified lipoteichoic acid activates neutrophil granulocytes and delays their spontaneous apoptosis via CD14 and TLR2. *J. Leukoc. Biol.*, **75**, 467–477.
89. Hashimoto, M., Furuyashiki, M., Kaseya, R., Fukada, Y., Akimaru, M., Aoyama, K., Okuno, T., Tamura, T., *et al.* (**2007**) Evidence of immunostimulating lipoprotein co-existing in natural lipoteichoic acid fraction. *Infect. Immun.*, **75**, 1926–1932.
90. Canepari, P. and Boaretti, M. (**1996**) Lipoteichoic acid as a target for antimicrobial action. *Microb. Drug Resist.*, **2**, 85–89.
91. Wootton, M., MacGowan, A.P. and Walsh, T.R (**2006**) Comparative bactericidal activities of daptomycin and vancomycin against glycopeptide-intermediate *Staphylococcus aureus* (GISA) and heterogeneous GISA isolates. *Antimicrob. Agents Chemother.*, **50**, 4195–4197.
92. Sader, H.S., Watters, A.A., Fritsche, T.R. and Jones, R.N (**2007**) Daptomycin antimicrobial activity tested against methicillin-resistant staphylococci and vancomycin-resistant enterococci isolated in European medical centers (2005). *BMC Infect. Dis.*, **2007** *Apr 18*, **7**, 29.
93. Schröder, N.W., Heine, H., Alexander, C., Manukyan, M., Eckert, J., Hamann, L., Göbel, U.B. and Schumann, R.R. (**2004**) Lipopolysaccharide binding protein binds to triacylated and diacylated lipopeptides and mediates innate immune responses. *J. Immunol.*, **173**, 2683–2691.
94. Hoebe, K., Georgel, P., Rutschmann, S., Du, X., Mudd, S., Crozat, K., Sovath, S., Shamel, L., Hartung, T., Zähringer, U. and Beutler, B. (**2005**) CD36 is a sensor of diacylglycerides. *Nature*, **433**, 523–527.
95. Schromm, A.B., Howe, J., Ulmer, A.J., Wiesmüller, K.H., Seyberth, T., Jung, G., Rössle, M., Koch, M.H., Gutsmann, T. and Brandenburg, K. (**2007**) Physicochemical and biological analysis of synthetic bacterial lipopeptides: validity of the concept of endotoxic conformation. *J. Biol. Chem.*, **282**, 11030–11037.

96. Akira, S. and Takeda, K. (**2004**) Toll-like receptor signalling. *Nat. Rev. Immunol.*, **4**, 499–511.
97. Lee, M.S. and Kim, Y.J. (**2007**) Signaling pathways downstream of pattern-recognition receptors and their cross talk. *Annu. Rev. Biochem.*, **76**, 447–480.
98. Liew, F.Y., Xu, D., Brint, E.K. and O'Neill, L.A. (**2005**) Negative regulation of Toll-like receptor-mediated immune responses. *Nat. Rev. Immunol.*, **5**, 446–458.
99. Horng, T., Barton, G.M., Flavell, R.A. and Medzhitov, R. (**2002**) The adaptor molecule TIRAP provides signalling specificity for Toll-like receptors. *Nature*, **420**, 329–333.
100. Yamamoto, M., Sato, S., Hemmi, H., Sanjo, H., Uematsu, S., Kaisho, T., Hoshino, K., Takeuchi, O., *et al.* (**2002**) Essential role for TIRAP in activation of the signalling cascade shared by TLR2 and TLR4. *Nature*, **420**, 324–329.
101. Kagan, J.C. and Medzhitov, R. (**2006**) Phosphoinositide-mediated adaptor recruitment controls Toll-like receptor signaling. *Cell*, **125**, 943–955.
102. Kotb, M. (**1998**) Superantigens of gram-positive bacteria: structure-function analyses and their implications for biological activity. *Curr. Opin. Microbiol.*, **1**, 56–65.
103. Dinges, M.M., Orwin, P.M. and Schlievert, P.M. (**2000**) Exotoxins of *Staphylococcus aureus*. *Clin. Microbiol. Rev.*, **1**, 16–34.
104. McCormick, J.K., Yarwood, J.M. and Schlievert, P.M. (**2001**) Toxic shock syndrome and bacterial superantigens: an update. *Annu. Rev. Microbiol.*, **55**, 77–104.
105. Proft, T. and Fraser, J.D. (**2003**) Bacterial superantigens. *Clin. Exp. Immunol.*, **133**, 299–306.
106. White, J., Herman, A., Pullen, A.M., Kubo, R., Kappler, J.W. and Marrack, P. (**1989**) The V-beta specific superantigen staphylococcal enterotoxin B: stimulation of mature T cells and clonal deletion in neonatal mice. *Cell*, **56**, 27–35.
107. Jupin, C., Anderson, S., Damais, C., Alouf, J.E. and Parant, M. (**1988**) Toxic shock syndrome toxin as an inducer of human tumor necrosis factors and γ-interferon. *J. Exp. Med.*, **167**, 752–761.
108. Miethke, T., Wahl, C., Heeg, K., Echtenacher, B., Krammer, P.H. and Wagner, H. (**1992**) T cell-mediated lethal shock triggered in mice by the superantigen staphylococcal enterotoxin B: critical role of tumor necrosis factor. *J. Exp. Med.*, **175**, 91–98.
109. Hackett, S.P. and Stevens, D.L. (**1993**) Superantigens associated with staphylococcal and streptococcal toxic shock syndrome are potent inducers of tumor necrosis factor-beta synthesis. *J. Infect. Dis.*, **168**, 232–235.
110. Rink, L., Nicklas, W., Luhm, J., Kruse, R. and Kirchner, H. (**1996**) Induction of a proinflammatory cytokine network by *Mycoplasma arthritidis*-derived superantigen (MAS). *J. Interferon Cytokine Res.*, **16**, 861–868.
111. Fraser, J., Arcus, V., Kong, P., Baker, E. and Proft, T. (**2000**) Superantigens – powerful modifiers of the immune system. *Mol. Med. Today*, **6**, 125–132.
112. Marrack, P. and Kappler, J. (**1990**) The staphylococcal enterotoxins and their relatives. *Science*, **248**, 705–711.
113. Manders, S.M. (**1998**) Toxin-mediated streptococcal and staphylococcal disease. *J. Am. Acad. Dermatol.*, **39**, 383–398.
114. Wieneke, A.A., Roberts, D. and Gilbert, R.J. (**1993**) Staphylococcal food poisoning in the United Kingdom. *Epidemiol. Infect.*, **110**, 519–531.
115. Swaminathan, S., Furey, W., Pletcher, J. and Sax, M. (**1992**). Crystal structure of staphylococcal enterotoxin B, a superantigen. *Nature*, **359**, 801–805.
116. Wang, L., Zhao, Y., Li, Z., Guo, Y., Jones, J.L., Kranz, D.M., Mourad, W.

and Li, H. (**2007**). Crystal structure of a complete ternary complex of TCR, a superantigen and peptide-MHC. *Nat. Struct. Mol. Biol.*, **14**, 169–171.

117. Haffner, A.C., Zepter, K. and Elmerts, C.A. (**1996**) Major histocompatibility complex class I molecules serves as a ligand for presentation of the superantigen staphylococcal enterotoxin B to T cells. *Proc. Natl. Acad. Sci. USA*, **93**, 3037–3042.
118. Gregory, S., Zilber, M.T., Choqueux, C., Mooney, N., Charron, D. and Gelin, C. (**2000**) Role of the CD1a molecules in the superantigen-induced activation of MHC class II negative human thymocytes. *Hum. Immunol.*, **61**, 427–437.
119. Brown, J.H., Jardetzky, T.S., Gorga, J.C., Stern, L.J., Urban, R.G., Strominger, J.L. and Wiley, D.C. (**1993**). Three-dimensional structure of the human class II histocompatibility antigen HLA-DR1. *Nature*, **364**, 33–39.
120. Omoe, K., Imanishi, K., Hu., D.-L., Kato, H., Takahashi-Omoe, H., Nakane, A., Uchiyama, T. and Shinagawa, K. (**2004**) Biological properties of staphylococcal enterotoxin-like toxin type R. *Infect. Immun.*, **72**, 3664–3667.
121. Williams, R.J., Ward, J.M., Henderson, B., Poole, S., O'Hara, B.P., Wilson, M. and Nair, S.P. (**2000**) Identification of a novel gene cluster encoding staphylococcal exotoxin-like proteins: characterization of the prototypic gene and its protein product, SET1. *Infect. Immun.*, **68**, 4407–4415.
122. Bestebroer, J., Poppelier, M.J., Ulfman, L.H., Lenting, P.J., Denis, C.V., van Kessel, K.P., van Strijp, J.A. and de Haas, C.J. (**2006**) Staphylococcal superantigen-like 5 binds PSGL-1 and inhibits P-selectin-mediated neutrophil rolling. *Blood*, **109**, 2936–2943.
123. Bessen, D.E., Izzo, M.W., Fiorentino, T.R., Caringal, R.M., Hollingshead, S.K. and Beall, B. (**1999**) Genetic linkage of exotoxin alleles and emm gene markers for tissue tropism in group A streptococci. *J. Infect. Dis.*, **179**, 627–636.
124. Kline, J.B. and Collins, C.M. (**1996**) Analysis of the superantigenic activity of mutant and allelic forms of streptococcal pyrogenic exotoxin A. *Infect. Immun.*, **64**, 861–869.
125. Kamezawa, Y., Nakahara, T., Nakano, S., Abe, Y., Nozaki-Renard, J. and Isono, T. (**1997**) Streptococcal mitogenic exotoxin Z, a novel acidic superantigenic toxin produced by a T1 strain of *Streptococcus pyogenes*. *Infect. Immun.*, **65**, 3828–3833.
126. Proft, T., Moffatt, S.L., Weller, K.D., Paterson, A., Martin, D. and Fraser, J.D. (**2000**) The streptococcal superantigen SMEZ exhibits wide allelic variation, mosaic structure, and significant antigenic variation. *J. Exp. Med.*, **191**, 1765–1776.
127. Yang, L., Thomas, M., Woodhouse, A., Martin, D., Fraser, J.D. and Proft, T. (**2006**) Involvement of streptococcal mitogenic exotoxin Z in streptococcal toxic shock syndrome. *J. Clin. Microbiol.*, **43**, 3570–3573.
128. Hudson, K.R., Robinson, H. and Fraser, J.D. (**1993**) Two adjacent residues in staphylococcal enterotoxins A and E determine T cell receptor Vβ specificity. *J. Exp. Med.*, **177**, 175–184.
129. Bergdoll, M.S., Crass, B.A., Reiser, R.F., Robbins, R.N. and Davis, P.J. (**1981**) A new staphylococcal enterotoxin, enterotoxin F, associated with toxic shock syndrome *Staphylococcus aureus* isolates. *Lancet*, **1**, 1017–1021.
130. Musser, J., Hauser, A., Kim, M., Schlievert, P.M., Nelson, K. and Selander, R.K. (**1991**) *Streptococcus pyogenes* causing toxic-shock-like syndrome and other invasive diseases: clonal diversity and pyrogenic exotoxin expression. *Proc. Natl. Acad. Sci. USA*, **88**, 2668–2672.

131. Holtfreter, S. and Broker, B.M. (**2005**) Staphylococcal superantigens: do they play a role in sepsis? *Arch. Immunol. Ther. Exp.*, **53**, 13–27.
132. Abe, J., Kotzin, B.L., Meissner, C., Melish, M.E., Takahashi, M., Fulton, D., Romagne, F., Malissen, B., *et al.* (**1993**) Characterization of T cell repertoire changes in acute Kawasaki disease. *J. Exp. Med.*, **177**, 791–796.
133. Smoot, L.M., McCormick, J.K., Smoot, J.C., Hoe, N.P., Strickland, I., Cole, R.L., Barbian, K.D., Earhart, C.A., *et al.* (**2002**) Characterization of two novel pyrogenic toxin superantigens made by an acute rheumatic fever clone of *Streptococcus pyogenes* associated with multiple disease outbreaks. *Infect. Immun.*, **70**, 7095–7104.
134. Faulkner, L., Cooper, A., Fantino, C., Altmann, D.M. and Sriskandan, S. (**2005**) The mechanism of superantigen-mediated toxic shock: not a simple Th1 cytokine storm. *J. Immunol.*, **175**, 6870–6877.
135. Matthys, P., Mitera, T., Heremans, H., van Damme, J. and Billiau, A. (**1995**) Anti-γ interferon and anti-interleukin-6 antibodies affect staphylococcal enterotoxin B-induced weight loss, hypoglycaemia, and cytokine release in D-galactosamine-sensitized and unsensitized mice. *Infect. Immun.*, **63**, 1158–1164.
136. Florquin, S., Amraoui, Z. and Goldman, M. (**1995**) T cells made deficient in interleukin-2 production by exposure to staphylococcal enterotoxin B *in vivo* are primed for interferon-γ and interluekin-10 secretion. *Eur. J. Immunol.*, **25**, 1148–1153.
137. Ferry, T., Thomas, D., Genestier, A.-L., Bes, M., Lina, G., Vandenesch, F. and Etienne, J. (**2005**) Comparative prevelance of superantigen genes in *Staphylococcus aureus* isolates causing sepsis with and without septic shock. *Clin. Infect. Dis.*, **41**, 771–777.
138. Dauwalder, O., Thomas, D., Ferry, T., Debard, A.-L., Badiou, C., Vandenesch, F., Etienne, J., Lina, G., *et al.* (**2006**) Comparative inflammatory properties of staphylococcal superantigenic enterotoxins SEA and SEG: implications for septic shock. *J. Leukoc. Biol.*, **80**, 753–758.
139. Todd, J.K., Kapral, F.A., Fishaut, M. and Welch, T.R. (**1978**) Toxic shock syndrome associated with phage group I staphylococci. *Lancet*, **2**, 1116–1118.
140. Schlievert, P.M., Shands, K.N., Dan, B.B., Schmid, G.P. and Nishimura, N.D. (**1981**) Identification and characterization of an exotoxin from *Staphylococcus aureus* associated with toxic-shock syndrome. *J. Infect. Dis.*, **143**, 509–516.
141. Altemeier, W.A., Lewis, S.A., Schlievert, P.M., Bergdoll, M.S., Bjornson, H.J., Taneck, L. and Crass, B.A. (**1982**) *Staphylococcus aureus* associated with toxic shock syndrome: phage typing and toxin capability tested. *Ann. Intern. Med.*, **96**, 978–982.
142. Chesney, P.J. (**1989**) Clinical aspects and spectrum of illness of toxic shock syndrome: overview. *Rev. Infect. Dis.*, **11**, S1–7.
143. Bohach, G.A., Fast, D.J., Nelson, R.D. and Schlievert, P.M. (**1990**) Staphylococcal and streptococcal pyrogenic toxins involved in toxic shock syndrome and related illnesses. *Crit. Rev. Microbiol.*, **17**, 251–272.
144. Hajjeh, R.A., Reingold, A., Weil, A., Shutt, K., Schuchat, A. and Perkins, B.A. (**1999**) Toxic shock syndrome in the United States: surveillance update: 1979–1996. *Emerg. Infect. Dis.*, **5**, 807–810.
145. MacDonald, K.L., Osterholm, M.T., Hedberg, C.W., Schrock, C.G., Peterson, G.F., Jentzen, J.M., Leonard, S.A. and Schlievert, P.M. (**1987**) Toxic shock syndrome: a newly recognized complication of

influenza and influenzalike illness. *JAMA*, **257**, 1053–1058.
146. Boventre, P.F., Heeg, H., Cullen, C. and Lian, C.J. (**1993**) Toxicity of recombinant toxic shock syndrome toxin 1 and mutant toxins produced by *Staphylococcus aureus* in a rabbit infection model of toxic shock syndrome. *Infect. Immun.*, **61**, 793–799.
147. de Azavedo, J.C., Foster, T.J., Hartigan, P.J., Arbuthnott, J.P., O'Reilly, M., Kreiswirth, B.N. and Novick, R.P. (**1985**) Expression of the cloned toxic shock syndrome toxin 1 gene (tst) *in vivo* with a rabbit uterine model. *Infect. Immun.*, **50**, 304–309.
148. Dinges, M.M., Jessurun, J. and Schlievert, P.M. (**1998**) Comparisons of mouse and rabbit models of toxic shock syndrome. *Int. Congr. Symp. Ser.*, **229**, 167–168.
149. Fremont, D.H., Rees, W.A. and Kozono, H. (**1996**) Biophysical studies of T-cell receptors and their ligands. *Curr. Opin. Immunol.*, **8**, 93–100.
150. Welcher, B.C., Carra, J.H., DaSilva, L., Hanson, J., David, C.S., Aman, M.J. and Bavari, S. (**2002**) Lethal shock induced by streptococcal pyrogenic exotoxin A in mice transgenic for human leukocyte antigen-DQ8 and human CD4 receptors: Implications for development of vaccines and therapeutics. *J. Infect. Dis.*, **186**, 501–510.
151. Taneja, V. and David, C.S. (**1999**) HLA class II transgenic mice as models of human diseases. *Immunol. Rev.*, **169**, 67–79.
152. Parsonnet, J., Gillis, Z.A., Richter, A.G. and Pier, G.B. (**1987**) A rabbit model of toxic shock syndrome that uses a constant, subcutaneous infusion of toxic syndrome toxin 1. *Infect. Immun.*, **55**, 1070–1076.
153. Basma, H., Norrby-Teglund, A., Guedez, Y., McGeer, A., Low, D.E., El-Ahmedy, O., Schwartz, B. and Kotb, M. (**1999**) Risk factors in the pathogenesis of invasive group A streptococcus infections: role of protective humoral immunity. *Infect. Immun.*, **67**, 1871–1877.
154. Childs, C., Edwards-Jones, V., Heathcote, D.M., Dawson, M. and Davenport, P.J. (**1994**) Patterns of *Staphylococcus aureus* colonization, toxin production, immunity and illness in burned children. *Burns*, **20**, 514–521.
155. Stolz, S.J., Davis, J.P., Vergeront, J.M., Crass, B.A., Chesney, P.J., Wand, P.J. and Bergdoll, M.S. (**1985**) Development of serum antibody to toxic shock toxin among individuals with toxic shock syndrome in Wisconsin. *J. Infect. Dis.*, **151**, 883–889.
156. Vergeront, J.M., Stolz, S.J., Crass, B.A., Nelson, D.B., Davis, J.P. and Bergdoll, M.S. (**1983**) Prevelance of serum antibody to staphylococcal enterotoxin F among Wisconsin residents: implications for toxic-shock syndrome. *J. Infect. Dis.*, **148**, 692–698.
157. Schlievert, P.M. (**1983**) Alteration of immune function by staphylococcal pyrogenic exotoxin type C: possible role in toxic-shock syndrome. *J. Infect. Dis.*, **147**, 391–398.
158. Krakauer, T. (**1995**) Inhibition of toxic shock syndrome toxin-1-induced cytokine production and T cell activation by interleukin-10, interleukin-4, and dexamethasone. *J. Infect. Dis.*, **172**, 988–992.
159. Norrby-Teglund, A., Ihendyane, N., Kansal, R., Basma, H., Kotb, M., Andersson, J. and Hammarstrom, L. (**2000**) Relative neutralizing activity in polyspecific IgM, IgA, and IgG preparations against group A streptococcal superantigens. *Clin. Infect. Dis.*, **31**, 1175–1182.
160. Hamad, A.R.A., Marrack, P. and Kappler, J.W. (**1997**) Transcytosis of staphylococcal superantigen toxins. *J. Exp. Med.*, **185**, 1447–1454.
161. Hashikawa, S., Iinuma, Y., Furushita, M., Ohkura, T., Nada, T., Torii, K., Hasegawa, T. and Ohta, M. (**2004**) Characterization of group C and G streptococcal strains that cause

streptococcal toxic shock syndrome. *J. Clin. Microbiol.*, **42**, 186–192.

162. Yu, C. and Ferretti, J. (**1989**) Molecular epidemiologica analysis of the type A streptococcal exotoxin (erythrogenic toxin) gene (speA) in clinical *Streptococcus pyogenes* strains. *Infect. Immun.*, **57**, 3715–3719.
163. Sriskandan, S., Moyes, D. and Cohen, J. (**1996**) Detection of circulating bacterial superantigen and lymphotoxin-a in patients with streptococcal toxic shock syndrome. *Lancet*, **348**, 1315–1316.
164. Herman, A., Croteau, G., Sekaly, R.P., Kappler, J. and Marrack, P. (**1990**) HLA-DR alleles differ in their ability to present staphylococcal enterotoxins to T cells. *J. Exp. Med.*, **172**, 709–717.
165. Kotb, M., Norrby-Teglund, A., McGeer, A., El-Sherbini, H., Dorak, M.T., Khurshid, A., Green, K., Peeples, J., *et al.* (**2002**) An immunogenetic and molecular basis for differences in outcomes of invasive group A streptococcal infections. *Nat. Med.*, **8**, 1398–1404.
166. Nooh, M.M., El-Gengehi, N., Kansal, R., David, C.S. and Kotb, M. (**2007**) HLA transgenic mice provide evidence for a direct and dominant role of HLA class II variation in modulating the severity of streptococcal sepsis. *J. Immunol.*, **178**, 3076–3083.
167. Mehrotra, M., Wang, G. and Johnson, W.M. (**2000**) Multiplex PCR for detection of genes for *Staphylococcus aureus* enterotoxins, exfoliative toxins, toxic shock syndrome toxin 1, and methicillin resistance. *J. Clin. Microbiol.*, **38**, 1032–1035.
168. Monday, S.R. and Bohach, G.A. (**1999**) Use of a multiplex PCR to detect classical and newly described pyrogeneic toxin genes in staphylococcal isolates. *J. Clin. Microbiol.*, **37**, 3411–3414.
169. Lintges, M., Arlt, S., Uciechowski, P., Plümäkers, B., Reinert, R.R., Al-Lahham, A., Lütticken, R. and Rink, L. (**2007**) A new closed-tube multiplex real-time PCR to detect eleven superantigens of *Streptococcus pyogenes* identifies a strain without superantigen activity. *Int. J. Med. Microbiol.*, [Epub ahead of print].
170. Schrage, B., Duan, G., Yang, L.P., Fraser, J.D. and Proft, T. (**2006**) Different preparations of intravenous immunoglobulin vary in their efficacy to neutralize streptococcal superantigens: implications for treatment of streptococcal toxic shock syndrome. *Clin. Infect. Dis.*, **43**, 743–746.
171. Wang, J., Soisson, S.M., Young, K., Shoop, W., Kodali, S., Galgoci, A., Painter, R., Parthasarathy, G., *et al.* (**2006**) Platensimycin is a selective FabF inhibitor with potent antibiotic properties. *Nature*, **441**, 358–361.
172. Shinefield, H.R. and Black, S. (**2005**) Prevention of *Staphylococcus aureus* infections: advances in vaccine development. *Expert. Rev. Vaccines*, **4**, 669–676.
173. Opal, S.M. and Huber, C.E. (**2002**) Bench-to-bedside review: Toll-like receptors and their role in septic shock. *Crit. Care*, **6**, 125–136.
174. Cavaillon, J.M., Adrie, C., Fitting, C. and Adib-Conquy, M. (**2005**) Reprogramming of circulatory cells in sepsis and SIRS. *J. Endotoxin. Res.*, **11**, 311–320.
175. Pugin, J. (**2007**) Immunostimulation is a rational therapeutic strategy in sepsis. *Novartis Found. Symp.*, **280**, 21–27.
176. Blank, C., Luz, A., Bendigs, S., Erdmann, A., Wagner, H. and Heeg, K. (**1997**) Superantigen and endotoxin synergize in the induction of lethal shock. *Eur. J. Immunol.*, **27**, 825–833.
177. Luhm, J., Kirchner, H. and Rink, L. (**1997**) One-way synergistic effect of low superantigen concentrations on lipopolysaccharide-induced cytokine production. *J. Interferon Cytokine Res.*, **17**, 229–238.

7
Pathogens in Sepsis: Fungi, Parasites

Frédéric Méchaï, Fabrice Bruneel, Fanny Lanternier, Stephane Legriel, Jean-Pierre Bédos and Olivier Lortholary

7.1
Fungi

7.1.1
Introduction

Sepsis represents a substantial health care burden and is often lethal, killing 20 to 50% of severely ill patients. It is the second leading cause of death among patients in intensive care units (ICUs) [1]. By consensus, sepsis is defined as the combination of infection and pathophysiological changes known collectively as the systemic inflammatory response syndrome. Furthermore, nosocomial bloodstream infections (BSIs) represent an important cause of morbidity and mortality in the United States. In one of the largest multicenter studies involving 49 hospitals over a 7-year period from March 1995 through September 2002, Gram-positive organism were found to cause 65% of the BSIs, Gram-negative organisms 25%, and yeasts 9.5%. In ICU patients, the crude mortality rate for Candida species was 27% (26 and 34% for coagulase-negative staphylococci and *E. Coli*, respectively) [2]. Thus, the relative frequency of specific causative organisms has shifted over time with the emergence of fungal pathogens. Candida species represent the most common cause of bloodstream fungal infections. Candidemia is defined by at least a positive blood culture for Candida spp. and always requires prompt appropriate antifungal therapy [3]. Proven invasive candidiasis is defined by any situation in which a biopsy of a normally sterile site shows Candida species by culture or histopathologic examination [4]. In this chapter, we will use invasive Candida infections as a pertinent model of fungal sepsis and will review recent acquisitions on this topic.

7.1.2
Epidemiology of Candidemia in Intensive Care Unit

The genus Candida comprises approximately 200 species. Candida spp. are yeasts and are of a variety of shapes and biochemical abilities, both

Sepsis and Non-infectious Systemic Inflammation. Edited by J.-M. Cavaillon and C. Adrie

ISBN: 978-3-527-31935-0

assimilative and fermentative. Candida, and particularly the major pathogenic species *C. albicans*, is unique among opportunistic pathogens because they are ubiquitous organisms which commonly colonize the oropharynx, intestine, vagina, and skin of humans. The ability of these yeasts to cause human infectious disease relates more to the immunosuppresed status of the host than to any obvious and notorious virulence factors produced, maintained, or exhibited by the fungus. Candida species are responsible for invasive infections among patients suffering from immunosuppression, but currently they are also responsible for the more frequent occurrences of nosocomial bloodstream infections among ICUs patients. Because blood cultures are an insensitive diagnostic test, estimates of disease frequency based on positive blood culture results are artificially low. The NEMIS SICU study is the largest prospective multicenter study to evaluate risk factors for the development of Candida spp. bloodstream infections (CBSIs) among surgical ICU patients [5]. After multivariate analysis, risk factors that were independently associated with an increased risk of CBSI included recent surgery, acute renal failure and receipt of parenteral nutrition [5]. In previous studies, potential risk factors for developing candidemia have included receipt of broad-spectrum antimicrobial agents, corticosteroids, antineoplastic chemotherapy, parenteral nutrition, hematological or solid organ malignancy, neutropenia, extensive surgery or burns, mechanical ventilation, ICU stay, an indwelling central venous catheter or hemodialysis and prior fungal colonization of the mucosa [6, 7].

More specifically in the ICU, patients are subjected to a number of therapeutic and supportive interventions which interfere with the normal barriers to the entry by microorganisms. Examples include mechanical ventilation and intravascular catheters. In one study, multiple logistic regression analysis has identified the use of Hickman catheters as an independent predictor of Candida infection while in others, use of Swan-Ganz catheters, parenteral nutrition, multiple transfusions and artificial ventilatory support were also found to be significant risk factors after univariate analysis [7–10].

Defining risk factors for developing candidemia and/or invasive candidiasis might include the selection of high-risk patients who could potentially benefit from early antifungal therapy to reduce infection-related deaths [11]. The mean interval between admission and infection has been estimated to be 22 days for Candida species in one study [2].

In the latter study, the authors found a significant increase in the proportion of Candida species among isolates from blood cultures, ranging from 8% in 1995 to 12% in 2002. The most common organisms to cause BSIs were coagulase-negative staphylococci (31%), *Staphylococcus aureus* (20%), enterococci (9%), and Candida species (9%). Furthermore, the proportion of *C. albicans* and *C. parapsilosis* among these isolates increased from 1995 to 2002, whereas the proportions of *C. tropicalis* and *C. glabrata* decreased [2].

Thus, Candida is now the fourth most commonly isolated pathogen in patients with nosocomial bloodstream infections [2, 12, 13]. In the NEMIS prospective multicenter study, the mortality rate was significantly higher among surgical intensive care patients who developed Candida bloodstream infections (41%) in comparison with data obtained from previous reports [5]. In another prospective, multicenter observational study of adults and children with candidemia in tertiary care centers in the United States, *Candida albicans* was the most frequent bloodstream isolate (45% among adults and 49% among children). In adults, mortality was high and similar among subjects with *Candida albicans* or non-*albicans* candidemia (46%). *Candida parapsilosis* fungemia has been associated with a lower mortality rate than sepsis due to other Candida species (24%) [14]. In the surveillance program (SENTRY) for BSI in the United States, Canada, Latin America, and Europe from 1997 through 1999, 1184 episodes of candidemia were detected in 71 medical centers [14]. Candida spp. were again the fourth most-common nosocomial BSI isolates [2, 12, 14]. Among these BSI, 75% were nosocomially acquired and 50% occurred in patients hospitalized in an ICU. Of the 1184 BSI in which the causative organism was identified, 55% were due to *C. albicans*, 15% to *C. glabrata*, 15% to *C. parapsilosis*, 9% to *C. tropicalis*, and 6% to miscellaneous Candida spp. [14] (Table 7.1). In a study on incidence rates and characteristics of candidemia in 25 French hospitals, from January 1995 through December 1995, *C. albicans* was found to be the most common species isolated, representing 53% of the isolates, followed by *C. parapsilosis* (16%), *C. glabrata* (11.5%), and *C. tropicalis* (9.5%) [15].

The results of the National Nosocomial Infections Surveillance System study (NNISS) which took place from January 1989 through December 1999, showed that there was a significant decrease in the incidence of *C. albicans* BSI and a significant increase in the incidence of *C. glabrata* BSIs [16]. The fact that prophylactic use of fluconazole was more frequently prescribed among high risk patients may have contributed to this decrease. Non-*albicans* Candida species are found more frequently in patients with hematological malignancies, and *C. parapsilosis* is more frequent in children, especially in neonates.

Isolates of *C. albicans*, *C. parapsilosis*, and *C. tropicalis* were all highly susceptible to fluconazole in the SENTRY program undertaken from 1997 through 1999 [14]. A trend of increased susceptibility of *C. glabrata* to fluconazole was noted over the 3-year period. The percentage of *C. glabrata* isolates susceptible to fluconazole increased from 48% in 1997 to 84% in 1999, and MIC50s decreased from 16 to 4 μg/ml [14]. In another study, only 1.2% of *C. albicans* isolates were resistant to fluconazole (MIC, ≥64 μg/ml, NCCLS method), compared to 7% of *C. glabrata* isolates [17]. Only 4.3% of *C. albicans* isolates were resistant to flucytosine (MIC, ≤32 μg/ml), compared to <1% of *C. parapsilosis* and *C. tropicalis* isolates and no *C. glabrata* isolates. As determined by E-test, the MICs of amphotericin B were ≥0.38 μg/ml for 10% of *Candida* isolates, ≥1 μg/ml for 1.7% of isolates, and ≥2 μg/ml for 0.4% of isolates [17]. In a multi-institutional survey of Candida BSIs carried out in Europe by the European Confederation of Medical Mycology (ECMM), it was found that there were a limited number of species with decreased susceptibility to azoles that were still able to cause BSIs and a low proportion of antifungal resistance (2089 episodes from seven countries) [18]. In a national survey of fungemia initiated in 2003 in Denmark, 303 episodes were registered. *C. albicans* was the predominant species (63%), *C. glabrata* was the second most frequently isolated species (20%) and *C. krusei* was rarely isolated (3%). Decreased azole susceptibility, defined as a fluconazole MIC of $>8\,\mu g/ml$ and/or an itraconazole MIC of $>0.125\,\mu g/ml$, was detected for 11 Candida isolates that were neither *C. glabrata* nor *C. krusei* [19]. The Barcelona candidemia project study group has conducted a population-based surveillance for Candida bloodstream infections in Spain over 2 years. Overall mortality was 44%, *C. albicans* was the most frequent causative species (51% of cases) followed by *C. parapsilosis* (23%) and 24 isolates (7%) had decreased susceptibility to fluconazole (MIC $\geq 16\,\mu g/ml$) [20]. An active surveillance program called YEASTS (Yeasts Ecosystems and Antifungal susceptibility Trends Study) was implemented in October 2002 by the French National Reference Center for Mycoses (NRCM). A total of 1164 episodes of incident BSI and 41 recurrences were recorded. *C. albicans* was the most frequent species (49%) followed by *C. glabrata* (14.5%). Fluconazole MICs $>8\,\mu g/ml$ were observed for 14% of the isolates including 55% of *C. glabrata* and 35% of the rare species. This active surveillance program in the Paris area demonstrates that non-neutropenic ICU-hospitalized patients represent the major group of patients currently developing yeast BSI (Dromer *et al.*, unpublished data). The epidemiology of sepsis in European care units was evaluated by the SOAP study which was a prospective, observational study carried out in 198 intensive care units [21]. Cultures were positive in 60% of the patients with sepsis. Gram-positive organisms were isolated from 40% of patients, Gram-negative from 38% and fungi from 17%. *Candida albicans* was thought to be involved in 13% of infections.

Table 7.1 Summary of the current epidemiology of candidemia from a worldwide perspective: recent studies.

	Tortorano (n = 569) J Hosp Infect 2002 (Italy)	Trick (n = 2759) CID 2002 (USA)	Diekema (n = 254) J Clin Microbiol 2002 (USA)	Richet (n = 377) CMI 2002 (France)	Pfaller (n = 1134) J Clin Microbiol 2002 (USA)	Marchetti (n = 1137) CID 2004 (Switzerland)
C. albicans	58.50%	59%	58%	53%	55%	66%
C. glabrata	12.80%	12%	20%	11%	15%	15%
C. parapsilosis	14.60%	11%	7%	16%	15%	1%
C. tropicalis	6.10%	10%	11%	9%	9%	9%
C. krusei	0.90%	1.20%	2%	4%	1%	2%
Miscellaneous	7.10%	7%	2%	6%	1%	7%

7.1.3
Pathogenesis of Candida Sepsis

Candida spp. are able to colonize mucosal epithelial surfaces, particularly that of the digestive tract. Under some circumstances (particularly changes in the digestive tract following surgery or antineoplastic chemotherapy), Candida spp. adhere to mucosal epithelial cells and cross this natural barrier, resulting in endogenous candidemia. Candida spp. can also cross the cutaneous barrier during use of intravenous catheters and are responsible for fungemia of exogenous origin.

The first mechanism of defence encountered by Candida spp. is innate immunity.

Like several other microorganisms, Candida spp. exhibit invariant molecular structures called PAMPs (pathogen-associated molecular patterns) which are recognized by recognition receptors, particularly Toll-like receptors (TLR). TLR are expressed by phagocytes and dentritic cells (DCs) and its intracellular domain shares similarities with the intracellular domain of IL1-RI. *Candida albicans* is recognized by IL-1R and TLR2. β-Glucan of *C. albicans* is recognized by dectin-1 and Candida spp. mannan antigen by TLR4. TLR activates a stereotyped response. TLR provokes a MYD88 (myeloid differentiation primary response 88) dependent or independent signaling pathway and induces synthesis of pro-inflammatory cytokines such as TNF-α, IL1β. TLR and IL 1R activate antifungal effectors; mostly respiratory burst and degranulation of neutrophils. Quantitative and qualitative functions of neutrophils are also crucial for anti-*Candida* defense as well as macrophages which contribute to their phagocytosis. Complement and antibodies are known to promote the opsonization of fungi. However, the role of antibodies in anti-*Candida* defense is marginal [22]. In addition, TNF-α and IL-1β synthesis by mononuclear human cells induced by Candida is reduced by anti-TLR2 antibodies [23]. Human leucocytes produce cytokines IL-1, 6, 8, 10, 12, 18, IFN, RANTES, MCP-1, MIP-1 and GM-CSF in presence of Candida [24–31].

In vivo studies have demonstrated the major role of TLR4 in innate anti-Candida immunity. Indeed, the growth of *Candida albicans* is 10 time greater in TLR4−/− mice compared to wild-type mice [32]. Macrophages of these mice exhibit a lower expression of chemokines and MIP2 and a lower recruitment of neutrophils induced by *Candida albicans.* Macrophages of MYD88-deficient mice exhibit reduced Candida phagocytosis and reduced cytokine production [33]. Anti-*Candida* yeasts and the hyphae activity of neutrophils is reduced in MYD88−/− and IL1R−/− mice. A decrease in Th1 response and an increase in Th2 response have been observed. Overall, in TLR2−, TLR4−, MYD88− and IL1R− mice infected with *Candida* spp., there is a reduction in the Th1 response and an increase in Th2 response [32]. Murine models have shown the major contribution of inflammatory cytokines such as IL-1 and IL-6 [34, 35].

A unique study of 43 patients with candidemia has suggested that the prevalence of TLR4 polymorphism Asp299Gly was more frequent in patients with candidemia compared with their respective controls [36].

7.1.4
Diagnosis of Candidemia

Disseminated candidiasis is a major cause of death in compromised patients, therefore early recognition of this infection is essential. Occasionally, patients present with severe myalgia, peripheral nodular lesions of the skin or endophtalmitis suggesting emboli due to candidemia. Other organs can be involved [37]. Unfortunately, the diagnosis of systemic candidiasis remains difficult due to the non-specific clinical signs. Otherwise, due to the frequent colonization of predisposed patients, interpretation of positive samples from various sites is still controversial and varies with the species isolated and the type of unit in which the patient is hospitalized. As for most systemic infectious diseases, the current "gold standard" for this diagnosis is either a positive culture specimen from a sterile site or characteristic histopathology. The lack of an early diagnosis leads to a delay in the institution of appropriate antifungal therapy. Unfortunately, conventional microbiological techniques for diagnosis of invasive candidiasis often fail to detect the disease. Moreover, the use of invasive diagnostic techniques for histopathological studies is frequently impossible due to the underlying conditions of critically ill patients, while biopsy of skin lesions may lead to a diagnosis in a minority of patients.

As previously stated, isolation of Candida species from a single blood culture is now considered to be sufficient evidence for the immediate initiation of systemic antifungal therapy [38]. However, the problem is that blood culture lacks sensitivity, widely reported to be less than 50%, particularly for deep-seated infections, and usually take several days to obtain a positive result [38, 39]. Improvements in blood culture techniques have increased the sensitivity to only 70% at best [40]. Furthermore, the availability of specific fungal media and rapid *in situ* hybridization/immunofluorescent techniques has significantly shortened the time to identification at the species level. For Candida blood cultures, an automated broth-based system appears to equal the sensitivity of lysis–centrifugation methods. Identification of species other than *C. albicans* is now easier and quicker than ever, as a result of the development of new agars which produce different colony colors for different species [41]. One study analyzed the performance of mycosis IC/F selective fungal medium for the diagnosis of fungemia [42]. This medium was shown to be more effective than the Plus Aerobic/F medium for the diagnosis of fungemia in terms of the number of positive results and detection speed. Moreover, there was a high level of disparity between species. The mean time saved using Mycosis IC/F medium was 8.8 h for *C. albicans* and 43.7 h for *C. glabrata*.

Efforts have been made to find either antibodies against *Candida albicans* molecules or Candida-derived molecules whose presence in patient's sera could indicate deep-tissue invasion. The use of mannan antigenemia (mannanemia) detection for the immunodiagnosis of systemic candidiasis was suggested about two decades ago [39]. In contrast to mannanemia detection, tests based on antimannan antibody detection have been used less frequently in clinical diagnostic mycology laboratories because they have been described as lacking in both specificity and sensitivity. The elevated antibody titers in heavily colonized but uninfected hospitalized patients and the possible lack of antibody response in infected immunocompromised patients could explain this weak interest. Some antigen detection assays are based on the recognition of mannan polysaccharide, a major heat-stable component of the yeast cell wall. Bougnoux *et al.* have analyzed the performance of candida antigen assays. They calculated the sensitivity and specificity of the mannan antigen assay to be 29% and 97% respectively, for the diagnosis of disseminated candidiasis [43]. Furthermore, it was suggested that the combined use of both enzyme immunoassays (mannan catabolite and antimannan antibody response) may increase the accuracy of routine diagnosis of candidiasis [44]. The evaluation of the utility of both tests in prospective studies enrolling large numbers of patients at risk for candidiasis is necessary. Furthermore, repeated serum sampling may also improve the reliability of antigen detection tests for the diagnosis of candidiasis [45]. The increasing incidence of fungal infections in immunocompromised patients has focused attention on the rapid and accurate diagnosis of invasive fungal infections using molecular biological techniques. The methods of nucleic acid hybridization and amplification can lead to both high detection rates and identification of specific fungal pathogens. A clinical trial on blood samples taken from 72 patients with hematological malignancies, neutropenia and fever, showed the greater sensitivity of the molecular approach compared to that of conventional blood cultures. Of note, the authors reported a negative predictive value of 98% for disseminated candidiasis in neutropenic patients [46].

7.1.5
Complications and Prognosis of Candidemia

Disseminated candidiasis is a major cause of death in immunocompromised patients. The organs commonly involved are the kidneys, liver, spleen, lungs and muscles [37]. Candida endocarditis is relatively rare but the incidence has increased recently, probably as a result of more aggressive therapeutic approaches in various severe underlying diseases. Candida endophtalmitis is a relatively rare condition however, being found in 9–15% of candidemic patients. It is associated with a mortality rate estimated at 40–80%. Other manifestations of disseminated infection are even less common, skin lesions and septic arthritis being rare in ICU patients.

The prognosis of disseminated candidiasis is extremely poor, particularly in granulocytopenic patients and prompt administration of antifungal therapy is necessary. In a study involving non-neutopenic critically ill patients, candidemia was diagnosed in 46 patients. The overall mortality was 56% and the attributable mortality 21.7%. Using univariate analysis, mortality was significantly associated with the time elapsed between the episode of candidemia and the start of antifungal therapy 48 h or more later ($p < 0.02$). Patients given "early" antifungal therapy (≤48 h between the onset of candidemia and the start of antifungal therapy) tended to have a higher probability of survival compared with patients whose therapy was delayed [47]. In another study, 134 patients were administered empiric antifungal treatment after the results of fungal cultures were known. From the time that the first blood sample which proved to be positive was taken, nine (5.7%) patients received antifungal treatment within 12 h, 10 (6.4%) patients between 12 and 24 h, 86 (54.8%) patients between 24 and 48 h, and 52 (33.1%) patients after 48 h. Multiple logistic regression analysis identified the administration of antifungal treatment 12 h after the first positive-culture blood sample as an independent determinant of hospital mortality. Increased use of empiric antifungal treatment in selected patients at high risk for fungal bloodstream infection also reduced delays in treatment [48]. Several studies have found crude mortality rates for candidemia to be in the range 25–60% [10]. One investigation from a large teaching hospital found a mortality of 38% directly attributable to candidemia [8]. In addition to mortality, candidemia was associated with considerable morbidity and a median hospital stay 8 days longer than that of controls. However, when only survivors were considered, the median length of stay was 30 days longer for cases compared to controls [8, 10]. In the EPIC study, infections caused by fungi alone were associated with a 6% mortality rate [49]. In the study by Zaoutis *et al.*, the mortality attributable to candidemia was 10.1% among children and 14.5% in adults [50]. In France, data from the National Reference Center for Mycology and Antifungals obtained in collaboration with the YEASTS group (CNRMA, Pasteur Institute), about 1024 candida bloodstream infections over a 4-year period (from 2002 through 2005), showed a mortality rate of 38.5% after 30 days (Dromer *et al.*, unpublished data). Moreover, treatment with antifungals and removal of central venous catheters were protective factors against early death [20]. Intravenous catheters are a well known risk factor for candidemia. The influence of the catheter management was studied by Rex *et al.* for the treatment of candidemia in 206 non-neutropenic patients [51]. For the subset of patients with a catheter in place at the time of their first positive blood culture, removal of all intravascular catheters was associated with a significant reduction in the subsequent mean duration of candidemia, from 5.6 ± 0.8 days to 2.6 ± 0.5 days.

7.1.6 Treatment of Candidemia

For decades, only one class of antifungal agent, the polyenes, was available for the treatment of disseminated candidiasis. The first useful alternatives emerged in the late 1980s with the triazoles, fluconazole and itraconazole [52].

There have been many advances in antifungal treatment during the last decade with the availability of more potent and less toxic antifungal agents: new active broad-spectrum azoles, safer lipid formulations of amphotericin and the novel echinocandin class of antifungals. Amphotericin B has a broad spectrum of antifungal activity *in vitro* and has been the standard treatment for most invasive fungal infections in immunocompromised patients for more than 30 years. Its main side-effect is significant nephrotoxicity with increased cost and prolongation of hospital stay [53]. In order to reduce toxicity, lipid formulations of amphotericin B with fewer side-effects were developed such as AMB lipid complex (ABLC) and liposomal AMB (LAMB) [52]. An unpublished comparative trial has shown that ABLC was as effective but less toxic than amphotericin B deoxycholate (Anaissie, ICAAC 95). A recent study has shown that liposomal amphotericin B was associated with a 90.8% cure in patients with candidemia and 81.5% cure in those with invasive candidiasis [54].

Flucytosine should only be used in combination with other antifungal agents as monotherapy can rapidly induce resistance. It shows inhibitory activity against many yeasts including Candida but *C. krusei* is often resistant (70%) to 5FC [14]. Flucytosine is rapidly and almost completely absorbed from the gastrointestinal tract. In IDSA guidelines, oral flucytosine can be used for eradicating candiduria in patients with urologic infection due to non-*albicans* species of Candida and in association with amphotericin B for Candida endocarditis and endophtalmitis [55].

The triazoles, fluconazole and itraconazole and the new triazoles, voriconazole and posaconazole are systemically acting azoles. Available since 1990, fluconazole is well established as a leading drug despite its fungistatic action. It displays a bioavailability of over 80% and an excellent tolerance profile. The drug may be given orally or by the intravenous route and is easily absorbed from the gastrointestinal tract. Fluconazole is active against most pathogenic Candida spp. with the exception of *C. krusei*, which is intrinsically resistant and *C. glabrata* against which it is not consistently effective [56]. Thus, the emergence of azole-resistant strains has raised important questions about its use as a first line drug. In neutropenic patients, the IDSA is careful not to recommend the use of fluconazole as the first line treatment if the patient's condition is not stable and/or if the strain has not been identified [57]. In theory at least, only patients infected with a strain that is usually susceptible to fluconazole and who have not received azole prophylaxis should be given fluconazole as a first line treatment. In contrast however, fluconazole has often been used for the treatment of fungal infections in non-neutropenic patients. A multicenter randomized trial has compared amphotericin B (0.6 mg/kg/day) with fluconazole (400 mg/day) as treatment for candidemia in patients without neutropenia and without major immunodeficiency [58]. Of the 237 patients included in the study, there was no statistically significant difference in outcome. The bloodstream infection failed to resolve in 12 patients in the amphotericin group and in 15 in the fluconazole group. Fluconazole was less toxic than amphotericin

B [58]. The Canadian Candidemia Study Group has also compared these two antifungal therapies in a randomized trial in non-neutropenic patients with candidemia to compare the safety and efficacy of fluconazole (800 mg intravenous loading dose then 400 mg/day) versus that of amphotericin B (0.6 mg/kg/day). A total of 106 patients were enrolled in the trial. Fluconazole and amphotericin B were associated with similar clinical response rates and survival in the treatment of candidemia among non-neutropenic patients; however, drug-related adverse events were more frequent with amphotericin B [59]. Itraconazole has potent activity against Candida spp. *in vitro* but there are no published clinical trials of its efficacy for the treatment of invasive candida infections. Voriconazole, and posaconazole belong to the second generation of systemic antifungal triazoles. Voriconazole appears to be active against all Candida spp., but less so against those with cross reduced susceptibility to fluconazole. The MICS for *C. glabrata* and *C. krusei* are higher than those for other species, but they are still in the presumed susceptible range [60]. Data collected on 137 487 isolates of Candida spp. tested with voriconazole from 2001 through 2005 showed that 94.8% were susceptible and 3.1% were resistant. Less than 30% of fluconazole-resistant isolates of *C. albicans, C. glabrata, C. tropicalis,* and *C. rugosa* remained susceptible to voriconazole [59]. Moreover, voriconazole is interesting because it is available in both intravenous and oral formulations. A randomized non-inferiority trial has recently compared voriconazole with a regimen of amphotericin B followed by fluconazole for the treatment of candidemia in non-neutropenic patient [61]. A total of 370 patients were included in the modified intention-to-treat population. Voriconazole cleared blood cultures as quickly as amphotericin B/fluconazole (median time to negative blood culture, 2.0 days). At the last evaluable assessment, outcome was successful in 162 (65%) patients assigned voriconazole and 87 (71%) assigned amphotericin B/fluconazole. Thus, voriconazole was as effective as the regimen of amphotericin B followed by fluconazole with fewer toxic effects [61]. Posaconazole has a broad spectrum of activity but is not any more effective than voriconazole against Candida spp. and has not been evaluated for the treatment of systemic candidiasis [62].

The echinocandins represent a novel class of antifungals, which act as rapid fungicidal agents against Candida spp. Caspofungin is the first agent of this class to be licensed for the treatment of invasive candidiasis in both non-neutropenic and neutropenic patients. Its use as well as that of other echinocandins is restricted to parenteral administration because of poor oral absorption [52]. Caspofungin is less effective against *C. parapsilosis* and *C. guilliermondii.* The Infectious Diseases Society of America (IDSA) has published evidence-based guidelines for the management of candidemia [55]. Knowledge of the infecting species is highly predictive of likely susceptibility and can be used as a guide for therapy. To a greater extent than for other fungi, treatment of candidiasis can now be guided by *in vitro* susceptibility testing. In the case of systemic infections, *C. albicans* usually remains susceptible to all major agents. Azole resistance (<3%) for this species is reported sporadically in critically ill adults with invasive candidiasis [14, 63]. However, non-*albicans* species of Candida are becoming increasingly associated with invasive candidiasis which is problematic in patients with acute life-threatening invasive candidal infections. Therefore, susceptibility testing for azole resistance is used to guide the management of candidiasis in patients, especially in situations where a failure to respond to the initial empirical therapy is observed. Furthermore, most Candida isolates remain susceptible to amphotericin B. Initial medical therapy should involve caspofungin, fluconazole, voriconazole or an amphotericin B preparation. The choice between these agents depend on many factors: the clinical status of the patient, the species and/or antifungal susceptibility of the infecting isolate, the toxicity of the drug and the patient's prior exposure to antifungal agents [55]. Two large randomized studies and two large observational studies have demonstrated that fluconazole (400 mg/day) and amphotericin B deoxycholate (0.5–0.6 mg/kg/day) are similarly effective as therapy. The use of caspofungin (a 70-mg loading dose followed by 50 mg daily) is now widespread based on its excellent clinical activity, its broad-spectrum activity against Candida species, and a low toxicity which make it a suitable choice for initial therapy in adults. For clinically stable patients who have not recently received azole therapy, fluconazole is the most appropriate choice. For clinically unstable patients infected with an unspeciated isolate, fluconazole has been used successfully, but many authorities prefer amphotericin B deoxycholate (0.7 mg/kg/day) because of its broader spectrum. When a lipid-associated formulation of amphotericin B is selected, a dosage of 3 mg/kg/day is recommended for LAMB and 5 mg/kg/day for ABLC.

For candidemia, therapy should be continued for 2 weeks after the last positive blood culture result and resolution of signs and symptoms of infection. For neutropenic patients, recombinant cytokine (granulocyte colony-stimulating factor) can be used to assist recovery from neutropenia. Since these guidelines, two other compounds have been approved for the treatment of Invasive Candida infections by the US FDA and/or EMEA: micafungin and anidulafungin [64]. The former has been compared with liposomal amphotericin B for the treatment of adult patients with candidemia [54]. A total 264 individuals were randomly assigned to treatment with micafungin (100 mg/day) and 267 were randomly assigned to receive liposomal amphotericin B (3 mg/kg/day) in a double-blind, randomized, multi-national non-inferiority study. Treatment success was observed for 181 (89.6%) patients treated with micafungin and 170 (89.5%) patients treated with liposomal amphotericin B. Efficacy was independent of the Candida spp, primary site infection, as well as neutropenic status. There were fewer treatment-related adverse events with micafungin than there were with liposomal amphotericin B [54]. The latter was studied in a double-blind, non-inferiority study where adults were randomly assigned to receive either intravenous amidulafungin (200 mg on day 1 and then 100 mg daily) or intravenous fluconazole (800 mg on day 1 and then 400 mg daily) [65]. Treatment was successful in 76.5% of patients treated with anidulafungin, as compared with 60.2% of those treated with fluconazole. The frequency and types of adverse events were similar in the two groups. Thus, anidulafungin was shown to be non-inferior to fluconazole in the treatment of invasive candidiasis [65].

The initial or empirical choice between a polyene, an echinocandin, or an azole depends on two critical parameters: the epidemiologic characteristics of the clinical unit and host factors such as severity of illness, infection site, neutropenia, coexisting renal dysfunction and potential drug/drug interactions. Furthermore, general principles of therapy include removing all compromised vascular lines, devices, and implants when possible, as this procedure is actually correlated with better patient outcomes [66]. In patients with neutropenia, most experts recommend using a fungicidal agent that covers all the Candida species (such as polyenes or echinochandins), rather than fungistatic agents (such as the azoles) that may not be effective against all species observed in such immunocompromised patients [66].

7.1.7 Conclusion

Candida species are the fourth most common cause of bloodstream infection and are the leading cause of invasive fungal infection among hospitalized patients. Acute disseminated candidiasis remains a life-threatening disease that now occurs mainly in ICU hospitalized immunocompromised patients. Delayed treatment of bloodstream infections caused by Candida could be minimized by the development of more rapid and susceptible diagnostic techniques for the identification of Candida spp. in bloodstream infections. Current guidelines for the management of these diseases recommend amphotericin B, fluconazole, or caspofungin as the primary therapeutic option. The optimal choice of antifungal agent should depend on local epidemiology and patient factors. For further development of antifungal therapeutics such as immunotherapies with cytokines, a better understanding of the protective immune mechanisms against candidemia is warranted.

7.2 Severe Falciparum Malaria: An Interesting Example of Parasite-induced Sepsis

7.2.1 Introduction

Severe falciparum malaria (SFM) is one of the most serious infectious emergencies and usually requires management in the intensive care unit (ICU). Efforts to unravel the complex pathophysiology of SFM have shed valuable light on severe sepsis induced by parasites.

7.2.2
Epidemiology of Falciparum Malaria

In endemic areas, *Plasmodium falciparum* malaria exacts a huge public health toll, causing about 1.5 to 2.7 million deaths each year. Over 90% of patients who die from falciparum malaria live in tropical Africa, and many are children younger than 5 years of age [67].

In non-endemic industrialized areas, imported malaria may develop in travellers or military personnel, as well as in immigrants from endemic countries. The number of imported cases is rising and can be expected to increase further throughout the industrialized world. About 16 000 cases are diagnosed in Europe each year [68], including more than 6000 in France [69], and about 1300 in the United States [70]. SFM contributes about 2 to 5% of the cases of imported malaria.

7.2.3
Definition of Severe Falciparum Malaria and Relevance to Severe Imported Malaria

In the absence of antimalarial treatment, uncomplicated falciparum malaria can lead to SFM and death, especially in non-immune individuals. The criteria for defining severe and complicated falciparum malaria were established by the World Health Organization (WHO) in 1990 [71] and revised in 2000 [72, 73]. In an adult with *P. falciparum* asexual parasitemia, the presence of one or more of the clinical or laboratory features reported in Table 7.2 indicates SFM. A definition for children has been published by the WHO [72, 73].

These definitions were developed based on studies carried out in tropical endemic areas. Few studies have focused on imported SFM in patients admitted to the ICU, and little is known about the relevance of the WHO criteria in populations living in non-endemic industrialized areas [74]. To better describe the clinical spectrum of imported malaria, we conducted the largest study to date of imported SFM in a non-endemic industrialized country, in which we retrospectively evaluated 188 consecutive patients who were admitted to our ICU in France in 1988–1999 [75]. Among them, 93 met criteria for severe malaria and 95 for less severe malaria, according to the 1990 WHO report [71]. The mean age of patients was 38 years, 51% of whom were non-immune, 94% acquired *P. falciparum* in sub-Saharan Africa, and 96% had taken inadequate antimalarial chemoprophylaxis. Mortality was 12% (11 patients) in the severe malaria group, whereas no patients died in the less-severe malaria group ($P = 0.002$). Bivariable analysis conducted in the severe malaria group suggested that, in our population in an industrialized country, the most relevant WHO defining criteria were neurological, cardio-circulatory, and pulmonary failures together with metabolic acidosis ($P < 0.001$ for each). Other significant parameters were the Simplified Acute Physiology Score (SAPS II), Glasgow

Table 7.2 Severe manifestations of *P. falciparum* malaria in adults (WHO 2000) [6, 7, 72, 73].

Prognostic value	Clinical manifestations and laboratory findings	Frequency
(?) no data	Prostration	+++
+	Impaired consciousness (score <11 on the Glasgow Coma Scale)	++
+++	Acute respiratory distress	+
++	Multiple seizures	+
+++	Circulatory collapse (systolic blood pressure <80 mmHg with features of peripheral circulatory failure)	+
+++	Pulmonary edema (radiological)	+
++	Abnormal bleeding (clinically defined)	+
+	Jaundice (clinically defined or serum bilirubin >50 μmol/l	+++
+	Macroscopic hemoglobinuria	+
+	Severe anemia (haemoglobin <5 g/dl or hematocrit <15%)	+
+++	Hypoglycemia (blood glucose concentration <2.2 mmol/l)	++
+++	Acidosis (pH < 7.35 or plasma bicarbonate <15 mmol/l)	++
+++	High plasma lactate (>5 mmol/l)	++
++	High parasitemia (especially ≥4% in non-immune patients)	+
++	Acute renal failure (serum creatinine >265 μmol/l and 24-h urine output <400 ml)	+++

Coma Scale score, need for mechanical ventilation, arterial pH, arterial lactate, and coagulation disorders ($P \leq 0.002$). Renal failure, high parasitemia, jaundice, and bleeding were fairly common defining criteria, but their prognostic value was not significant. Finally, hypoglycemia, profound anemia, seizures, and hemoglobinuria were rare and of questionable significance [75].

7.2.4
Pathophysiology of Severe Falciparum Malaria

The pathophysiology of SFM is extremely complex. It combines both specific features and those also seen in sepsis/severe sepsis/septic shock. The three main mechanisms of infection are sequestration of parasitized red blood cells (pRBC) which bind to the vascular endothelium; activation of the systemic inflammatory response; and hemostasis dysfunction [72, 76, 77]. None of these mechanisms alone fully accounts for the pathogenesis of SFM [77]. Dynamic interactions occur among the three mechanisms, explaining the complexity of this potentially fatal infection. The variability in presentation of SFM involves numerous other factors related to the parasite, host, and setting.

7.2.4.1 The Sequestration Hypothesis

Sequestration of pRBC in organ capillaries is mediated by the phenomenon of adherence and/or rolling of pRBC on endothelial cells. The main parasite adhesins concentrated at the surface of pRBC are *P. falciparum* Erythrocyte Membrane Protein-1 (PFEMP-1), Histidin Rich Proteins (HRP), Ring Erythrocyte Surface Antigen (RESA Ag), and rifins. The main endothelial receptors are CD36, ICAM-1, thrombospondin, ELAM-1, V-CAM-1, PECAM-1, and chondroitin sulfate A (CS-A) [72]. Sequestered pRBC partially obstruct the capillaries, thereby severely reducing tissue blood flow, particularly in the brain during cerebral malaria. Capillary obstruction may be worsened by decreased RBC deformability and by the rosetting phenomenon, in which pRBC aggregate with non-parasitized RBC [72, 76, 78]. Nevertheless, RBC sequestration does not seem to obstruct the capillaries completely (in contradiction to stroke, where ischemic damage is usually not reversible). The effect of RBC sequestration is probably patchy impairment of microvascular blood flow, explaining why patients with profound coma during cerebral malaria can achieve a full recovery [79]. Sequestration occurs in all patients with falciparum malaria, whereas cerebral malaria develops in only about 1% of these patients. The number of sequestered pRBC (which probably constitute the pRBC population most relevant to the pathophysiology of the disease) is difficult to quantitate. Parasitemia reflects circulating pRBC, whose pathogenic role may be limited. Therefore, when used in isolation as a severity criterion, parasitemia may be of limited prognostic value. A recent study showed that plasma *P. falciparum* HRP2 concentration provided a useful estimate of the total parasite burden [80]. Finally, pRBC, non-parasitized RBC, endothelial cells, mononuclear cells, and platelets come into close contact with each other within the capillaries, thereby creating ideal conditions for the immune/humoral and hemostasis mechanisms described below.

7.2.4.2 The Immunological (or Inflammation, or Humoral) Hypothesis

As with the bacterial model of sepsis, the humoral response includes not only immune activation of macrophages and T-lymphocytes by *P. falciparum*, but also activation of monocytes by malarial toxins (probably glycophosphatidylinositol). The result is a systemic inflammatory response with secretion of proinflammatory cytokines including TNF, IL-1, IFNγ, IL-6, IL-8, IL-12, M-CSF, and lymphotoxin. These cytokines, most notably TNF, generate a vicious circle of macrophage activation, increased cytoadherence and sequestration, and up-regulated production of reactive oxygen species (ROS) including nitric oxide (NO) in the general circulation and *in situ* [72, 76, 77, 79]. However, levels of proinflammatory cytokines are high in non-fatal *P. vivax* malaria, suggesting that this mechanism may be necessary but not sufficient to induce SFM. Furthermore, the inability of anti-TNF monoclonal antibodies, steroids, and pentoxyfylline to protect against severe malaria and death indicates that other mechanisms are involved [77].

7.2.4.3 The Role of Hemostasis and Platelets

Activation of coagulation and platelets occurs during SFM. The coagulation cascade is initiated by activated monocytes, which express tissue factor; this results in consumption of clotting factors, presence of fibrin dimers in plasma and, more generally, activation of coagulation. Nevertheless, disseminated intravascular coagulation (DIC) is rare during SFM. Platelets are actively involved in sequestration, humoral responses, and hemostasis disorders; and some authors have suggested that platelets may act as a link between these three mechanisms [77]. Platelets bind to the endothelial cells through overlapping adhesins (ICAM-1, P-selectin) and can also bind to rosettes [81]. Moreover, platelets contribute to immune activation by binding parasite-derived molecules via their toll-like receptors. In turn, platelets are activated by *P. falciparum* to express cytokines and chemokines. All these mechanisms can contribute to vascular leakage and microvascular damage, which may explain the observation of retinal hemorrhage upon funduscopy and brain petechiae at autopsy in patients with cerebral malaria [72].

Finally, all these activated cells (pRBC, non-parasitized RBC, endothelial cells, platelets, mononuclear cells) are in close contact with each other within the host capillaries, where they generate a storm of inflammation and coagulation. They may release microparticles which constitute a marker for cellular activation and a mechanism of intercellular signaling that may contribute to the activation of the above-described mechanisms. Although the role for circulating microparticles in SFM requires further investigation, microparticles hold promise as a biological marker for disease severity [82].

7.2.4.4 Genetic Polymorphisms

Individuals vary in their response to malaria. Resistance and susceptibility to malaria is influenced by both host and parasite genes and the manner in which they interact. All host molecules involved in the pathogenesis of SFM may be subject to genetic polymorphism. Many genetic studies have been conducted, chiefly in endemic areas. They have investigated genes for erythrocyte surface molecules, hemoglobin, adhesion molecules, haptoglobin, HLA groups, cytokines, NO synthase, complement receptors, and the coagulation pathway [83]. A role was established for a few genes, such as the Duffy blood group antigen, sickle hemoglobin gene, and TNF promoter gene. Other genes may have weak effects (IL-1, complement receptor-1, mannose-binding protein). For some genes, the results varied across geographic areas (ICAM-1, HLA-B53) or proved conflicting (inducible NO synthase, CD36).

Genetic polymorphisms also occur in the *P. falciparum* genome and may affect all the main parasite-derived molecules involved in the pathogenesis of falciparum malaria. Recently developed tools can be used to analyze the *P. falciparum* transcriptome. This approach provides information regarding the expression of *P. falciparum* genes and its variability, which may lead to advances in the identification of virulence factors [83–85].

7.2.4.5 Other Factors

Finally, many other factors may contribute in varying degrees to the severity of falciparum malaria. They include the initial size of the plasmodial inoculum, clonal population dynamics [86], chemosusceptibility of *P. falciparum* isolates [87], intensity of specific anti-malarial immunity, co-infection with other microorganisms, nutritional status, age, co-morbidities, access to health services, and available healthcare resources (most notably in ICUs) [72].

7.2.5 Important Clinical Issues

7.2.5.1 From Uncomplicated to Severe Malaria

The symptoms of uncomplicated falciparum malaria are non-specific and include asthenia, fever, chills, headache, myalgia, abdominal pain, diarrhea, cough, and a broad range of neurological symptoms. In the absence of adequate antimalarial treatment, especially in non-immune patients, worsening of the disease is inevitable. Therefore, although the symptoms are non-specific, their occurrence in a patient returning from an endemic area should prompt immediate parasitological investigations for falciparum malaria. If the diagnosis is confirmed, antimalarial treatment must be given immediately.

7.2.5.2 Populations at Risk

Some populations are particularly vulnerable to SFM, and to death from the disease. Children are at high risk, most notably those younger than 5 years of age who live in endemic areas. Other high-risk populations include non-immune travellers, pregnant women, older patients, and patients with co-morbidities including those with HIV infection [72, 88].

7.2.5.3 Clinical Aspects of SFM

Nearly all the SFM-defining criteria shown in Table 7.1 can occur during severe sepsis and/or septic shock due to any cause. Furthermore, in the most severe form of SFM characterized by multiorgan failure, both the parasite and/or the non-specific systemic inflammatory response can contribute to the failure of each organ.

The neurological manifestations range from simple delirium to profound coma. Coma is common, but usually resolves completely. The duration of coma is 4 days on average but may be longer, and sedative agents may contribute to prolong the coma. Neurological deterioration requires immediate testing for hypoglycemia. In addition, in comatose patients, some authors recommend tests for bacterial meningitis or locally prevalent viral encephalitis. [72, 88]. Cerebral edema, focal neurological signs, and seizures are rare in adults [72, 88]. Brain CT or MRI usually shows mild cerebral swelling with a slight increase in brain volume, the mechanism of which is thought to involve an increase in cerebral blood volume due to the presence of sequestered pRBC within the blood vessels [89]. Focal lesions are uncommon. Nevertheless, in a study of 28 patients who underwent brain imaging, abnormalities were detected in 10 cases; they included cortical infarcts in three patients, cerebral edema in three, hematoma in two, abnormalities of the deep white matter in three, and meningeal enhancement in one [75].

Cardio-circulatory failure during SFM (algid malaria) carries an extremely poor prognosis. Coma, acute respiratory failure, and profound acidosis are often present, especially in imported cases, which rarely manifest as isolated cerebral malaria [75]. A hyperdynamic pattern is the rule in patients with shock during SFM; myocardial failure and cardiac arrhythmia are extremely rare, despite pRBC sequestration within the myocardial vessels and cardiac side-effects associated with many antimalarial drugs [72, 75, 90]. Initial hypovolemia and bacterial co-infection may contribute to the development of shock, although shock can also be induced by *P. falciparum* alone. Bacterial co-infection is found in 30 to 50% of patients with SFM and shock. SFM is associated with increased susceptibility to bacterial infection. The most common infections are community acquired or early nosocomial pneumonia and/or bacteremia [72, 75, 90]. A few cases of fatal invasive aspergillosis have been reported, and malaria is known to induce transient but profound immunodepression [91]. Lactic acidosis is very common in patients with algid malaria. It results from numerous interacting factors including anaerobic glycolysis, bacterial and/or plasmodial septic shock, decreased oxygen delivery secondary to anemia, mitochondrial dysfunction, seizures, altered redox state, and impaired clearance by the liver [72, 88, 92]. Finally, metabolic acidosis results from both lactic acidosis and acute renal failure.

The cause of acute lung injury during SFM is unknown. This manifestation is associated with the most severe form of falciparum malaria, and it carries a higher than 30% mortality rate in patients with imported SFM [93]. Falciparum malaria is among the established causes of ARDS (Adult Respiratory Distress Syndrome). However, many factors may contribute to worsen the respiratory failure, such as aspiration pneumonia in comatose patients, community-acquired or nosocomial bacterial pneumonia, other co-infections, shock, overhydration, and anuria [72, 75, 93].

Acute renal failure (ARF) during SFM is virtually confined to adults. The absence of hypertension and the rapid and complete resolution of ARF during SFM are consistent with acute tubular necrosis. The main mechanisms of this tubular necrosis are hypovolemia due to dehydration and/or associated septic shock, hemolysis, and increased blood viscosity with reduced blood flow due to sequestered pRBC in renal vessels [72]. When malaria-related ARF is isolated, complete recovery is the rule. In contrast, ARF in the setting of falciparum-induced multi-organ failure carries a poor prognosis.

Thrombocytopenia is a feature of SFM. However, although profound thrombocytopenia may occur, this criterion is not sufficiently consistent to be of prognostic significance [72]. Possible causes include reduced platelet survival, intravascular lysis, splenomegaly, decreased central production, and trapping at sites of pRBC sequestration. Although coagulation is activated, DIC is rare. Consequently, significant bleeding may occur in the most severe forms of falciparum malaria but is less common than expected taking into account the severe hemostasis abnormalities [72].

The relevance of parasitemia during SFM is hotly debated, and few data are available on parasitemia in imported SFM [72, 75]. Parasitemia reflects only the number of circulating pRBC and correlates poorly with sequestered pRBC [80]. In addition, the relationship between parasitemia and malaria severity varies across populations. In non-immune patients, parasitemia values of 4% or more indicate an increased risk of death and should be taken as a sign of severe malaria (Table 7.1). However, in patients with partial specific immunity the thresholds indicating increased risk should be determined based on local experience. In vast studies of adults in Asia, parasitemia values of 10% or more were considered to indicate severe malaria [94, 95]. In regions of high and stable endemicity in Africa, parasitemia values greater than 20% (extreme parasitemia) consistently indicate severe malaria in adults [72, 88]. Data available from patients with imported SFM are too scant to enable determination of the parasitemia threshold associated with an increased risk of death in adults and children [75]. In adults, when parasitemia is the only severity criterion, 10 to 15% is probably a better severity threshold than 4%. When a combination of severity criteria is used, parasitemia is less relevant than other criteria such as coma, shock, ARDS, or acidosis.

Profound anemia and severe hypoglycemia are extremely rare in adults but more common in children. Macroscopic hemoglobinuria in the absence of G6PD deficiency is very infrequent during SFM. When present, especially in Caucasian expatriates, blackwater fever must be ruled out first, since quinine may increase the massive intravascular hemolysis observed during this disease [96].

7.2.6 The Management of Severe Falciparum Malaria

The mortality rate in patients with untreated SFM, most notably those with imported malaria, is probably close to 100% [97].

7.2.6.1 Specific Antimalarial Treatments

Early full-dose antimalarial treatment is crucial in patients with SFM. Two drug classes are available for the parenteral treatment of SFM: quinine (and quinidine) and artemisinin derivatives (chiefly artesunate and artemether).

In Europe, quinine treatment reduces mortality by up to 90% [75] and intravenous (IV) quinine is still the treatment of choice in SFM. A loading dose of 16 mg base/kg is given by IV infusion over 4 h. The infusion is then stopped for 4 h, after which a continuous IV infusion of 24 mg base/kg/24 h is given for 7 days. The goal of the loading dose is to produce quinine concentrations with schizonticidal activity (usually about 12 mg/l) as promptly as possible [72, 88]. A recent meta-analysis of four randomized studies found that a loading dose had no significant effect on mortality but decreased the times to parasite clearance and defervescence without increasing the rate of hypoglycemia [98]. Most experts currently recommend a loading dose [72, 88], which was administered in the recent extensive studies of SFM [94, 95]. The loading dose should be reduced only when there is reliable evidence of adequate previous treatment with quinine, mefloquine or halofantrine, to minimize the risk of cardiac toxicity. During treatment, blood glucose and ECG should be closely monitored. The dosage should be reduced by one-third in patients with hepatic dysfunction or acute renal failure, but only after the second day of treatment [88].

Parenteral artemisinin derivatives are at least as effective as IV quinine [94, 95, 99], and their safety profile is better [95]. Artesunate exhibits better pharmacokinetic properties compared to artemether and artemotil, as it is water-soluble and can be given intravenously, intramuscularly, or orally. In recent multicenter trials conducted chiefly in Asian adults with SFM, IV artesunate was clearly superior over IV quinine, producing significantly less mortality and a better safety profile [95]. Consequently, recent WHO guidelines recommend artesunate as the first choice in SFM. Data is lacking in patients with imported SFM, essentially because artemisinin derivatives are not actually authorized and consequently only available for this indication in industrialized countries. IV artesunate would probably be as effective as quinine in adults with imported SFM; it has been suggested that IV artesunate therapy may decrease mortality from imported SFM [97].

A number of antibiotics with parasitistatic activity (as opposed to parasiticidal activity) on *P. falciparum*, such as doxycycline and clindamycin, have been considered. However, concomitant administration of a schizonticidal drug is indispensable. These antibiotics should be reserved for patients with SFM due to multidrug resistant strains (such as those encountered in the jungles of Amazonia, Myanmar, Cambodia, and Thailand).

7.2.6.2 **Continuing Supportive Care**

Concomitantly with antimalarial treatment, intensive supportive management must be provided in the ICU to patients with imported SFM. Comatose patients should be immediately intubated by the orotracheal route. Standard measures for protecting the brain should be implemented, including semi-recumbent position, maintenance of pCO_2 around 35–40 mmHg and serum sodium around 145 mmol/l, blood pressure control, and maintenance of core temperature under 38 °C. Seizures should be promptly treated with diazepam and phenobarbital, but preventive anticonvulsant treatment is not recommended. Blood glucose should be carefully monitored throughout treatment, especially during quinine loading, which is associated with a high risk of hypoglycemia. If focal neurological signs or seizures develop, the coma is abnormally prolonged, or the clinician has any doubts whatsoever, an EEG and brain imaging should be obtained [72, 75, 88].

At presentation, patients with SFM may be dehydrated. Fluid resuscitation relies on crystalloids. When shock persists, a hyperdynamic hemodynamic profile is the rule and requires early goal-directed therapy according to recent guidelines published in industrialized countries [100], including fluid loading, vasopressors, inotropic agents, blood transfusion, and scheduled hemodynamic monitoring. The WHO recommendations concerning limitation of fluid loading to maintain central venous pressure lower than 5 cm of water, are relevant only to endemic areas that lack critical-care resources (especially mechanical ventilation and renal replacement therapies). These recommendations do not apply to patients with imported SFM managed in industrialized countries. The key concern in patients with shock/acidosis during SFM is bacterial co-infection. Specimens for blood cultures and other relevant microbiological tests should be obtained and parenteral broad-spectrum antimicrobials should be initiated as soon as possible [75, 88, 90]. Low-dose steroid therapy should be given, as with septic shock [100]. Activated protein C should be considered, most notably in SFM with multi-organ failure [101]; however, severe thrombocytopenia and/or significant bleeding contraindicate the administration of activated protein C, whose usefulness is therefore severely limited in SFM [102].

Acute lung injury/ARDS during SFM should be managed according to recent guidelines on lung-protective mechanical ventilation for patients with ARDS [100]. Community-acquired and nosocomial pneumonia are common and require adequate antimicrobial therapy.

Acute renal failure, especially with anuria, requires prompt extra-renal replacement therapy. Sequential hemodialysis or continuous hemo(dia)filtration are better than peritoneal dialysis, which is no longer used for managing SFM in industrialized countries [88].

Monitoring in the ICU requires careful attention, as abrupt deterioration may occur, most notably during the first 48 h of treatment. However, the only issue that is specific for SFM is the need for close monitoring of blood glucose levels.

7.2.6.3 **Other Adjuvant Treatments**

Anecdotal reports suggest that exchange blood transfusion (EBT) may be beneficial in SFM. However, no robust comparative trials are available. There is no consensus regarding whether EBT reduces mortality, and which mechanisms may be involved [88, 103]. A 2002 meta-analysis not only failed to identify any beneficial effects, but also highlighted the potential risks of EBT [104]. In our opinion, EBT no longer has a role in the management of SFM. In our study of imported SFM, in which conventional treatment without EBT was used, mortality was 11%, which is one of the lowest mortality rates reported to date in SFM of similar severity [75].

Although many other adjuvant treatments have been suggested for SFM, few of them produced convincing benefits, and many caused side effects. Heparin, prostacyclin, pentoxifylline, deferoxamine, low-molecular-weight dextran, high-dose steroids, acetylsalicylic acid, anti-TNF antibody, cyclosporine, dichloroacetate, and hyperimmune serum have all been suggested, but none is recommended [72, 76, 88]. Erythropoietin was recently shown to prevent death from cerebral malaria in mice [105]. No data are available in humans, although a study is ongoing in Africa.

7.2.7 Conclusion

SFM is a fascinating and complex illustration of parasite-induced sepsis. In endemic areas, it is among the main priorities selected by the WHO. In industrialized countries, the mortality rate remains about 10%, making imported SFM a challenge for intensivists. To progress, we must both increase the quality of prevention and improve the management of uncomplicated and severe falciparum malaria. Finally, experimental and clinical research has a key role to play in unraveling the pathophysiology of this complex and potentially fatal infection.

References

1. Martin, G.S., Mannino, D.M., Eaton, S. and Moss, M. (**2003**) The epidemiology of sepsis in the United States from 1979 through 2000. *N. Engl. J. Med.*, **348**, 1546–1554.
2. Wisplinghoff, H., Bischoff, T., Tallent, S.M., Seifert, H., Wenzel, R.P. and Edmond, M.B. (**2004**) Nosocomial bloodstream infections in US hospitals: analysis of 24,179 cases from a retrospective nationwide surveillance study. *CID*, **39**, 309–317.
3. Edwards, J.E., Bodey, G.P., Bowden, R.A., Büchner, T., de Pauw, B.E., Filler, S.G. and Ghannoum, M.A. (**1997**) International Conference for the Development of a Consensus on the Management and Prevention of Severe Candidal Infections. *CID*, **25**, 43–59.
4. Ascioglu, S., Rex, J.H., de Pauw, B., Bennett, J.E., Crokaert, F., Denning, D.W., *et al.* (**2002**) Defining opportunistic invasive fungal infections in immunocompromised patients with cancer and hematopoietic stem cell transplants: an international consensus. *CID*, **34**, 7–14.
5. Blumberg, H.M., Jarvis, W.R., Soucie, J.M., Edwards, J.E., Patterson, J.E., Pfaller, A., *et al.* (**2001**) Risk factors for candidal bloodstream infections in surgical intensive care unit patients: the

NEMIS Prospective Multicenter Study. *CID*, **33**, 177–186.
6. Karabinis, A., Hill, C., Leclercq, B., Tancrede, C., Baume, D. and Andremont, A. (**1988**) Risk factors for candidemia in cancer patients: a case study. *J. Clin. Microbiol.*, **26**(3), 429–432.
7. Bross, J., Talbot, G.H., Maislin, G., Hurwitz, S. and Strom, B.L. (**1989**) Risk factors for nosocomial candidemia: a case–control study in adults without leukemia. *Am. J. Med.*, **87**(6), 614–620.
8. Wey, S.B., Mori, M., Pfaller, M.A., Woolson, R.F. and Wenzel, R.P. (**1989**) Risk factors for hospital-acquired candidemia. A matched case–control study. *Arch. Intern. Med.*, **149**, 2349–2353.
9. McKinnon, P.S., Goff, D.A., Kern, J.W., Devlin, J.W., Barletta, J.F., Sierawski, S.J., *et al.* (**2001**) Temporal assessment of candida risk factors in the surgical intensive care unit. *Arch. Surg.*, **136**, 1401–1408.
10. Vincent, J.L., Anaissie, E., Bruining, H., Demajo, W., El-Ebiary, M., Haber, J., *et al.* (**1998**) Epidemiology, diagnosis and treatment of systemic Candida infection in surgical patients under intensive care. *Intensive Care Med.*, **24**, 206–216.
11. Wenzel, R.P. and Gennings, C. (**2005**) Bloodstream infections due to candida species in the intensive care unit: identifying especially high-risk patients to determine prevention strategies. *CID*, **41**, S389–93.
12. Marchetti, O., Bille, J., Fluckiger, U., Eggimann, P., Ruef, C., Garbino, J., Calandra, T., Glauser, M.P., Täuber, M.G. and Pittet, D. (**2004**) Epidemiology of candidemia in Swiss tertiary care hospitals: secular trends, 1991–2000. *CID*, **38**, 311–320.
13. Pappas, P.G., Rex, J.H., Lee, J., Hamill, R.H., Larsen, R.A., Powderly, W., Kauffman, C.A., Hyslop, N., Mangino, J.E., Chapman, S., Horowitz, H.W., Edwards, J.E. and Dismukes, W.E. (**2003**) for the NIAID Mycoses Study Group. A prospective observational study of candidemia: epidemiology, therapy, and influences on mortality in hospitalized adult and pediatric patients. *CID*, **37**, 634–643.
14. Pfaller, M.A., Diekema, D.J., Jones, R.N., Sader, H.S., Fluit, A.C., Hollis, R.J., Messer, S.A. and the SENTRY participant group. (**2001**) International surveillance of bloodstream infections due to Candida species: frequency of occurrence and *in vitro* susceptibilities to fluconazole, ravuconazole, and voriconazole of isolates collected from 1997 through 1999 in the SENTRY Antimicrobial Surveillance Program. *J. Clin. Microbiol.*, **39**, 3254–3259.
15. Richet, H., Roux, P., Des Champs, C., Esnault, Y., Andremont, A. and the French Candidemia Study Group. (**2002**) Candidemia in French hospitals: incidence rates and characteristics. *Clin. Microbiol. Infect.*, **8**, 405–412.
16. Trick, W.E., Fridkin, S.K., Edwards, J.R., Hajjeh, R.A. and Gaynes, R.P. (**2002**) The National Nosocomial Infections Surveillance System Hospitals. Secular trends of hospital-acquired candidemia among intensive care unit patients in the United States during 1989–1999. *CID*, **35**, 627–630.
17. Hajjeh, R.A., Sofair, A.N., Harrison, L.H., Lyon, G.M., Arthington-Skaggs, B.A., Mirza, S.A., Phelan, M., Morgan, J., Lee-Yang, W., Ciblak, M.A., Benjamin, L.E., Sanza, L.T., Huie, S., Yeo, S.F., Brandt, M.E. and Warnock, D.W. (**2004**) Incidence of bloodstream infections due to Candida species and *in vitro* susceptibilities of isolates collected from 1998 to 2000 in a population-based active surveillance program. *J. Clin. Microbiol.*, **42**, 1519–1527.

18. Tortorano, A.M., Kibbler, C., Peman, J., Bernhardt, H., Klingspor, L. and Grillot, R. (**2006**) Candidemia in Europe: epidemiology and resistance. *Int. J. Antimicrob. Agents*, **27**, 359–366.
19. Arendrup, M.C., Fuursted, K., Gahrn-Hansen, B., Jensen, I.M., Knudsen, J.D., Lundgren, B., *et al.* (**2005**) Seminational surveillance of fungemia in Denmark: notably high rates of fungemia and numbers of isolates with reduced azole susceptibility. *J. Clin. Microbiol.*, **43**, 4434–4440.
20. Almirante, B., Rodriguez, D., Park, B.J., Cuenca-Estrella, M., Planes, A.M., Almela, M., *et al.* (**2005**) Epidemiology and predictors of mortality in cases of *candida* bloodstream infection: results from population-based surveillance, Barcelona, Spain, from 2002 to 2003. *J. Clin. Microbiol.*, **43**, 1829–1835.
21. Vincent, J.L., Sakr, Y., Sprung, C.L., Ranieri, V.M., Reinhart, K., *et al.* (**2006**) Sepsis in European intensive care units: results of the SOAP study. *Crit. Care Med.*, **34**, 344–353.
22. Romani, L. (**2004**) Immunity to fungal infections. *Nat. Rev. Immunol.*, **4**(1), 1–23.
23. Netea, M.G., Van Der Graaf, C.A., Vonk, A.G., Verschueren, I., Van Der Meer, J.W. and Kullberg, B.J. (**2002**) The role of toll-like receptor (TLR) 2 and TLR4 in the host defense against disseminated candidiasis. *J. Infect. Dis.*, **185**(10), 1483–1489.
24. Cassone, A., Palma, C., Djeu, J.Y., Aiuti, F. and Quinti, I. (**1993**) Anticandidal activity and interleukin 1 beta and interleukin-6 production by polymorphonuclear leukocytes are preserved in subjects with AIDS. *J. Clin. Microbiol.*, **31**(5), 1354–1357.
25. Castro, M., Bjoraker, J.A., Rohrbach, M.S. and Limper, A.H. (**1996**) *Candida albicans* induces the release of inflammatory mediators from human peripheral blood monocytes. *Inflammation*, **20**(1), 107–122.
26. Kobrynski, L.J., Tanimune, L., Kilpatrick, L., Campbell, D.E. and Douglas, S.D. (**1996**) Production of T-helper cell subsets and cytokines by lymphocytes from patients with chronic mucocutaneous candidiasis. *Clin. Diagn. Lab. Immunol.*, **3**(6), 740–745.
27. Xiong, J., Kang, K., Liu, L., Yoshida, Y., Cooper, K.D. and Ghannoum, M.A. (**2000**) *Candida albicans* and *Candida krusei* differentially induce human blood mononuclear cell interleukin-12 and gamma interferon production. *Infect. Immun.*, **68**(5), 2464–2469.
28. Netea, M.G., Stuyt, R.J., Kim, S.H., Van der Meer, J.W., Kullberg, B.J. and Dinarello, C.A. (**2002**) The role of endogenous interleukin (IL)-18, IL-12, IL-1beta, and tumor necrosis factor-alpha in the production of interferon-gamma induced by Candida albicans in human whole-blood cultures. *J. Infect. Dis.*, **185**(7), 963–970.
29. Levitz, S.M. and North, E.A. (**1996**) Gamma interferon gene expression and release in human lymphocytes directly activated by *Cryptococcus neoformans* and *Candida albicans*. *Infect. Immun.*, May **64**(5), 1595–1599.
30. Jiang, Y., Russell, T.R., Graves, D.T., Cheng, H., Nong, S.H. and Levitz, S.M. (**1996**) Monocyte chemoattractant protein 1 and interleukin-8 production in mononuclear cells stimulated by oral microorganisms. *Infect. Immun.*, **64**(11), 4450–4455.
31. Jeremias, J., Kalo-Klein, A. and Witkin, S.S. (**1991**) Individual differences in tumour necrosis factor and interleukin-1 production induced by viable and heat-killed *Candida albicans*. *J. Med. Vet. Mycol.*, **29**(3), 157–163.
32. Bellocchio, S., Montagnoli, C., Bozza, S., *et al.* (**2004**) The contribution of the Toll-like/IL-1 receptor superfamily to innate and

adaptive immunity to fungal pathogens *in vivo*. *J. Immunol.*, **172**(5), 3059–3069.

33. Marr, K.A., Balajee, S.A., Hawn, T.R., *et al.* (**2003**) Differential role of MyD88 in macrophage mediated responses to opportunistic fungal pathogens. *Infect. Immun.*, **71**(9), 5280–5286.
34. Minami, A., Fujimoto, K., Ozaki, Y. and Nakamura, S. (**1988**) Augmentation of host resistance to microbial infections by recombinant human interleukin-1 alpha. *Infect. Immun.*, **56**(12), 3116–3120.
35. Van Enckevort, F.H., Netea, M.G., Hermus, A.R., *et al.* (**1999**) Increased susceptibility to systemic candidiasis in interleukin-6 deficient mice. *Med. Mycol.*, Dec **37**(6), 419–426.
36. Van der Graaf, C.A., Netea, M.G., Morre, S.A., *et al.* (**2006**) Toll-like receptor 4 Asp299Gly/Thr399Ile polymorphisms are a risk factor for Candida bloodstream infection. *Eur. Cytokine Netw.*, **17**(1), 29–34.
37. Meunier, F. (**1989**) Candidiasis. *Eur. J. Clin. Microbiol. Infect. Dis.*, **8**, 438–447.
38. Year, H., Sendid, B., Francois, N., Camus, D. and Poulain, D. (**2001**) Contribution of serological tests and blood culture to the early diagnosis of systemic candidiasis. *Eur. J. Microbiol. Infect. Dis.*, **20**, 864–870.
39. Weiner, M.H. and Coats-Stephen, M. (**1979**) Immunodiagnosis of systemic candidiasis: mannan antigenemia detected by radioimmunoassay in experimental and human infections. *J. Infect. Dis.*, **140**, 989–993.
40. Ostrosky-Zeichner, L., Rex, J.H., Bennett, J., *et al.* (**2002**) Deeply invasive candidiasis. *Infect. Dis. Clin. North Am.*, **16**, 821–835.
41. Stevens, D.A. (**2002**) Diagnosis of fungal infections: current status. *JAC*, **49**, 11–19.
42. Meyer, M.H., Letscher-Bru, V., Jaulhac, B., Waller, J. and Candolfi, Ermanno (**2004**) Comparison of mycosis IC/F and plus aerobic/F media for diagnosis of fungemia by the Bactec 9240 system. *J. Clin. Microbiol.*, **42**, 773–777.
43. Bougnoux, M.E., Hill, C., Moissenet, D., Feuilhade de Chauvin, M., Bonnay, M., Vicens-Sprauel, I., *et al.* (**1990**) Comparison of antibody, antigen, and metabolite assays for hospitalized patients with disseminated or peripheral candidiasis. *J. Clin. Microbiol.*, **28**, 905–909.
44. Sendid, B., Tabouret, M., Poirot, J.L., Mathieu, D., Fruit, J. and Poulain, D. (**1999**) Nex enzyme immunoassays for sensitive detection of circulating *Candida albicans* mannan and antimannan antibodies: useful combined test for diagnosis of systemic candidiasis. *J. Clin. Microbiol.*, **37**, 1510–1517.
45. Yeo, S.F. and Wong, B. (**2002**) Current status of nonculture methods for diagnosis of invasive fungal infections. *Clin. Microbiol. Rev.*, **15**, 465–484.
46. Morace, G., Pagano, L., Sanguinetti, M., Posteraro, B., Mele, L. and Equitani, F. (**1999**) PCR-restriction enzyme analysis for detection of Candida DNA in blood from febrile patients with haematological malignancies. *J. Clin. Microbiol.*, **37**, 1871–1875.
47. Nolla-Salas, J., Sitges-Serra, A., Leon-Gil, C., Martinez-Gonzalez, J., Leon-Regidor, M.A., Ibanez-Lucia, P., *et al.* (**1997**) Candidemia in non-neutropenic critically ill patients: analysis of prognostic factors and assessment of systemic antifungal therapy. *Intensive Care Med.*, **23**, 23–30.
48. Morrell, M., Fraser, V.J. and Kollef, M.H. (**2005**) Delaying the empiric treatment of Candida bloodstream infection until positive blood culture results are obtained: a potential risk factor for hospital mortality. *Antmicrob. Agents Chemother.*, **49**, 3640–3645.
49. Vincent, J.L., Bihari, D.J., Suter, P.M., Bruining, H.A., White, J. and Nicolas-Chanoin, M.H. (**1995**)

The prevalence of nosocomial infection in intensive care units in Europe. Results of the European Prevalence of Infection in Intensive Care (EPIC) Study. *JAMA*, **274**, 639–644.

50. Zaoutis, T.E., Argon, J., Chu, J., Berlin, J.A., Walsh, T.J. and Feudtner, C. (**2005**) The epidemiology and attributable outcomes of candidemia in adults and children hospitalized in the United States: a propensity analysis. *CID*, **41**, 1232–1239.
51. Rex, J.H., Bennett, J.E., Sugar, A.M., Pappas, P.G., Serody, J., Edwards, J.E. and Washburn, R.G. (**1995**) Intravascular catheter exchange and duration of candidemia. NIAID Mycoses Study Group and the Candidemia Study Group. *CID*, **21**, 994–996.
52. Willinger, B. (**2006**) Laboratory Diagnosis and Therapy of Invasive Fungal Infections. *Curr. Drug Targets*, **7**, 513–522.
53. Bates, D.W., Su, L., Yu, D.T., Chertow, G.M., Seger, D.L., Gomes, D.R.J., *et al.* (**2001**) Mortality and costs of acute renal failure associated with amphotericin B therapy. *CID*, **32**, 686–693.
54. Kuse, E.R., Chetchotisakd, P., da Cuhna, C.A., Ruhnke, M., Barrios, C., Raghunadharao, D., *et al.* (**2007**) Micafungin versus liposomal amphotericin B for candidaemia and invasive candidosis: a phase 3 randomised double-blind trial. *Lancet.*, **369**, 1519–1527.
55. Pappas, P.G., Rex, J.H., Sobel, J.D., Filler, S.G., Dismukes, W.E., Walsh, T.J. and Edwards, J.E. (**2004**) Guidelines for treatment of candidiasis. *CID*, **38**, 161–189.
56. Charlier, C., Hart, E., Lefort, A., Ribaud, P., Dromer, F., Denning, D.W., *et al.* (**2006**) Fluconazole for the management of invasive candidiasis: where do we stand after 15 years? *JAC*, **57**, 384–410.
57. Pfaller, M.A., Diekema, D.J., Gibbs, D.L., Newell, V.A., Meis, J.F., Gould, I.M., *et al.* (**2007**) Results of the ARTEMIS DISK global antifungal surveillance study, 1997–2005: an 8.5 year analysis of susceptibilities of candida and other yeast species to fluconazole and voriconazole by CLSI standardized disk diffusion testing. *J. Clin. Microbiol.*, 18 [Epub ahead of print].
58. Rex, J.H., Bennett, J.E., Sugar, A.M., Pappas, P.G., van der Horst, C.M., Edwards, J.E., *et al.* (**1994**) A randomized trial comparing fluconazole with amphotericin B for the treatment of candidemia in patients without neutropenia. *NEJM*, **331**, 1325–1330.
59. Phillips, P., Shafran, S., Garber, G., Rotstein, C., Smaill, F., Fong, I., *et al.* (**1997**) Multicenter randomized trial of fluconazole versus amphotericin B for treatment of candidemia in non-neutropenic patients. Canadian candidemia study group. *Eur. J. Clin. Microbiol. Infect. Dis.*, **16**, 337–345.
60. Johnson, L.B. and Kauffman, C.A. (**2003**) Voriconazole: a new Triazole antifungal Agent. *CID*, **36**, 630–637.
61. Kullberg, J., Sobel, J.D., Ruhnke, M., Pappas, P.G., Viscoli, C., Rex, J.H., *et al.* (**2005**) Voriconazole versus a regimen of amphotericin B followed by fluconazole for candidaemia in non-neutropenic patients: a randomised non-inferiority trial. *Lancet.*, **366**, 1435–1442.
62. Lortholary, O., Dannaoui, E., Raoux, D., Hoinard, D., Datry, A., Paugam, A., *et al.* (**2007**) *In vitro* susceptibility to posaconazole of 1903 yeast isolates recovered in France (2003–2006) and tested by the EUCAST method. *Antimicrob. Agents Chemother.*, 18 [Epub ahead of print].
63. Marr, K.A., White, T.C., van Burik, J.A.H. and Bowden, R.A. (**1997**) Development of fluconazole resistance in Candida albicans causing disseminated infection in a patient undergoing marrow transplantation. *CID*, **25**, 908–910.

64. Ostrosky-Zeichner, L. and Pappas, P.G. (**2006**) Invasive candidiasis in the intensive care unit. *Crit. Care. Med.*, **34**, 857–863.
65. Reboli, A.C., Rotstein, C., Pappas, P.G., Chapman, S.W., Kett, D.H., Kumar, D., *et al.* (**2007**) Anidulafungin versus Fluconazole for invasive candidiasis. *N. Engl. J. Med.*, **356**, 2472–2482.
66. Nucci, M. and Anaissie, E. (**2002**) Should vascular catheters be removed from all patients with candidemia? An evidence-based review. *CID*, **34**, 591–599.
67. Sachs, J. and Malaney, P. (**2002**) The economic and social burden of malaria. *Nature*, **415**, 680–685.
68. Jelinek, T., Schulte, C., Behrens, R., *et al.* (**2002**) Imported falciparum malaria in Europe: sentinel surveillance data from the European network on surveillance of imported infectious diseases. *Clin. Infect. Dis.*, **34**, 572–576.
69. Legros, F., Arnaud, A., El Mimouni, B. and Danis, M. (**2006**) Paludisme d'importation en France métropolitaine: données épidémiologiques 2001–2004. *Bull. Epidémiol. Hebdomadaire*, **32**, 235–236.
70. Holtz, T.H., Kachur, S.P., MacArthur, J.R., Roberts, J.M., Barber, A.M., Steketee, R.W. and Parise, M.E. (**2001**) Malaria surveillance – United States, **1998**. *Mor. Mortal. Wkly. Rep. CDC. Surveill. Summ.*, **50**, 1–20.
71. World Health Organization. (**1990**) Severe and complicated malaria. *Trans. R. Soc. Trop. Med. Hyg.*, **84**, 1–65.
72. World Health Organization. (**2000**) Severe falciparum malaria. *Trans. R. Soc. Trop. Med. Hyg.*, **94**, 1–90.
73. Correction. World Health Organization. (**2000**) Severe falciparum malaria. *Trans. R. Soc. Trop. Med. Hyg.*, **94**, 3.
74. Bruneel, F., Hocqueloux, L., Chevret, S., Regnier, P. and Vachon, F. (**1999**) Paludisme d'importation à *P. falciparum*. Quelle est la pertinence des critères de gravité de l'Organisation Mondiale de la Santé? *Méd. Malad. Infect.*, **29**(Suppl. 3), 345–355.
75. Bruneel, F., Hocqueloux, L., Alberti, C., *et al.* (**2003**) The clinical spectrum of severe imported falciparum malaria in the intensive care unit. *Am. J. Respir. Crit. Care. Med.*, **167**, 684–689.
76. White, N.J. and Ho, M. (**1992**) The pathophysiology of malaria. *Adv. Parasitol.*, **31**, 83–173.
77. Van der Heyde, H.C., Nolan, J., Combes, V., Gramaglia, I. and Grau, G.E. (**2006**) A unified hypothesis for the genesis of cerebral malaria: sequestration, inflammation and hemostasis leading to microcirculatory dysfunction. *Trends Parasitol.*, **22**, 503–508.
78. Dondorp, A.M., Kager, P.A., Vreeken, J. and White, N.J. (**2000**) Abnormal blood flow and red blood cell deformability in severe malaria. *Parasitol. Today*, **16**, 228–232.
79. Idro, R., Jenkins, N.E. and Newton, C.R. (**2005**) Pathogenesis, clinical faetures and neurological outcome of cerebral malaria. *Lancet. Neurol.*, **4**, 827–840.
80. Dondorp, A.M., Desakorn, V., Pongtavornpinyo, W., *et al.* (**2005**) Estimation of the total parasite biomass in acute falciparum malaria from plasma PfHRP2. *PloS. Med.*, **2**, e204.
81. Pain, A., Ferguson, D.J., Kai, O., *et al.* (**2001**) Platelet-mediated clumping of *Plasmodium falciparum*-infected erythrocytes is a common adhesive phenotype and is associated with severe malaria. *Proc. Natl. Acad. Sci. USA.*, **98**, 1805–1810.
82. Combes, V., Taylor, T.E., Juhan-Vague, I., *et al.* (**2004**) Circulating endothelial microparticles in Malawian children with severe falciparum malaria complicated with coma. *JAMA*, **291**, 2542–2544.
83. Kwiatkowski, D.P. and Luoni, G. (**2005**) Host genetic factors in resistance and susceptibility to

malaria, in *Molecular Approaches to Malaria*. ASM Press, USA.

84. Daily, J.P., Le Roch, K.G., Sarr, O., *et al.* (**2004**) *In vivo* transcriptional profiling of *Plasmodium falciparum*. *Malaria J.*, **3**, 30.
85. Daily, J.P., Le Roch, K.G., Sarr, O., *et al.* (**2005**) *In vivo* transcriptome of *Plasmodium falciparum* reveals overexpression of transcripts that encode surface proteins. *J. Infect. Dis.*, **191**, 1196–1203.
86. Jafari, S., Le Bras, J., Bouchaud, O. and Durand, R. (**2004**) *Plasmodium falciparum* clonal population dynamics during malaria treatment. *J. Infect. Dis.*, **189**, 195–203.
87. Olumese, P.E., Amodu, O.K., Björkman, A., *et al.* (**2002**) Chloroquine resistance of *Plasmodium falciparum* is associated with severity of disease in Nigerian children. *Trans. R. Soc. Trop. Med. Hyg.*, **96**, 418–420.
88. World Health Organization. *Guidelines for the Treatment of Malaria*, WHO, Geneva, **2006**.
89. Looareesuwan, S., Wilairatana, P., Krishna, S., *et al.* (**1995**) Magnetic resonance imaging of the brain in cerebral malaria. *Clin. Infect. Dis.*, **21**, 300–309.
90. Bruneel, F., Gachot, B., Timsit, J.F., *et al.* (**1997**) Shock complicating severe falciparum malaria in European adults. *Int. Care. Med.*, **23**, 698–701.
91. Hocqueloux, L., Bruneel, F., Lagorce Pages, C. and Vachon, F. (**2000**) Fatal invasive aspergillosis complicating severe Plasmodium falciparum malaria. *Clin. Infect. Dis.*, **30**, 940–942.
92. Maitland, K. and Newton, C.R. (**2005**) Acidosis of severe falciparum malaria: heading for a shock? *Trends Parasitol.*, **21**, 11–16.
93. Gachot, B., Wolff, M., Nissack, G., *et al.* (**1995**) Acute lung injury complicating imported Plasmodium falciparum malaria. *Chest*, **108**, 746–749.
94. Hien, T.T., Day, N.P.J., Phu, N.H., *et al.* (**1996**) A controlled trial of artemether or quinine in Vietnamese adults with severe falciparum malaria. *N. Engl. J. Med.*, **335**, 76–83.
95. Dondorp, A.M., Nosten, F., Stepniewska, K., *et al.* (**2005**) Artesunate versus quinine for treatment of severe falciparum malaria: a randomised trial. *Lancet.*, **366**, 717–725.
96. Bruneel, F., Gachot, B., Wolff, M., *et al.* (**2001**) Resurgence of blackwater fever in long-term European expatriates in Africa. Report of 21 cases and review. *Clin. Infect. Dis.*, **32**, 1133–1140.
97. White, N.J. (**2003**) The management of severe falciparum malaria. *Am. J. Resp. Crit. Care. Med.*, **167**, 673–674.
98. Lesi, A. and Meremikwu, M. (**2004**) High first dose quinine regimen for treating severe malaria. *Cochrane. Database Syst. Rev.*, (3), CD003341.
99. Artemether-quinine Meta-analysis Study Group. (**2001**) A meta-analysis using individual patient data of trials comparing artemether with quinine in the treatment of severe falciparum malaria. *Trans. R. Soc. Trop. Med. Hyg.*, **95**, 637–650.
100. Dellinger, R.P., Carlet, J.M., Masur, H., *et al.* (**2004**) Surviving Sepsis Campaign guidelines for management of severe sepsis and septic shock. *Crit. Care Med.*, **32**, 858–873.
101. Nau, A., Peytel, E. and Petrognani, R. (**2006**) Is there a place for activated protein C in the treatment of severe imported falciparum malaria in adult? *Ann. Fr. Anesth. Reanim.*, **25**, 1085–1086.
102. Bruneel, F., Hocqueloux, L., Legriel, S., *et al.* (**2007**) Activated protein C during severe imported falciparum malaria: look at benefit/risk ratio. *Ann. Fr. Anesth. Reanim.*, **26**, 263–264.
103. Pasvol, G. (**2005**) Management of severe malaria: interventions and controversies. *Infect. Dis. Clin. N. Am.*, **19**, 211–240.
104. Riddle, M.S., Jackson, J.L., Sanders, J.W. and Blazes, D.L. (**2002**) Exchange transfusion as an adjunct

therapy in severe *Plasmodium falciparum* malaria: a meta-analysis. *Clin. Infect. Dis.*, **34**, 1192–1198.

105. Kaiser, K., Texier, A., Ferrandiz, J., *et al.* (**2006**) Recombinant human erythropoietin prevents the death of mice during cerebral malaria. *J. Infect. Dis.*, **193**, 987–995.

PART III
Physiopathology of Sepsis and SIRS

Sepsis and Non-infectious Systemic Inflammation. Edited by J.-M. Cavaillon and C. Adrie

ISBN: 978-3-527-31935-0

8
Inflammatory Mediators

François Philippart, Jean-Marc Cavaillon and Charles Dinarello

8.1
Inflammatory Mediators in Sepsis

During sepsis, microbial products (see Chapters 5 and 6) induce gene expression, synthesis and release of inflammatory mediators by host cells (see Table 8.1). Each of these mediators can affect the production and activities of each other. For example, one of the most potent of these is prostaglandin E2 (PGE2), which is a vasodilator that inhibits the production of TNFα but augments the production of IL-6. Although the primary function of each of these mediators is to assist the host to eliminate life-threatening infections, these same mediators are also responsible for several deleterious effects on the host and eventual death in sepsis. Fortunately, nature provided mechanisms by which the agonists in sepsis are countered by inhibitors of inflammation in an attempt to limit the damage that occurs during the inflammatory storm of sepsis. For example, the cytokine IL-10 suppresses IL-1, TNFα and IFNγ production. As a result, the effect on the immune status is a result of the pro- and anti-inflammatory balance of the mediators.

8.1.1
Cytokines, Chemokines, and Growth Factors

Cytokines play a central role in the initiation of any inflammatory response. In response to pathogens and endogenous alarm signals, production of cytokines and chemokines rapidly takes place and facilitates activation of the immune cells that are required to fight infection. Most cytokines when given 24 h before lethal LPS were protective. These same cytokines and chemokines also contribute to the pathogenesis and lethal consequences of endotoxemia or septic shock. In septic patients, cytokine production can be high, to a level that is detectable in the circulation, where they are normally absent [1]. The concept of cytokine storm has been proposed to illustrate a devastating inflammatory

Sepsis and Non-infectious Systemic Inflammation. Edited by J.-M. Cavaillon and C. Adrie

ISBN: 978-3-527-31935-0

Table 8.1 Immune dysregulation during sepsis is characterized by an exacerbated production of pro-inflammatory mediators that lead to deleterious effects and lethality, and an exacerbated production of anti-inflammatory mediators that contribute to the induction of an immune suppressive status.

	Mediators contributing to tissue injury, organ or system dysfunction, and eventual to lethality[a]	Mediators favoring immune suppression
Cytokines	Tumor necrosis factor (TNF)	IL-10
	Interleukin-1 (IL-1)	IL-13
	IL-12, IL-15, IL-18, IL-27, IL-33	Transforming growth factor-β
	Gamma interferon (IFNγ), IFNβ	
	Granulocyte-macrophage colony-stimulating factor	
	Leukemia inhibitory factor (LIF)	
	Macrophage migration inhibitory factor (MIF)	
	Some chemokines: CXCL8 (IL-8), CCL5,	
	CXCR1 & 2 ligands, CCR1 ligands, CCR4 ligands[b]	
Growth factors	Vascular endothelial growth factor (VEGF)	–
Cell markers of stress	High Mobility Group Box 1 protein (HMGB1)	Heat shock proteins (HSP)
	Crystal of uric acid	
	S100	
Plasma factors	Ligand of TREM-1[c]	Ligand of TREM-2
	Anaphylatoxin C5a	
	Mannose-binding lectin (MBL)	
Lipid Mediators	Prostaglandins	Prostaglandins
	Leukotrienes	
	Platelet activating factor (PAF)	
	Oxidized phospholipids	
Hormones	–	Glucocorticoids
Neuromediator	Substance P	Adrenalin, Acetylcholine
	Neurokinin	α-melanocyte stimulating hormones
	Noradrenalin	Vasoactive intestinal peptide (VIP)
		Urocortin, Cortistatin, Adrenomodulin

(continued overleaf)

Table 8.1 *(continued).*

Enzymes	Cyclooxygenase-2, 5-Lipoxygenase Phospholipase A2, Metaloproteinase 9 Elastase, mast cell dipeptidyl peptidase I Glycogen synthase kinase-3 (GSK-3) Inducible nitrite oxide synthase (iNOS) Nicotinamide adenine dinucleotide phosphate (NADPH) oxidase	–
Coagulation factor & finibrolysis	Tissue factor, Thrombin Thrombin activable Fibrinolysis inhibitor	–
Free radicals	Nitric oxide (NO), O_2^-	
Purine nucleoside	Adenosine (via A_{2A} receptor)	Adenosine (via A_{2A} receptor)

[a] As demonstrated in animal models with the help of specific antibodies, inhibitors or antagonists, or with KO-mice.
[b] Either CCL17 or CCL22.
[c] Triggering receptor expressed on myeloid.

situation that is associated with critical illness and organ dysfunction. Similar observations have been documented in the absence of an infectious process in clinical or experimental settings where high concentrations of inflammatory cytokines are released [2]. For example, patients with acute pancreatitis can succumb to lethal cytokinemia in the absence of infection [3]. However, circulating cytokines are only the tip of the iceberg [4], and undetectable levels in the circulation does not indicate their absence [5]. Indeed, numerous circulating cells can effectively carry any cytokine present in the environment via their high affinity receptors, removing them from the fluid phase. The levels of most cytokines are correlated with each other, other markers of inflammation (e.g. C3a, lactate, C-reactive protein, creatinine, α1-anti-trypsin), circulating endotoxin, clinical status (shock, organ failure, fever), clinical scores and even with outcome. However, the most predictive value associated with clinical status is rather the persistence of detectable elevated levels of cytokines over time than an absolute high level: sustained levels of tumor necrosis factor-α (TNFα) are predictive of multiple organ failure [6], and sustained levels of interleukin-8 (IL-8) predict acute respiratory distress syndrome [7].

IL-1β and TNFα are synergistic in orchestrating the inflammatory response. TNF and IL-1 enhance procoagulant activity of vascular endothelial cells, activate neutrophils, and increase gene expression of adhesion molecules, which in turn worsens tissue injury during sepsis. TNFα has been shown to contribute to mortality in animal models infected with either Gram-positive or Gram-negative bacteria [8], and numerous studies in animal models have established that neutralization of TNFα was beneficial prior to the administration of lipopolysaccharide (LPS) or bacteria. Studies performed with knock-out mice may lead to different conclusions. For example, the major role of TNFα in LPS-induced mortality in TNFα-KO mice was only achieved when LPS was injected together with the hepatotoxic agent galactosamine [9]. Injection of LPS into IL-1β-deficient mice produces similar observations as in wild-type mice, probably because other IL-1-like cytokines (i.e. IL-1α, IL-18, IL-33) can compensate for the absence of IL-1β. Indeed, in contrast, in the absence of caspase-1, the maturation enzyme required to produce biologically active IL-1β, IL-18, and IL-33, mice are resistant to endotoxic shock. In the latter case, due to role of IL-18 in the production of IFNγ, mice deficient in capsase-1 are incapable of producing IFNγ. Mice deficient in IFNγ are also resistant to lethal endotoxemia. Although administration of LPS to healthy human subjects results in low but detectable levels of IL-1β and TNFα [10], IL-1β levels are rarely found in patients with sepsis and serum TNFα levels are poorly associated with outcome in septic patients. In some studies the long-lasting presence of detectable circulating TNFα levels correlates with mortality. In meningococcal sepsis, circulating TNFα correlates with pejorative prognosis [11].

IL-12, a heterodimeric cytokine composed of p40 and p35 chains, shares with IL-15 and IL-18 the capacity to induce IFNγ production by T- and NK cells. IL-12 is known to play a major role in defense against bacterial infection, and IL-12 deficiency decreases resistance to polymicrobial sepsis due to cecal ligature and puncture (CLP), a model of peritonitis and sepsis [12]. Transgenic mice that over-express IL-15 are resistant to an otherwise lethal challenge with *Escherichia coli*, although bacterial burden and serum levels of TNFα are similar in non-transgenic mice [13]. Neutralization of IL-18 protected mice against lethal *E. coli* or *Salmonella typhimurium* endotoxemia [14], and IL-18-deficient mice exhibit decreased sensitivity toward LPS-induced shock [15]. Most interestingly, it has been suggested that high levels of circulating IL-18 are associated with Gram-positive infections whereas lower levels were reported for Gram-negative infection [16, 17]. IL-23 shares with IL-12 its p40 subunit. IL-23 is involved in chronic inflammatory response to infection that involves adaptive immune cells. Interestingly, IL-23 seems to be produced early in response to *Pseudomonas aeruginosa*, and its blockade by antibodies is associated with dose-dependent improvement in survival [18]. Similarly to IL-23, IL-27 is a heterodimeric cytokine consisting of an IL-12 p40-related EBV-induced gene 3 (EBI3) and a p28 chain. IL-27 mRNA is expressed after CLP in lungs and spleen. EBI3-deficient mice are resistant to CLP-induced lethality, and neutralization of IL-27 protects against sepsis [19]. The last member of the IL-1 family, IL-33 also contributes to the LPS-induced lethality. The receptor for IL-33 is ST2 (an orphan receptor for a while) and soluble IL-33 receptor (sST2) protects against injection of LPS and reduces the levels of circulating cytokines; in contrast to agonistic anti-ST2 antibodies, that mimic IL-33 activity, aggravate outcome and enhance levels of plasma cytokines [20].

IFNγ possesses well-known pro-inflammatory and antibacterial properties. For example, IFNγ enhances phagocytosis and free radical production and increases bactericidal activity of macrophages and neutrophils. However, blockade of IFNγ is associated with reduced lethality in animal models of sepsis, associated with a decreased bacterial load and lower systemic inflammation [21]. This phenomenon was partially attributed to an increase in fibrin deposition that could be a protective factor against pathogen dissemination [21]. Echtenacher *et al.* [22] reported that concomitant injection of IFNγ at the time of CLP increases lethality. Indeed, simultaneous injection of a low dose of IFNγ together with a non-lethal dose of LPS or TNFα results in lethality. Of note, IFNγ stimulates cells to release high mobility group box-1 (HMGB-1) from monocytes/macrophages *in vitro* (see below). In a CLP model of sepsis, inhibition of IFNγ is associated with decreased HMGB-1 expression in the peritoneum and increased survival of experimental animals [23].

Macrophage migration inhibitory factor (MIF), produced in response to TNFα and IFNγ, is also induced by glucocorticoids. MIF is preformed within leukocytes and can also be produced by the pituitary gland. MIF directly or indirectly promotes the production or expression of a large panel of pro-inflammatory cytokines. MIF-deficient mice display a significant improvement in survival after challenge with either LPS or *Staphylococcus aureus* enterotoxin B. During peritonitis, MIF increases rapidly in the peritoneal space, and its inhibition protects mice from death [24]. In a two-hit model, with chemically-induced pancreatitis followed by an injection of endotoxin, MIF is associated with a pejorative evolution of acute lung injury [25]. Increased levels of MIF have been detected in the blood of patients with severe sepsis or septic shock and correlate with severity [24].

Chemokines control leukocyte trafficking during homeostasis as well as inflammation [26, 27]. Clinical studies identified elevated levels of both CXC (e.g. CXCL8 [IL-8]), and CC chemokines such as monocyte chemotactic protein (MCP)-1 (CCL2) associated with human sepsis and acute lung injury. Elevated plasma levels of nearly all chemokines except RANTES (CCL5) correlate with poor outcome [1]. In sepsis models, macrophage inflammatory protein-2 (CXCL2) increases severity, whereas blocking its activity decreases mortality. Similarly, CXCL-receptor inhibitor improves survival in sepsis [28] and reduces HMGB-1-induced lung inflammation and injury. In a sepsis model of peritonitis, the neutralization of CCL2 decreases the production of IL-13 and IL-12 and increases the production of TNFα and IL-10. In an LPS-induced lethality model, administration of MCP-1 protects mice, whereas the neutralization of MCP-1 with antibodies increases mortality. Indeed, the relative contribution of chemokines to the pathophysiologic events associated with sepsis is linked to their capacity to recruit inflammatory cells within tissues. In contrast, their presence within the bloodstream may limit the inflammatory process by desensitizing circulating cells to further chemoattractant signals.

The relationship between the production of cytokines in sepsis or after injection of LPS and the requirement of various transcription factors has been addressed using mice in which a specific gene has been deleted at the level of the germline. As such, mice deficient in certain genes can provide a great deal of information regarding the role of a particular gene in the pathogenesis of sepsis, but such mice can have a dysfunctional immune and endocrine system as shown in IL-18 deficient mice [29]. Nevertheless, as summarized in Table 8.2, it appears that enhanced survival after LPS administration, following the deletion of certain transcription factors (e.g. IRF-2, HIF-1α) can been associated with either enhanced or reduced levels of TNFα. Similarly, the deletions of other transcription factor (e.g. HSF-1, ATF-2) that are associated with a reduced survival to LPS can be either associated with enhanced or reduced expression of TNFα. Enhanced expression of TNFα does not always parallel an increase in levels of other cytokines (e.g. IRF-2 −/−). To add to the complexity, different results can be obtained with different experimental models. For example, the deletion of STAT-4 and STAT-6 transcription factors allows an enhanced survival in the CLP model but a reduced survival after LPS. These latter results illustrate that the same cytokines required to fight infectious process can be associated with severe side-effects and death.

From gene deletion studies it seems that the detrimental effect of cytokines on the host during an infectious event is due to overproduction or alternatively, receptor hypersensitivity. For example, IFNγ increases the receptors for many cytokines and it remains possible that the role of IFNγ in sepsis is due to an increase in the surface expression of pro-inflammatory cytokines.

At the time of writing, little has been demonstrated concerning the role of growth factors in sepsis, except for vascular endothelial growth factor (VEGF). Yano *et al.* [30] showed an increased level of VEGF in patients with sepsis. They also demonstrated that over-expression of soluble Flt-1 (VEGF receptor) limited inflammatory markers in a mouse model of endotoxemia and increased survival after CLP or LPS injection.

8.1.2 Complement System

Without question, the complement system plays a pivotal role during sepsis. The major purpose of the activation of the complement system through its three pathways is to assist the host in removal and/or destruction of pathogens. However, some of the complement-derived compounds released after activation of complement favor the inflammatory process, as illustrated by their ability to increase vascular permeability and contribute to granulocyte and monocyte recruitment within inflamed tissues.

Table 8.2 Role of transcription factor in cytokine induction and survival in experimental endotoxinemia or sepsis assessed with KO mice.

Transcription Factor	Experimental Model	Survival KO versus WT	TNF Levels plasma	TNF Levels tissues	Other cytokines	References
IRF-2 −/−	LPS challenge	↑	↑	↑ (liver)	IFNγ, IL-1β, ↓ IL-6, IL-12 ↓	Cuesta *et al.* J. Immunol. 2003, 170, 5739 [126]
STAT-1 −/−	LPS challenge	↑	nd		IFNβ ↓	Karaghiosoff *et al.* Nature Immunol. 2003, 4, 473 [127]
HIF-1α − /−	LPS challenge	↑	↓		IL-1β, IL-6, IL-12 ↓ IL-10, IFNγ ⇔	Peyssonnaux *et al.* J. Immunol. 2007, 178, 7516 [128]

(*continued overleaf*)

Table 8.2 *(continued)*.

Transcription Factor	Experimental Model	Survival KO versus WT	TNF Levels plasma	tissues	Other cytokines	References
STAT-4 −/−	CLP	↑		nd	IL-10 (liver) ↑	Matsukawa *et al.* J. Exp. Med. 2001, 193, 679 [129]
					MIP-2, KC ↓ (liver, lung, kidney)	
	LPS challenge	↓	⇔		IL-6, MCP-1, MIP1α ⇔	Lentsch *et al.* J. Clin. Invest. 2001, 108, 1475 [130]
					(liver, lung)	
STAT-6 −/−	CLP	↑		↑	IL-12, MDC, C10 ↑	Matsukawa *et al.* J. Exp. Med. 2001, 193, 679 [129]
				(peritoneum)	(peritoneum)	
	LPS challenge	↓	↑		IL-6, MCP-1, MIP1α ↑	Lentsch *et al.* J. Clin. Invest. 2001, 108, 1475 [130]

					(liver, lung)	
STAT-3 −/−	CLP	↓	↑	↑	IL-1β, IL-6, IL-10 ↑	Matsukawa *et al.* J. Immunol. 2003, 171, 6198 [131]
(MØ + PMN)				(peritoneum)	IL-12, IFNγ ↑	
ATF-2 −/−	LPS challenge	↓		↓	IL-1β, IL-6, KC ↓	Reimold *et al.* Intern. Immunol. 2001, 13, 241 [132]
				(lung, kidney, spleen)		
HSF-1 −/−	LPS challenge	↓	↑		IL-10 ⇔	Xiao *et al* EMBO J. 1999, 18, 5943 [133]
NF-κB	LPS challenge	↓	↑		IFNγ ↑	Gadjeva *et al.* J. Immunol. 2004, 173, 5786 [134]
p50−/− or p50−/− p65+/−					(spleen)	

Anaphylatoxins C3a and C5a are involved in infection control and inflammatory regulation. Both pro- and anti-inflammatory properties have been attributed to C3a [31]. Anaphylatoxin C5a favors the synthesis and release of pro-inflammatory cytokines such as TNFα, IL-1β, IL-6, and IL-8 from human leukocytes [32], and can enhance phagocytosis, induce oxidative burst, and favor the release of granular enzymes from neutrophils. In the lung, C5a enhances inflammatory response, particularly by its action on alveolar epithelial cells. C5a has also been found to be a vasodilator, to enhance the expression of adhesion molecules, and to activate the coagulation system. Furthermore, C5a has a major role in cardiac dysfunction during sepsis, participating in suppressive cardiomyopathy [33]. Of note, IL-1β and TNFα do the same but are independent of C5a action. Excessive production of C5a may compromise the host defenses, and a high level of plasma C5a in sepsis patients correlates with poor outcomes [34]. In some models of shock, C5-deficient animals are more resistant to the effects of TNFα and LPS infusions than controls. Studies of sepsis in animal models have shown that blockade of C5a attenuates physiologic perturbations and prevents the development of acute respiratory distress syndrome and multiple organ failure. After CLP, an increased expression of C5a receptor occurs in most tissues including lungs, liver, kidneys and heart. Most importantly, blockade of C5a receptor is highly protective [34].

C4- or C3-deficient mice are more susceptible to LPS, with an increased consumption of C1q inhibitor. C1q inhibitor is a plasma glycoprotein, which participates in both complement regulation and contact system activation. Interestingly, supplementation with C1q inhibitor prevents the endotoxin-induced increase in vascular permeability [35] and protects against LPS-induced mortality. In addition, C1q inhibitor protects mice by interacting directly with LPS [36]. Thus, despite the fact that the circulating inactivated form of C1q inhibitor is increased in sepsis and is associated with a pejorative evolution of sepsis, animal models have suggested a protective role for this molecule.

8.1.3
Coagulation and Fibrinolysis

Disorders of coagulation are common in sepsis, and 30 to 50% of patients with the most severe clinical manifestations have disseminated intravascular coagulation. Tissue factor is the link between inflammation and coagulation by interacting with factor VII and increasing the production of fibrin through activation of thrombin. In healthy volunteers challenged with LPS and in patients with sepsis, tissue factor expression is enhanced, and its increased membrane expression on monocytes appears to be a prognostic factor of poor outcome in sepsis [37]. Factor VII co-localizes with tissue factor, which correlates with fibrin deposition, mainly at bifurcations of blood vessels. Thrombin signaling in endothelial cells results in changes in cell shape, cell permeability, proliferative responses, and leukocyte adhesion. Thrombin is inactivated by anti-thrombin, but during sepsis, levels of anti-thrombin are reduced. Numerous animal models of sepsis have revealed the beneficial effects of anti-thrombin. Factor XIII is involved in fibrin stabilization. Whereas factor XIII subunit A and its cross-linking are decreased in septic patients, the specific activity of factor XIII is increased in these patients [38]. Neither subunit A nor cross-linking activity is associated with severity of the patient's condition; nevertheless, the specific activity of factor XIII is strongly associated with severity and fatality in septic groups [38].

The study by Bernard *et al.* [39] reported that activated protein C (APC) displays beneficial effects in human septic shock and resulted in increased studies on APC and its role in sepsis. However, clinical studies subsequent to regulatory approval have been inconsistent with respect to the role of APC in human sepsis. A clear role for APC has yet to be established and it is also clear that more than one mechanism may account for the action of APC in human sepsis. Protein C is synthesized by the liver and circulates as an inactive zymogene. During sepsis, protein C synthesis is reduced, and endothelial shedding is associated with a decrease in thrombomodulin expression. These phenomena lead to a significant decrease in APC, which eventually results in a prothrombotic state. Decreased levels of protein C during sepsis are associated with a poor outcome in patients [40]. It has been postulated that protein C plays a central role in linking inflammatory and coagulation processes. APC seems to possess anti-inflammatory properties and counteracts the inhibition of fibrinolysis [41, 42].

Although coagulation is increased during sepsis, the mechanisms that favor fibrinolysis are reduced. Thrombin-activatable fibrinolysis inhibitor (TAFI) is an inhibitor of the fibrinolytic system. Much of the regulation of TAFI is due to IL-1β [43]. TAFI is able to inhibit C3a and C5a. In mice, administration of LPS or *Escherichia coli* induces an enhanced plasma TAFI activity. On the other hand, TAFI deficiency modifies neither coagulation markers during sepsis nor fibrin deposition in liver and lung tissue. However, TAFI-deficient mice are protected from liver necrosis during peritonitis. In humans, TAFI was shown to be decreased in plasma from healthy subjects injected with LPS [44]. Plasminogen activator inhibitor-1 is another inhibitor of fibrinolysis, which acts by inhibiting the transformation of plasminogen into active plasmin. Sepsis is associated with increased levels of plasminogen activator inhibitor-1, which seem to be a pejorative factor in patients with sepsis complicated by intravascular coagulation [45] but a good prognosis marker during pneumonia [46].

8.1.4
Proteases

During sepsis, proteases are released by activated leukocytes and play an important role in the host inflammatory response. Their role seems dependent on the pathogen. For example, mice deficient in neutrophil elastase are more susceptible to Gram-negative peritoneal infections, whereas no difference in mortality was shown during peritoneal Gram-positive infection. On the other hand, elastase is associated with organ dysfunction, because its inhibition decreases this phenomenon [47]. Among proteases, matrix metalloproteinases (MMPs) are known to be involved in tissue remodeling (i.e. the degradation and remodeling of all components of extracellular matrix) and inflammatory processes. Among them, MMP-9 is of paramount importance during sepsis. Serum levels of MMP-9 have been shown to increase rapidly in an *E. coli*-induced bacteremia in baboons and following injection of endotoxin in healthy humans. These increases are observed not only in serum, but also in lung and liver tissues as well as in the peritoneal fluid of CLP peritonitis [48]. Moreover, MMP-9 levels correlate with severity and mortality of sepsis in humans [49]. Interestingly, whereas MMP-9 deficiency protects mice against mortality in an endotoxin model, the deficiency is associated with an enhancement of bacterial growth and bacterial dissemination following CLP [48]. Other enzymes can also contribute to the deleterious effects observed during sepsis (Table 8.1).

8.1.5 Lipid Mediators

Pro-inflammatory cytokines induce the synthesis of phospholipase A2 (PLA2), inducible cyclooxygenase-2, 5′-lipooxygenase, and acetyltransferase, which contribute to eicosanoids synthesis. Leukotrienes, 5′-lipoxygenase products, are increased in endotoxemic and septic animals. These factors promote inflammation, alter vasomotor tone, and increase blood flow and vascular permeability. Blocking the synthesis of leukotrienes or using 5′-lipoxygenase-deficient mice allowed researchers to demonstrate the deleterious effect of this eicosanoid in a mouse sepsis model [50]. Mice deficient in PLA2 are resistant to lethal endotoxemia, and *in vivo* inhibition of PLA2 decreases neutrophil infiltration of the lungs and deterioration of gas exchanges during endotoxin or zymosan challenge in mice. Cyclooxygenase-2-deficient mice are also resistant to endotoxin-induced inflammation and death [51]. There is extensive literature on the ability of cyclooxygenase inhibitors to reduce LPS-induced side effects and even lethality. A clinical study in patients with sepsis, showed that treatment with ibuprofen reduces levels of prostacyclin and thromboxane and decreases fever, tachycardia, oxygen consumption, and lactic acidosis, but it did not prevent the development of shock or acute respiratory distress syndrome and did not improve survival [52].

Platelet-activating factor (PAF) is released by a large number of cell types, such as platelets, endothelial cells, macrophages, and neutrophils. Blood levels of PAF are elevated during septic shock, whereas the activity of PAF-acetylhydrolase, its inhibitor, decreases in humans during experimental endotoxemia or sepsis [53]. Inhibition of PAF before endotoxin challenge in healthy humans is associated with a decreased intensity of symptoms and decreased pro-inflammatory cytokine levels. Conversely, transgenic mice that over-express PAF are more susceptible to LPS challenge, but surprisingly, mice lacking PAF receptors display a similar susceptibility to endotoxin as their wild-type counterparts. In human sepsis, targeting PAF has failed to demonstrate any beneficial effects.

8.1.6
Nitric Oxide

Nitric oxide (NO) involved in sepsis is produced by inducible NO synthase (iNOS), an enzyme produced in response to endotoxin or inflammatory cytokines. Enhanced iNOS activity in inflamed tissues and vessel walls of sepsis patients has been demonstrated [54]. NO production has been involved in many pathophysiologic processes during sepsis or in endotoxemia models, and NO is rendered responsible for multi-organ failure [55]. On the other hand, NO was shown to reduce neutrophil migration to infected tissues by inhibiting leukocyte rolling. iNOS-deficient mice exhibit a high mortality rate after *S. aureus* infection or during polymicrobial sepsis [56]. iNOS deficiency is sometimes associated with improvement in metabolic perturbations, such as acidosis and hypotension [57]. In contrast, iNOS-deficient mice and the wild-type group have similar sensitivities to LPS. NO and coagulation are interrelated during inflammation and sepsis. Inhibition of iNOS was associated with a decrease in tissular plasminogen activator (tPA). One of the major deleterious effects of NO is probably its ability to alter epithelial tight junctions, responsible for pulmonary, liver, and gut dysfunction in endotoxemic mice [58].

8.1.7
Cellular Markers of Stress and Soluble Cell Surface Markers

HMGB-1 is a nuclear protein present in almost all eukaryotic cells, acting as a transcription factor-like protein regulating the expression of several genes associated with neural growth and differentiation. When released from the cell, likely by cell death, HMGB-1 binds to RAGE and TLR4, inducing IL-1β and TNFα. As such, HMGB-1 is also a late mediator in endotoxemia and sepsis [59]. Levels of HMGB-1 are correlated with severity during pneumonia [60], and other sepsis in humans. Passive immunization with neutralizing anti-HMGB-1 antibodies prevents lethality from sepsis, even when administrated 24 h after the induction of peritonitis [59]. HMGB-1 mediates hepatic injury after murine liver ischemia–reperfusion in a Toll-like receptor 4-dependent fashion [61]. Importantly, HMGB-1 appears to be the link between the occurrence of apoptosis, organ damage and lethality in sepsis [62].

Triggering receptor expression on myeloid cells (TREM)-1 is expressed on the surface of monocytes and neutrophils. Its expression is upregulated by LPS but its ligand remains to be identified. Activation of TREM-1 leads to synergy with different microbial-derived products [63]. This protein may be released from the cell surface, and large amounts of soluble TREM-1 are found in the body fluids of sepsis patients. Soluble TREM-1 has been suggested to be a marker of infection, although recent studies have questioned that specificity [64].

CD163 is a hemoglobin scavenger receptor, which exists exclusively on monocytes and macrophages and which is shed after stimulation with LPS. Accordingly, the soluble form of CD163 is increased in plasma during sepsis. It plays a role in the anti-inflammatory response [65], as do heat shock proteins released by stressed cells.

8.1.8 Lipoproteins and Lipopolysaccharide Inhibitors

Besides their function in lipid transport and metabolism, lipoproteins play a role during sepsis. LPS interacts with lipoproteins and is neutralized and shuttled to the liver. Apolipoprotein (apo) E-deficient mice are more susceptible to LPS or *Klebsiella pneumonia* challenge. In addition, apoA-I can protect mice against mortality [66].

CD14 exists as a membrane form and is part of the LPS receptor. It also exists as a soluble form known to favor the responsiveness to LPS of cells that lack membrane CD14. However, the role of soluble CD14 remains controversial. In a transgenic mouse model expressing different copy numbers of the human CD14 gene, it was reported that animals expressing the highest levels of soluble CD14 were the less sensitive to LPS-induced lethality [67]. In human sepsis, several studies have shown that increased levels of circulating soluble CD14 correlates with mortality.

LPS-binding protein (LBP) is an acute phase protein, levels of which are enhanced during sepsis. Despite its ability to facilitate the binding of LPS to CD14, LBP seems to protect mice from septic shock during Gram-negative bacterial challenge, whereas LBP-deficient mice are more susceptible to *Salmonella* challenge than wild-type mice. Bactericidal/permeability-increasing protein (BPI) shares more than 40% homology with LBP. It attenuates LPS-mediated endothelial damage and IL-6 production, as well as LPS-mediated NO, IL-1β, IL-6, IL-8, and TNFα production by macrophages or in cultures of whole blood. Lactoferrin, lysozyme, hemoglobin, surfactant proteins A and D, antimicrobial cationic peptide (CAP18 and CAP37) also bind endotoxins and modify their capacity to initiate an inflammatory process.

8.1.9 Anti-inflammatory Mediators

Anti-inflammatory cytokines downregulate the production of pro-inflammatory cytokines, and favor the release of soluble TNFα receptors and the production of IL-1 receptor antagonist (IL-1Ra). These effects are evident for IL-10, IL-4, IL-13, transforming growth factor-β, and IFNα. Each has been shown to protect animals from sepsis or endotoxin-induced shock. However, they probably also contribute to the altered immune status observed in septic patients. Although sepsis is associated with a large production of pro-inflammatory cytokines at sites of infection, the systemic response could be principally anti-inflammatory [68]. Indeed, high levels of anti-inflammatory cytokines such as IL-10 [69] or TGFβ [70] are found in the plasma of sepsis patients. Other anti-inflammatory cytokines such as IL-4 or IL-13 remained undetectable in the plasma of sepsis patients. High serum levels of soluble TNF receptor [71], soluble IL-1 receptor [72], IL-1Ra [73], and IL-18 binding protein [74] are found in plasma of such patients, with the highest levels often associated with the poorest of outcomes. Enhanced levels of circulating IL-1Ra are detected in human volunteers injected with LPS and exceed IL-1β levels by a factor of 100, a molar ratio sufficient to counteract the activity of IL-1 [75]. IL-1Ra-deficient mice are highly susceptible to endotoxin-induced death, and conversely, mice that over-express IL-1Ra are relatively protected. IL-1Ra prevents *E. coli*-induced shock in rabbits [76]. Unfortunately, human trials have failed to find any beneficial effects for IL-1Ra. Because of its capacity to block the IL-1 receptor and thus counteract the effect of IL-1 that contributes to its role in the infectious process as well to the deleterious inflammatory process, IL-1Ra can be both beneficial or deleterious depending upon the model [77].

IL-6 is unquestionably one of the best cytokine markers of disease severity. In both infectious as well as non-infectious conditions, levels of IL-6 correlate with disease indicators. IL-6 induces the release of cortisol, IL-10, IL-1Ra, and hepatic acute-phase proteins. Accordingly, it can be considered as an anti-inflammatory cytokine. In fact, mice deficient in IL-6 have increased levels of TNFα and, in general, IL-6-deficient mice are not protected from inflammatory processes. It is unclear whether IL-6 is a direct agonist in sepsis. However, IL-6 also possesses some pro-inflammatory properties, as illustrated by its role in myocardial depression [78]. Yet, IL-6 administered to humans in high doses induces fever, leukocytosis and thombocytosis, yet has no effect on blood pressure. In contrast, IL-1β or TNFα administered to humans at exceedingly low doses of nanograms/kg (1000-fold less than IL-6) results in frank hypotension and hemodynamic shock.

Heme oxygenase (HO)-1 is induced by IL-10 and plays a protective role against oxidative stress. HO-1-deficient mice are more sensitive to endotoxin than wild-type mice. HO-1 has been shown to exert beneficial effects in ischemia/reperfusion and hemorrhagic shock models.

Angiopoietins are a class of angiogenic growth factors that act selectively on endothelial cells. Angiopoietin (Ang)-1 possesses anti-inflammatory properties reducing leukocyte–endothelial cell adhesion and transmigration, and it protects mice from endotoxin-induced mortality [79]. Ang-2 is another angiopoietin, which recognizes the same receptor as Ang-1 but antagonizes this latter molecule by blocking signal transduction. Ang-2 levels are also increased during sepsis in humans, and interestingly, these levels correlate with disease severity scores and with levels of inflammatory cytokines, C-reactive protein, and procalcitonin [80].

Angiotensin converting enzyme (ACE)-2 inactivates angiotensin II and is a negative regulator of the renin–angiotensin system. Interestingly, ACE-2 is involved in the limitation of lung injury in a peritonitis model of sepsis, in contrast with ACE, which seems to worsen lung disease in the same conditions [81]. It is worth mentioning that ACE-inhibitors suppress IL-1β and TNFα synthesis at the posttranscriptional level [82].

Adenosine acts on specific A2 receptors and inhibits numerous neutrophil functions such as phagocytosis, generation of superoxide anion, arachidonic acid release, leukotriene B4 biosynthesis, and adhesion to endothelial cells. Adenosine also inhibits IL-12 and TNF production by macrophages, and the nucleotide seems to play a protective role against tissue damage. Finally, leptin, a homeostatic protein involved in body weight and satiety, has been shown to increase during LPS challenge, probably in response to IL-1β. Leptin has a protective role in mice injected with LPS or TNFα [83].

8.1.10 Neuromediators

With the exception of substance P, neurokinins and norepinephrine, most neuromediators dampen the inflammatory process. Substance P and neurokinin A derive from the preprotachykinin-A gene. Mortality due to CLP is decreased and the onset of mortality is delayed in preprotachykinin-A-deficient mice compared to wild-type mice [84]. Norepinephrine increases TNFα production via α2-adrenergic receptors, whereas epinephrine via β2-adrenergic receptors decreases production of TNF, IL-6, and NO in response to LPS and potentiates IL-10 production [85].

Adrenomedullin is a neuropeptide structurally related to corticotropin-releasing factor, which increases in response to LPS injection with an enhanced expression in both plasma and viscera. Adrenomedullin inhibits TNFα expression [86] and stimulates IL-6 production by macrophages in response to LPS. Enhanced levels of plasma of calcitonin gene-related peptide (CGRP), neuropeptide Y (NPY) are elevated in sepsis patients and controversial data have been reported for substance P. Increasingly, observations have implicated the vagus nerve in the downregulation of inflammation. Borovikova *et al.* [87] established the key role played by acetylcholine and showed that the α7-nicotinic receptor contributes to the central modulation and integration of vagal regulation during inflammation. On the other hand, peripheral muscarinic receptors do not have such anti-inflammatory properties [88].

8.2 Inflammatory Mediators in Non-infectious SIRS

8.2.1 Sepsis and Non-infectious SIRS share Similar Inflammatory Mediators

Polly Matzinger revisited the definition of immunology when she introduced the concept of danger [89]. She postulated that *endogenous* danger signals generated by host cells in response to different type of stress could initiate the immune response. This concept has now been widely accepted, particularly after it was demonstrated that the same pattern recognition receptors (PRRs) first identified by their specific recognition of pathogen associated molecular patterns (PAMPs), were similarly capable of sensing endogenous signal dangers. As summarized in Figure 8.1, necrotic cells and injured tissues release endogenous molecules that directly interact with PRRs. These endogenous mediators are known as either "alarmins" [90] or "damage associated molecular patterns" (DAMPs) [91], new names given to an old concept. As a consequence, it not surprising that most cytokines induced after the stimulation of PRRs are similar, whatever the stimuli, a microbial ligand or an endogenous molecule. Then, because cytokines orchestrate the inflammatory process, most other inflammatory mediators previously detected during sepsis are similarly found during non-infectious SIRS.

In ICU patients many situations can mimic infection, although no pathogen is involved. Indeed, many clinical settings associated with tissue damage, hypoxia, or ischemia will lead to systemic inflammatory response. Thus, many

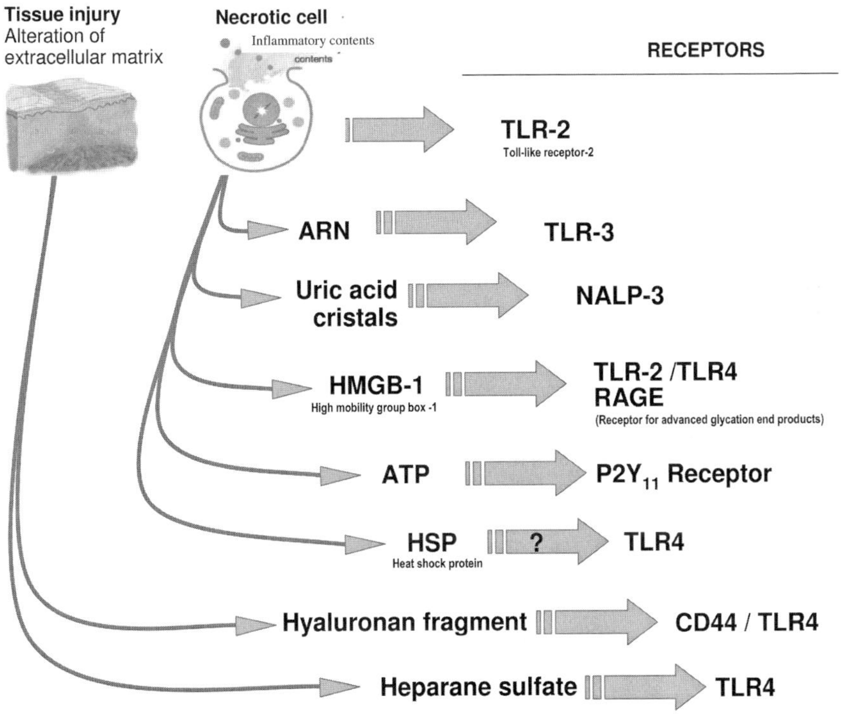

Figure 8.1 Sterile inflammation: numerous "alarmins" or "damage associated molecular patterns" are released after tissue injury and cell necrosis. These molecules share similar sensors (pattern recognition receptors) with microbial compounds. Their ligation to their respective receptors activates the release of inflammatory cytokines.

non-infectious stressful conditions could be responsible for local manifestations of inflammation and fever, and can lead to shock, organ failure and even multi-organ failure known as multi-organ dysfunction syndrome (MODS) [3]. Similar systemic complications accompany all these "sepsis-like" inflammatory responses [3, 92, 93]. Among these clinical situations, burns and trauma should be mentioned, in which inflammatory process respond to an external source of damage and will take part in wound healing; ischemia (e.g. stroke, acute myocardial infarction, peripheral vascular obstruction) during which tissular anoxia will trigger inflammatory response, is another frequent situation as are pancreatitis or surgery. Surgical intervention is associated with tissular injury and can be associated with significant blood loss [94]. It is worth noting, that in contrast to other stressful situations, surgery is performed under anesthesia which may interfere with the inflammatory response. As expected, length and extent of tissular aggression is correlated with the intensity of the inflammatory response [95–97].

8.2.2 Are Inflammatory Mediators Markers of Evolution and/or Infection?

Similarly to sepsis, the intensity of the inflammatory process is associated with high levels of inflammatory mediators that correlate with mortality [98–100], clinical evolution [101–103], and organ dysfunction [104–110]. In certain types of surgery the length of aortic crossclamping in vascular or cardiac surgery [95, 96, 111], or length of extracorporeal circulation [112] are additional events that contribute to the amplitude of the inflammatory process. In sepsis, additional parameters interfere with the prediction of outcome. These are linked to the type of pathogen and adequate antibiotherapy. It is noteworthy that in most ICU patients, inflammatory response is associated with an increased chance of developing an infection [113–115].

It has been postulated that certain inflammatory mediators are markers of infection [116]. Among those, C-reactive protein [117], procalcitonin [118] and soluble TREM-1 [119] are the most popular. However, none of them seems to be really characteristic of infectious process [3, 92, 93, 120–123]. Of particular interest, among mechanisms responsible for inflammatory response is the notion of translocation, a process that allows bacteria from the digestive flora or fragments of these microbes to move from the digestive lumen to the blood circulation through the epithelial gut barrier or via the lymphatic system. Under insufficient vascular perfusion or under the effect of inflammatory mediators (e.g. NO, HMGB-1), the alteration in the epithelial barrier is responsible for an impermeability weakness allowing bacterial fragment dissemination. LPS is the most frequently identified PAMPs within the blood stream. Cabié *et al.* [124] showed that gut manipulation and aortic clamping during vascular surgery led to the detection of LPS in the plasma of non-septic patients. During abdominal vascular surgery, in resuscitated cardiac arrest patients, and other types of ICU patients, LPS and probably numerous other PAMPs can trigger inflammatory response, in conditions identical to those reported during sepsis. These mechanisms may explain the absence of notable differences between infectious and non-infectious situations.

8.3 Conclusion

Syndromes associated with sepsis are due to the infection itself as well as the host response to the pathogens. Interestingly, during non-infectious inflammatory processes, "alarmins" and most probably endogenous PAMPs seem to be the cornerstones of the host response. Thus, in both situations the host response is mediated through a similar complex network of pro- and anti-inflammatory mediators. Although each pro-inflammatory mediator has its own potential deleterious effect, the synergy between them and with the microorganisms (or possibly the endogenous compounds) is responsible for the occurrence of organ dysfunction and eventually death. Despite positive results in animal models, the therapeutic targeting of these mediators has been disappointing in human settings. A meta-analysis by Natanson *et al.* examined the large number of placebo-controlled studies in sepsis [125]. The conclusions were that only patients with the highest likelihood of death benefited in the anti-mediator trials. Clearly, the trials were comprised of patients with low, moderate and high disease severity and thus the patient population was heterogeneous. The analysis by Natanson supports the concept that animal models show benefit because they are also severe (most are dead within 1 or 2 days) (See also Chapter 17). Therefore, the failure of these trials does not indicate that these mediators are not involved in the syndromes associated with human sepsis and non-infectious SIRS. Further knowledge is required to better appreciate the respective contribution of pro- and anti-inflammatory mediators to the side effects associated with the natural host response. The control of sepsis, and non-inflammatory SIRS, with respect to inflammation equilibrium should be kept in mind for the elaboration of therapeutic interventions in the future.

References

1. Cavaillon, J.M., Adib-Conquy, M., Fitting, C., Adrie, C. and Payen, D. (**2003**) Cytokine cascade in sepsis. *Scand. J. Infect. Dis.*, **35**(9), 535–44.
2. Suntharalingam, G., Perry, M.R., Ward, S., Brett, S.J., Castello-Cortes, A., Brunner, M.D., *et al.* (**2006**) Cytokine storm in a phase 1 trial of the anti-CD28 monoclonal antibody TGN1412. *N. Engl. J. Med.*, **355**(10), 1018–28.
3. Granger, J. and Remick, D. (**2005**) Acute pancreatitis: models, markers, and mediators. *Shock*, **24**(Suppl 1), 45–51.

4. Cavaillon, J.M., Munoz, C., Fitting, C., Misset, B. and Carlet, J. (**1992**) Circulating cytokines: the tip of the iceberg?. *Circ. Shock*, **38**(2), 145–52.
5. Munoz, C., Misset, B., Fitting, C., Bleriot, J.P., Carlet, J. and Cavaillon, J.M. (**1991**) Dissociation between plasma and monocyte-associated cytokines during sepsis. *Eur. J. Immunol.*, **21**(9), 2177–84.
6. Pinsky, M.R., Vincent, J.L., Deviere, J., Alegre, M., Kahn, R.J. and Dupont, E. (**1993**) Serum cytokine levels in human septic shock. Relation to multiple-system organ failure and mortality. *Chest*, **103**(2), 565–75.
7. Hildebrand, F., Stuhrmann, M., van Griensven, M., Meier, S., Hasenkamp, S., Krettek, C., *et al.* (**2007**) Association of IL-8-251A/T polymorphism with incidence of Acute Respiratory Distress Syndrome (ARDS) and IL-8 synthesis after multiple trauma. *Cytokine.*, **37**(3), 192–9.
8. Sappington, P.L., Yang, R., Yang, H., Tracey, K.J., Delude, R.L. and Fink, M.P. (**2002**) HMGB1 B box increases the permeability of Caco-2 enterocytic monolayers and impairs intestinal barrier function in mice. *Gastroenterology*, **123**(3), 790–802.
9. Amiot, F., Fitting, C., Tracey, K.J., Cavaillon, J.M. and Dautry, F. (**1997**) Lipopolysaccharide-induced cytokine cascade and lethality in LT alpha/TNF alpha-deficient mice. *Mol Med.*, (12), 864–75.
10. Michie, H.R., Manogue, K.R., Spriggs, D.R., Revhaug, A., O'Dwyer, S. and Dinarello, C.A., *et al.* (**1988**) Detection of circulating tumor necrosis factor after endotoxin administration. *N. Engl. J. Med.*, **318**(23), 1481–6.
11. Vermont, C.L., Hazelzet, J.A., de Kleijn, E.D., van den Dobbelsteen, G.P. and de Groot, R. (**2006**) CC and CXC chemokine levels in children with meningococcal sepsis accurately predict mortality and disease severity. *Crit. Care.*, **10**(1), R33.
12. Moreno, S.E., Alves-Filho, J.C., Alfaya, T.M., da Silva, J.S., Ferreira, S.H. and Liew, F.Y. (**2006**) IL-12, but not IL-18, is critical to neutrophil activation and resistance to polymicrobial sepsis induced by cecal ligation and puncture. *J. Immunol.*, **177**(5), 3218–24.
13. Hiromatsu, T., Yajima, T., Matsuguchi, T., Nishimura, H., Wajjwalku, W. and Arai, T., *et al.* (**2003**) Overexpression of interleukin-15 protects against Escherichia coli-induced shock accompanied by inhibition of tumor necrosis factor-alpha-induced apoptosis. *J. Infect. Dis.*, **187**(9), 1442–51.
14. Netea, M.G., Fantuzzi, G., Kullberg, B.J., Stuyt, R.J., Pulido, E.J. and McIntyre, R.C., Jr., *et al.* (**2000**) Neutralization of IL-18 reduces neutrophil tissue accumulation and protects mice against lethal Escherichia coli and Salmonella typhimurium endotoxemia. *J. Immunol.*, **164**(5), 2644–9.
15. Hochholzer, P., Lipford, G.B., Wagner, H., Pfeffer, K. and Heeg, K. (**2000**) Role of interleukin-18 (IL-18) during lethal shock: decreased lipopolysaccharide sensitivity but normal superantigen reaction in IL-18-deficient mice. *Infect. Immun.*, **68**(6), 3502–8.
16. Tschoeke, S.K., Oberholzer, A. and Moldawer, L.L. (**2006**) Interleukin-18: a novel prognostic cytokine in bacteria-induced sepsis. *Crit. Care. Med.*, **34**(4), 1225–33.
17. Feezor, R.J., Oberholzer, C., Baker, H.V., Novick, D., Rubinstein, M., Moldawer, L.L., *et al.* (**2003**) Molecular characterization of the acute inflammatory response to infections with gram-negative versus gram-positive bacteria. *Infect. Immun.*, **71**(10), 5803–13.
18. Belladonna, M.L., Vacca, C., Volpi, C., Giampietri, A., Fioretti, M.C., Puccetti, P., *et al.* (**2006**) IL-23 neutralization protects

mice from Gram-negative endotoxic shock. *Cytokine.*, **34**(3–4), 161–9.
19. Wirtz, S., Tubbe, I., Galle, P.R., Schild, H.J., Birkenbach, M. and Blumberg, R.S., *et al.* (**2006**) Protection from lethal septic peritonitis by neutralizing the biological function of interleukin 27. *J. Exp. Med.*, **203**(8), 1875–81.
20. Sweet, M.J., Leung, B.P., Kang, D., Sogaard, M., Schulz, K. and Trajkovic, V., *et al.* (**2001**) A novel pathway regulating lipopolysaccharide-induced shock by ST2/T1 via inhibition of Toll-like receptor 4 expression. *J. Immunol.*, **166**(11), 6633–9.
21. Qiu, G., Gribbin, E., Harrison, K., Sinha, N. and Yin, K. (**2003**) Inhibition of gamma interferon decreases bacterial load in peritonitis by accelerating peritoneal fibrin deposition and tissue repair. *Infect. Immun.*, **71**(5), 2766–74.
22. Echtenacher, B., Freudenberg, M.A., Jack, R.S. and Mannel, D.N. (**2001**) Differences in innate defense mechanisms in endotoxemia and polymicrobial septic peritonitis. *Infect. Immun.*, **69**(12), 7271–6.
23. Yin, K., Gribbin, E. and Wang, H. (**2005**) Interferon-gamma inhibition attenuates lethality after cecal ligation and puncture in rats: implication of high mobility group box-1. *Shock.*, **24**(4), 396–401.
24. Calandra, T., Echtenacher, B., Roy, D.L., Pugin, J., Metz, C.N. and Hultner, L., *et al.* (**2000**) Protection from septic shock by neutralization of macrophage migration inhibitory factor. *Nat. Med.*, **6**(2), 164–70.
25. Matsuda, N., Nishihira, J., Takahashi, Y., Kemmotsu, O. and Hattori, Y. (**2006**) Role of macrophage migration inhibitory factor in acute lung injury in mice with acute pancreatitis complicated by endotoxemia. *Am. J. Respir. Cell. Mol. Biol.*, **35**(2), 198–205.
26. VanOtteren, G.M., Strieter, R.M., Kunkel, S.L., Paine, R, 3rd, Greenberger M.J, Danforth J.M., *et al.* (**1995**) Compartmentalized expression of RANTES in a murine model of endotoxemia. *J. Immunol.*, **154**(4), 1900–8.
27. Imhof, B.A. and Aurrand-Lions, M. (**2004**) Adhesion mechanisms regulating the migration of monocytes. *Nat. Rev. Immunol.*, **4**(6), 432–44.
28. Kaneider, N.C., Agarwal, A., Leger, A.J. and Kuliopulos, A. (**2005**) Reversing systemic inflammatory response syndrome with chemokine receptor pepducins. *Nat. Med.*, **11**(6), 661–5.
29. Netea, M.G., Joosten, L.A., Lewis, E., Jensen, D.R., Voshol, P.J. and Kullberg, B.J., *et al.* (**2006**) Deficiency of interleukin-18 in mice leads to hyperphagia, obesity and insulin resistance. *Nat. Med.*, **12**(6), 650–6.
30. Yano, K., Liaw, P.C., Mullington, J.M., Shih, S.C., Okada, H., Bodyak, N., *et al.* (**2006**) Vascular endothelial growth factor is an important determinant of sepsis morbidity and mortality. *J. Exp. Med.*, **203**(6), 1447–58.
31. Haeffner-Cavaillon, N., Cavaillon, J.M., Laude, M. and Kazatchkine, M.D. (**1987**) C3a(C3adesArg) induces production and release of interleukin 1 by cultured human monocytes. *J. Immunol.*, **139**(3), 794–9.
32. Cavaillon, J.M., Fitting, C. and Haeffner-Cavaillon, N. (**1990**) Recombinant C5a enhances interleukin 1 and tumor necrosis factor release by lipopolysaccharide-stimulated monocytes and macrophages. *Eur. J. Immunol.*, **20**(2), 253–7.
33. Niederbichler, A.D., Hoesel, L.M., Westfall, M.V., Gao, H., Ipaktchi, K.R., Sun, L., *et al.* (**2006**) An essential role for complement C5a in the pathogenesis of septic cardiac dysfunction. *J. Exp. Med.*, **203**(1), 53–61.
34. Ward, P.A. (**2004**) The dark side of C5a in sepsis. *Nat. Rev. Immunol.*, **4**(2), 133–42.
35. Liu, D., Zhang, D., Scafidi, J., Wu, X., Cramer, C.C., Davis, A.E, 3rd. (**2005**)

C1 inhibitor prevents Gram-negative bacterial lipopolysaccharide-induced vascular permeability. *Blood.*, **105**(6), 2350–5.
36. Liu, D., Cai, S., Gu, X., Scafidi, J., Wu, X., Davis, A.E, 3rd. (**2003**) C1 inhibitor prevents endotoxin shock via a direct interaction with lipopolysaccharide. *J. Immunol.*, **171**(5), 2594–601.
37. Osterud, B., Flaegstad, T. (**1983**) Increased tissue thromboplastin activity in monocytes of patients with meningococcal infection: related to an unfavourable prognosis. *Thromb. Haemost.*, **49**(1), 5–7.
38. Zeerleder, S., Schroeder, V., Lammle, B., Wuillemin, W.A., Hack, C.E. and Kohler, H.P. (**2007**) Factor XIII in severe sepsis and septic shock. *Thromb. Res.*, **119**(3), 311–8.
39. Bernard, G.R., Vincent, J.L., Laterre, P.F., LaRosa, S.P., Dhainaut, J.F., Lopez-Rodriguez, A., *et al.* (**2001**) Efficacy and safety of recombinant human activated protein C for severe sepsis. *N. Engl. J. Med.*, **344**(10), 699–709.
40. Yan, S.B., Helterbrand, J.D., Hartman, D.L., Wright, T.J. and Bernard, G.R. (**2001**) Low levels of protein C are associated with poor outcome in severe sepsis. *Chest.*, **120**(3), 915–22.
41. Murakami, K., Okajima, K., Uchiba, M., Johno, M., Nakagaki, T., Okabe, H., *et al.* (**1997**) Activated protein C prevents LPS-induced pulmonary vascular injury by inhibiting cytokine production. *Am. J. Physiol.*, **272**(2 Pt 1), L197–202.
42. Esmon, C. (**2000**) The protein C pathway. *Crit. Care. Med.*, **28**(9 Suppl), S44–8.
43. Becker, R.C. (**2002**) Antithrombotic therapy after myocardial infarction. *N. Engl. J. Med.*, **347**(13), 1019–22.
44. Verbon, A., Meijers, J.C., Spek, C.A., Hack, C.E., Pribble, J.P., Turner, T., *et al.* (**2003**) Effects of IC14, an anti-CD14 antibody, on coagulation and fibrinolysis during low-grade endotoxemia in humans. *J. Infect. Dis.*, **187**(1), 55–61.
45. Madoiwa, S., Nunomiya, S., Ono, T., Shintani, Y., Ohmori, T., Mimuro, J., *et al.* (**2006**) Plasminogen activator inhibitor 1 promotes a poor prognosis in sepsis-induced disseminated intravascular coagulation. *Int. J. Hematol.*, **84**(5), 398–405.
46. Renckens, R., Roelofs, J.J., Bonta, P.I., Florquin, S., de Vries, C.J., Levi, M., *et al.* (**2007**) Plasminogen activator inhibitor type 1 is protective during severe Gram-negative pneumonia. *Blood.*, **109**(4), 1593–601.
47. Nakamura, A., Mori, Y., Hagiwara, K., Suzuki, T., Sakakibara, T., Kikuchi, T., *et al.* (**2003**) Increased susceptibility to LPS-induced endotoxin shock in secretory leukoprotease inhibitor (SLPI)-deficient mice. *J. Exp. Med.*, **197**(5), 669–74.
48. Renckens, R., Roelofs, J.J., Florquin, S., de Vos, A.F., Lijnen, H.R., van't Veer, C., *et al.* (**2006**) Matrix metalloproteinase-9 deficiency impairs host defense against abdominal sepsis. *J. Immunol.*, **176**(6), 3735–41.
49. Yassen, K.A., Galley, H.F., Webster, N.R. (**2001**) Matrix metalloproteinase-9 concentrations in critically ill patients. *Anaesthesia.*, **56**(8), 729–32.
50. Benjamim, C.F., Canetti, C., Cunha, F.Q., Kunkel, S.L., Peters-Golden, M. (**2005**) Opposing and hierarchical roles of leukotrienes in local innate immune versus vascular responses in a model of sepsis. *J. Immunol.*, **174**(3), 1616–20.
51. Ejima, K., Layne, M.D., Carvajal, I.M., Kritek, P.A., Baron, R.M., Chen, Y.H., *et al.* (**2003**) Cyclooxygenase-2-deficient mice are resistant to endotoxin-induced inflammation and death. *FASEB J.*, **17**(10), 1325–7.
52. Bernard, G.R W.A., Russell, J.A., Schein, R., Summer, W.R., Steinberg, K.P., Fulkerson, W.J., Wright, P.E., Christman, B.W., Dupont, W.D., Higgins, S.B.,

Swindell, B.B. (**1997**) The effects of ibuprofen on the physiology and survival of patients with sepsis. The Ibuprofen in Sepsis Study Group. *N. Engl. J. Med.*, **336**(13), 912–8.
53. Gomes, R.N., Bozza, F.A., Amancio, R.T., Japiassu, A.M., Vianna, R.C., Larangeira, A.P., *et al.* (**2006**) Exogenous platelet-activating factor acetylhydrolase reduces mortality in mice with systemic inflammatory response syndrome and sepsis. *Shock.*, **26**(1), 41–9.
54. Annane, D., Sanquer, S., Sebille, V., Faye, A., Djuranovic, D., Raphael, J.C., *et al.* (**2000**) Compartmentalised inducible nitric-oxide synthase activity in septic shock. *Lancet.*, **355**(9210), 1143–8.
55. Asakura, H., Asamura, R., Ontachi, Y., Hayashi, T., Yamazaki, M., Morishita, E., *et al.* (**2005**) Selective inducible nitric oxide synthase inhibition attenuates organ dysfunction and elevated endothelin levels in LPS-induced DIC model rats. *J. Thromb. Haemost.*, **3**(5), 1050–5.
56. Benjamim, C.F., Silva, J.S., Fortes, Z.B., Oliveira, M.A., Ferreira, S.H., Cunha, F.Q. (**2002**) Inhibition of leukocyte rolling by nitric oxide during sepsis leads to reduced migration of active microbicidal neutrophils. *Infect. Immun.*, **70**(7), 3602–10.
57. Matejovic, M., Krouzecky, A., Radej, J., Rokyta, R., Jr., Kralova H, Radermacher P., *et al.* (**2007**) Coagulation and endothelial dysfunction during longterm hyperdynamic porcine bacteremia–effects of selective inducible nitric oxide synthase inhibition. *Thromb. Haemost.*, **97**(2), 304–9.
58. Han, X., Fink, M.P., Uchiyama, T., Yang, R., Delude, R.L. (**2004**) Increased iNOS activity is essential for pulmonary epithelial tight junction dysfunction in endotoxemic mice. *Am. J. Physiol. Lung. Cell. Mol. Physiol.*, **286**(2), L259–67.
59. Yang, H., Ochani, M., Li, J., Qiang, X., Tanovic, M., Harris, H.E., *et al.* (**2004**) Reversing established sepsis with antagonists of endogenous high-mobility group box 1. *Proc. Natl. Acad. Sci. USA.*, **101**(1), 296–301.
60. Angus, D.C., Yang, L., Kong, L., Kellum, J.A., Delude, R.L., Tracey, K.J., *et al.* (**2007**) Circulating high-mobility group box 1 (HMGB1) concentrations are elevated in both uncomplicated pneumonia and pneumonia with severe sepsis. *Crit. Care. Med.*, **35**(4), 1061–7.
61. Tsung, A., Sahai, R., Tanaka, H., Nakao, A., Fink, M.P., Lotze, M.T., *et al.* (**2005**) The nuclear factor HMGB1 mediates hepatic injury after murine liver ischemia-reperfusion. *J. Exp. Med.*, **201**(7), 1135–43.
62. Qin, S., Wang, H., Yuan, R., Li, H., Ochani, M., Ochani, K., *et al.* (**2006**) Role of HMGB1 in apoptosis-mediated sepsis lethality. *J. Exp. Med.*, **203**(7), 1637–42.
63. Bouchon, A., Facchetti, F., Weigand, M.A., Colonna, M. (**2001**) TREM-1 amplifies inflammation and is a crucial mediator of septic shock. *Nature.*, **410**(6832), 1103–7.
64. Adib-Conquy, M., Monchi, M., Goulenok, C., Laurent, I., Thuong, M., Cavaillon, J.M. and Adrie, C. (**2007**) Enhanced plasma levels of soluble triggering expressed on myeloid cells-1 and procalcitonin after cardiac surgery and cardiac arrest in the absence of infection. *Shock.*
65. Moestrup, S.K., Moller, H.J. (**2004**) CD163: a regulated hemoglobin scavenger receptor with a role in the anti-inflammatory response. *Ann. Med.*, **36**(5), 347–54.
66. Imai, T., Fujita, T., Yamazaki, Y. (**2003**) Beneficial effects of apolipoprotein A-I on endotoxemia. *Surg. Today.*, **33**(9), 684–7.
67. Jacque, B., Stephan, K., Smirnova, I., Kim, B., Gilling, D., Poltorak, A. (**2006**) Mice expressing high levels of soluble CD14 retain LPS in the

circulation and are resistant to LPS-induced lethality. *Eur. J. Immunol.*, **36**(11), 3007–16.
68. Munford, R.S., Pugin, J. (**2001**) Normal responses to injury prevent systemic inflammation and can be immunosuppressive. *Am. J. Respir. Crit. Care. Med.*, **163**(2), 316–21.
69. Marchant, A., Bruyns, C., Vandenabeele, P., Ducarme, M., Gerard, C., Delvaux, A., *et al.* (**1994**) Interleukin-10 controls interferon-gamma and tumor necrosis factor production during experimental endotoxemia. *Eur. J. Immunol.*, **24**(5), 1167–71.
70. Marie, C., Cavaillon, J.M., Losser, M.R. (**1996**) Elevated levels of circulating transforming growth factor-beta 1 in patients with the sepsis syndrome. *Ann. Intern. Med.*, **125**(6), 520–1.
71. Girardin, E., Roux-Lombard, P., Grau, G.E., Suter, P., Gallati, H., Dayer, J.M. (**1992**) Imbalance between tumour necrosis factor-alpha and soluble TNF receptor concentrations in severe meningococcaemia. The J5 Study Group. *Immunology.*, **76**(1), 20–3.
72. van Deuren, M., van der Ven-Jongekrijg, J., Vannier, E., van Dalen, R., Pesman, G., Bartelink, A.K., *et al.* (**1997**) The pattern of interleukin-1beta (IL-1beta) and its modulating agents IL-1 receptor antagonist and IL-1 soluble receptor type II in acute meningococcal infections. *Blood.*, **90**(3), 1101–8.
73. Fischer, E., Van Zee, K.J., Marano, M.A., Rock, C.S., Kenney, J.S., Poutsiaka, D.D., *et al.* (**1992**) Interleukin-1 receptor antagonist circulates in experimental inflammation and in human disease. *Blood.*, **79**(9), 2196–200.
74. Novick, D., Schwartsburd, B., Pinkus, R., Suissa, D., Belzer, I., Sthoeger, Z., *et al.* (**2001**) A novel IL-18BP ELISA shows elevated serum IL-18BP in sepsis and extensive decrease of free IL-18. *Cytokine.*, **14**(6), 334–42.
75. Granowitz, E.V., Santos, A.A., Poutsiaka, D.D., Cannon, J.G., Wilmore, D.W., Wolff, S.M., *et al.* (**1991**) Production of interleukin-1-receptor antagonist during experimental endotoxaemia. *Lancet.*, **338**(8780), 1423–4.
76. Wakabayashi, G., Gelfand, J.A., Jung, W.K., Connolly, R.J., Burke, J.F., Dinarello, C.A. (**1991**) *Staphylococcus epidermidis* induces complement activation, tumor necrosis factor and interleukin-1, a shock-like state and tissue injury in rabbits without endotoxemia. Comparison to *Escherichia coli*. *J. Clin. Invest.*, **87**(6), 1925–35.
77. Mancilla, J., Garcia, P., Dinarello, C.A. (**1993**) The interleukin-1 receptor antagonist can either reduce or enhance the lethality of *Klebsiella pneumoniae* sepsis in newborn rats. *Infect. Immun.*, **61**(3), 926–32.
78. Pathan, N., Hemingway, C.A., Alizadeh, A.A., Stephens, A.C., Boldrick, J.C., Oragui, E.E., *et al.* (**2004**) Role of interleukin 6 in myocardial dysfunction of meningococcal septic shock. *Lancet.*, **363**(9404), 203–9.
79. Witzenbichler, B., Westermann, D., Knueppel, S., Schultheiss, H.P., Tschope, C. (**2005**) Protective role of angiopoietin-1 in endotoxic shock. *Circulation.*, **111**(1), 97–105.
80. Orfanos, S.E., Kotanidou, A., Glynos, C., Athanasiou, C., Tsigkos, S., Dimopoulou, I., *et al.* (**2007**) Angiopoietin-2 is increased in severe sepsis: correlation with inflammatory mediators. *Crit. Care. Med.*, **35**(1), 199–206.
81. Imai, Y., Kuba, K., Rao, S., Huan, Y., Guo, F., Guan, B., *et al.* (**2005**) Angiotensin-converting enzyme 2 protects from severe acute lung failure. *Nature.*, **436**(7047), 112–6.
82. Schindler, R.DC., Koch, K.M. (**1995**) Angiotensin-converting-enzyme inhibitors suppress synthesis of tumour necrosis factor and interleukin 1 by human peripheral

blood mononuclear cells. *Cytokine.*, 7(6), 526–33.
83. Faggioni, R., Fantuzzi, G., Gabay, C., Moser, A., Dinarello, C.A., Feingold, K.R., *et al.* (**1999**) Leptin deficiency enhances sensitivity to endotoxin-induced lethality. *Am. J. Physiol.*, **276**(1 Pt 2), R136–42.
84. Puneet, P., Hegde, A., Ng, S.W., Lau, H.Y., Lu, J., Moochhala, S.M., *et al.* (**2006**) Preprotachykinin-A gene products are key mediators of lung injury in polymicrobial sepsis. *J. Immunol.*, **176**(6), 3813–20.
85. Deng, J., Muthu, K., Gamelli, R., Shankar, R., Jones, S.B. (**2004**) Adrenergic modulation of splenic macrophage cytokine release in polymicrobial sepsis. *Am. J. Physiol. Cell. Physiol.*, **287**(3), C730–6.
86. Wong, L.Y., Cheung, B.M., Li, Y.Y., Tang, F. (**2005**) Adrenomedullin is both proinflammatory and antiinflammatory: its effects on gene expression and secretion of cytokines and macrophage migration inhibitory factor in NR8383 macrophage cell line. *Endocrinology.*, **146**(3), 1321–7.
87. Borovikova, L.V., Ivanova, S., Zhang, M., Yang, H., Botchkina, G.I., Watkins, L.R., *et al.* (**2000**) Vagus nerve stimulation attenuates the systemic inflammatory response to endotoxin. *Nature.*, **405**(6785), 458–62.
88. Pavlov, V.A., Ochani, M., Gallowitsch-Puerta, M., Ochani, K., Huston, J.M., Czura, C.J., *et al.* (**2006**) Central muscarinic cholinergic regulation of the systemic inflammatory response during endotoxemia. *Proc. Natl. Acad. Sci. USA.*, **103**(13), 5219–23.
89. Matzinger, P. (**1994**) Tolerance, danger, and the extended family. *Annu. Rev. Immunol.*, **12**, 991–1045.
90. Oppenheim, J.J., Yang, D. (**2005**) Alarmins: chemotactic activators of immune responses. *Curr. Opin. Immunol.*, **17**(4), 359–65.
91. Bianchi, M.E. (**2007**) DAMPs, PAMPs and alarmins: all we need to know about danger. *J. Leukoc. Biol.*, **81**(1), 1–5.
92. Muir, K.W., Tyrrell, P., Sattar, N., Warburton, E. (**2007**) Inflammation and ischaemic stroke. *Curr. Opin. Neurol.*, **20**(3), 334–42.
93. Tousoulis, D., Antoniades, C., Koumallos, N., Stefanadis, C. (**2006**) Pro-inflammatory cytokines in acute coronary syndromes: from bench to bedside. *Cytokine Growth Factor Rev.*, **17**(4), 225–33.
94. Miki, C., Iriyama, K., Gunson, B.K., Mayer, A.D., Buckels, J.A., McMaster, P. (**1997**) Influence of intraoperative blood loss on plasma levels of cytokines and endotoxin and subsequent graft liver function. *Arch. Surg.*, **132**(2), 136–41.
95. Hennein, H.A., Ebba, H., Rodriguez, J.L., Merrick, S.H., Keith, F.M., Bronstein, M.H., *et al.* (**1994**) Relationship of the proinflammatory cytokines to myocardial ischemia and dysfunction after uncomplicated coronary revascularization. *J. Thorac. Cardiovasc. Surg.*, **108**(4), 626–35.
96. Khabar, K.S., el Barbary, M.A., Khouqeer, F., Devol, E., al-Gain, S., al-Halees, Z. (**1997**) Circulating endotoxin and cytokines after cardiopulmonary bypass: differential correlation with duration of bypass and systemic inflammatory response/multiple organ dysfunction syndromes. *Clin. Immunol. Immunopathol.*, **85**(1), 97–103.
97. Cruickshank, A.M., Fraser, W.D., Burns, H.J., Van Damme, J., Shenkin, A. (**1990**) Response of serum interleukin-6 in patients undergoing elective surgery of varying severity. *Clin. Sci. (Lond).*, **79**(2), 161–5.
98. Gueugniaud, P.Y., Bertin-Maghit, M., Hirschauer, C., Bouchard, C., Vilasco, B., Petit, P. (**1997**) In the early stage of major burns, is there a correlation between survival, interleukin-6 levels and oxygen delivery and consumption? *Burns.*, **23**(5), 426–31.
99. Geppert, A., Dorninger, A., Delle-Karth, G., Zorn, G., Heinz, G., Huber, K. (**2006**) Plasma

concentrations of interleukin-6, organ failure, vasopressor support, and successful coronary revascularization in predicting 30-day mortality of patients with cardiogenic shock complicating acute myocardial infarction. *Crit. Care. Med.*, **34**(8), 2035–42.

100. Waydhas, C., Nast-Kolb, D., Jochum, M., Trupka, A., Lenk, S., Fritz, H., *et al.* (**1992**) Inflammatory mediators, infection, sepsis, and multiple organ failure after severe trauma. *Arch. Surg.*, **127**(4), 460–7.
101. Fassbender, K., Rossol, S., Kammer, T., Daffertshofer, M., Wirth, S., Dollman, M., *et al.* (**1994**) Proinflammatory cytokines in serum of patients with acute cerebral ischemia: kinetics of secretion and relation to the extent of brain damage and outcome of disease. *J. Neurol. Sci.*, **122**(2), 135–9.
102. de Beaux, A.C., Ross, J.A., Maingay, J.P., Fearon, K.C., Carter, D.C. (**1996**) Proinflammatory cytokine release by peripheral blood mononuclear cells from patients with acute pancreatitis. *Br. J. Surg.*, **83**(8), 1071–5.
103. Riche, F., Dosquet, C., Panis, Y., Valleur, P., Laisne, M.J., Briard, C., *et al.* (**1995**) Levels of portal and systemic blood cytokines after colectomy in patients with carcinoma or Crohn's disease. *J. Am. Coll. Surg.*, **180**(6), 718–24.
104. Miyao, Y., Miyazaki, S., Goto, Y., Itoh, A., Daikoku, S., Morii, I., *et al.* (**1999**) Role of cytokines and adhesion molecules in ischemia and reperfusion in patients with acute myocardial infarction. *Jpn. Circ. J.*, **63**(5), 362–6.
105. Nathan, N., Denizot, Y., Cornu, E., Jauberteau, M.O., Chauvreau, C., Feiss, P. (**1997**) Cytokine and lipid mediator blood concentrations after coronary artery surgery. *Anesth. Analg.*, **85**(6), 1240–6.
106. Saatvedt, K., Lindberg, H., Michelsen, S., Pedersen, T., Seem, E., Geiran, O. (**1996**) Release of soluble tumour necrosis factor alpha receptors during and after paediatric cardiopulmonary bypass. Correlation with haemodynamic and clinical variables. *Cytokine.*, **8**(12), 944–8.
107. Ikei, S., Ogawa, M., Yamaguchi, Y. (**1998**) Blood concentrations of polymorphonuclear leucocyte elastase and interleukin-6 are indicators for the occurrence of multiple organ failures at the early stage of acute pancreatitis. *J. Gastroenterol. Hepatol.*, **13**(12), 1274–83.
108. Gross, V., Andreesen, R., Leser, H.G., Ceska, M., Liehl, E., Lausen, M., *et al.* (**1992**) Interleukin-8 and neutrophil activation in acute pancreatitis. *Eur. J. Clin. Invest.*, **22**(3), 200–3.
109. Pezzilli, R., Billi, P., Miniero, R., Fiocchi, M., Cappelletti, O., Morselli-Labate, A.M., *et al.* (**1995**) Serum interleukin-6, interleukin-8, and beta 2-microglobulin in early assessment of severity of acute pancreatitis. Comparison with serum C-reactive protein. *Dig. Dis. Sci.*, **40**(11), 2341–8.
110. Kossmann, T., Hans, V.H., Imhof, H.G., Stocker, R., Grob, P., Trentz, O., *et al.* (**1995**) Intrathecal and serum interleukin-6 and the acute-phase response in patients with severe traumatic brain injuries. *Shock.*, **4**(5), 311–7.
111. Kawamura, T., Wakusawa, R., Okada, K. and Inada, S. (**1993**) Elevation of cytokines during open heart surgery with cardiopulmonary bypass: participation of interleukin 8 and 6 in reperfusion injury. *Can. J. Anaesth.*, **40**(11), 1016–21.
112. Deng, M.C., Dasch, B., Erren, M., Mollhoff, T., Scheld, H.H. (**1996**) Impact of left ventricular dysfunction on cytokines, hemodynamics, and outcome in bypass grafting. *Ann. Thorac. Surg.*, **62**(1), 184–90.
113. Struzyna, J., Pojda, Z., Braun, B., Chomicka, M., Sobiczewska, E., Wrembel, J. (**1995**) Serum cytokine levels (IL-4, IL-6, IL-8, G-CSF,

GM-CSF) in burned patients. *Burns.*, **21**(6), 437–40.
114. Haupt, W., Hohenberger, W., Mueller, R., Klein, P., Christou, N.V. (**1997**) Association between preoperative acute phase response and postoperative complications. *Eur. J. Surg.*, **163**(1), 39–44.
115. Vindenes, H., Ulvestad, E., Bjerknes, R. (**1995**) Increased levels of circulating interleukin-8 in patients with large burns: relation to burn size and sepsis. *J. Trauma.*, **39**(4), 635–40.
116. Silva-Mejias, C., Gamboa-Antinolo, F., Lopez-Cortes, L.F., Cruz-Ruiz, M., Pachon, J. (**1995**) Interleukin-1 beta in pleural fluids of different etiologies: its role as inflammatory mediator in emphysema. *Chest.*, **108**(4), 942–5.
117. Povoa, P. (**2002**) C-reactive protein: a valuable marker of sepsis. *Intensive Care. Med.*, **28**(3), 235–43.
118. Meisner, M. (**2005**) Biomarkers of sepsis: clinically useful? *Curr. Opin. Crit. Care.*, **11**(5), 473–80.
119. Gibot, S., Cravoisy, A., Levy, B., Bene, M.C., Faure, G., Bollaert, P.E. (**2004**) Soluble triggering receptor expressed on myeloid cells and the diagnosis of pneumonia. *N. Engl. J. Med.*, **350**(5), 451–8.
120. van der Meer, V., Neven, A.K., van den Broek, P.J., Assendelft, W.J. (**2005**) Diagnostic value of C reactive protein in infections of the lower respiratory tract: systematic review. *BMJ.*, **331**(7507), 26.
121. Tang, B.M., Eslick, G.D., Craig, J.C., McLean, A.S. (**2007**) Accuracy of procalcitonin for sepsis diagnosis in critically ill patients: systematic review and meta-analysis. *Lancet. Infect. Dis.*, **7**(3), 210–7.
122. Wang, D.Y., Qin, R.Y., Liu, Z.R., Gupta, M.K., Chang, Q. (**2004**) Expression of TREM-1 mRNA in acute pancreatitis. *World J. Gastroenterol.*, **10**(18), 2744–6.
123. Koussoulas, V., Vassiliou, S., Demonakou, M., Tassias, G., Giamarellos-Bourboulis, E.J., Mouktaroudi, M., *et al.* (**2006**) Soluble triggering receptor expressed on myeloid cells (sTREM-1): a new mediator involved in the pathogenesis of peptic ulcer disease. *Eur. J. Gastroenterol. Hepatol.*, **18**(4), 375–9.
124. Cabie, A., Farkas, J.C., Fitting, C., Laurian, C., Cormier, J.M., Carlet, J., *et al.* (**1993**) High levels of portal TNF-alpha during abdominal aortic surgery in man. *Cytokine.*, **5**(5), 448–53.
125. Eichacker, P.Q P.C., Kalil, A., Esposito, C., Cui, X., Banks, S.M., Gerstenberger, E.P., Fitz, Y., Danner, R.L., Natanson, C. (**2002**) Risk and the efficacy of antiinflammatory agents: retrospective and confirmatory studies of sepsis. *Am. J. Respir. Crit. Care. Med.*, **166**(9), 1197–205.
126. Cuesta, N., Salkowski C.A., Thomas K.E., Vogel S.N. Regulation of lipopolysaccharide sensitivity by IFN regulatory factor-2. *J Immunol.* **2003** Jun 1; **170**(11), 5739–47.
127. Karaghiosoff M., Steinborn R., Kovarik P., Kriegshäuser G., Baccarini M., Donabauer B., Reichart U., Kolbe T., Bogdan C., Leanderson T., Levy D., Decker T., Müller M. Central role for type I interferons and Tyk2 in lipopolysaccharide-induced endotoxin shock. *Nat Immunol.* **2003** May; **4**(5), 471–7. Epub 2003 Apr 7.
128. Peyssonnaux C., Cejudo-Martin P., Doedens A., Zinkernagel A.S., Johnson R.S., Nizet V. Cutting edge: Essential role of hypoxia inducible factor-1alpha in development of lipopolysaccharide-induced sepsis. *J Immunol.* **2007** Jun 15; **178**(12), 7516–9.
129. Matsukawa A., Kaplan M.H., Hogaboam C.M., Lukacs N.W., Kunkel S.L. Pivotal role of signal transducer and activator of transcription (Stat)4 and Stat6 in the innate immune response during sepsis. *J Exp Med.* **2001** Mar 19; **193**(6), 679–88.

130. Lentsch A.B., Kato A., Davis B., Wang W., Chao C., Edwards M.J. STAT4 and STAT6 regulate systemic inflammation and protect against lethal endotoxemia. *J Clin Invest.* **2001** Nov; **108**(10), 1475–82.
131. Matsukawa A., Takeda K., Kudo S., Maeda T., Kagayama M., Akira S. Aberrant inflammation and lethality to septic peritonitis in mice lacking STAT3 in macrophages and neutrophils. *J Immunol.* **2003** Dec 1; **171**(11), 6198–205.
132. Reimold A.M., Kim J., Finberg R., Glimcher L.H. Decreased immediate inflammatory gene induction in activating transcription factor-2 mutant mice. *Int Immunol.* **2001** Feb; **13**(2), 241–8.
133. Xiao X., Zuo X., Davis A.A., McMillan D.R., Curry B.B., Richardson J.A., Benjamin I.J. HSF1 is required for extra-embryonic development, postnatal growth and protection during inflammatory responses in mice. *EMBO J.* **1999** Nov 1; **18**(21), 5943–52.
134. Gadjeva M., Tomczak M.F., Zhang M., Wang Y.Y., Dull K., Rogers A.B.,Erdman S.E., Fox J.G., Carroll M., Horwitz B.H. A role for NF-kappa B subunits p50 and p65 in the inhibition of lipopolysaccharide-induced shock. *J Immunol.* **2004** Nov 1; **173**(9), 5786–93.

9
Apoptosis: A Potential Therapeutic Target in Sepsis and Non-Infectious Systemic Inflammation

Chun-Shiang Chung, Ryan Swan, Mario Perl, Doreen E. Wesche-Soldato and Alfred Ayala

9.1
Introduction

With approximately 700 000 cases annually and a mortality of nearly 30%, sepsis remains a major health problem in the United States [1]. The principal hypothesis has been that sepsis induces an overwhelming inflammatory response, which then causes significant cell/tissue injury in the host; however, while numerous treatments based on anti-inflammatory or anti-coagulant therapies showed promise in the experimental setting, they have frequently failed to provide a survival benefit in randomized human clinical trials [2, 3]. The exception, recombinant human activated protein C [4], low-dose corticosteroids [5], and intensive insulin therapy [6] have been proven to reduce mortality; however, the survival benefit provided is relatively modest. As these drugs (with the exception of corticosteroids) are not classic anti-inflammatory agents, their mechanisms of salutary benefit remain to be clarified. This also implies that the pathology behind the septic condition in the patient/animal may not just be simply about inflammation. Since human sepsis is a complex and evolving disease, defining both the patient population who may benefit from a potential therapy and the timing of delivery of the therapy is also critical. For example, it is well documented that the immune system in a septic individual exhibits a biphasic response, with an early hyper-inflammatory response and a late stage generalized immunosuppressive state. Hence, the application of anti-inflammatory therapy without knowledge of the patient's inflammatory/immune status may mitigate the benefit of these anti-inflammatory treatments [7]. Thus, more information is needed to understand the pathology and processes underpinning the development of sepsis if we are to optimize these present clinical therapies and/or develop novel approaches.

In this respect, over the last decade, developing evidence in sepsis research suggests that increased apoptosis may play a crucial role in the outcome of experimental animals and possibly septic patients. The induction of increased

Sepsis and Non-infectious Systemic Inflammation. Edited by J.-M. Cavaillon and C. Adrie

ISBN: 978-3-527-31935-0

apoptosis during sepsis is suggested to result in loss of immune and/or non-immune cell populations that may contribute to immunosuppression and/or subsequent multiple organ failure [8, 9]. Here, we will overview not only the evidence supporting the concept that changes in the septic animal/patient's apoptotic process and/or the interaction of the immune system/body with these apoptotic materials can contribute to septic morbidity and mortality, but also discuss which components of the apoptotic process may serve as potential novel therapeutic targets.

9.2 Apoptosis

9.2.1 Overview

The word "apoptosis" (*apo* = from and *ptosis* = falling) is of Greek origin and was introduced by Kerr and colleagues [10] to describe a specific form of cell death. Classic apoptosis or programmed cell death (PCD) is a genetically regulated process that can be initiated by intra- as well as extracellular events and is characterized by deformation of the cell membrane, cell shrinkage, condensation of the nuclear chromatin, activation of endonucleases which cleave DNA into oligonucleotide fragments, and disruption of the mitochondrial transmembrane potential [11]. Typically, the dying cell maintains its plasma membrane integrity, yet changes in the cell plasma membrane antigen/receptor expression also serve as a signal to local phagocytes to clear the developing apoptotic cells. This latter process is largely a non-inflammatory event. Classic apoptosis can be differentiated from necrosis (in Greek = death), a pathologic form of non-programmed and energy-independent cell death, by rapid loss of cell viability, cell swelling, loss of membrane integrity, and induction of an inflammatory response by the release of cytoplasmic contents including proteases, toxic proteins and oxidizing molecules [10, 12]. However, recent evidence suggests that at least two intermediate patterns of cell death appear to exist, which fall between apoptosis and necrosis [13].

During normal embryogenesis, apoptosis is involved in maintaining tissue homeostasis in the process of tissue differentiation/development and organogenesis, as well as in optimization of biological functions in the immune and central nervous systems [10]. In the immune system, apoptosis plays a pivotal role in the selection of lymphocyte populations and maintenance of normal functional immune responses. Strict control of cell death ensures the deletion of excessive, improperly developed, genetically damaged and immunologically autoreactive cells, thereby protecting the host. On the other hand, it is also now clear that overt activation or suppression of the apoptotic pathway can contribute to a variety of pathological conditions, such as

HIV immune depression, cancer, autoimmune disorders, neurodegenerative diseases, inflammatory bowel disease and ischemic injury [14–17]. Apoptosis can be triggered by various stimuli from outside or inside the cell, e.g. by ligation of cell surface receptors, by DNA damage resulting from defects in DNA repair mechanisms, treatment with cytotoxic drugs or irradiation, by a lack of survival signals, or by developmental death signals. All these different apoptotic signals appear to activate common cell death machinery leading to the characteristic features of programmed cell death. The importance of this normal physiologic process is reflected in a substantial increase in publications in the literature over the past decade. Today, entering the term *apoptosis* into the PubMed search engine yields about 130 000 peer reviewed articles; more than 17 000 papers were published in 2006 alone.

9.2.2 Pathways of Apoptosis

There are primarily three general apoptotic pathways in mammalian cells: the extrinsic or death receptor (DR) pathway, the intrinsic or mitochondrial pathway and the endoplasmic reticulum (ER) or stress-induced pathway [18, 19]. The death receptor pathway is triggered by specific ligands and their death receptors, a family of membrane protein receptors including Fas (CD95 or Apo1), tumor necrosis factor receptor 1 (TNFR1 or p55), DR3 (TNF receptor apoptosis-mediating protein [TRAMP]), DR4 (Apo2 or TNF related apoptosis-inducing ligand [TRAIL]-R1) and DR5 (TRAIL-R2). Engagement of these receptors and ligands causes them to cluster and recruits Fas associated death domain (FADD) or TNF Receptor I associated death domain (TRADD), procaspase-8, FADD-like IL-1β-converting enzyme (FLICE), etc. to form a death-inducing signaling complex (DISC). This complex can activate caspase-8, which in turn cleaves and activates a signaling cascade of downstream proteases including caspase-3, -6 or -7 and leads to apoptosis (Figure 9.1).

The intrinsic mitochondrial pathway is driven by changes in the interaction of B-cell CCl/lymphoma 2 (Bcl-2) family members including anti-apoptotic Bcl-2, Bcl-extra long (Bcl-x_L), myeloid cell leukemia sequence 1 (Mcl-1), and pro-apoptotic Bcl-2 associated X (Bax), Bcl/Bcl-x_L-associated death promoter (Bad), Bcl-2 interacting mediator of cell death (Bim), BH3 interacting domain death agonist (Bid), Bcl-2 antagonistic killer (Bak), etc. In a viable cell, anti-apoptotic Bcl-2 and Bcl-x_L proteins counteract the pro-apoptotic Bax, Bak, and BH3-only proteins such as Bid, etc. In response to the apoptotic stimuli, BH3-only members are activated by transcriptional upregulation (Bax, Noxa, Puma), subcellular relocalization (Bim, Bmf), dephosphorylation (Bad), or proteolysis (Bid). Bax-like factors then undergo a conformational change (possibly assisted by some BH3-only proteins), insert into the outer mitochondrial membrane where they provoke permeability transition (PT) and the release of apoptogenic factors [20]. When changes in the relationship of these Bcl-2 family members

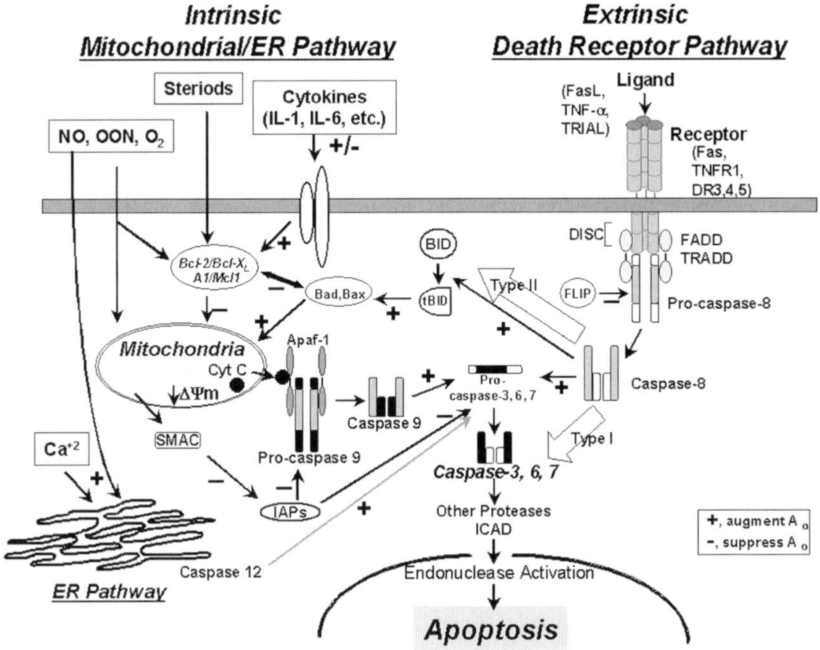

Figure 9.1 Schematic representation of the three signaling pathways of apoptosis: the death receptor pathway through ligation of TNFR or Fas (extrinsic signaling), the mitochondrial pathway through Bcl-2 family members (intrinsic signaling), and the ER stress pathway. The two-pathway model of Fas/FasL signaling (Type I and Type II) is also illustrated.

occur, cells undergo apoptosis: mitochondrial membrane potential ($\Delta\Psi_m$) collapses due to disruption of normal ion flux PT and cytochrome c and other downstream mitochondrial pro-apoptotic factors are released into the cytosol [21]. Cytosolic cytochrome c binds to apoptosis protease activating factor-1 (Apaf-1), which undergoes an ATP-dependent conformational change, oligomerizes to assemble the apoptosome that activates procaspase-9 [22]. Activated caspase-9 subsequently initiates a downstream caspase cascade resulting in cell death. Besides cytochrome c, many other apoptotic factors are released from mitochondria when cells undergo apoptosis, such as apoptosis-inducing factor (AIF), which can activate caspase-3 *in vitro* [23], as well as the inhibitors of apoptosis (IAPs), which are a family of anti-apoptotic proteins that directly inhibit caspases. IAP expression can be upregulated in response to survival signals to suppress apoptosis signaling, such as by activation of the transcription factor NF-κB through growth factor receptors. Another factor released from mitochondria is Smac/Diablo, which acts as a pro-apoptotic mediator by releasing the blockade on caspase activity by IAPs [24]. In addition, the endonuclease endoG from mitochondria can cleave DNA resulting in an apoptotic phenotype [25]. Recently, it has been suggested that there is cross-talk between the intrinsic and extrinsic apoptotic pathways. Pro-apoptotic Bcl-2 member Bid protein can promote cell death and is also a substrate for caspase-8 during apoptosis induced by the extrinsic death receptor pathway. Caspase-8 cleaves cytoplasmic Bid into truncated Bid (tBid), which then translocates to the mitochondria where it works together with Bax and Bak to activate the intrinsic death pathway [26]. Studies showed that Bid knockout hepatocytes are resistant to Fas and TNF-α induced cell death [27, 28]. Whether cell death is activated through the extrinsic or intrinsic pathway is dependent on not only the type of cells, but also the type of death signals seen (Figure 9.1).

The third and the least well understood apoptotic pathway is the ER stress-induced pathway, activated by oxidants or calcium dyshomeostasis. Studies using gene deficient mice suggest that caspase-12 plays a major role in the ER apoptotic pathway and in the development of Alzheimer's disease [29]. Phylogenetic analysis reveals that caspase-12 most resembles caspase-1 and -11, which are often referred to as inflammatory caspases as they play an important role in processing and maturation of cytokines such as IL-1β and IL-18 [30]. However, this pathway appears to be a species- and cell type-specific apoptotic effecter. Recent studies by Saleh *et al.* showed that caspase-12 deficiency increased bacterial clearance and survival in a colon ascendens stent peritonitis model [31]. However, more studies are needed to clarify the role of the ER stress-induced apoptotic pathway in sepsis (Figure 9.1).

9.2.3
Biochemistry of Apoptosis

The key components of cell death controlling the apoptotic signaling network in mammalian cells are caspases, which are homologous to the ced-3 gene originally identified in the soil nematode, *Caenorhabiditis elegans* [32]. Three *C. elegans* genes involved in the regulation of apoptosis during development, CED-3, CED-4 and CED-9, were found to be functionally and structurally similar to the mammalian caspase family, Apaf-1, and the Bcl-2 gene family, respectively [33, 34]. This suggests that the regulation of apoptosis is an evolutionarily conserved mechanism.

Caspases are a family of cysteine-dependent aspartate-specific proteases, thus named caspase, which specifically cleave their substrates after Asp residues and share similar amino acid sequence, structure, and substrate specificity, and exist in an inactive zymogen form that must be cleaved to generate the active form [35]. Once activated, caspases can catalyze other members of the family, resulting in a sequential activation and amplification of a proteolytic cascade. So far, 14 different members of the caspase-family have been described in mammals. According to their function in cell death, caspases can be divided into three groups. Initiator caspases (i.e. caspase-2, -8, -9 and -10) are activated in response to apoptotic signals and then cleave and activate effector caspases (i.e. caspase-3, -6 and -7), which serve as the major players in apoptosis by catalyzing many key proteins associated with apoptosis [36, 37]. The last group of caspases (i.e. caspase-1, -4, -5, -11 and -12), as mentioned briefly earlier, appear to be involved in the cleavage of pro-inflammatory cytokines such as pro-IL-1β and pro-IL-18 rather than apoptosis. A number of caspase substrates have been identified which include proteins involved in cellular structure (such as nuclear lamins, gelsolin and P21-activated kinase 2 (PAK2)), DNA repair (such as poly-ADP ribose polymerase (PARP), DNA fragmentation factor (DFF), inhibitory subunit caspase-activated DNase (ICAD) and DNA-dependent protein kinase), cell cycle control (such as retinoblastoma protein, cyclin dependent kinase and tyrosine kinase c-Abl), and precursors of pro-inflammatory cytokines, IL-1β and IL-18, etc. [34].

Although caspases are thought to be the central component of cell death machinery, inhibition of caspase activity does not always prevent cell death. Studies indicated that z-VAD-fmk, a pancaspase inhibitor, treatment did not prevent cell death of cortical neurons [38] and photoreceptor cells [39]. This implies that there is a caspase-independent cell death pathway. AIF, a mitochondrial intermembrane flavoprotein, is one of the main agents of caspase-independent apoptosis. Upon an apoptotic stimulus, AIF is released from mitochondria and translocates to the nucleus, where it induces chromatin condensation and large-scale DNA fragmentation (50 Kbp) [40]. Additionally, EndoG and Omi/HtrA2 (a serine protease) also can induce apoptotic cell death in a caspase-independent manner [41].

9.3
Pathological Role of Apoptotic Cell Death in Sepsis

The lesson from the failed anti-inflammatory-based therapeutic trials in sepsis has led to a query of the hypothesis that the major issue/pathologic process in sepsis is uncontrolled inflammation [42, 43]. Increasing evidence has suggested that during sepsis an impaired immune response develops and may also contribute to sepsis-induced multiple organ failure. This is supported by a study in which Docke *et al.* observed that immune-enhancing interferon-γ treatment improved outcome in septic patients [44]. It has also been documented that mice lacking functional lymphocytes i.e. RAG−/− [45] or γδT − /− mice [46] showed a much higher mortality when subjected to sepsis induced by cecal ligation and puncture (CLP) as compared to their background controls. In this regard, evidence has begun to point towards pathogenic changes in the apoptotic process as another possible explanation for the morbidity/mortality observed [8, 47]. Here, we review the information that has formed the basis for this concept in sepsis.

With respect to the type of immune cells lost, lymphocytes appear to be the most overt cell type to be the target of increased apoptosis seen during sepsis. Studies from a number of laboratories show that increased lymphocyte apoptosis was readily observed in the thymus, spleen, and gut-associated lymphoid tissues (GALT) after CLP [47–49]. The increased thymocyte apoptosis in CLP mice appears to be driven by glucocorticoids or nitric oxide but not endotoxin or TNF [50]. In addition, complement C5a has been reported to contribute to thymocyte apoptosis in early sepsis [51]. However, sepsis-induced apoptosis in lymphocytes from GALT, such as T- and B-cells from Peyer's patches, small intestinal intraepithelial lymphocytes, and lamina propria B-cells appears to be directed primarily through the Fas-FasL death receptor pathway [52]. Besides the thymus and mucosal lymphoid cells, a number of other lymphoid tissues also show evidence of increased apoptosis after injury. Mixed bone marrow cells from mice show increased apoptosis following CLP [47]. In the spleen, studies have indicated that increased apoptosis was present in splenocytes derived from mice following septic insult [53] as well as from patients who had recently succumbed to the lethal effects of sepsis, shock, and multiple organ failure [8]. In addition, a broad range of blood lymphocyte subsets (CD4 and CD8 T-cells, B-cells and NK cells) taken from septic patients show an increase in apoptosis; furthermore, these cells also exhibit evidence of activation of both intrinsic and extrinsic pathways [54] (Table 9.1).

Interestingly, unlike lymphocytes (where apoptosis must typically, be actively induced), neutrophils, once mature, exhibit a spontaneous (constitutive) programmed cell death process. Normally, the life span of mature blood neutrophils is between 6 and 12 h in the circulation, following which apoptosis is spontaneously initiated; they are then recognized as dying cells and removed by phagocytes, thus limiting neutrophil-induced tissue injury. However, many

Table 9.1 The levels of apoptosis (Ao) seen in experimental septic mice and septic patients in various immune and none-immune cell types.

Cell population	Septic mice	Septic human
Thymus	⇑	?
Bone marrow	⇑	?
Splenic lymphoid	⇑	⇑
Macrophage		
Peritoneal macrophage	⇑	?
Blood monocyte	?	⇑
Dendritic cell	⇑	⇑
Mucosal GALT		
Lymphocyte (T and B cell)	⇑	⇑
Epithelia	⇑	⇑
Blood		
Lymphocyte	⇑	⇑
Neutrophil	⇓	⇓
Endothelial cells	⇑	?
Hepatocytes	?	?

inflammatory agents reported to be released during sepsis, such as LPS, TNF, IL-8, IL-6, IL-1, GM-CSF, etc., have also been observed to stimulate the arrest or inhibition of neutrophil apoptosis. This is associated with a decrease in caspase-3 and -9 activities as well as a prolonged maintenance of the mitochondrial membrane potential [55]. The delayed apoptotic response provides the neutrophils with a longer life span, which in turn allows them to accumulate at the local tissue sites. The result of this persistence is leukocyte influx, which is speculated to contribute to tissue injury. In this respect, many studies have indicated that neutrophils from patients following major surgery, burn injury, sepsis, and acute respiratory distress syndrome (ARDS), as well as from mice subjected to CLP, showed evidence of decreased apoptosis [56]. However, controversy persists as to whether this actually contributes to organ injury (Table 9.1).

Other immune cell types that have been reported to show potentially important septic changes in their apoptotic process are the antigen presenting cell sub-populations. For example, macrophages have been shown to increase apoptosis after sepsis [57, 58]. The significance of this apoptotic myeloid event is illustrated by the observation that transgenic mice which overexpress Bcl-2 under myeloid restriction showed a survival benefit [58]. However, evidence of phagocytic cell apoptosis can be controversial, as these cells also actively phagocytose apoptotic material, which can make the delineation of macrophage apoptosis at times difficult. Hotchkiss and colleagues have reported the loss of $CD8^+$ lymphoid-derived dendritic cells from the spleen, after $CD4^+$ T-cell activation, via apoptosis during sepsis in both experimental and clinical settings [59]. As these are critical cells in mediating the adaptive immune response in the presentation of foreign antigens, their loss would be significant for fighting chronic infection. To this end, Scumpia *et al.* showed that loss of this cell population significantly compromised survival of septic mice [60] (Table 9.1).

Studies have also indicated that non-immune cells such as gut mucosal epithelial cells [61], lung epithelial cells [62], hepatocytes [63], and endothelial cells [64] exhibit apoptotic changes in clinical or experimental sepsis. While the significance of these apoptotic events in sepsis is yet to be clarified, it is tempting to speculate how this potentiated cell death might contribute not only to loss of innate host defense/barrier function but also to organ dysfunction/damage (Table 9.1).

Together, these studies have documented the changes in apoptosis in both immune and non-immune cells in both experimental and clinical settings of sepsis. However, the significance of the observation from a pathologic perspective, as well as the value of targeting biochemical events of the apoptotic process in sepsis, have been defined by the capacity to arrest experimental septic morbidity and mortality by inhibiting this process. The residual discussion of this review overviews these findings, while considering their value as therapeutic targets in sepsis.

9.4
Potential Anti-apoptotic Therapies for the Treatment of Sepsis

9.4.1
Recombinant Human Activated Protein C

Despite many anti-inflammatory and anti-thrombotic drugs showing promise in the treatment of sepsis in animal models and reaching clinical trials for the treatment of sepsis in humans, only recombinant human activated Protein C (rhaPC) has shown any efficacy in the clinical setting [2]. Recombinant activated protein C was found to reduce 28-day mortality by 6%, a relative reduction of 19%, in patients with a documented or suspected infectious source, signs of the systemic inflammatory response syndrome, and evidence of organ dysfunction [4].

The successful application of rhaPC and the concurrent failure of other anticoagulant trials led to the investigation of the mechanism of action of rhaPC, with the discovery that although rhaPC does affect both the coagulation cascade and the mechanisms of the inflammatory response, it also has a significant anti-apoptotic effect on multiple cell types. The effects of rhaPC on coagulation, inflammation, as well as apoptosis will be briefly reviewed here.

Recombinant activated protein C inhibits thrombin formation directly and via inactivation of factors Va and VIIIa, which attenuates thrombin-induced inflammatory cytokine release and endothelial dysfunction, as well as directly attenuating the inflammatory response of human endothelial cells via binding to protease-activated receptor-1 (PAR-1) in an endothelial protein C receptor (EPCR)-dependent manner [65, 66]. Joyce *et al.* [67] demonstrated that rhaPC alone down-regulated the expression of nuclear factor-kappaB (NF-κB) and attenuated NF-κB expression following TNF-α induction. Furthermore, TNF-α induced expression of NF-κB regulated adhesion molecules, such as ICAM, VCAM, and E-selectin, and was inhibited by rhaPC, potentially inhibiting leukocyte adherence *in vivo*.

The effect of rhaPC on microcirculation *in vivo* was recently demonstrated in a rat model utilizing intravital microscopy following LPS versus LPS plus rhaPC injection [68]. At 1, 2, and 3 h post-injection of LPS alone or LPS plus either low or high-dose rhaPC, the mesenteric microcirculation was examined under intravital microscopy. Both treatment groups had a significantly lower number of adherent leukocytes per field than the control group. The low-dose rhaPC group was found to have a significantly lower number of microvascular bleeding events and a higher arteriole and venule red blood cell velocity than both the control and high-dose rhaPC group. In addition, LPS also induced an early rise in serum TNF and a later rise in serum IL-6, both of which were attenuated by rhaPC treatment. A possible correlation with ensuing organ damage was also demonstrated by rise in alanine transaminase and blood urea nitrogen in the LPS group, which was blocked by rhaPC treatment.

Activated protein C has been shown to counter the induction of apoptosis in animal studies as well as in human endothelial and monocyte cell lines *in vitro* [65, 67, 69, 70]. Recombinant human activated protein C prevented staurosporine-induced apoptosis in human endothelial cells and up-regulated expression of multiple anti-apoptotic genes, including endothelial nitric oxide synthase (eNOS), Al, Bcl-2 homolog, and inhibitor of apoptosis-1 [67]. Apoptosis in a human monocyte cell line is also inhibited by rhaPC, an effect that is mediated by EPCR [65]. In human brain endothelial cells subjected to hypoxic injury, Cheng *et al.* [69] found a 60% reduction in apoptosis with the addition of rhaPC. In addition, hypoxia increased levels of the pro-apoptotic molecules p53, Bax, and caspase-3, and decreased levels of the anti-apoptotic protein Bcl-2. All of these changes were attenuated by the addition of rhaPC.

In an *in vivo* murine model of ischemic stroke, the application of rhaPC decreased the size of infarction and edema formation, and decreased neutrophil infiltration and endothelial ICAM-1 expression [71]. In a similar model, Cheng *et al.* [69] demonstrated that the rescue effect of rhaPC is mediated by EPCR and PAR-1. A direct neuroprotective effect of aPC was recently shown *in vitro* and *in vivo* following N-methyl-D-aspartate (NMDA) and staurosporine induction of apoptosis [72]. In NMDA-induced apoptosis, aPC was found to decrease translocation of apoptosis-inducing factor (AIF) into the nucleus, block caspase-3 and p53 induction, and attenuate the increase in Bax and decrease in Bcl-2. Activated protein C also blocked caspase-8 activation, thus blocking staurosporine-induced apoptosis.

These insights into the anti-apoptotic effects of the only drug developed explicitly for the treatment of sepsis, combined with current insights into the association of apoptosis with morbidity and mortality in sepsis, may shed some light on why rhaPC has had an effect on mortality where other therapies have not. Furthermore, these observations indicate that altering the apoptotic response during sepsis may potentially be of benefit.

9.4.2
Targeting the Extrinsic Apoptotic Pathway

9.4.2.1 Targeting Fas

Manipulation of the extrinsic apoptotic pathway has shown promise in the mouse model of polymicrobial sepsis cecal ligation and puncture (CLP). Utilizing C3H/HeJ (endotoxin-tolerant) and C3H/HeJ-FasL^{gld} (endotoxin-tolerant/FasL-deficient) mice, our laboratory showed that the Fas/FasL pathway, rather than endotoxin, is a mediator of apoptosis in the CD220+ B-cells of the Peyer's patches as well as the CD4+ and CD8+ T-cells of the intestinal intraepithelial lymphocyte (IEL) population during sepsis. FasL deficiency was also associated with a better survival following CLP as compared to the endotoxin-tolerant mice [73, 74]. Similarly, splenocytes harvested 24 h after CLP and stimulated with concavalin A showed an increase in CD4+ T-cell apoptosis as compared to Sham, which was associated with an increase in Fas expression. This increase in apoptosis seen following sepsis was found to be dependent on Fas/FasL when these studies were repeated in FasL-deficient mice [75].

Following these promising results in the FasL-deficient mouse, methods for reversibly inhibiting the Fas/FasL pathway after septic injury were examined. In a follow-up series of experiments, Fas-receptor fusion protein (FasFP) was injected 12 h after the induction of sepsis. Radioactive micro-beads were then injected 24 h after CLP/Sham, and the hemodynamic profile was measured prior to euthanasia and organ harvest. Sepsis caused a decrease in indicators of tissue perfusion such as cardiac index and delivery as well as consumption of oxygen, which was reversed with the administration of FasFP. Organ-specific blood flow, as measured by the radioactivity of whole organs, was decreased by sepsis and restored by FasFP administration. Serum alanine transaminase (ALT), aspartate transaminase (AST), and lactate dehydrogenase (LDH) levels were similarly decreased during sepsis and restored by FasFP, indicating an attenuation of liver injury [76]. Injection of FasFP 12 h after sepsis was also associated with a significant decrease in mortality following CLP, as well as decreased Kupffer cell apoptosis and markers of liver injury [77].

Small interfering RNA (siRNA) is another potential reversible inhibitor of the Fas/FasL pathway. SiRNA consists of short segments (20–25 nucleotides) of double-stranded RNA targeted to a specific genetic sequence, with the effect of disrupting gene expression. When given by hydrodynamic injection (2 ml given over 5 s) 30 min after CLP, siRNA has the capability of knocking down mRNA and protein levels of Fas for up to 10 days. The administration of Fas siRNA reduced the levels of active caspase-3 and pro-caspase-9 (markers of apoptosis) in the liver and spleen, reduced liver toxicity as measured by serum AST and ALT, and improved survival following CLP. Significantly, the administration of Fas siRNA was still effective in improving survival when given 12 h after CLP, likely making any transition to a clinical setting more feasible [78]. A major limitation to the clinical application of siRNA is the method of delivery, as the hydrodynamic injection method employed in the laboratory setting to ensure rapid delivery of the naked siRNA constructs into cells, is not feasible in the clinical setting. Methods of packaging the siRNA for targeted delivery, including cationic liposomes or viral vectors are a current topic of research [79]. Another limitation is the potential for sequence-independent activation of the innate immune system by siRNA through PKR or TLR3, 7, 8 or 9, with subsequent elevation in interferon (IFN)-α and -β as well as TNF-α and IL-6 [80–82]. In the above study, serum IFN-α and IL-6 levels in naïve mice following injection of Fas siRNA were not elevated, indicating that delivery of Fas siRNA via this method does not induce inflammation; however, as future delivery mechanisms are explored, this potential for toxicity must continue to be explored.

9.4.2.2 Targets Downstream of Fas

The extrinsic apoptotic pathway has also been disrupted downstream from the Fas receptor at the Fas-associated Death Domain (FADD), which, as stated earlier, is an adapter protein that associates with Fas and caspase-8 to form the Death-inducing Signaling Complex (DISC) responsible for the initiation of further caspase activation and the initiation of programmed cell death. Transgenic expression of a FADD dominant negative (FADD-DN) mutation in mice significantly reduced thymocyte as well as splenic CD3+ T-cell and CD20+ B-cell apoptosis seen in wild type mice following CLP. Mortality following CLP was also significantly reduced in the FADD-DN mutant mice, indicating that this reduction in lymphocyte apoptosis correlates with the outcome of sepsis [83]. This transgenic model illustrates the potential for targeting the extrinsic pathway at a number of points downstream from Fas.

Our laboratory has also used siRNA to target caspase-8 during sepsis. As mentioned above, caspase-8 associates with FADD to form the DISC, thus initiating a sequence of caspase activation and the process of programmed cell death. As was the case for Fas siRNA, caspase-8 siRNA delivered 30 min after CLP resulted in long-lasting suppression of caspase-8 mRNA and protein levels (up to 10 days), as well as decreased indices of apoptosis in the liver and spleen. Similarly, caspase-8 siRNA reduced indices of liver toxicity and significantly improved survival following CLP [78].

Manipulation of the extrinsic pathway of programmed cell death has proven to be beneficial in a mouse model of polymicrobial sepsis; however, as was seen in previous trials for the treatment of human sepsis, patient selection and timing of delivery of these potential therapies will be crucial to their success in clinical trials (Table 9.2).

Table 9.2 Targets for intervention within the apoptotic pathways and the current therapies that have been used to manipulate these targets are listed.

Target	Treatment
Fas death receptor	Fas fusion protein [76, 77] Fas siRNA [78]
Fas-associated Death Domain (FADD)	Expression of FADD-DN (gene therapy) [83]
Bcl-2 family	Overexpression of anti-apoptotic members (gene therapy) [45, 61, 84, 85] TAT-Bcl-x_L construct [88], TAT-BH4 construct [84] Adrenomedullin/AMBP-1 [95, 96]
Caspases	Pan-caspase inhibitor (z-VAD) [90] Caspase-3 inhibitor [91] Caspase-8 siRNA [78]
Proteases	Protease Inhibitors [92]
Akt	Overexpression of Akt (gene therapy) [94]
Apoptotic cell clearance	Dendritic cell-derived MFG-E8 exosomes [105]

9.4.3
Targeting the Intrinsic Apoptosis Pathway

9.4.3.1 Targeting the BCL-2 Family

Manipulation of the intrinsic, or mitochondrial, apoptotic pathway has also proven to be beneficial in murine sepsis. The anti-apoptotic protein Bcl-2 was one of the first targets for manipulation during experimental sepsis. Selective transgenic overexpression of T lymphocyte Bcl-2 completely protected thymocytes and splenic T-cells from sepsis-induced apoptosis. T lymphocyte Bcl-2 overexpression also significantly improved survival following CLP [45]. Similarly, selective Bcl-2 overexpression targeted to intestinal epithelial cells decreased intestinal epithelial apoptosis and improved survival following induction of sepsis via CLP or intratracheal instillation of *Pseudomonas aeruginosa* [61, 84]. A similar result was found when Bcl-x_L was selectively overexpressed in T lymphocytes, with a decrease in thymocyte and splenic T lymphocyte and splenic B lymphocyte apoptosis, as well as a significant improvement in survival following CLP [85].

The anti-apoptotic effects of Bcl-2 (and others in the Bcl-2 family such as Bcl-x_L) are inhibited by a group of proteins termed the BH3-only proteins, which in turn are activated by a variety of stimuli, as discussed above. This inhibition of the Bcl-2 family then allows pro-apoptotic proteins such as Bax to initiate the loss of mitochondrial membrane potential, leading to the release of cytochrome c into the cytoplasm and the activation of caspase-9 [86]. It is with this in mind that mice deficient in Bim, a BH3-only protein, were studied in the setting of sepsis. Bim−/− mice had near complete rescue of sepsis-induced apoptosis in the thymus and spleen as well as a significant survival advantage over wild type septic mice [83].

The above studies demonstrate that a therapy that could reversibly manipulate the Bcl-2 pathway at a variety of levels after a septic insult could potentially be beneficial in the treatment of sepsis. Finding a suitable means of delivery of a drug therapy to the cytosol to affect these proteins has been a challenge; however, recently the HIV-1 TAT basic protein has been conjugated to proteins for rapid, receptor-independent uptake into cells [87]. When conjugated to anti-apoptotic molecules such as Bcl-x_L [88] or FLICE inhibiting protein (FLIP) [89], these constructs have been shown to inhibit apoptosis *in vivo*. Hotchkiss and colleagues [85] utilized the TAT-Bcl-x_L construct as well as a TAT-conjugated peptide of the anti-apoptotic BH4 domain of Bcl-x_L to examine the effects of these constructs in experimental sepsis. The TAT-BH4 construct or a control construct with a two-amino acid substitution (as a control) was then infused via a subcutaneous continuous infusion pump and i.p. injection following CLP, and organs were harvested for evaluation of apoptosis 18 h post-CLP or sham operation. The TAT-BH4 construct decreased sepsis-induced splenocyte T- and B-cell apoptosis to near sham levels, and there was a trend toward decreased apoptosis in the blood and thymus. Survival using this novel construct has yet to be evaluated; nevertheless, the novel method of stable delivery of an anti-apoptotic therapy constitutes a major advancement (Table 9.2).

9.4.4 Targeting Both Pathways

9.4.4.1 Caspase Inhibitors

As mentioned above, inhibition of caspase-8 by caspase-8 siRNA reduced sepsis-induced apoptosis and improved survival after CLP [78]. Other caspases, within both the intrinsic and extrinsic pathways, have been targeted as well. The first caspase inhibitor examined in the mouse CLP model of sepsis was the pan-caspase inhibitor N-benzyloxycarbonyl-Val-Ala-Asp(O-methyl) fluoromethyl ketone (z-VAD) [90]. When administered 1 h after CLP, z-VAD significantly reduced sepsis-induced apoptosis in the thymus and spleen. When survival was assessed, z-VAD administration (6 mg/kg) immediately after and 24 h after CLP resulted in 100% survival as compared to 40% in controls. Interestingly, when the same dose was given 1 and 24 h post-CLP, the survival of the z-VAD group was only 72% as compared to 44% for the control group, and when high dose z-VAD (30 mg/kg) was given at 1 and 24 h, there was no difference from controls. When an intermediate dose (20 mg/kg) was given at 3, 24 and 48 h post-CLP, survival was improved to 82.4%. These differences in survival indicate that immediate administration of z-VAD was more beneficial than even a 1-h delay; however, by increasing the dose and adding a third dose at 48 h, survival could be restored even after a 3-h delay in treatment, an important consideration for clinical application. Survival in the high dose group was at the level of the controls, indicating also that there appears to be dose-limiting toxicity associated with the administration of z-VAD. In a similar study, a pan-caspase inhibitor L-826,920 (M-920) and a caspase-3 selective caspase inhibitor L-826,791 (M-791) were tested in the CLP model [91]. When administered 90 min after CLP by i.p. injection, both M-920 and M-791 significantly improved survival as compared to controls; furthermore, both caspase inhibitors effectively inhibited sepsis-induced apoptosis in the spleen and thymus. To prove that this survival benefit was the result of the rescue of lymphocytes from apoptosis, a survival study was then performed in Rag−/− mice, which resulted in no improvement in survival with M920.

The administration of caspase inhibitors in sepsis is not without its potential drawbacks. As only small amounts of caspases are needed to initiate the cell death pathway, large amounts of caspase inhibitor must reach the cell cytosol to inhibit apoptosis, while at the same time, the caspase inhibitors themselves can have detrimental effects within the cell, such as dampening the inflammatory response to infection [49]. Nevertheless, the administration of caspase inhibitors remains a potential future therapy in the treatment of sepsis (Table 9.2).

9.4.4.2 Bid Protein

There is also considerable cross-talk between the extrinsic and intrinsic pathways, which can result in cross-activation of caspases, potentially requiring inhibition of apoptosis at multiple sites to truly block the progression of cell death. Our laboratory and others have looked at the role of Bid in apoptosis during sepsis. As mentioned above, Bid is a link between the extrinsic and intrinsic pathways. Using Bid−/− mice, our laboratory has shown a decrease in sepsis-induced apoptosis in the spleen, thymus and Peyer's patch, and an increase in survival following CLP (unpublished data). Similarly, Chang *et al.* [83] have demonstrated a decrease in sepsis-induced apoptosis in CD3+ T-cells of the thymus and spleen as well as improved survival in Bid−/− mice. This data indicates that Bid, as a mediator of both apoptotic pathways, is a potential target for inhibition during sepsis.

9.4.5 Other Potential Anti-apoptotic Therapies

9.4.5.1 Protease Inhibitors in Sepsis

The HIV protease inhibitors (PI) that are currently used to treat HIV have been shown to decrease apoptosis in CD4+ T lymphocytes; thus, applying these drugs to the treatment of sepsis is potentially promising. A mixture of nelfinavir and ritonavir at doses that produced a serum level equivalent to that seen in humans during HIV treatment, was given as an oral lavage to mice subjected to CLP [92]. When given 24 h prior to CLP and every 8 h thereafter, this PI mixture improved survival to 67% as compared to 17% for controls. More pertinent for any clinical scenario, when the PI lavage protocol was started 4 h post-CLP, survival was increased to 50% from 17%. This PI mixture was also found to reduce sepsis-induced apoptosis in the spleen and thymus, as well as decrease the bacterial load in the blood. Furthermore, when given to Rag−/− mice they found no survival difference, indicating that this reduction in lymphocyte apoptosis is responsible for the noted survival effect. These clinically tested and readily available drugs could rapidly move to clinical trials for sepsis (Table 9.2).

9.4.5.2 Akt

Cell survival and programmed cell death are also regulated by trophic factors via the binding to membrane receptors with tyrosine kinase activity or seven transmembrane G protein-coupled receptors and the subsequent activation of the phosphatidylinositide 3′-OH kinase (PI3K)/c-Akt pathway. In regards to the intrinsic apoptotic pathway, Akt has been shown to phosphorylate Bad, a pro-apoptotic member of the Bcl-2 family, resulting in the sequestration of Bad in the cytoplasm away from the mitochondria, thus preventing the induction of apoptosis. Downstream from the Bcl-2/mitochondria interaction, Akt has also been found to inactivate caspase-9. Beyond the intrinsic and extrinsic apoptotic machinery, Akt has also been shown to inactivate the forkhead family of transcription factors, which target many of the genes for proteins associated with cell death, including the gene for FasL. In addition, Akt has been shown to enhance degradation of IκB, allowing NF-κB to translocate to the nucleus. NF-κB targets multiple genes, including those for the Bcl-2 family of proteins. Thus Akt has a broad effect on the apoptotic process [93].

In this regard, Akt has proven to be important in the regulation of apoptosis during sepsis. Transgenic mice that overexpress Akt in T lymphocytes have an improved survival, from 47 to 94%, following CLP [94]. This improvement in survival was associated with a three-fold increase in the release of IFN-γ by CD4+ splenocytes from septic transgenic Akt-overexpressing mice as compared to background septic controls, in response to anti-CD3/anti-CD28 stimulation. Additionally, Akt overexpression was associated with a reduction in sepsis-induced splenocyte CD3+ T-cell and CD19+ B-cell apoptosis. Clearly, manipulation of the PI3K/c-Akt pathway holds promise for future treatment of sepsis. In fact, the successful administration of intensive insulin therapy for glucose control in the intensive care unit may indeed prove to have anti-apoptotic effects through the PI3K/c-Akt pathway [6] (Table 9.2).

9.4.5.3 Adrenomedullin

Adrenomedullin (AM), a potent endogenous vasodilator, and its binding protein adrenomedullin binding protein-1 (AMBP-1) have been implicated in the progression to the hypodynamic phase of late sepsis. Administration of AM and AMBP-1 together at the onset of sepsis has been shown to prevent this transition, increase organ blood flow, decrease liver damage, and improve survival in CLP [95]. Adrenomedullin and AMBP-1, when administered together to septic mice, have also been shown to reduce aortic endothelial cell apoptosis as well as prevent the decrease in Bcl-2 protein and the increase in Bax gene expression in aortic tissue during sepsis [96]. These data suggest that the combined administration of AM and AMBP-1 may prove to be a beneficial in human sepsis (Table 9.2).

9.4.5.4 Clearance of Apoptotic Cells

There is gathering evidence that the pathology of apoptosis in sepsis may be more complex than just the loss of competent immune cells in the setting of infection. Another potential mechanism implicates the body's mechanisms for disposing of apoptotic material. As mentioned earlier, this is a normal physiologic process that is necessary for the resolution of inflammation and tissue remodeling [97–99]. Once a cell becomes apoptotic, it must then be recognized as such and disposed of by phagocytes without the inflammatory cytokine release typically observed after contact with pathogens [99]. In fact, *in vitro* studies have demonstrated that exposure of macrophages to apoptotic cells causes the macrophages to acquire an anti-inflammatory phenotype [100–103].

It is thus possible, yet so far untested, that prevention of lymphocyte apoptosis alters the outcome of sepsis by subsequently decreasing the load of apoptotic material present to be cleared in an anti-inflammatory or anergic fashion by phagocytes.

Recent evidence also indicates that which receptors and ligands are involved in the clearance of apoptotic cells will have an effect on the phenotypic alterations of the phagocyte. Multiple receptors and ligands have been implicated in the interaction between apoptotic cells and phagocytes [97–99]. Some of these interactions have been demonstrated to be distinctly anti-inflammatory, such as those involving phosphatidylserine (PS), while others, such as calreticulin/CD91, may induce inflammation; furthermore, it is likely that a balance between pro-inflammatory and anti-inflammatory signals influences the overall macrophage phenotype [97].

9.4.5.5 Milk Fat Globule Epidermal Growth Factor 8 (MFG-E8)

MFG-E8 is a bridging protein that opsonizes apoptotic cells by binding to $\alpha_v\beta_3$ integrins and PS to facilitate phagocytosis [104]. When administered immediately after CLP in the form of immature dendritic cell-derived exosomes containing MFG-E8, this molecule improved clearance of apoptotic cells and also improved survival [105]. At 20 h post-CLP, levels of serum IL-6 and TNF-α, both classic inflammatory cytokines, were reduced following the administration of MFG-E8, indicating the induction of a shift from an inflammatory to an anti-inflammatory state. The authors suggest that the increased clearance of apoptotic cells prior to them becoming secondarily necrotic and inducing organ damage by spilling their intracellular contents, is responsible for this shift and confers the survival benefit; however, it would also be intriguing to see what effect the administration of MFG-E8 had on classic anti-inflammatory serum cytokines, as well as the cytokine production and immune phenotype of the phagocytes that handle apoptotic cells. As with any anti-inflammatory therapy, timing of delivery and patient selection would be crucial in a clinical scenario, as patients who are outside of the pro-inflammatory phase of sepsis would likely not benefit from treatment. In an intriguing study by Hotchkiss and colleagues [106], apoptotic or necrotic cells were adoptively transferred to mice 5 days prior to CLP. The mice receiving necrotic cells survived significantly better than controls with an enhanced splenocyte IFN-γ response, whereas those receiving apoptotic cells survived significantly worse, with significantly less IFN-γ production. When these experiments were performed immediately following CLP, there was no effect on survival. This would seem to argue that there is an immune suppressive effect of apoptotic cells *in vivo*, and that an excess of necrotic material, rather than inducing inflammation and worsening survival, may act more as an immune adjuvant when present prior to sepsis. In either case, the clearance of apoptotic cells by a multitude of receptors undoubtedly plays a key role in the immune response during sepsis, and manipulation of these receptors/bridging molecules/ligands is a potential target of future therapies (Table 9.2).

9.5 Conclusions

In recent years there have been promising candidates for therapeutic intervention in sepsis; however, almost all of these trials have failed due to a lack of appreciation of the full spectrum of inflammatory and immunologic alterations that confront a patient with sepsis. Ongoing research into this complex disease process has shown that apoptosis is clearly associated with the morbidity and mortality of sepsis, and that it may indeed provide us with a means to change the outcome of this disease. Here we have considered how rhaPC, an anti-thrombotic therapy for sepsis, has proven beneficial over other anti-inflammatory and anti-thrombotic therapies, and has subsequently been proven to have substantial anti-apoptotic effects itself. We have also reviewed several experimental studies focusing on altering the outcome of sepsis through altering the apoptotic process at multiple intervention sites, revealing several targets that may be useful in designing stand-alone and/or adjuvant therapies. Together, these findings suggest that therapies which alter the apoptotic process during sepsis may prove effective in altering the outcome of this disease.

Acknowledgments

This work was supported in part by funds from NIH-RO1s GM53209 and HL73525 (to A.A.), as well as fellowship support from NIH-T32 GM65085 (for R.S.) and GAANN P200A03100 (for D.E.W-S.).

References

1. Angus, D.C. and Wax, R.S. (**2001**) Epidemiology of sepsis: an update. *Crit. Care Med.*, **29**, S109–S116.
2. Rice, T.W. and Bernard, G.R. (**2005**) Therapeutic interventions and targets for sepsis. *Ann. Rev. Med.*, **56**, 225–248.
3. Deans, K.J, Haley, M., Natanson, C, Eichacker, P.Q. and Minneci, P.C. (**2005**) Novel therapies for sepsis: a review. *J. Trauma*, **58**, 867–874.
4. Bernard, G.R., Vincent, J-L., Laterre, P-F., LaRosa, S.P., Dhainaut, J-F., Lopez-Rodriguez, A., Steingrub, J.S., Garber, G.E., Helterbrand, J.D., Ely, E.W. and Fisher, C.J. (**2001**) Efficacy and safety of recombinant human activated protein C for severe sepsis. *N. Engl. J. Med.*, **344**, 699–709.
5. Annane, D., Sebille, V., Charpentier, C., Bollaert, P-E,

Francois, B., Korach, J-M., Capellier, G., Cohen, Y., Azoulay, E., Troche, G., Chaumet-Riffaut, P. and Bellissant, E. (**2002**) Effect of treatment with low doses of hydrocortisone and fluorocortisone on mortality in patients with septic shock. *JAMA*, **288**, 862–871.

6. van den Berghe, G., Wouters, P., Weekers, F., Verwaest, C., Bruyninckx, F., Schetz, M., Vlasselaers, D., Ferdinande, P., Lauwers, P. and Bouillon, R. (**2001**) Intensive Insulin Therapy in Critically Ill Patients. *N. Engl. J. Med.*, **345**, 1359–1367.
7. Riedemann, N.C., Guo, R-F. and Ward, P. (**2003**) The enigma of sepsis. *J. Clin. Invest.*, **112**, 460–467.
8. Hotchkiss, R.S., Swanson, P.E., Freeman, B.D., Tinsley, K.W., Cobb, J.P., Matuschak, G.M., Buchman, T.G. and Karl, I.E. (**1999**) Apoptotic cell death in patients with sepsis, shock and multiple organ dysfunction. *Crit. Care Med.*, **27**, 1230–1251.
9. Ayala, A., Chung, C.S. and Chaudry, I.H. (**1998**) Lymphocyte apoptosis in sepsis. *Sepsis*, **2**, 55–71.
10. Kerr, J.F., Wyllie, A.H. and Currie, A.R. (**1972**) Apoptosis: a basic biological phenomenon with wide-ranging implications in tissue kinetics. *Br. J. Cancer*, **26**, 239–257.
11. Cohen, J.J., Duke, R.C., Fadok, V.A. and Sellins, K.S. (**1992**) Apoptosis and programmed cell death in immunity. *Annu. Rev. Immunol.*, **10**, 267–293.
12. Hirsch, T., Marchetti, P., Susin, S.A., Dallaporta, B., Zamzami, N., Marzo, I., Geuskens, M. and Kroemer, G. (**1997**) The apoptosis-necrosis paradox. Apoptogenic proteases activated after mitochondrial permeability transition determine the mode of cell death. *Oncogene.*, **15**, 1573–1581.
13. Leist, M. and Jaattela, M. (**2001**) Four deaths and a funeral: from caspases to alternative mechanisms. *Nat. Rev. Mol. Cell Biol.*, **2**, 589–598.
14. Li, C.J., Friedman, D.J., Wang, C., Metelev, V. and Pardee, A.B. (**1995**) Induction of apoptosis in uninfected lymphocytes by HIV-1 Tat protein. *Science*, **268**, 429–431.
15. Liles, W.C. (**1997**) Apoptosis-role in infection and inflammation. *Curr. Opin. Infect. Dis.*, **10**, 165–170.
16. Barinaga, M. (**1998**) Is apoptosis key in Alzheimer's disease? *Science*, **281**, 1303–1304.
17. Barr, P.J. and Tomei, L.D. (**1994**) Apoptosis and its role in human disease. *Bio/Technology*, **12**, 487–493.
18. Wesche, D.E., Lomas-Neira, J.L., Perl, M., Chung, C.S. and Ayala, A. (**2005**) Leukocyte apoptosis and its significance in sepsis and shock. *J. Leukocyte Biol.*, **25**, 325–337.
19. Hotchkiss, R.S., Tinsley, K.W. and Karl, I.E. (**2003**) Role of apoptotic cell death in sepsis. *Scand. J. Infect. Dis.*, **35**, 585–592.
20. Borner, C. (**2003**) The Bcl-2 protein family: sensors and checkpoints for life-or-death decisions. *Mol. Immunol.*, **39**, 615–647.
21. Luo, X., Budihardjo, I., Zou, H., Slaughter, C. and Wang, X. (**2007**) Bid, a Bcl2 interacting protein, mediates cytochrome c release from mitochondria in response to activation of cell surface death receptors. *Cell*, **94**, 481–490.
22. Acehan, D., Jiang, X., Morgan, D.G., Hauser, J.E., Wang, X. and Akey, C.W. (**2002**) Three-dimensional structure of the apoptosome: implications for assembly, procaspase-9 binding, and activation. *Mol. Cell*, **9**, 423–432.
23. Susin, S.A., Zamzami, N., Casado, M., Hirsch, T., Marchetti, P., Macho, A., Daugas, E., Geuskens, M. and Kroemer, G. (**1996**) Bcl-2 inhibits the mitochondrial release of an apoptogenic protease. *J. Exp. Med.*, **184**, 1331–1341.
24. Verhagen, A.M., Ekert, P.G., Pakusch, M., Silke, J., Connolly, L.M., Reid, G.E., Moritz, R.L., Simpson, R.J. and Vaux, D.L. (**2000**) Identification of DIABLO, a mammalian protein that

promotes apoptosis by binding to and antagonizing IAP proteins. *Cell*, **102**, 43–53.
25. Li, L.Y., Luo, X. and Wang, X. (**2001**) Endonuclease G is an apoptotic DNase when released from mitochondria. *Nature*, **412**, 95–99.
26. Scaffidi, C., Fulda, S., Srinivasan, A., Friesen, C., Li, F., Tomaselli, K.J., Debatin, K.M., Krammer, P.H. and Peter, M.E. (**1998**) Two CD95 (APO-1/Fas) signaling pathways. *EMBO J.*, **17**, 1675–1687.
27. Yin, X-M., Wang, K., Gross, A., Zhao, Y., Zinkel, S., Klocke, B., Roth, K.A. and Korsmeyer, S.J. (**1999**) Bid-deficient mice are resistant to Fas-induced hepatocellular apoptosis. *Nature*, **400**, 886–891.
28. Guicciardi, M.E., Bronk, S.F., Werneburg, N.W., Yin, X-M. and Gores, G.J. (**2005**) Bid is upstream of lysosome-mediated caspase 2 activation in tumor necrosis factor alpha-induced hepatocyte apoptosis. *Gastroenterology*, **129**, 269–284.
29. Nakagawa, T., Zhu, H., Morishima, N., Li, E., Yanker, B.A. and Yuan, J. (**2000**) Caspase-12 mediates endoplasmic-reticulum-specific apoptosis and cytotoxicity by amyloid-beta. *Nature*, **403**, 98–103.
30. Lamkanfi, M., Declercq, W., Kalai, M., Saelens, X. and Vandenabeele, P. (**2002**) Alice in caspase land. A phylogenetic analysis of caspases from worm to man. *Cell Death Differ.*, **9**, 358–361.
31. Saleh, M., Mathison, J.C., Wolinski, M.K., Bensinger, S.J., Fitzgerald, P., Droin, N., Ulevitch, R.J., Green, D.R. and Nicholson, D.W. (**2006**) Enhanced bacterial clearance and sepsis resistance in caspase-12-deficient mice. *Nature*, **440**, 1064–1068.
32. Yuan, J., Shaham, S., Ledoux, S., Ellis, H.M. and Horvitz, H.R. (**1993**) The C. elegans Cell Death Gene ced-3 encodes a protein similar to mammalian interleukin-1β-coverting enzyme. *Cell*, **75**, 641–652.
33. Hengartner, M.O. and Horvitz, H.R. (**1994**) *C. elegans* cell survival gene ced-9 encodes a functional homolog of the mammalian proto-oncogene bcl-2. *Cell*, **76**, 665–676.
34. Thompson, C.B. (**1999**) Apoptosis, in *Fundamental Immunology* (W.E. Paul), Lippincott-Raven, Philadelphia, pp. 813–829.
35. Kumar, S. (**1995**) ICE-like proteases in apoptosis. *Trends. Biochem. Sci.*, **20**, 198–202.
36. Miura, M., Zhu, H., Rotello, R., Hartwieg, E.A. and Junying, Y. (**1993**) Induction of apoptosis in fibroblasts by IL-1β-converting enzyme, a mammalian homolog of the *C. elegans* cell death gene *ced*3. *Cell*, **75**, 653–660.
37. Creagh, E.M. and Martin, S.J. (**2001**) Caspases, cellular demolition experts. *Biochem. Soc. Trans.*, **29**, 696–702.
38. Zhu, C., Qiu, L., Wang, X., Hallin, U., Cande, C., Kroemer, G., Hagberg, H. and Blomgren, K. (**2003**) Involvement of apoptosis-inducing factor in neuronal death after hypoxia-ischemia in the neonatal rat brain. *J. Neurochem.*, **86**, 306–317.
39. Hisatomi, T., Sakamoto, T., Murata, T., Yamanaka, I., Oshima, Y., Hata, Y., Ishibashi, T., Inomata, H., Susin, S.A. and Kroemer, G. (**2001**) Relocalization of apoptosis-inducing factor in photoreceptor apoptosis induced by retinal detachment *in vivo*. *Am. J. Pathol.*, **158**, 1271–1278.
40. Susin, S.A., Lorenzo, H.K., Zamzami, N., Marzo, I., Snow, B.E., Brothers, G.M., Mangion, J., Jacotot, E., Costantini, P., Loeffler, M., Larochette, N., Goodlett, D.R., Aebersold, R., Siderovski, D.P., Penninger, J.M. and Kroemer, G. (**1999**) Molecular characterization of mitochondrial apoptosis-inducing factor. *Nature*, **397**, 441–446.
41. Kroemer, G. and Martin, S.J. (**2005**) Caspase-independent cell death. *Nat. Med.*, **11**, 725–730.

42. Nasraway, S.A. (**1999**) Sepsis research: we must change course. *Crit. Care Med.*, **27**, 427–430.
43. Remick, D.G., Garg, S.J., Newcomb, D.E., Wollenberg, G., Huie, T.K. and Bolgos, G.L. (**1998**) Exogenous interleukin-10 fails to decrease the mortality or morbidity of sepsis. *Crit. Care Med.*, **26**, 895–904.
44. Docke, W.D., Randow, F., Syrbe, U., Krausch, D., Asadullah, K., Reinke, P., Volke, H.D. and Kox, W.J. (**1997**) Monocyte deactivation in septic patients: restoration by IFN-gamma treatment. *Nat. Med.*, **3**, 678–681.
45. Hotchkiss, R.S., Swanson, P.E., Knudson, C.M., Chang, K.C., Cobb, J.P., Osborne, D.F., Zollner, K.M., Buchman, T.G., Korsmeyer, S.J. and Karl, I.E. (**1999**) Overexpression of Bcl-2 in transgenic mice decreases apoptosis and improves survival in sepsis. *J. Immunol.*, **162**, 4148–4156.
46. Chung, C.S., Watkins, L., Funches, A., Lomas-Neira, J.L., Cioffi, W.G. and Ayala, A. (**2006**) Deficiency of gamma-delta T-lymphocytes contributes to mortality and immunosuppression in sepsis. *Am. J. Physiol.*, **291**, R1338–R1343.
47. Ayala, A., Herdon, C.D., Lehman, D.L., Ayala, C.A. and Chaudry, I.H. (**1996**) Differential induction of apoptosis in lymphoid tissues during sepsis: variation in onset, frequency, and the nature of the mediators. *Blood*, **87**, 4261–4275.
48. Oberholzer, C., Oberholzer, A., Clare-Salzler, M. and Moldawer, L.L. (**2001**) Apoptosis in sepsis: a new target for therapeutic exploration. *FASEB J.*, **15**, 879–892.
49. Hotchkiss, R.S. and Nicholson, D.W. (**2006**) Apoptosis and caspases regulate death and inflammation in sepsis. *Nat. Rev. Immunol.*, **6**, 813–822.
50. Ayala, A., Herdon, C.D., Lehman, D.L., DeMaso, C.M., Ayala, C.A. and Chaudry, I.H. (**1995**) The induction of accelerated thymic programmed cell death during polymicrobial sepsis: Control by corticosteroids but not tumor necrosis factor. *Shock*, **3**, 259–267.
51. Guo, R-F., Huber-Lang, M., Wang, X., Sarma, V., Padgaonkar, V.A., Craig, R.A., Riedemann, N.C., McClintock, S.D., Hlaing, T., Shi, M.M. and Ward, P.A. (**2000**) Protective effects of anti-C5a in sepsis-induced thymocyte apoptosis. *J. Clin. Invest.*, **106**, 1271–1280.
52. Ayala, A., Lomas, J.L., Grutkoski, P.S. and Chung, C.S. (**2003**) Pathological aspects of apoptosis in severe sepsis and shock? *Int. J. Biochem. Cell Biol.*, **35**, 7–15.
53. Hotchkiss, R.S., Swanson, P.E., Cobb, J.P., Jacobson, A., Buchman, T.G. and Karl, I.E. (**1997**) Apoptosis in lymphoid and parenchymal cells during sepsis: findings in normal and T and B cell deficient mice. *Crit. Care Med.*, **25**, 1298–1307.
54. Hotchkiss, R.S., Osmon, S.B., Chang, K.C., Wagner, T.H., Coopersmith, C.M. and Karl, I.E. (**2005**) Accelerated lymphocyte death in sepsis occurs by both the death receptor and mitochondrial pathway. *J. Immunol.*, **174**, 5110–5118.
55. Taneja, R., Parodo, J., Jia, S.H., Kapus, A. and Rotstein, O.D. (**2004**) (**1999**) Delayed neutrophil apoptosis in sepsis is associated with maintenance of mitochondrial transmembrane potential and reduced caspase-9 activity. *Crit. Care Med.*, **32**, 1460–1469.
56. Taneja, R., Yue, L., Marshall, J.C. Programmed cell death (apoptosis) and the immunologic derangements of critical illness, in *Immune Response in the Critically Ill* (eds J.C. Marshall and J. Cohen), Springer, Berlin Heidelberg New York, pp. 264–279.
57. Williams, M.A., Withington, S., Newland, A.C. and Kelsey, S.M. (**1998**) Monocyte anergy in septic shock is associated with a predilection to apoptosis and is reversed by granulocyte-macrophage

colony stimulating factor *ex vivo*. *J. Infect. Dis.*, **178**, 1421–1433.
58. Iwata, A., Stevenson, V.M., Minard, A., Tasch, M., Tupper, J., Lagasse, E., Weissman, I., Harlan, J.M. and Winn, R.K. (**2003**) Over-expression of Bcl-2 provides protection in septic mice by a trans effect. *J. Immunol.*, **171**, 3136–3141.
59. Hotchkiss, R.S., Tinsley, K.W., Swanson, P.E., Grayson, M.H., Osborne, D.F., Wagner, T.H., Cobb, J.P., Coopersmith, C. and Karl, I.E. (**2002**) Depletion of dendritic cells, but not macrophages, in patients with sepsis. *J. Immunol.*, **168**, 2493–2500.
60. Scumpia, P.O., McAuliffe, P.F., O'Malley, K.A., Ungaro, R., Uchida, T., Matsumoto, T., Remick, D.G., Clare-Salzler, M.J., Moldawer, L.L. and Efron, P.A. (**2005**) CD11c+ dendritic cells are required for survival in murine polymicrobial sepsis. *J. Immunol.*, **175**, 3282–3286.
61. Coopersmith, C.M., Stromberg, P.E., Dunne, W.M., Davis, C.G., Amiot, I.DM., Buchman, T.G., Karl, I.E. and Hotchkiss, R.S. (**2002**) Inhibition of intestinal epithelial apoptosis and survival in a murine model of pneumonia-induced sepsis. *JAMA*, **287**, 1716–1721.
62. Perl, M., Chung, C.S., Lomas-Neira, J., Rachel, T.M., Biffl, W.L., Cioffi, W.G. and Ayala, A. (**2005**) Silencing of Fas- but not caspase-8 in lung epithelial cells ameliorates experimental acute lung injury. *Am. J. Pathol.*, **167**, 1545–1559.
63. Kim, Y-M., Kim, T-H., Chung, H-T., Talanian, R.V., Yin, X-M. and Billiar, T.R. (**2000**) Nitric oxide prevents tumor necrosis factor α-induced rat hepatocyte apoptosis by the interruption of mitochondrial apoptotic signaling through S-nitrosylation of caspase-8. *Hepatology*, **32**, 770–778.
64. Mutunga, M., Fulton, B., Bullock, R., Batchelor, A., Gascoigne, A., Gillespie, J.I. and Baudouin, S.V. (**2001**) Circulating endothelial cells in patients with septic shock. *Am. J. Respis. Crit. Care Med.*, **163**, 195–200.
65. Joyce, D.E., Nelson, D.R. and Grinnell, B.W. (**2004**) Leukocyte and endothelial cell interactions in sepsis: relevance of the protein C pathway. *Crit. Care Med.*, **32**, S280–S286.
66. Riewald, M., Petrovan, R.J., Donner, A. and Ruf, W. (**2003**) Activated protein C signals through the thrombin receptor PAR1 in endothelial cells. *J. Endotox. Res.*, **9**, 317–321.
67. Joyce, D.E., Gelbert, L., Ciaccia, A., DeHoff, B. and Grinnell, B.W. (**2001**) Gene expression and profile of antithrombotic protein c defines new mechanisms modulating inflammation and apoptosis. *J. Biol. Chem.*, **276**, 11199–11203.
68. Iba, T., Kidokoro, A., Fukunaga, M., Nagakari, K., Shirahama, A. and Ida, Y. (**2005**) Activated protein C improves the visceral microcirculation by attenuating the leukocyte-endothelial interaction in a rat lipopolysaccharide model. *Crit. Care Med.*, **33**, 368–372.
69. Cheng, T., Liu, D., Griffin, J.H., Fernandez, J.A., Castellino, F., Rosen, E.D., Fukudome, K. and Zlokovic, B.V. (**2003**) Activated protein C blocks p53-mediated apoptosis in ischemic human brain endothelium and is neuroprotective. *Nat. Med.*, **9**, 338–342.
70. Joyce, D.E. and Grinnell, B.W. (**2002**) Recombinant human activated protein C attenuates the inflammatory response in endothelium and monocytes by modulating nuclear factor-kB. *Crit. Care Med.*, **30**(Suppl), S288–S292.
71. Shibata, M., Kumar, R., Amar, A., Fernandez, J.A., Hofman, F., Griffin, J.H. and Zlokovic, B.V. (**2001**) Anti-inflammatory, antithrombotic, and neuroprotective effects of activated protein C in a murine model of focal ischemic stroke. *Circulation*, **103**, 1799–1805.
72. Guo, H., Liu, D., Gelbard, H., Cheng, T., Insalaco, R., Fernandez, J.A., Griffin, J.H. and

Zlokovic, B.V. (**2004**) Activated protein C prevents neuronal apoptosis via protease activated receptors 1 and 3. *Neuron*, **41**, 563–572.

73. Ayala, A., Xu, Y.X., Ayala, C.A., Sonefeld, D.E., Karr, S.M., Evans, T.A. and Chaudry, I.H. (**1998**) Increased mucosal B-lymphocyte apoptosis during polymicrobial sepsis is a Fas ligand but not an endotoxin mediated process. *Blood*, **91**, 1362–1372.
74. Chung, C.S., Xu, Y.X., Wang, W., Chaudry, I.H. and Ayala, A. (**1998**) Is Fas ligand or endotoxin responsible for mucosal lymphocyte apoptosis in sepsis? *Arch. Surg.*, **133**, 1213–1220.
75. Ayala, A., Chung, C.S., Xu, Y.X., Evans, T.A., Redmond, K.M. and Chaudry, I.H. (**1999**) Increased inducible apoptosis in CD4+ T-lymphocytes during polymicrobial sepsis is mediated by Fas ligand (FasL) and not endotoxin. *Immunology*, **97**, 45–55.
76. Chung, C.S., Yang, S.L., Song, G.Y., Lomas, J., Wang, P., Simms, H.H., Chaudry, I.H.. and Ayala, A. (**2001**) Inhibition of Fas signaling prevents hepatic injury and improves organ blood flow during sepsis. *Surgery*, **130**, 339–345.
77. Chung, C.S., Song, G.Y., Lomas, J., Simms, H.H., Chaudry, I.H. and Ayala, A. (**2003**) Inhibition of Fas/Fas ligand signaling improves septic survival: differential effects on macrophage apoptotic and functional capacity. *J. Leukoc. Biol.*, **74**, 344–351.
78. Wesche-Soldato, D.E., Chung, C.S., Lomas-Neira, J.L., Doughty, L.A., Gregory, S.H. and Ayala, A. (**2005**) *In vivo* delivery of caspase 8 or Fas siRNA improves the survival of septic mice. *Blood*, **106**, 2295–2301.
79. Ayala, A., Wesche-Soldato, D.E., Perl, M., Lomas-Neira, J., Swan, R. and Chung, C.S. (**2007**) Blockade of apoptosis as a rational therapeutic strategy for the treatment of sepsis. *Novartis. Found Symp.*, **280**, 37–164.
80. Sledz, C.A., Holko, M., de Veer, M.J., Silverman, R.H. and Williams, B.RG. (**2003**) Activation of the interferon system by short-interfering RNAs. *Nat. Cell Biol.*, **5**, 834–839.
81. Kariko, K., Bhuyan, P., Capodici, J. and Weissman, D. (**2004**) Small interfering RNAs mediate sequence-independent gene suppression and induce immune activation by signaling through toll-like receptor 3. *J. Immunol.*, **172**, 6545–6549.
82. Sioud, M. (**2007**) RNA interference and innate immunity. *Adv. Drug Del. Rev.*, **59**, 153–163.
83. Chang, K.C., Unsinger, J., Davis, C.G., Schwulst, S.J., Muenzer, J.T., Strasser, A. and Hotchkiss, R.S. (**2007**) Multiple triggers of cell death in sepsis: death receptor and mitochondrial-mediated apoptosis. *FASEB J.*, **21**, 708–719.
84. Coopersmith, C.M., Chang, K.C., Swanson, P.E., Tinsley, K.W., Stromberg, P.E., Buchman, T.G., Karl, I.E. and Hotchkiss, R.S. (**2002**) Overexpression of Bcl-2 in the intestinal epithelium improves survival in septic mice. *Crit. Care Med.*, **30**, 195–201.
85. Hotchkiss, R.S., McConnell, K.W., Bullock, K., Davis, C.G., Chang, K.C., Schwulst, S.J., Dunne, J.C., Dietz, G.P., Bahr, M., McDunn, J.E., Karl, I.E., Wagner, T.H., Cobb, J.P., Coopersmith, C.M. and Piwnica-Worms, D. (**2006**) TAT-BH4 and TAT-Bcl-xL peptides protect against sepsis-induced lymphocyte apoptosis *in vivo*. *J. Immunol.*, **179**, 5471–5477.
86. Strasser, A. (**2005**) The role of BH3-only proteins in the immune system. *Nat. Rev. Immunol.*, **5**, 189–200.
87. Schwarze, S.R., Ho, A., Vocero-Akbani, A. and Dowdy, S.F. (**1999**) *In vivo* protein transduction: delivery of a biologically active protein into the mouse. *Science*, **285**, 1569–1572.
88. Kilic, E., Deitz, G.PH., Hermann, D.M. and Bahr, M. (**2002**) Intravenous TAT-Bcl-XL is protective

after middle cerebral artery occlusion in mice. *Ann. Neurol.*, **52**, 617–622.

89. Krautwald, S., Ziegler, E., Tiede, K., Pust, R. and Kunzendorf, U. (**2004**) Transduction of the TAT-FLIP fusion protein results in transient resistance to Fas-induced apoptosis *in vivo*. *J. Biol. Chem.*, **279**, 44005–44011.
90. Hotchkiss, R.S., Tinsley, K.W., Swanson, P.E., Chang, K.C., Cobb, J.P., Buchman, T.G., Korsmeyer, S.J. and Karl, I.E. (**1999**) Prevention of lymphocyte cell death in sepsis improves survival in mice. *Proc. Natl. Acad. Sci. USA*, **96**, 14541–14546.
91. Hotchkiss, R.S., Chang, K.C., Swanson, P.E., Tinsley, K.W., Hui, J.J., Klender, P., Xanthoudakis, S., Roy, S., Black, C., Grimm, E., Aspiotis, R., Han, Y., Nicholson, D.W. and Karl, I.E. (**2001**) Caspase inhibitors improve survival in sepsis: a critical role of the lymphocyte. *Nat. Immunol.*, **1**, 496–501.
92. Weaver, J.GR., Rouse, M.R., Steckelburg, J.M. and Badley, A.M. (**2004**) Improved survival in experimental sepsis with an orally administered inhibitor of apoptosis. *FASEB J.*, **18**, 1185–1191.
93. Datta, S.R., Brunet, A. and Greenberg, M.E. (**1999**) Cellular survival: a play in three Akts. *Genes. Dev.*, **13**, 2905–2927.
94. Bommhardt, U., Chang, K.C., Swanson, P.E., Wagner, T.H., Tinsley, K.W., Karl, I.E. and Hotchkiss, R.S. (**2004**) Akt decreases lymphocyte apoptosis and improves survival in sepsis. *J. Immunol.*, **172**, 7583–7591.
95. Yang, S., Zhou, M., Chaudry, I.H. and Wang, P. (**2002**) Novel approach to prevent the transition from the hyperdynamic phase to the hypodynamic phase of sepsis, role of adrenomedullin and adrenomedullin binding protein-1. *Ann. Surg.*, **236**, 625–633.
96. Zhou, M., Simms, H.H. and Wang, P. (**2004**) Adrenomedullin and adrenomedullin binding protein - 1 attenuate vascular endothelial cell apoptosis in sepsis. *Ann. Surg.*, **240**, 321–330.
97. Henson, P.M. and Hume, D.A. (**2006**) Apoptotic cell removal in development and tissue homeostasis. *Trends Immunol.*, **27**, 244–250.
98. Gregory, C.D. and Devitt, A. (**2004**) The macrophage and the apoptotic cell: an innate immune interaction viewed simplistically? *Immunology*, **113**, 1–14.
99. Fadok, V.A., Bratton, D.L. and Henson, P.M. (**2001**) Phagocyte receptors for apoptotic cells: recognition, uptake and consequences. *J. Clin. Invest.*, **108**, 957–962.
100. Fadok, V.A., Bratton, D.L., Konowal, A., Freed, P.W., Westcott, J.Y. and Henson, P.M. (**1998**) Macrophages that have ingested apoptotic cells *in vitro* inhibit proinflammatory cytokine production through autocrine/paracrine mechanisms involving TGF-beta, PGE2, and PAF. *J. Clin. Invest.*, **101**, 890–898.
101. McDonald, P.P., Fadok, V., Bratton, D. and Henson, P.M. (**1999**) Transcriptional and translational regulation of inflammatory mediator production by endogenous TGF-β in macrophages that have ingested apoptotic cells. *J. Immunol.*, **163**, 6164–6172.
102. Lucas, M., Stuart, L.M., Savill, J. and Lacy-Hulbert, A. (**2003**) Apoptotic cells and innate immune stimuli combine to regulate macrophage cytokine secretion. *J. Immunol.*, **171**, 2610–2615.
103. Lucas, M., Stuart, L.M., Zhang, A., Hodivala-Dilke, K., Febbraio, M., Silverstein, R., Savill, J. and Lacy-Hulbert, A. (**2006**) Requirements for apoptotic cell contact in regulation of macrophage responses. *J. Immunol.*, **177**, 4047–4054.
104. Hanayama, R., Tanaka, M., Miyasaka, K., Aozasa, K., Koike, M., Uchiyama, Y. and Nagata, S. (**2004**) Autoimmune disease and impaired

uptake of apoptotic cells in MFG-E8-deficient mice. *Science*, **304**, 1147–1150.
105. Miksa, M., Wu, R., Dong, W., Das, P., Yang, D. and Wang, P. (**2006**) Dendritic cell-derived exosomes containing milk fat globule epidermal growth factor-factor VIII attenuate proinflammatory responses in sepsis. *Shock*, **25**, 586–593.
106. Hotchkiss, R.S., Chang, K.C., Grayson, M.H., Tinsley, K.W., Dunne, B.S., Davis, C.G., Osborne, D.F. and Karl, I.E. (**2003**) Adoptive transfer of apoptotic splenocytes worsens survival, whereas adoptive transfer of necrotic splenocytes improves survival in sepsis. *Proc. Natl. Acad. Sci. USA*, **100**, 6724–6729.

10
The Role of Endothelium

Siew-Yean Lau and Fumito Ichinose

10.1
Introduction

The endothelium forms the inner cellular lining of blood and lymphatic vessels in the human body and plays a number of functions that are important for normal homeostasis. These functions include prevention of coagulation, orchestration of migration of blood cells into the tissues (by expression of adhesion molecules that specifically recognize and bind counterstructures on blood cells), regulation of the microcirculation (by dictating the tonus of the arterioles through vasodilating and vasoconstricting agents), and the regulation of vascular permeability [1] (Figure 10.1). Endothelium phenotype is modulated by systemic inflammatory syndrome (SIRS) in a number of different ways and the alteration of endothelial function plays a central role in sepsis pathogenesis. The goal of this chapter is to present an up-to-date review of the function of endothelium in health and during SIRS and sepsis and to explore the potential of endothelium as a therapeutic target.

10.2
Endothelial Cell Functions and their Changes in SIRS and Sepsis

Under physiological conditions, endothelial cells are highly active and constantly sensing and responding to changes in the local environment. The hallmark of the physiological function of endothelium is its spatial and temporal heterogeneity [2]. For instance, minor trauma and/or infection may trigger local endothelial cells to release inflammatory mediators that recruit leukocytes to wall off infection without affecting the function of endothelial cells in the remote vascular bed. The local activation of endothelium, that tends to benefit the host, may be considered physiologic or adaptive. In contrast, SIRS or sepsis induces generalized maladaptive endothelial dysfunction in which local physiological control of vascular function by endothelium is largely disrupted.

Sepsis and Non-infectious Systemic Inflammation. Edited by J.-M. Cavaillon and C. Adrie

ISBN: 978-3-527-31935-0

10.2.1 Regulation of Coagulation

10.2.1.1 Regulation of Coagulation in Normal Endothelium

Under normal conditions, endothelial cells inhibit blood coagulation by various mechanisms. They express thrombomodulin (TM), which not only binds thrombin but also shifts the specificity of this clotting enzyme from fibrin to protein C [3]. Activated protein C (APC), in the presence of its cofactor protein S, catalytically inactivates activated factors V and VIII of the clotting system [4]. Endothelial cells also have proteoglycans, such as heparin sulfate [5], on their surface, which can bind and potentiate the inhibitors anti-thrombin (AT III) [6] and tissue factor pathway inhibitor (TFPI) [7]. ATIII-heparin neutralizes the serine proteases in the clotting cascade, while TFPI limits activation of the extrinsic pathway. Endothelial cells also release low amounts of the tissue type plasminogen activator (tPA), leading to the formation of plasmin that degrades preformed fibrin. Additionally, endothelial cells inhibit platelet aggregation by producing prostacyclin (PGI2) [8] and nitric oxide (NO) [9], and by expressing a surface-bound adenosine diphosphatase, which hydrolyzes an important agonist of platelets, adenosine diphosphate [10]. Together, these natural anticoagulants dampen coagulation, enhance fibrinolysis, and remove microthrombi.

On the other hand, endothelium also produces procoagulant factors at low controlled levels including von Willebrand factor (vWF), thrombin receptor, tissue factor (TF) and plasminogen activator inhibitor-1 (PAI-1) to maintain vascular homeostasis.

Production of each of these factors is differentially regulated in time and space. For example, vWF, a glycoprotein that mediates primary hemostasis, is expressed at high levels within the heart and lung and in low levels within the kidney of mice [11]. Also, within a given organ, vWF expression is higher on the venous side of the circulation compared with the arterial side [11]. In contrast to vWF, tPA expression is highest in the brain, whereas PAI-1 levels are greatest in the aorta [12]. Finally, TM is expressed in most organs, with the notable exception of the brain [13].

10.2.1.2 Procoagulant Properties of Endothelium in Severe Sepsis

Endothelium triggered by inflammatory stimuli may lose its anticoagulant properties and become a procoagulant surface. Under *in vitro* conditions, the addition of bacterial endotoxin or cytokines to endothelial cells has been shown to decrease synthesis of TM, tPA, and heparin, to increase expression of TF and PAI-1 [14], and to generate procoagulant microparticles. The extent to which these changes occur in the intact endothelium is not entirely clear. In a recent study of patients with meningococcemia, TM levels were reduced in dermal microvessels, an effect that would be predicted to yield decreased levels of APC [15]. In a mouse model of endotoxemia, the administration of LPS resulted in reduction in total tissue TM antigen in the lung and brain, but not in the kidney [16], suggesting that sepsis-associated changes in TM expression may vary between organs. While sepsis is associated with increased levels of PAI-1 [17], an endothelial source of PAI-1 has not been established. With few exceptions, sepsis studies have consistently failed to demonstrate TF in the intact endothelium.

Other factors induced by severe sepsis further enhance procoagulant properties of endothelium. For instance, activated endothelial cells attract platelets, monocytes, and neutrophils – cells that are capable of initiating or amplifying coagulation. Endothelial activation may result in translocation of cell surface phospholipids that enhance binding of coagulation complexes [18]. Endothelial cells undergoing apoptosis may express an increasingly procoagulant phenotype. The development of a low blood-flow state in sepsis, whether secondary to reduced cardiac output, vasoconstriction, or occlusive lesions, may reduce clearance of activated serine proteases, thus promoting additional clotting.

10.2.2 Regulation of Leukocyte Adhesion/Migration and Vascular Permeability

Under physiologic conditions, the endothelium expresses few adhesion molecules. However, upon stimulation with a variety of agonists, such as cytokines, this changes dramatically. The cells express P-selectin, E-selectin, intercellular adhesion molecule-1 (ICAM), and other adhesion molecules such as vascular cell adhesion molecule-1 (VCAM) [19]. As a result, leukocytes and platelets may accumulate near the endothelium to bind to adhesion molecules [19]. Migration of leukocyte into tissues involves rolling of the leukocytes over the endothelium, followed by strong adherence, and finally transmigration into the tissues.

"Rolling" of leukocytes on endothelium involves the selectin family. Selectins are molecules on leukocytes (L-selectin) or even on platelets (P-selectin) and endothelial cells (E-selectin) that act as receptors to provide loose binding for rolling [20]. Selectins allow leukocytes to roll in the direction of flow into the proximity of activating signals displayed by endothelial cells.

The second step, adherence to the endothelium, involves receptors of the integrin family (β2-integrin) and immunoglobulin-like receptors [21]. These receptors allow leukocyte arrest and adherence to the endothelium. Simultaneously, this process requires some activation of the leukocytes as well. This is achieved by production of leukocyte agonists such as platelet-activating factor and chemokines by the endothelium. Studies in genetically-modified mice have confirmed the essential role of adhesion molecules in the process of migration of leukocytes into the tissues [19]. Meanwhile, activated endothelial cells also recruit increased numbers of platelets. In turn, platelet/endothelium interactions induce endothelial cells to produce specific neutrophil and mononuclear cell chemoattractants thus perpetuating inflammation [22].

A growing body of evidence suggests that the expression of endothelial-derived cell adhesion molecules such as E-selectin and ICAM are increased in septic shock and that these adhesion molecules may have prognostic value because the degree of expression tends to correlate with hemodynamic instability and the future development of multiple organ dysfunction syndrome [23]. Sessler and coworkers measured blood levels of the adhesion molecule ICAM-1 as a potential marker of endothelial cell activation in septic adults and healthy volunteers [24]. They established a relationship between increased ICAM-1 levels and the consequences of sepsis (i.e. multiple organ failure and death). Meanwhile, Watanabe and coworkers prevented endotoxin shock in rabbits by administering a specific monoclonal antibody against CD18 (integrin β2) [25]. In another study, Xu and coworkers observed that mice deficient in ICAM-1 were markedly protected against death in septic shock [26].

As such, by expressing adhesion molecules, the endothelium plays an important role in leukocyte adhesion and migration into the tissues, which leads to the development of capillary leakage throughout the body, resulting in systemic hypotension, peripheral edema, adult respiratory distress syndrome, and multiple organ dysfunction.

10.2.3 Regulation of the Microcirculation/Vasomotor Properties

The normal endothelium produces a number of vasoactive compounds that regulate the tonus of the arterioles and hence have a great influence on blood pressure. These compounds include the vasodilators NO and PGI_2 and the vasoconstrictors endothelin (ET), and thromboxane A_2 (TXA_2).

10.2.3.1 Nitric Oxide

Nitric oxide (NO) is a free radical with a short half-life in biological fluids [27]. NO reacts with a variety of cellular targets leading to a plethora of responses including vasorelaxation [9] and bronchodilation, as well as inhibition of mitochondrial respiration [28], inhibition of platelet and leukocyte activation, and modulation of vascular smooth muscle cell proliferation. NO is synthesized by a family of enzymes referred to as nitric oxide synthases (NOSs). There are three known isoforms (NOS1, NOS2, and NOS3), that are obligate homodimers that catalyze NADPH-dependent oxidation of L-arginine to NO and L-citrulline [29]. NOS1 is constitutively expressed in neuronal cells, including those in the heart. NOS2 or inducible NOS was first identified in macrophages but has since been detected in a wide variety of cells, typically after exposure to LPS and/or cytokines. NOS3 is constitutively expressed in the endothelial cells, endocardial cells, and cardiac myocytes [27]. There are important differences in the mechanisms regulating NOS isoform expression and activity [30]. NOS2 is predominantly regulated at the levels of transcription and protein stability. NOS1 and NOS3 are stimulated by increased intracellular calcium concentrations via activation and binding of calmodulin. NOS3 is also regulated by post-translational events that control its distribution within the cell (membrane-bound versus cytosolic) and its specific activity.

Under normal conditions, a small amount of NO is produced in endothelial cells by NOS3 in response to receptor-mediated and shear stress vasodilatory stimuli [31]. As a small gaseous molecule, NO diffuses down its concentration gradient, from endothelial cell to smooth muscle cell, where it stimulates soluble guanylate cyclase (sGC) to synthesize cGMP [32] that, in turn, activates cGMP-dependent protein kinase (PKG) leading to vascular relaxation [9]. The actions of cGMP are limited via its catabolism by phosphodiesterases (PDEs). NO can also elicit effects via cGMP-independent mechanisms including interactions with heme-containing molecules (in addition to sGC) and proteins containing reactive thiol groups [33]. NO also interacts with superoxide radical (O_2^-), thereby limiting NO bioavailability and resulting in the formation of the potent oxidant peroxynitrite ($ONOO^-$) [34]. In the presence of oxygenated hemoglobin (Hb), NO is rapidly metabolized to nitrate with formation of met-Hb. Met-Hb in erythrocytes is rapidly reduced to ferrous-Hb by met-Hb reductase.

NO produced by endothelial cells also diffuses into the microvascular lumen where it regulates RBC [35] and leukocyte deformability [36], leukocyte–endothelial adhesion in postcapillary venules [37], and platelet adhesion and aggregation [38]. As such, NO is an important factor in maintaining the integrity of blood flow through the microcirculation by regulating resistance vessel diameter, blood rheology, interaction between cellular blood elements and the vascular wall, and blood volume.

High levels of NO produced by NOS2 have been implicated in the pathogenesis of cardiac dysfunction in sepsis [39]. For instance, increased levels of nitrite and nitrate were measured in the plasma of septic patients [40], and selective NOS2 inhibitors restored blood pressure in experimental models of sepsis and reversed hypotension in human endotoxemia [41]. Despite the prominent role of NOS2 in cardiovascular dysfunction of sepsis, NOS inhibitors that are not isoform-selective, such as N^G-monomethyl-L-arginine (L-NMMA), significantly decrease cardiac output and oxygen delivery [42]. A recent phase III clinical trial with L-NMMA for sepsis was terminated because of a higher mortality rate in the treatment group [42]. Of note, the serious adverse events in the treatment group appeared to be primarily cardiac in origin (decreased cardiac index and potential myocardial ischemia). These observations may relate to the deleterious effects of inhibition of NOS1 and/or NOS3 by nonselective NOS inhibitors in the setting of sepsis. We and others have previously reported that agents which inhibit all three isoforms of NOS have detrimental effects on cardiac function in endotoxemic animals [43, 44].

Drugs that generate NO, such as nitroglycerin and sodium nitroprusside, have long been used to reduce blood pressure and treat angina pectoris. Systemically-administered NO-donor compounds can dilate the pulmonary vasculature, but their efficacy is limited by systemic hypotension [45]. In lung injury, these drugs can impair matching of ventilation with perfusion leading to systemic arterial hypoxemia. Frostell and colleagues reasoned that NO administered via inhalation would relax the pulmonary vasculature but, upon reaching the bloodstream, would be scavenged by Hb thereby preventing systemic vasodilation [46]. These investigators studied conscious lambs with pulmonary vasoconstriction induced by intravenous administration of U46619, a thromboxane mimetic, or by breathing low oxygen concentrations. Inhalation of NO gas dose-dependently decreased PAP and pulmonary vascular resistance (PVR) in sheep with PAH but not in sheep with normal pulmonary vascular tone. The pulmonary vasodilator effects of breathing NO were readily reversible upon discontinuation of the gas. Breathing NO up to 80 parts per million (ppm) did not alter systemic blood pressure, a finding which has been consistently observed over a wide range of species, including man.

In clinical studies of patients with severe acute respiratory distress syndrome (ARDS), inhaled NO has been shown to produce selective pulmonary vasodilation and improve systemic oxygenation [47]. Although subsequent clinical studies, as well as a number of studies in animal models of acute lung injury, have confirmed a physiological benefit of inhaled NO therapy, subsequent randomized clinical trials reported disappointing outcome results. Inhaled NO therapy did not affect mortality rate, duration of mechanical ventilation, or the number of days alive and off mechanical ventilation in the two single-center pilot trials of small sample sizes ($n = 40$ [48] and 30 [49]) and two larger multi-center randomized trials ($n = 177$ [50] and 286 [51]). Since the majority of patients dying with ARDS suffer from multiple organ failure, beneficial effects of a lung-selective therapy such as inhaled NO (e.g. improvement of gas exchange and reduced PAP) may not alter the overall survival rate. Whether or not a subgroup of severely hypoxemic ARDS patients will respond more favorably to inhaled NO with improved clinical outcome has not yet been conclusively resolved.

10.2.3.2 **Prostacyclin (PGI_2) and Thromboxane A_2 (TXA_2)**

The endothelium also produces various eicosanoids including prostaglandins, PGI_2 and TXA_2. They are synthesized from arachidonic acid by endothelial cells via the enzyme cyclooxygenase (COX) [52]. Two isoforms of COX have been identified. COX-1 is a constitutively expressed isoform, whereas COX-2 is inducible [53]. COX-1 is continuously expressed in tissues and plays a role in normal homeostatic processes. Meanwhile, COX-2 synthesis is induced by inflammatory stimuli such as LPS and by cytokines.

Like NO, PGI_2 is a potent vasodilator, an inhibitor of platelet aggregation and thrombosis, and may synergize with NO in that respect [54]. PGI_2

spontaneously hydrolyzes to form the stable but inactive metabolite 6-keto-PGF1α. PGI_2 mediates its effects through the prostacyclin (IP) receptor. On the other hand, TXA_2 was so named because it induces platelet aggregation and thrombosis [55]. It is unstable and spontaneously hydrolyzes to form the stable but inactive product TXB_2. TXA_2 is a potent vasoconstrictor, bronchoconstrictor, and promoter of platelet aggregation [56]. Increased synthesis of eicosanoids in response to endotoxemia and sepsis occurs in several animal species. The relative amounts of TXB_2 and 6-keto-PGF1α are influenced by the experimental model and the frequency of endotoxin administration [57]. More direct evidence that eicosanoids mediate endotoxin-induced sequelae is provided by observations that inhibition of eicosanoid synthesis or blockade of specific receptors protects animals from shock sequelae. Numerous NSAIDs (COX inhibitors) have been evaluated for potential therapeutic benefit in endotoxemia and sepsis in animal models [57]. These compounds, when used in experimental sepsis or endotoxemia, generally improve survival or survival time and reduce cardiopulmonary dysfunction and indices of tissue injury. More selective inhibition of specific arachidonic acid metabolites has also been examined. Pretreatment with TXA_2 synthesis inhibitors or TXA_2 receptor antagonists ameliorated endotoxin-induced pulmonary hypertension, reduced cardiac output and hypotension, and decreased renal blood flow and renal glomerular microthrombi [57]. However, clinical trials with selective TXA_2 synthesis inhibitors have shown no survival benefit in patients with ARDS or sepsis. Bernard and associates conducted a double-blind, placebo-controlled trial of short-term intravenous ibuprofen in 455 patients with a diagnosis of sepsis [58]. The ibuprofen-treated group did not experience any increased incidence of renal dysfunction, gastrointestinal bleeding, or other adverse effects. Short-term treatment with ibuprofen did not significantly affect the duration of shock or 30-day survival rate. However, treatment with ibuprofen significantly decreased urinary 6-keto-PGF1α and TXB_2 excretion, temperature, heart rate, oxygen consumption, and lactic acidosis. In a subsequent analysis of a subset of hypothermic septic patients in this study, ibuprofen treatment was demonstrated to improve 30-day survival [59].

In contrast to the deleterious effects of TXA_2, other prostanoids such as PGI_2, and more recently 15-deoxy-Δ(12, 14)-PGJ_2, have been shown to be beneficial in endotoxemia or septic shock [60, 61]. In human septic conditions, a decrease in PGI_2 is thought to contribute to the reduction in capillary density in the microcirculation and heterogeneity of their spatial distribution, leading to impaired perfusion and local tissue oxygenation. Infusion of PGI_2 is protective in canine endotoxic shock. 15-Deoxy-Δ(12, 14)-PGJ_2 is protective in endotoxemic and septic animal models [60, 61]. Aerosolized PGI_2 has also been tested in ARDS patients [62]. The effect of PGI_2 was shown to be similar to that of inhaled NO by inducing pulmonary vasodilation and improving ventilation/perfusion matching [62]. Finally, newly evolving concepts about the roles that specific eicosanoids play in the resolution of inflammation [63, 64] may direct future therapeutic interventions.

10.2.3.3 Endothelin (ET)

In addition to the generation of vasodilating agents, endothelial cells also form vasoconstricting compounds, including ET. There are three types of ET, but vascular endothelial cells produce mainly ET-1 [65]. It is secreted constitutively and thereby participates in the regulation of vascular tone. ET-1 exerts vasoconstrictor effects through stimulation of ET_A receptors in vascular smooth muscle and vasodilator actions through stimulation of ET_B receptors in endothelial cells. However, ET_B receptors also contribute to vasoconstriction in some blood vessels [66].

Among the pathophysiological conditions known to involve the ET system, sepsis presents with the highest plasma levels of ET [67]. The possible involvement of the ET system in human septic shock is supported by a correlation between ET plasma levels and morbidity and mortality in septic patients [68]. In the experimental setting, endotoxin induces the expression of prepro ET-1 mRNA in the lung and heart [69]. Infusion of ET-1 to humans causes cardiovascular changes in part resembling those seen in sepsis, i.e. decreased cardiac output and vasoconstriction in the pulmonary, renal and splanchnic circulations [70].

The mechanisms responsible for the cardiac dysfunction observed during SIRS and sepsis have been the object of a number of studies. *In vivo* effects of ET-1 on the heart include coronary vascular constriction, decreased cardiac output, and arrhythmias [71]. The mixed ET receptor antagonist, bosentan, has been shown to increase coronary blood flow in a porcine model of endotoxin shock. NO and cytokines, such as TNFα and IL-1β, are released during sepsis and endotoxemia, and are suggested to exert cardiodepressant effects [72]. Endotoxin-induced coronary vasoconstriction in rats depends on TNFα-mediated ET-1 release [73] and antagonizing ET receptors might, therefore, influence the effects of cytokines, such as TNFα, by modulating their cardiodepressant effects. Nevertheless, clinical studies aimed at preventing or opposing the effects of TNFα, IL-1β and NO have so far not shown favorable effects on cardiac performance [42].

ET-1 is also believed to play an important role in the regulation of pulmonary vascular tone, and the main vascular effects of ET-1 during pathological conditions within the lung are believed to be constrictor [74]. ET-1 has also been shown to increase pulmonary microvascular permeability in experimental models [75]. The pulmonary dysfunction during sepsis includes pulmonary hypertension, hypoxemia and low lung compliance and may progress to ARDS [76]. The involvement of the ET system in the late phase of endotoxin-induced pulmonary hypertension, has been shown, and mixed as well as selective ET_A receptor antagonism can counteract the late changes in the experimental setting [77]. ET-1 can also increase the expression of neutrophil adhesive molecules [78] and promote leukocyte migration into the alveoli [79]. Significantly lower levels of protein and numbers of white blood cells in broncheo-alveolar lavage fluid after endotoxemia are seen with mixed ET receptor antagonism [80], indicating a lower degree of permeability disturbances in the lungs.

Because of the various detrimental effects of ET during sepsis, many studies using mainly combined ET_A and ET_B receptor antagonism in experimental settings, have been carried out and have shown beneficial effects with improvements in central, regional and metabolic parameters [81]. Hypotension due to vasodilation often accompanies human septic shock and further vasodilatation as a result of ET receptor antagonism may be of some concern for the clinician. However, in the experimental setting, the vasodilating properties of ET antagonists have resulted in increased perfusion rather than hypotension, even though human data are still absent.

10.3
Endothelium-centered Therapeutic Approaches in Severe Sepsis

How do all of these altered endothelial properties contribute to altered perfusion and organ dysfunction? The endothelium is not the sole, but is a critical component of the host response to sepsis; therefore it is a potentially valuable target for sepsis therapy. While the endothelium can be activated under physiological conditions, generalized systemic activation of endothelium that characterizes severe sepsis is considered to be dysfunctional or pathological. It is therefore critical to distinguish physiological activation of endothelium from endothelial dysfunction when considering therapeutic strategies targeting endothelial function in sepsis. An approach that impairs normal physiological function of endothelium can be harmful to the host. Excessive inhibition of endothelial reaction to sepsis may be one of the reasons that many therapeutic strategies have failed to improve outcome in septic patients in the past. For instance, inhibition of all three NOSs by nonselective NOS inhibitors is likely to adversely affect endothelial function in severe sepsis [42].

Over the past decade, enormous resources have been expended on sepsis trials. However, the vast majority of these trials have failed to show reduced mortality in patients with severe sepsis with a few exceptions including APC. In 2004, consensus guidelines for the management of sepsis were published. Based on these guidelines, a therapeutic plan consisting of early, goal-directed therapy, lung-protective ventilation, broad-spectrum antibiotics, and APC has been advocated [82]. In this section, therapeutic approaches will be discussed from the perspective of endothelial function, potential differences between successful and unsuccessful therapeutic strategies, as well as emerging novel approaches. Readers are referred to a more comprehensive review of sepsis treatment in Chapter 17 of this book.

10.3.1
Soluble Inflammatory Mediators as Therapeutic Targets

Considerable efforts have been made to target LPS or inflammatory mediators that directly activate endothelial cells. Unfortunately, in large clinical trials, the use of specific anti-mediator therapy, including anti-endotoxin, anti-cytokine, anti-prostaglandin, anti-bradykinin, and anti-platelet activating factor, have consistently failed to decrease mortality in patients with severe sepsis [83]. This is perhaps because the approach was narrowly focused, pathways are redundant, or cytokines are critical to host defense and their blockade is excessively immunosuppressive. For instance, in the MONARCS trial (Monoclonal anti-TNF, A Randomized Controlled Sepsis) [84], 3000 patients were randomly assigned to receive one of several doses of a Fab component of a monoclonal antibody against TNF or placebo. There was a very small improvement in mortality, and this suggested that it would be unlikely that any single inflammatory mediator modulation would provide the "golden bullet" therapy.

10.3.2
Therapeutic Strategies Targeting Coagulation/Anti-coagulation Balance

Several anticoagulant molecules have been studied in nonhuman primate models of sepsis. While heparin and inactivated factor Xa inhibited activation of coagulation, they did not protect against organ dysfunction or mortality [85], suggesting that inhibition of coagulation is not sufficient to decrease mortality in sepsis. In contrast to agents that inhibit activity or generation of thrombin, administration of inactivated factor VIIa, ATIII, APC, or TFPI blocked activation of the coagulation and inflammatory pathways, reduced organ damage, and improved survival in a baboon model of sepsis [86]. It is of note that the anti-inflammatory effect of these anti-coagulant agents is related, at least in part, to their ability to block protease-activated receptor-mediated signaling and/or to activate protective signaling in the endothelial cells [87]. Taken together with the results of the failed anti-mediator/anti-cytokine trials, these observations suggest that death in severe sepsis is associated with simultaneous activation of the coagulation and pro-inflammatory pathways.

The therapeutic potential of activated protein C was evidenced in the phase 3 Protein C Worldwide Evaluation in Severe Sepsis (PROWESS) trial, in which the administration of human recombinant APC (drotrecogin alfa) to patients with severe sepsis resulted in reduced mortality [88]. A total of 1690 patients with a diagnosis of severe sepsis were randomized to receive either drotrecogin alfa or placebo. There was a statistically significant reduction in 28-day all-cause mortality (24.7 versus 30.8% in the treatment and placebo groups, respectively, $P < 0.005$) [88]. Based on these results, APC was approved for administration to patients with severe sepsis and an increased risk of death (as indicated by an

Acute Physiology and Chronic Health Evaluation (APACHE) II score greater than or equal to 25 or dysfunction of two or more organs).

Although the PROWESS trial was large and multicentered, and although APC treatment provided significant survival benefit in that trial, the use of APC as standard therapy in sepsis remains controversial [89, 90]. A subsequent trial of APC in patients with a low risk of death (the Administration of Drotrecogin Alfa [Activated] in Early Stage Severe Sepsis [ADDRESS] trial) was halted after an interim analysis for lack of effectiveness [91]. This outcome suggests that APC is not beneficial in low-risk patients. Criticism of the PROWESS trial includes patient selection criteria and the bleeding risk. Recent trauma or surgery, active hemorrhage, concurrent therapeutic anticoagulation, thrombocytopenia, and recent stroke were exclusion criteria for safety reasons in the PROWESS trial of APC. In the PROWESS trial, there was a trend toward a higher rate of serious bleeding among patients receiving APC than among patients in the placebo group (3.5 versus 2%, $P = 0.06$), especially during infusion of APC (2.4 versus 1%). Also, APC is very expensive, costing approximately $7000 per course [92]. Further randomized clinical trials will be needed to define the role of APC in the treatment of sepsis.

10.3.3 Nitric Oxide

Sepsis is associated with increased NOS2 activity and decreased NOS3 activity [93]. However, the relative role of NOS2 and NOS3 in mediating the sepsis phenotype remains unclear. In some studies, the use of NOS inhibitors yielded beneficial results [94], whereas other studies reported the opposite findings [42].

From 1997 to 1998, a multiple-center, randomized, placebo-controlled, double-blind study of the NOS inhibitor 546C88 was carried out to assess the safety and efficacy of this drug in patients with septic shock [42]. The predefined primary efficacy objective was survival at day 28. Patients with septic shock were allocated to receive 546C88 or placebo for up to 7 or 14 days in addition to conventional therapy. Unfortunately, the trial was stopped early after review by the independent Data Safety Monitoring Board. Day-28 mortality was 59% (259/439) in the 546C88 group and 49% (174/358) in the placebo group ($p < 0.001$).

The increase in mortality was evident early in the course of treatment. There was a higher proportion of cardiovascular deaths but a reduced proportion of deaths due to multiple organ failure. This may be explained by the fact that multiple organ failure tends to occur later in the course of septic shock, and the earlier deaths were due to cardiovascular causes. The mechanism underlying the overall adverse effect of 546C88 in the phase 3 study remains unknown. It is interesting to note the apparent absence of adverse effects on noncardiovascular organ function, the preponderance of possibly attributable cardiovascular

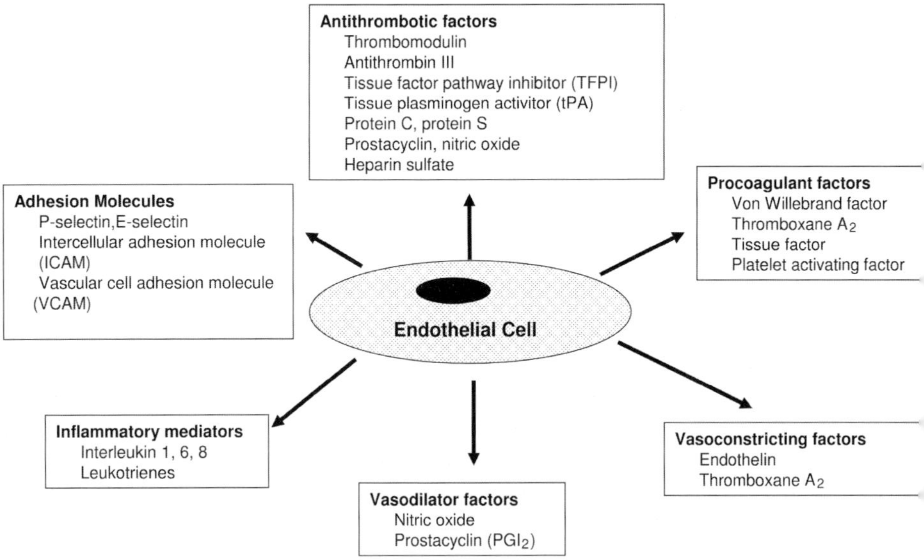

Figure 10.1 Multiple product of endothelial cells that confer physiological and pathophysiological functions.

adverse events (e.g. low cardiac output, pulmonary hypertension), and the increased proportion of deaths due to refractory shock in the 546C88 treatment group. These observations suggest that in some patients, treatment with 546C88 provoked a paradoxical worsening of their acute circulatory failure or perhaps, more specifically, myocardial dysfunction. This may have been due to either overcorrection of vascular tone resulting in an excessive increase in ventricular afterload compounded by inadequate inotropic support or a previously unrecognized primary effect of 546C88 on myocardial performance. Of note, in mouse models of septic shock, overexpression of NOS3 either in endothelial cells [95] or in cardiomyocyte attenuates cardiovascular dysfunction and improves survival [96]. Given the importance of NO-dependent signaling in endothelial cells, further exploitation of the NO/cGMP pathway may lead to novel therapeutic strategies in cardiovascular dysfunction of sepsis.

10.3.4 Statins

Statins or hydroxy methyl glutaryl-CoA (HMG-CoA) reductase inhibitors are widely used clinically as cholesterol-lowering agents because of their ability to block hepatic conversion of HMG-CoA to L-mevalonate and the production of isoprenoid geranylgeranylpyrophosphate (GGPP) [97]. Administration of statins has been shown to decrease risk of coronary and cerebrovascular events and increase survival rates in patients with coronary artery disease [97].

In vitro studies using human endothelial cells have shown that statins increase the expression of NOS3 [98], while decreasing the expression of NOS2 [99]. Statins may therefore restore vascular responsiveness during sepsis via re-establishment of a favorable balance between NOS3 and NOS2. In addition, statins also inhibits leukocyte–endothelial rolling, adherence, and transmigration in post-capillary venules in response to inflammatory stimuli [100]. They also exert anti-thrombotic effects by inhibiting the production of TXA_2 [101], and upregulate the expression of the COX-2 enzyme with a consequent increase in the synthesis of the anti-aggregant, vasodilating agent PGI_2 [102].

While no data from randomized trials of statins and sepsis are available, observational studies lent support to a potentially important salutary effect of statins. The largest study to date is a population-based cohort study involving the linked administrative databases in Ontario, Canada, and included a matched cohort of 69 168 patients [103]. The incidence of sepsis was substantially lower among patients receiving statins (hazard ratio (HR) 0.81; 95% CI 0.72–0.91). The protective association between statins and sepsis persisted in high-risk subgroups including patients with diabetes mellitus, malignancy, and those receiving oral steroids. Significant reductions in severe sepsis (HR 0.83; 95% CI 0.70–0.97) and fatal sepsis (HR 0.75; 95% CI 0.61–0.93) were also observed.

10.4
Role of Endothelium and Future Therapeutic Challenges in Sepsis and SIRS

Despite new information about the pathophysiology and treatment of severe sepsis, this disorder continues to be associated with an unacceptably high mortality rate. Many reasons have been postulated to explain the long history of failed clinical trials in sepsis. Future breakthroughs will require a conceptual shift that emphasizes relationships between the various mediators and cells involved in host response, rather than focusing on a single mediator or pathway of inflammation or coagulation. In that sense, endothelium presents itself as a very attractive target for the treatment of sepsis and SIRS. Additional studies promise to provide new insight into the endothelium, not as an isolated mechanism of sepsis pathophysiology, but rather as the coordinator of a far more expansive, spatially and temporally orchestrated reaction of the human body.

References

1. Hack, C.E. and Zeerleder, S. (**2001**) The endothelium in sepsis: source of and a target for inflammation. *Crit. Care Med.*, **29**, S21–S27.
2. Aird, W.C. (**2006**) Mechanisms of endothelial cell heterogeneity in health and disease. *Circ. Res.*, **98**, 159–162.

3. Esmon, C.T. (**1995**) Thrombomodulin as a model of molecular mechanisms that modulate protease specificity and function at the vessel surface. *FASEB J.*, **9**, 946–955.
4. Faust, S.N., Heyderman, R.S. and Levin, M. (**2001**) Coagulation in severe sepsis: a central role for thrombomodulin and activated protein C. *Crit. Care Med.*, **29**, S62–S67.
5. Ihrcke, N.S., Wrenshall, L.E., Lindman, B.J. and Platt, J.L. (**1993**) Role of heparan sulfate in immune system-blood vessel interactions. *Immunol. Today*, **14**, 500–505.
6. Mertens, G., Cassiman, J.J., Van den, B.H., Vermylen, J. and David, G. (**1992**) Cell surface heparan sulfate proteoglycans from human vascular endothelial cells. Core protein characterization and antithrombin III binding properties. *J. Biol. Chem.*, **267**, 20435–20443.
7. Iversen, N., Sandset, P.M., Abildgaard, U. and Torjesen, P.A. (**1996**) Binding of tissue factor pathway inhibitor to cultured endothelial cells-influence of glycosaminoglycans. *Thromb. Res.*, **84**, 267–278.
8. Weksler, B.B., Marcus, A.J. and Jaffe, E.A. (**1977**) Synthesis of prostaglandin I2 (prostacyclin) by cultured human and bovine endothelial cells. *Proc. Natl. Acad. Sci. USA*, **74**, 3922–3926.
9. Rees, D.D., Palmer, R.M. and Moncada, S. (**1989**) Role of endothelium-derived nitric oxide in the regulation of blood pressure. *Proc. Natl. Acad. Sci. USA*, **86**, 3375–3378.
10. Marcus, A.J., Broekman, M.J., Drosopoulos, J.H., Islam, N., Alyonycheva, T.N., Safier, L.B., Hajjar, K.A., Posnett, D.N., Schoenborn, M.A., Schooley, K.A., Gayle, R.B. and Maliszewski, C.R. (**1997**) The endothelial cell ecto-ADPase responsible for inhibition of platelet function is CD39. *J. Clin. Invest.*, **99**, 1351–1360.
11. Yamamoto, K., de Waard, V., Fearns, C. and Loskutoff, D.J. (**1998**) Tissue distribution and regulation of murine von Willebrand factor gene expression *in vivo*. *Blood*, **92**, 2791–2801.
12. Yamamoto, K. and Loskutoff, D.J. (**1996**) Fibrin deposition in tissues from endotoxin-treated mice correlates with decreases in the expression of urokinase-type but not tissue-type plasminogen activator. *J. Clin. Invest.*, **97**, 2440–2451.
13. Ishii, H., Salem, H.H., Bell, C.E., Laposata, E.A. and Majerus, P.W. (**1986**) Thrombomodulin, an endothelial anticoagulant protein, is absent from the human brain. *Blood*, **67**, 362–365.
14. Aird, W.C. (**2001**) Vascular bed-specific hemostasis: role of endothelium in sepsis pathogenesis. *Crit. Care Med.*, **29**, S28–S34.
15. Faust, S.N., Levin, M., Harrison, O.B., Goldin, R.D., Lockhart, M.S., Kondaveeti, S., Laszik, Z., Esmon, C.T. and Heyderman, R.S. (**2001**) Dysfunction of endothelial protein C activation in severe meningococcal sepsis. *N. Engl. J. Med.*, **345**, 408–416.
16. Weiler, H., Lindner, V., Kerlin, B., Isermann, B.H., Hendrickson, S.B., Cooley, B.C., Meh, D.A., Mosesson, M.W., Shworak, N.W., Post, M.J., Conway, E.M., Ulfman, L.H., von Andrian, U.H. and Weitz, J.I. (**2001**) Characterization of a mouse model for thrombomodulin deficiency. *Arterioscler. Thromb. Vasc. Biol.*, **21**, 1531–1537.
17. Mesters, R.M., Florke, N., Ostermann, H. and Kienast, J. (**1996**) Increase of plasminogen activator inhibitor levels predicts outcome of leukocytopenic patients with sepsis. *Thromb. Haemost.*, **75**, 902–907.
18. Combes, V., Simon, A.C., Grau, G.E., Arnoux, D., Camoin, L., Sabatier, F., Mutin, M., Sanmarco, M., Sampol, J. and Dignat-George, F. (**1999**) *In vitro* generation of endothelial microparticles and possible

prothrombotic activity in patients with lupus anticoagulant. *J. Clin. Invest.*, **104**, 93–102.
19. Springer, T.A. (**1995**) Traffic signals on endothelium for lymphocyte recirculation and leukocyte emigration. *Annu. Rev. Physiol.*, **57**, 827–872.
20. Kuebler, W.M., Borges, J., Sckell, A., Kuhnle, G.E., Bergh, K., Messmer, K. and Goetz, A.E. (**2000**) Role of L-selectin in leukocyte sequestration in lung capillaries in a rabbit model of endotoxemia. *Am. J. Respir. Crit. Care Med.*, **161**, 36–43.
21. Ma, X.L., Tsao, P.S. and Lefer, A.M. (**1991**) Antibody to CD-18 exerts endothelial and cardiac protective effects in myocardial ischemia and reperfusion. *J. Clin. Invest.*, **88**, 1237–1243.
22. Thienel, U., Loike, J. and Yellin, M.J. (**1999**) CD154 (CD40L) induces human endothelial cell chemokine production and migration of leukocyte subsets. *Cell Immunol.*, **198**, 87–95.
23. Cummings, C.J., Sessler, C.N., Beall, L.D., Fisher, B.J., Best, A.M. and Fowler, A.A., III (**1997**) Soluble E-selectin levels in sepsis and critical illness. Correlation with infection and hemodynamic dysfunction. *Am. J. Respir. Crit. Care Med.*, **156**, 431–437.
24. Sessler, C.N., Windsor, A.C., Schwartz, M., Watson, L., Fisher, B.J., Sugerman, H.J. and Fowler, A.A., III (**1995**) Circulating ICAM-1 is increased in septic shock. *Am. J. Respir. Crit. Care Med.*, **151**, 1420–1427.
25. Watanabe, S., Mukaida, N., Ikeda, N., Akiyama, M., Harada, A., Nakanishi, I., Nariuchi, H., Watanabe, Y. and Matsushima, K. (**1995**) Prevention of endotoxin shock by an antibody against leukocyte integrin beta 2 through inhibiting production and action of TNF. *Int. Immunol.*, **7**, 1037–1046.
26. Xu, H., Gonzalo, J.A., St Pierre, Y., Williams, I.R., Kupper, T.S., Cotran, R.S., Springer, T.A. and Gutierrez-Ramos, J.C. (**1994**) Leukocytosis and resistance to septic shock in intercellular adhesion molecule 1-deficient mice. *J. Exp. Med.*, **180**, 95–109.
27. Moncada, S. and Higgs, A. (**1993**) The L-arginine-nitric oxide pathway. *N. Engl. J. Med.*, **329**, 2002–2012.
28. Brown, G.C. (**2001**) Regulation of mitochondrial respiration by nitric oxide inhibition of cytochrome c oxidase. *Biochim. Biophy. Acta (BBA) - Bioenergetics*, **1504**, 46–57.
29. Gomez-Jimenez, J., Salgado, A., Mourelle, M., Martin, M.C., Segura, R.M., Peracaula, R. and Moncada, S. (**1995**) L-arginine: nitric oxide pathway in endotoxemia and human septic shock. *Crit. Care Med.*, **23**, 253–258.
30. Nathan, C. and Xie, Q.W. (**1994**) Nitric oxide synthases: roles, tolls, and controls. *Cell*, **78**, 915–918.
31. Macedo, M.P. and Lautt, W.W. (**1996**) Shear-induced modulation by nitric oxide of sympathetic nerves in the superior mesenteric artery. *Can. J. Physiol. Pharmacol.*, **74**, 692–700.
32. Vila-Petroff, M.G., Younes, A., Egan, J., Lakatta, E.G. and Sollott, S.J. (**1999**) Activation of distinct cAMP-dependent and cGMP-dependent pathways by nitric oxide in cardiac myocytes. *Circ. Res.*, **84**, 1020–1031.
33. Hare, J.M. and Stamler, J.S. (**2005**) NO/redox disequilibrium in the failing heart and cardiovascular system. *J. Clin. Invest.*, **115**, 509–517.
34. Beckman, J.S. and Koppenol, W.H. (**1996**) Nitric oxide, superoxide, and peroxynitrite: the good, the bad, and ugly. *Am. J. Physiol.*, **271**, C1424–C1437.
35. Bor-Kucukatay, M., Wenby, R.B., Meiselman, H.J. and Baskurt, O.K. (**2003**) Effects of nitric oxide on red blood cell deformability. *Am. J. Physiol. Heart Circ. Physiol.*, **284**, H1577–H1584.
36. Sato, Y., Walley, K.R., Klut, M.E., English, D., D'yachkova, Y., Hogg, J.C. and van Eeden, S.F. (**1999**) Nitric oxide reduces the sequestration

of polymorphonuclear leukocytes in lung by changing deformability and CD18 expression. *Am. J. Respir. Crit. Care Med.*, **159**, 1469–1476.

37. Kubes, P., Suzuki, M. and Granger, D.N. (**1991**) Nitric oxide: an endogenous modulator of leukocyte adhesion. *Proc. Natl. Acad. Sci. USA*, **88**, 4651–4655.
38. Radomski, M.W., Vallance, P., Whitley, G., Foxwell, N. and Moncada, S. (**1993**) Platelet adhesion to human vascular endothelium is modulated by constitutive and cytokine induced nitric oxide. *Cardiovasc. Res.*, **27**, 1380–1382.
39. Brady, A.J., Poole-Wilson, P.A., Harding, S.E. and Warren, J.B. (**1992**) Nitric oxide production within cardiac myocytes reduces their contractility in endotoxemia. *Am. J. Physiol.*, **263**, H1963–H1966.
40. Gomez-Jimenez, J., Salgado, A., Mourelle, M., Martin, M.C., Segura, R.M., Peracaula, R. and Moncada, S. (**1995**) L-arginine: nitric oxide pathway in endotoxemia and human septic shock. *Crit. Care Med.*, **23**, 253–258.
41. Wray, G.M., Millar, C.G., Hinds, C.J. and Thiemermann, C. (**1998**) Selective inhibition of the activity of inducible nitric oxide synthase prevents the circulatory failure, but not the organ injury/dysfunction, caused by endotoxin. *Shock*, **9**, 329–335.
42. Lopez, A., Lorente, J.A., Steingrub, J., Bakker, J., McLuckie, A., Willatts, S., Brockway, M., Anzueto, A., Holzapfel, L., Breen, D., Silverman, M.S., Takala, J., Donaldson, J., Arneson, C., Grove, G., Grossman, S. and Grover, R. (**2004**) Multiple-center, randomized, placebo-controlled, double-blind study of the nitric oxide synthase inhibitor 546C88: effect on survival in patients with septic shock. *Crit. Care Med.*, **32**, 21–30.
43. Ichinose, F., Hataishi, R., Wu, J.C., Kawai, N., Rodrigues, A.C., Mallari, C., Post, J.M., Parkinson, J.F., Picard, M.H., Bloch, K.D. and Zapol, W.M. (**2003**) A selective inducible NOS dimerization inhibitor prevents systemic, cardiac, and pulmonary hemodynamic dysfunction in endotoxemic mice. *Am. J. Physiol. Heart. Circ. Physiol.*, **285**, H2524–H2530.
44. Mitaka, C., Hirata, Y., Ichikawa, K., Uchida, T., Yokoyama, K., Nagura, T., Tsunoda, Y. and Amaha, K. (**1995**) Effects of nitric oxide synthase inhibitor on hemodynamic change and O2 delivery in septic dogs. *Am. J. Physiol.*, **268**, H2017–H2023.
45. Spronk, P.E., Ince, C., Gardien, M.J., Mathura, K.R., Straaten, H.M.O. and Zandstra, D.F. (**2002**) Nitroglycerin in septic shock after intravascular volume resuscitation. *Lancet*, **360**, 1395–1396.
46. Frostell, C., Fratacci, M.D., Wain, J.C., Jones, R. and Zapol, W.M. (**1991**) Inhaled nitric oxide. A selective pulmonary vasodilator reversing hypoxic pulmonary vasoconstriction. *Circulation*, **83**, 2038–2047.
47. Rossaint, R., Falke, K.J., Lopez, F., Slama, K., Pison, U. and Zapol, W.M. (**1993**) Inhaled nitric oxide for the adult respiratory distress syndrome. *N. Engl. J. Med.*, **328**, 399–405.
48. Michael, J.R., Barton, R.G., Saffle, J.R., Mone, M., Markewitz, B.A., Hillier, K., Elstad, M.R., Campbell, E.J., Troyer, B.E., Whatley, R.E., Liou, T.G., Samuelson, W.M., Carveth, H.J., Hinson, D.M., Morris, S.E., Davis, B.L. and Day, R.W. (**1998**) Inhaled nitric oxide versus conventional therapy: effect on oxygenation in ARDS. *Am. J. Respir. Crit. Care Med.*, **157**, 1372–1380.
49. Troncy, E., Collet, J.P., Shapiro, S., Guimond, J.G., Blair, L., Ducruet, T., Francoeur, M., Charbonneau, M. and Blaise, G. (**1998**) Inhaled nitric oxide in acute respiratory distress syndrome: a pilot randomized controlled study. *Am. J. Respir. Crit. Care Med.*, **157**, 1483–1488.

50. Dellinger, R.P., Zimmerman, J.L., Taylor, R.W., Straube, R.C., Hauser, D.L., Criner, G.J., Davis, K., Jr., Hyers, T.M. and Papadakos, P. (**1998**) Effects of inhaled nitric oxide in patients with acute respiratory distress syndrome: results of a randomized phase II trial. Inhaled Nitric Oxide in ARDS Study Group. *Crit. Care Med.*, **26**, 15–23.
51. Lundin, S., Mang, H., Smithies, M., Stenqvist, O. and Frostell, C. (**1999**) Inhalation of nitric oxide in acute lung injury: results of a European multicentre study. The European Study Group of Inhaled Nitric Oxide. *Intensive Care Med.*, **25**, 911–919.
52. Cook, J.A. (**2005**) Eicosanoids. *Crit. Care Med.*, **33**, S488–S491.
53. Turini, M.E. and DuBois, R.N. (**2002**) Cyclooxygenase-2: a therapeutic target. *Annu. Rev. Med.*, **53**, 35–57.
54. Zardi, E.M., Zardi, D.M., Dobrina, A. and Afeltra, A. (**2007**) Prostacyclin in sepsis: a systematic review. *Prostaglandins Other Lipid Mediat.*, **83**, 1–24.
55. Samuelsson, B., Goldyne, M., Granstrom, E., Hamberg, M., Hammarstrom, S. and Malmsten, C. (**1978**) Prostaglandins and thromboxanes. *Annu. Rev. Biochem.*, **47**, 997–1029.
56. Funk, C.D. (**2001**) Prostaglandins and leukotrienes: advances in eicosanoid biology. *Science*, **294**, 1871–1875.
57. Bulger, E.M. and Maier, R.V. (**2000**) Lipid mediators in the pathophysiology of critical illness. *Crit. Care Med.*, **28**, N27–N36.
58. Bernard, G.R., Wheeler, A.P., Russell, J.A., Schein, R., Summer, W.R., Steinberg, K.P., Fulkerson, W.J., Wright, P.E., Christman, B.W., Dupont, W.D., Higgins, S.B. and Swindell, B.B. (**1997**) The effects of ibuprofen on the physiology and survival of patients with sepsis. The Ibuprofen in Sepsis Study Group. *N. Engl. J. Med.*, **336**, 912–918.
59. Arons, M.M., Wheeler, A.P., Bernard, G.R., Christman, B.W., Russell, J.A., Schein, R., Summer, W.R., Steinberg, K.P., Fulkerson, W., Wright, P., Dupont, W.D. and Swindell, B.B. (**1999**) Effects of ibuprofen on the physiology and survival of hypothermic sepsis. Ibuprofen in Sepsis Study Group. *Crit. Care Med.*, **27**, 699–707.
60. Kaplan, J.M., Cook, J.A., Hake, P.W., O'Connor, M., Burroughs, T.J. and Zingarelli, B. (**2005**) 15-Deoxy-delta(12,14)-prostaglandin J(2) (15D-PGJ(2)), a peroxisome proliferator activated receptor gamma ligand, reduces tissue leukosequestration and mortality in endotoxic shock. *Shock*, **24**, 59–65.
61. Zingarelli, B., Sheehan, M., Hake, P.W., O'Connor, M., Denenberg, A. and Cook, J.A. (**2003**) Peroxisome proliferator activator receptor-gamma ligands, 15-deoxy-Delta(12,14)-prostaglandin J2 and ciglitazone, reduce systemic inflammation in polymicrobial sepsis by modulation of signal transduction pathways. *J. Immunol.*, **171**, 6827–6837.
62. Zwissler, B., Kemming, G., Habler, O., Kleen, M., Merkel, M., Haller, M., Briegel, J., Welte, M. and Peter, K. (**1996**) Inhaled prostacyclin (PGI2) versus inhaled nitric oxide in adult respiratory distress syndrome. *Am. J. Respir. Crit. Care Med.*, **154**, 1671–1677.
63. Levy, B.D., Clish, C.B., Schmidt, B., Gronert, K. and Serhan, C.N. (**2001**) Lipid mediator class switching during acute inflammation: signals in resolution. *Nat. Immunol.*, **2**, 612–619.
64. Lawrence, T., Willoughby, D.A. and Gilroy, D.W. (**2002**) Anti-inflammatory lipid mediators and insights into the resolution of inflammation. *Nat. Rev. Immunol.*, **2**, 787–795.
65. Inoue, A., Yanagisawa, M., Kimura, S., Kasuya, Y., Miyauchi, T., Goto, K. and Masaki, T. (**1989**) The human endothelin family: three structurally and pharmacologically

distinct isopeptides predicted by three separate genes. *Proc. Natl. Acad. Sci. USA*, **86**, 2863–2867.
66. Teerlink, J.R., Breu, V., Sprecher, U., Clozel, M. and Clozel, J.P. (**1994**) Potent vasoconstriction mediated by endothelin ETB receptors in canine coronary arteries. *Circ. Res.*, **74**, 105–114.
67. Tschaikowsky, K., Sagner, S., Lehnert, N., Kaul, M. and Ritter, J. (**2000**) Endothelin in septic patients: effects on cardiovascular and renal function and its relationship to proinflammatory cytokines. *Crit. Care Med.*, **28**, 1854–1860.
68. Weitzberg, E., Lundberg, J.M. and Rudehill, A. (**1991**) Elevated plasma levels of endothelin in patients with sepsis syndrome. *Circ. Shock*, **33**, 222–227.
69. Kaddoura, S., Curzen, N.P., Evans, T.W., Firth, J.D. and Poole-Wilson, P.A. (**1996**) Tissue expression of endothelin-1 mRNA in endotoxaemia. *Biochem. Biophys. Res. Commun.*, **218**, 641–647.
70. Weitzberg, E., Ahlborg, G. and Lundberg, J.M. (**1991**) Long-lasting vasoconstriction and efficient regional extraction of endothelin-1 in human splanchnic and renal tissues. *Biochem. Biophys. Res. Commun.*, **180**, 1298–1303.
71. Lerman, A., Hildebrand, F.L., Jr., Aarhus, L.L. and Burnett, J.C., Jr. (**1991**) Endothelin has biological actions at pathophysiological concentrations. *Circulation*, **83**, 1808–1814.
72. Kumar, A., Thota, V., Dee, L., Olson, J., Uretz, E. and Parrillo, J.E. (**1996**) Tumor necrosis factor alpha and interleukin 1beta are responsible for *in vitro* myocardial cell depression induced by human septic shock serum. *J. Exp. Med.*, **183**, 949–958.
73. Hohlfeld, T., Klemm, P., Thiemermann, C., Warner, T.D., Schror, K. and Vane, J.R. (**1995**) The contribution of tumour necrosis factor-alpha and endothelin-1 to the increase of coronary resistance in hearts from rats treated with endotoxin. *Br. J. Pharmacol.*, **116**, 3309–3315.
74. Cassin, S., Kristova, V., Davis, T., Kadowitz, P. and Gause, G. (**1991**) Tone-dependent responses to endothelin in the isolated perfused fetal sheep pulmonary circulation *in situ*. *J. Appl. Physiol*, **70**, 1228–1234.
75. Helset, E., Kjaeve, J. and Hauge, A. (**1993**) Endothelin-1-induced increases in microvascular permeability in isolated, perfused rat lungs requires leukocytes and plasma. *Circ. Shock*, **39**, 15–20.
76. Bigatello, L.M. and Zapol, W.M. (**1996**) New approaches to acute lung injury. *Br. J. Anaesth.*, **77**, 99–109.
77. Wanecek, M., Weitzberg, E., Alving, K., Rudehill, A. and Oldner, A. (**2001**) Effects of the endothelin receptor antagonist bosentan on cardiac performance during porcine endotoxin shock. *Acta Anaesthesiol. Scand.*, **45**, 1262–1270.
78. Ishizuka, T., Takamizawa-Matsumoto, M., Suzuki, K. and Kurita, A. (**1999**) Endothelin-1 enhances vascular cell adhesion molecule-1 expression in tumor necrosis factor alpha-stimulated vascular endothelial cells. *Eur. J. Pharmacol.*, **369**, 237–245.
79. Helset, E., Ytrehus, K., Tveita, T., Kjaeve, J. and Jorgensen, L. (**1994**) Endothelin-1 causes accumulation of leukocytes in the pulmonary circulation. *Circ. Shock*, **44**, 201–209.
80. Wanecek, M., Oldner, A., Rudehill, A., Sollevi, A., Alving, K. and Weitzberg, E. (**1997**) Cardiopulmonary dysfunction during porcine endotoxin shock is effectively counteracted by the endothelin receptor antagonist bosentan. *Shock*, **7**, 364–370.
81. Oldner, A., Goiny, M., Rudehill, A., Ungerstedt, U. and Sollevi, A. (**1999**) Tissue hypoxanthine reflects gut vulnerability in porcine endotoxin shock. *Crit. Care Med.*, **27**, 790–797.
82. Russell, J.A. (**2006**) Management of sepsis. *N. Engl. J. Med.*, **355**, 1699–1713.

83. Marshall, J.C. (**2000**) Clinical trials of mediator-directed therapy in sepsis: what have we learned? *Intensive Care Med.*, **26**(Suppl 1), S75–S83.
84. Panacek, E.A., Marshall, J.C., Albertson, T.E., Johnson, D.H., Johnson, S., MacArthur, R.D., Miller, M., Barchuk, W.T., Fischkoff, S., Kaul, M., Teoh, L., Van Meter, L., Daum, L., Lemeshow, S., Hicklin, G. and Doig, C. (**2004**) Efficacy and safety of the monoclonal anti-tumor necrosis factor antibody F(ab′)2 fragment afelimomab in patients with severe sepsis and elevated interleukin-6 levels. *Crit. Care Med.*, **32**, 2173–2182.
85. Taylor, F.B., Jr., Chang, A.C., Peer, G.T., Mather, T., Blick, K., Catlett, R., Lockhart, M.S. and Esmon, C.T. (**1991**) DEGR-factor Xa blocks disseminated intravascular coagulation initiated by *Escherichia coli* without preventing shock or organ damage. *Blood*, **78**, 364–368.
86. Welty-Wolf, K.E., Carraway, M.S., Miller, D.L., Ortel, T.L., Ezban, M., Ghio, A.J., Idell, S. and Piantadosi, C.A. (**2001**) Coagulation blockade prevents sepsis-induced respiratory and renal failure in baboons. *Am. J. Respir. Crit. Care Med.*, **164**, 1988–1996.
87. Oelschlager, C., Romisch, J., Staubitz, A., Stauss, H., Leithauser, B., Tillmanns, H. and Holschermann, H. (**2002**) Antithrombin III inhibits nuclear factor kappaB activation in human monocytes and vascular endothelial cells. *Blood*, **99**, 4015–4020.
88. Bernard, G.R., Vincent, J.L., Laterre, P.F., LaRosa, S.P., Dhainaut, J.F., Lopez-Rodriguez, A., Steingrub, J.S., Garber, G.E., Helterbrand, J.D., Ely, E.W. and Fisher, C.J., Jr. (**2001**) Efficacy and safety of recombinant human activated protein C for severe sepsis. *N. Engl. J. Med.*, **344**, 699–709.
89. Warren, H.S., Suffredini, A.F., Eichacker, P.Q. and Munford, R.S. (**2002**) Risks and benefits of activated protein C treatment for severe sepsis. *N. Engl. J. Med.*, **347**, 1027–1030.
90. Ely, E.W., Bernard, G.R. and Vincent, J.L. (**2002**) Activated protein C for severe sepsis. *N. Engl. J. Med.*, **347**, 1035–1036.
91. Abraham, E., Reinhart, K., Opal, S., Demeyer, I., Doig, C., Rodriguez, A.L., Beale, R., Svoboda, P., Laterre, P.F., Simon, S., Light, B., Spapen, H., Stone, J., Seibert, A., Peckelsen, C., De Deyne, C., Postier, R., Pettila, V., Artigas, A., Percell, S.R., Shu, V., Zwingelstein, C., Tobias, J., Poole, L., Stolzenbach, J.C. and Creasey, A.A. (**2003**) Efficacy and safety of tifacogin (recombinant tissue factor pathway inhibitor) in severe sepsis: a randomized controlled trial. *JAMA*, **290**, 238–247.
92. Manns, B.J., Lee, H., Doig, C.J., Johnson, D. and Donaldson, C. (**2002**) An economic evaluation of activated protein C treatment for severe sepsis. *N. Engl. J. Med.*, **347**, 993–1000.
93. Scott, J.A., Mehta, S., Duggan, M., Bihari, A. and McCormack, D.G. (**2002**) Functional inhibition of constitutive nitric oxide synthase in a rat model of sepsis. *Am. J. Respir. Crit. Care Med.*, **165**, 1426–1432.
94. Grover, R., Zaccardelli, D., Colice, G., Guntupalli, K., Watson, D. and Vincent, J.L. (**1999**) An open-label dose escalation study of the nitric oxide synthase inhibitor, N(G)-methyl-L-arginine hydrochloride (546C88), in patients with septic shock. Glaxo Wellcome International Septic Shock Study Group. *Crit. Care Med.*, **27**, 913–922.
95. Yamashita, T., Kawashima, S., Ohashi, Y., Ozaki, M., Ueyama, T., Ishida, T., Inoue, N., Hirata, K.I., Akita, H. and Yokoyama, M. (**2000**) Resistance to endotoxin shock in transgenic mice overexpressing endothelial nitric oxide synthase. *Circulation*, **101**, 931–937.
96. Ichinose, F., Buys, E.S., Neilan, T.G., Furutani, E.M., Morgan, J.G., Jassal, D.S., Graveline, A.R.,

Searles, R.J., Lim, C.C., Kaneki, M., Picard, M.H., Scherrer-Crosbie, M., Janssens, S., Liao, R. and Bloch, K.D. (**2007**) Cardiomyocyte-specific overexpression of nitric oxide synthase 3 prevents myocardial dysfunction in murine models of septic shock. *Circ. Res.*, **100**, 130–139.

97. Stalker, T.J., Lefer, A.M. and Scalia, R. (**2001**) A new HMG-CoA reductase inhibitor, rosuvastatin, exerts anti-inflammatory effects on the microvascular endothelium: the role of mevalonic acid. *Br. J. Pharmacol.*, **133**, 406–412.
98. Laufs, U., Gertz, K., Dirnagl, U., Bohm, M., Nickenig, G. and Endres, M. (**2002**) Rosuvastatin, a new HMG-CoA reductase inhibitor, upregulates endothelial nitric oxide synthase and protects from ischemic stroke in mice. *Brain Res.*, **942**, 23–30.
99. Madonna, R., Di Napoli, P., Massaro, M., Grilli, A., Felaco, M., De Caterina, A., Tang, D., De Caterina, R. and Geng, Y.J. (**2005**) Simvastatin attenuates expression of cytokine-inducible nitric-oxide synthase in embryonic cardiac myoblasts. *J. Biol. Chem.*, **280**, 13503–13511.
100. Scalia, R. and Stalker, T.J. (**2002**) Microcirculation as a target for the anti-inflammatory properties of statins. *Microcirculation*, **9**, 431–442.
101. Mazeaud, M.M., Driss, F., Quan Sang, K.H., Duranthon, V., Levenson, J., Simon, A. and Devynck, M.A. (**1992**) Biochemical and functional alterations associated with hypercholesterolemia in platelets from hypertensive patients. *Atherosclerosis*, **94**, 201–211.
102. Degraeve, F., Bolla, M., Blaie, S., Creminon, C., Quere, I., Boquet, P., Levy-Toledano, S., Bertoglio, J. and Habib, A. (**2001**) Modulation of COX-2 expression by statins in human aortic smooth muscle cells. Involvement of geranylgeranylated proteins. *J. Biol. Chem.*, **276**, 46849–46855.
103. Hackam, D.G., Mamdani, M., Li, P. and Redelmeier, D.A. (**2006**) Statins and sepsis in patients with cardiovascular disease: a population-based cohort analysis. *Lancet*, **367**, 413–418.

11 Coagulation

Marcel Levi

11.1 Introduction

Virtually all patients with sepsis have coagulation abnormalities. These abnormalities range from subtle activation of coagulation that can only be detected with sensitive markers for coagulation factor activation to somewhat stronger coagulation activation that may be detectable by a small decrease in platelet count and sub-clinical prolongation of global clotting times to fulminant disseminated intravascular coagulation (DIC), characterized by simultaneous widespread microvascular thrombosis and profuse bleeding from various sites [1]. Septic patients with severe forms of DIC may present with manifest thrombo-embolic disease or clinically less apparent microvascular fibrin deposition, that predominantly presents as multiple organ dysfunction [2–5]. Alternatively, severe bleeding may be the leading symptom [6], but quite often a patient with DIC has simultaneous thrombosis and bleeding (Figure 11.1). Bleeding is caused by consumption and subsequent exhaustion of coagulation proteins and platelets, due to the ongoing activation of the coagulation system [7]. In its most severe form this combination may present as the Waterhouse–Friderichsen syndrome, commonly seen during fulminant meningococcal septicemia, although many other microorganisms may cause this clinical state [8].

11.2 Relevance of Coagulation Abnormalities in Patients with Sepsis

There is ample evidence that activation of coagulation in concert with inflammatory activation can result in microvascular thrombosis and thereby contributes to multiple organ failure in patients with severe sepsis [9]. Firstly, extensive data has been reported on post-mortem findings of patients with coagulation abnormalities and DIC in patients with severe infectious

Sepsis and Non-infectious Systemic Inflammation. Edited by J.-M. Cavaillon and C. Adrie

ISBN: 978-3-527-31935-0

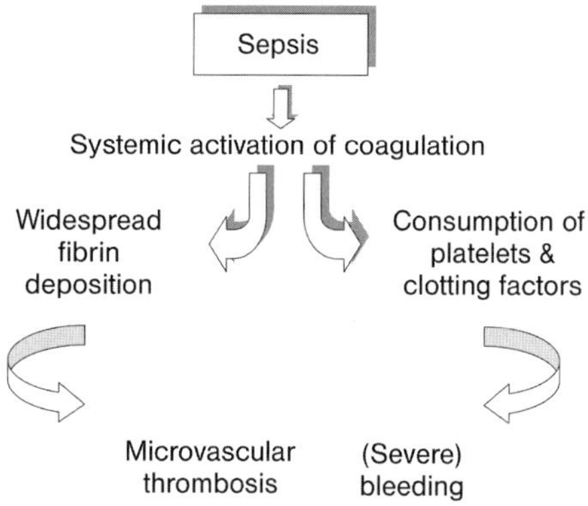

Figure 11.1 Schematic representation of coagulopathy in sepsis. Systemic activation of coagulation leads to microvascular thrombosis, contributing to organ dysfunction. Simultaneously ongoing and insufficiently compensated loss of platelets and coagulation factors may enhance the risk of bleeding.

diseases [10, 11]. These autopsy findings include diffuse bleeding at various sites, hemorraghic necrosis of tissue, microthrombi in small blood vessels and thrombi in mid-size and larger arteries and veins. The demonstration of ischemia and necrosis was invariably due to fibrin deposition in small and mid-size vessels of various organs [12]. Importantly, the presence of these intravascular thrombi appears to be clearly and specifically related to the clinical dysfunction of the organ. Secondly, experimental animal studies of DIC show fibrin deposition in various organs. Experimental bacteremia or endotoxemia causes intra- and extravascular fibrin deposition in kidneys, lungs, liver, brain and various other organs. Amelioration of the hemostatic defect by various interventions in these experimental models appears to improve organ failure and, in some but not all cases, mortality [13–16]. Interestingly, some studies indicate that amelioration of the activation of systemic coagulation will have a profound beneficial effect on the resolution of local fibrin deposition and improvement of organ failure [6, 17]. Lastly, clinical studies support the notion of coagulation as an important denominator of clinical outcome. DIC has shown to be an independent predictor of organ failure and mortality in patients with sepsis [2, 18]. In a consecutive series of patients with severe sepsis the mortality of patients with DIC was 43%, as compared with 27% in those without DIC. In this study, the severity of the coagulopathy was also directly related to mortality in septic patients [19].

Apart from microvascular thrombosis and organ dysfunction, coagulation abnormalities may also have other harmful consequences. The relevance of thrombocytopenia in patients with sepsis is in the first place related to an increased risk of bleeding. Indeed, in particular, critically ill patients with a platelet count of $<50 \times 10^9$/l have a 4- to 5-fold higher risk for bleeding as compared to patients with a higher platelet count [20, 21]. The risk of intracerebral bleeding in patients with sepsis during intensive care admission is relatively low (0.3–0.5%), but in 88% of patients with this complication the platelet count is less than 100×10^9/l [22]. Regardless of the cause, thrombocytopenia is an independent predictor of ICU mortality in multivariate analyses with a relative risk of 1.9 to 4.2 in various studies [20, 21, 23]. In particular, a sustained thrombocytopenia for more than 4 days after ICU admission or a drop in platelet count of >50% during ICU stay is related to a 4- to 6-fold increase in mortality [20, 24]. The platelet count was shown to be a stronger predictor for ICU mortality than composite scoring systems, such as the Acute Physiology and Chronic Evaluation (APACHE) II score or the Multiple Organ Dysfunction Score (MODS). Also low levels of coagulation factors in patients with sepsis, as reflected by prolonged global coagulation times, may be a risk factor for bleeding and mortality. A PT or aPTT ratio of >1.5 in critically ill patients was found to predict excessive bleeding and increased mortality [25, 26].

11.3 Incidence of Coagulation Abnormalities in Sepsis

Clinically relevant coagulation abnormalities may occur in 50–70% of patients with sepsis, whereas about 35% of patients will meet the criteria for DIC (see below) [27, 28]. In general, the incidence of thrombocytopenia (platelet count $<150 \times 10^9$/l) in critically ill medical patients is 35–50% [20, 21, 29]. Typically, the platelet count decreases during the first 4 days on the intensive care unit [24]. Sepsis is a clear risk factor for thrombocytopenia in critically ill patients and the severity of sepsis correlates with the decrease in platelet count [30]. Main factors that contribute to thrombocytopenia in patients with sepsis are impaired platelet production, increased consumption or destruction, or sequestration in the spleen or at the endothelial level. Impaired production of platelets in the bone marrow may seem contradictory to the high levels of platelet production-stimulating pro-inflammatory cytokines, such as tumor necrosis factor (TNF)-α and interleukin (IL)-6, and high concentration of circulating thrombopoietin in patients with sepsis, which theoretically should stimulate megakaryopoiesis in the bone marrow [31]. However, in a substantial number of patients with sepsis marked hemophagocytosis may occur, consisting of active phagocytosis of megakaryocytes and other hematopietic cells by monocytes and macrophages, hypothetically due to stimulation with high levels of macrophage colony stimulating factor (M-CSF) in sepsis [32]. Platelet consumption also plays an important role in patients with sepsis, due to the ongoing generation of thrombin. Platelet activation, consumption, and destruction may also occur at the endothelial site as a result of the extensive endothelial cell–platelet interaction in sepsis, which may vary between different vascular beds in various organs [33]. A prolonged global coagulation time (such as the prothrombin time (PT) or the activated partial thromboplastin time (aPTT)) occurs in 14 to 28% of patients [25, 26]. Other coagulation test abnormalities include high fibrin split products (in 99% of patients with sepsis) [34–36] and low levels of coagulation inhibitors, such as antithrombin and protein C (90% of sepsis patients) [36, 37].

11.4
Pathogenetic Pathways in the Coagulopathy of Sepsis

In recent years the mechanisms involved in the pathological derangement of coagulation in patients with sepsis have become increasingly clear. Apparently, various mechanisms at different sites in the hemostatic balance act simultaneously towards a procoagulant state (Figure 11.2). It has become clear that the most important mediators that orchestrate this imbalance of the coagulation system during sepsis are cytokines [38]. Increasing evidence points to extensive cross-talk between these two systems, whereby inflammation leads not only to activation of coagulation, but coagulation also considerably affects inflammatory activity [39]. Interestingly, systemic activation of coagulation and inflammation in sepsis can have some organ-specific manifestations, that are relevant for the specific organ dysfunction as a consequence of severe sepsis [40].

11.4.1
Initiation of Coagulation Activation

The tissue factor/factor VIIa pathway plays a central role in the initiation of coagulation activation in sepsis. Abrogation of the tissue factor/factor VII(a)

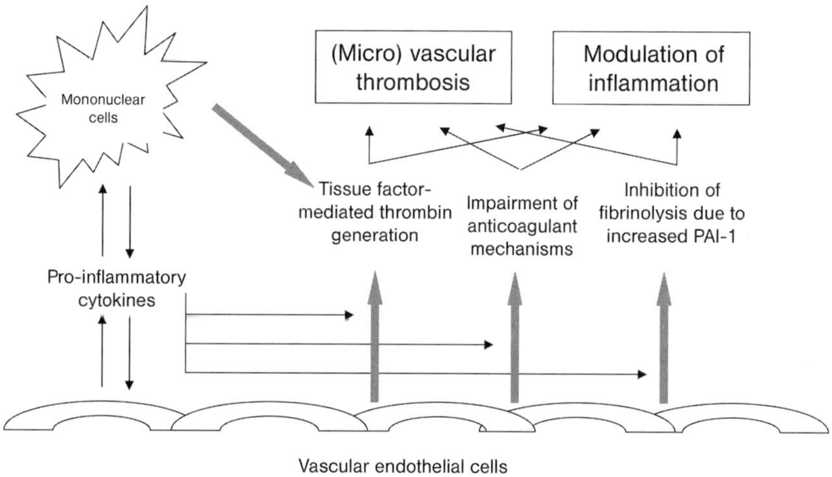

Figure 11.2 Schematic representation of pathogenetic pathways involved in the activation of coagulation in sepsis. During sepsis, both perturbed endothelial cells and activated mononuclear cells may produce pro-inflammatory cytokines that mediate coagulation activation. Activation of coagulation is initiated by tissue factor expression on activated mononuclear cells and endothelial cells. In addition, downregulation of physiological anticoagulant mechanisms and inhibition of fibrinolysis by endothelial cells will further promote intravascular fibrin deposition. PAI-1, plasminogen activator inhibitor, type 1.

pathway by monoclonal antibodies specifically directed against tissue factor or factor VIIa activity resulted in a complete inhibition of thrombin generation in endotoxin-challenged chimpanzees and prevented the occurrence of DIC and mortality in baboons that had been infused with *E. coli* [15, 41, 42]. On the other hand, studies of human endotoxemia or cytokinemia, did not show any change in markers for activation of the contact system [43, 44].

Tissue factor is a transmembrane 45-kD protein, that is constitutively expressed on a number of cells throughout the body [45]. The majority of these cells are in tissues not in direct contact with blood, such as the adventitial layer of large blood vessels. However, tissue factor comes into contact with blood upon disruption of the vascular integrity, or if cells present in the circulation start expressing tissue factor. In sepsis, circulating mononuclear cells that have been stimulated by pro-inflammatory cytokines, express tissue factor which leads to systemic activation of coagulation. However, other than in severe meningococcemia [46], it has proved difficult to demonstrate *ex vivo* tissue factor expression on monocytes of septic patients or experimental animals systemically exposed to microorganisms. It has been shown, however, that low dose endotoxemia in healthy subjects results in a 125-fold increase in tissue factor mRNA levels in blood monocytes [47]. Another source of tissue factor may be its localization on polymorphonuclear cells [48], although it is unlikely that these cells actually synthesize tissue factor in substantial quantities [49]. Based on the observation of transfer of tissue factor from leukocytes to activated platelets on a collagen surface in an *ex vivo* perfusion system, it is hypothesized that this "blood borne" tissue factor is transferred between cells through microparticles derived from activated mononuclear cells [50].

11.4.2 Impairment of the Antithrombin, Protein C, and TFPI Anticoagulant Pathways in Sepsis

In general, activation of coagulation is regulated by three major anticoagulant pathways: antithrombin, the protein C system and tissue factor pathway inhibitor (TFPI). During sepsis-induced activation of coagulation, the function of all three pathways can be impaired (Figure 11.3).

Antithrombin is a serine protease inhibitor and the main inhibitor of thrombin and factor Xa. During severe inflammatory responses, antithrombin levels are markedly decreased due to consumption (as a result of ongoing thrombin generation), impaired synthesis (as a result of a negative acute phase response) and degradation by elastase from activated neutrophils [51, 52]. A reduction in glycosaminoglycan availability at the endothelial surface (due to the influence of pro-inflammatory cytokines on endothelial synthesis) will also contribute to reduced antithrombin function, since glycosaminoglycans act as physiological heparin-like cofactors of antithrombin. Binding of glycosaminoglycans to

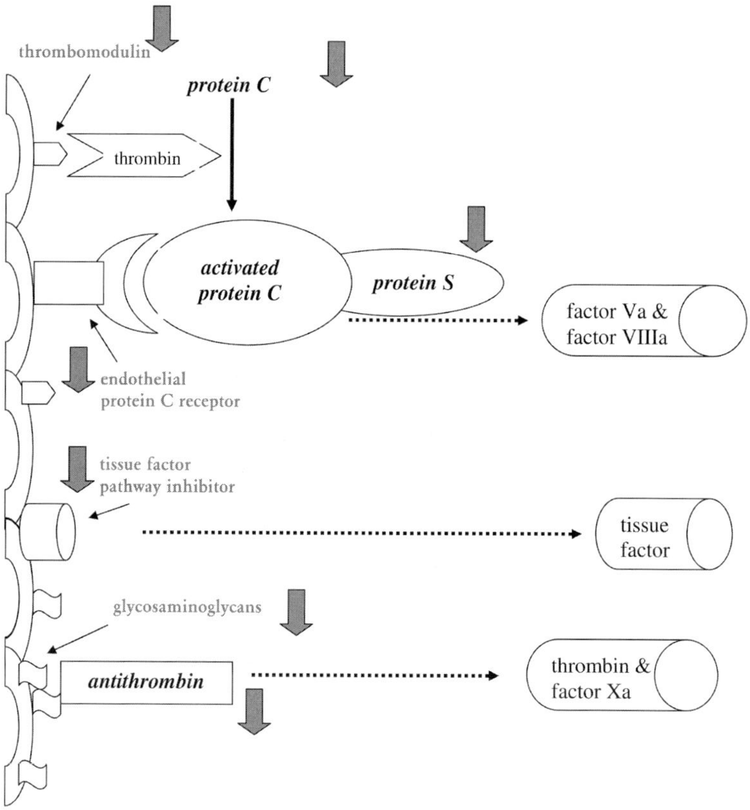

Figure 11.3 Schematic representations of the three important physiological antiocoagulant mechanisms and their point of impact on the coagulation system. In sepsis, these mechanisms are impaired by various mechanisms (red arrows). The protein C system is dysfunctional due to low levels of zymogen protein C, downregulation of thrombomodulin and the endothelial protein C receptor, and low levels of free protein S due to acute phase-induced high levels of its binding protein, i.e. C4b-binding protein. There is a relative insufficiency of the endothelial cell-associated tissue factor pathway inhibitor. The antithrombin system is defective due to low levels of antithrombin and impaired glycosaminoglycan expression on perturbed endothelial cells.

antithrombin induces a conformational change at the reactive center of the antithrombin molecule, thereby converting this protease inhibitor from a slow to a very efficient inhibitor of thrombin and other active coagulation factors [53]. Prospective clinical studies in patients at high risk for sepsis have shown that a marked decrease in levels of antithrombin precedes the clinical manifestation of the infection, which may indicate that antithrombin is involved in the early stages of coagulation activation during sepsis [54].

Endothelial dysfunction is even more important in the impairment of the protein C system during inflammation. Under physiologic conditions protein C is activated by thrombin bound to the endothelial cell membrane-associated thrombomodulin. Thrombomodulin is a membrane protein with several domains, including a lectin-like domain, six epidermal growth factor (EGF)-like repeats, a transmembrane domain and a short cytoplasmatic tail [55]. The binding of thrombin to thrombomodulin occurs at the site of the EGF-repeats [56]. This binding not only results in an about 100-fold increase in the activation of protein C, but also blocks the thrombin-mediated conversion of fibrinogen into fibrin and inhibits the binding of thrombin to other cellular receptors on platelets and inflammatory cells. In addition, thrombomodulin accelerates the activation of the plasma carboxypeptidase thrombin-activatable fibrinolysis inhibitor (TAFI), an important inhibitor of fibrinolysis [57]. Activated protein C regulates coagulation activation by proteolytic cleavage of the essential co-factors Va and VIIIa. Binding of protein C to the endothelial protein C receptor (EPCR) results in a 5-fold augmentation of the activation of protein C by the thrombomodulin–thrombin complex [58]. However, during severe inflammation, such as occurs in sepsis, in addition to low levels of protein C due to impaired synthesis [51] and degradation by neutrophil elastase (which has been described at least *in vitro*) [59], the protein C system is defective due to downregulation of thrombomodulin at the endothelial surface, mediated by the pro-inflammatory cytokines TNF-α and IL-1β [60]. Observations in patients with severe Gram-negative septicemia indeed confirmed the downregulation of thrombomodulin *in vivo* and impaired activation of protein C [61]. In this study histological analysis of skin biopsies from patients with meningococcal sepsis showed decreased endothelial expression of thrombomodulin, both in vessels with and without thrombosis. Low levels of free protein S (the co-factor of activated protein C) may further compromise the adequate functioning of the protein C system. In plasma, 60% of protein S is complexed to a complement regulatory protein, C4b binding protein (C4bBP). Increased plasma levels of C4bBP as a consequence of the acute phase reaction in inflammatory diseases may result in a relative free protein S deficiency. Although it has been shown that the β-chain of C4bBP (which mainly governs the binding to protein S) in not very much affected during the acute phase response [62], support for this hypothesis comes from studies showing that infusion of C4bBP increases organ dysfunction and mortality in septic baboons [63]. Animal experiments of severe inflammation-induced coagulation activation convincingly show that compromising the protein C system results in increased morbidity and mortality, whereas restoring the adequate function of activated protein C improves survival and organ failure [64]. Interestingly, experiments in mice with a one-allele targeted deletion of the protein C gene (resulting in heterozygous protein C deficiency) have more severe DIC and organ dysfunction and a higher mortality than wild type littermates [65].

A third inhibitory mechanism of thrombin generation involves TFPI, the main inhibitor of the tissue factor–factor VIIa complex. TFPI is a complex multi-domain Kunitz-type protease inhibitor, which binds to the tissue factor–factor VIIa complex and factor Xa [66]. The TFPI–factor Xa complex may bind to negatively-charged membrane surfaces, which may increase the local concentration of TFPI at cellular sites and facilitate inhibition of membrane-bound tissue factor–factor VIIa complex. The role of TFPI in the regulation of inflammation-induced coagulation activation is not completely clear. Experiments showing that administration of recombinant TFPI (and thereby achieving higher than physiological plasma concentrations of TFPI) blocks inflammation-induced thrombin generation in humans and the observation that pharmacological doses of TFPI are capable of preventing mortality during systemic infection and inflammation suggest that high concentrations of TFPI are capable of importantly modulating tissue factor-mediated coagulation [13, 67]. However, the endogenous concentration of TFPI is presumably insufficiently capable of regulating coagulation activation and downstream consequences during systemic inflammation, as has been confirmed in a clinical study of patients with sepsis [68, 69].

11.4.3
Inhibition of Endogenous Fibrinolysis in Sepsis

Experimental models indicate that at the time of maximal activation of coagulation in sepsis, the fibrinolytic system is largely shut off. The acute fibrinolytic response to inflammation is the release of plasminogen activators, in particular tissue-type plasminogen activator (t-PA) and urokinase-type plasminogen activator (u-PA), from storage sites in vascular endothelial cells. However, this increase in plasminogen activation and subsequent plasmin generation, is counteracted by a delayed but sustained increase in plasminogen activator inhibitor, type 1 (PAI-1) [70, 71]. The resulting effect on fibrinolysis is a complete inhibition and, as a consequence, inadequate fibrin removal, thereby contributing to microvascular thrombosis. Experiments in mice with targeted disruptions of genes encoding components of the plasminogen–plasmin system confirm that fibrinolysis plays a major role in inflammation-induced coagulation. Mice with a deficiency of plasminogen activators have more extensive fibrin deposition in organs when challenged with endotoxin, whereas PAI-1 knockout mice, in contrast to wild-type controls, have no microvascular thrombosis upon administration of endotoxin [72, 73]. Of interest, studies have shown that a functional mutation in the PAI-1 gene, the 4G/5G polymorphism, not only influenced the plasma levels of PAI-1, but was also linked to clinical outcome of meningococcal septicemia. Patients with the 4G/4G genotype had significantly higher PAI-1 concentrations in plasma and an increased risk of death [74]. Further investigations demonstrated that the PAI-1 polymorphism did not influence the risk of contracting meningitis as such, but probably increased the likelihood of developing septic shock from meningococcal infection [75].

11.4.4 Regulatory Role of Cytokines in the Coagulopathy of Sepsis

Similar to almost all systemic inflammatory responses to infection, the derangement of coagulation and fibrinolysis in sepsis is mediated by several cytokines. Most pro-inflammatory cytokines have been shown to activate coagulation *in vitro*. In patients with sepsis high levels of cytokines are detectable in the circulation and experimental bacteremia or endotoxemia results in the transient enhancement of serum levels of these cytokines [76]. Consecutively, tumor necrosis factor becomes first detectable, followed by an increase in circulating levels of interleukin-6 (IL-6), and interleukin 1 (IL-1). A number of experimental and clinical studies have focused on the roles of these cytokines in the pathogenesis of DIC.

Since TNF is the first cytokine to appear in the circulation after infusion of bacteria or endotoxin and exerts potent procoagulant effects *in vitro*, it was initially thought that activation of coagulation was mediated by TNF. The hypothesis that TNF played an important role in the induction of coagulation activation *in vivo* was strengthened by studies in which cancer patients or healthy human volunteers were injected with purified recombinant TNF [44, 77]. Following the injection of TNF the observed activation of the coagulation system was virtually identical to the endotoxin-induced effects on coagulation. However, in studies using various strategies to block TNF activity it became clear that the endotoxin-induced increase in TNF could be completely abolished whereas activation of coagulation was unchanged, although the effects on anticoagulant pathways and fibrinolysis seemed to be driven by TNF [78, 79]. Also, in baboons infused with a lethal dose of *E. coli*, treatment with an anti-TNF antibody had little or no effect on fibrinogen consumption [80]. Moreover, clinical studies in septic patients with an anti-TNF monoclonal antibody did not show a beneficial effect of this treatment [81]. These observations made it necessary to reconsider the role of TNF as principal mediator of endotoxin-induced activation of coagulation. In subsequent studies the role of IL-6 was investigated. It could be shown that infusion of a monoclonal anti-IL-6 antibody resulted in the complete abrogation of endotoxin-induced activation of coagulation in chimpanzees [82]. In addition, studies in cancer patients receiving recombinant IL-6 indicated that indeed thrombin is generated following the injection of this cytokine [83]. Thus, these data suggest that IL-6 rather than TNF is relevant as a mediator for the induction of the procoagulant response in DIC. While IL-1 is a potent agonist of tissue factor expression *in vitro*, its role has not been clarified *in vivo*. Administration of a IL-1 receptor antagonist partly blocked the procoagulant response in a sepsis model in baboons and treatment of patients with an IL-1 receptor inhibitor reduced thrombin generation [84–86]. However, most of the procoagulant changes after an endotoxin challenge occur well before IL-1 becomes detectable in the circulation, leaving a potential direct role of IL-1 in coagulation activation in sepsis an unresolved issue.

Anti-inflammatory cytokines, such as interleukin-10 (IL-10), may modulate the activation of coagulation. Infusion of recombinant human IL-10 was able to completely block the endotoxin-induced changes in coagulation and fibrinolysis in human volunteers [87]. However, the relevance of this regulatory role of anti-inflammatory cytokines in the pathogenesis of sepsis-associated coagulopathy remains to be established.

11.4.5 Cross-talk between Coagulation and Inflammation in Sepsis

Coagulation proteases and protease inhibitors not only interact with coagulation protein zymogens, but also with specific cell receptors to induce signaling pathways. In particular, protease interactions that affect inflammatory processes may be important in sepsis. Coagulation of whole blood *in vitro* results in a detectable expression of IL-1α mRNA in blood cells [88], and thrombin markedly enhances endotoxin-induced IL-1 activity in culture supernatants of guinea pig macrophages [89]. Similarly, clotting blood produces IL-8 *in vitro* [90]. Factor Xa, thrombin and fibrin can also activate endothelial cells, eliciting the synthesis of IL-6 and/or IL-8 [91, 92]. The most important mechanisms by which coagulation proteases influence inflammation is by binding to so-called protease activated receptors or PARs, of which four types (PAR 1–4) have been identified, all belonging to the family of transmembrane domain, G-protein coupled receptors [93]. A peculiar feature of PARs (in contrast to most other receptors of the superfamily) is that they serve as their own ligand. Proteolytic cleavage by an activated coagulation factor leads to exposure of a neo-amino terminus, that activates the same receptor (and possibly adjacent receptors), initiating transmembrane signaling. PARs 1, 3, and 4 are thrombin receptors whereas PAR-2 cannot bind thrombin but can be activated by the tissue factor–factor VIIa complex, factor Xa, and trypsin. PAR-1 can also serve as a receptor for the tissue factor–factor VIIa complex and factor Xa. PARs are localized in the vasculature on endothelial cells, mononuclear cells, platelets, fibroblasts, and smooth muscle cells [93]. *In vivo* evidence for a role of coagulation-protease stimulation of inflammation comes from recent experiments showing that the administration of recombinant factor VIIa to healthy human subjects causes a small but significant 3- to 4-fold rise in plasma levels of IL-6 and IL-8 [94].

There is also considerable cross-talk between physiological anticoagulant pathways and inflammatory mediators. Antithrombin can act as a mediator of inflammation, for example by direct binding to neutrophils and other leukocytes and thereby attenuating cytokine and chemokine receptor expression [95]. In addition, there is mounting evidence that the protein C system also has an important function in modulating inflammation [55, 96]. Indeed, activated protein C has been found to inhibit endotoxin-induced production of TNF-α, IL-1β, IL-6 and IL-8 by cultured monocytes/macrophages [97, 98].

Further, activated protein C abrogates endotoxin-induced cytokine release and leukocyte activation in rats *in vivo* [99]. Blocking the protein C pathway by a monoclonal antibody in septic baboons exacerbates the inflammatory response, as evidenced by increased levels of pro-inflammatory cytokines and more leukocyte infiltration and tissue destruction at histological analysis [100, 101]. Conversely, administration of activated protein C ameliorates the inflammatory activation in various models of severe systemic inflammation [64, 96]. Infusion of activated protein C abrogates inflammatory activity and improves organ function and survival in an experimental *E. coli* sepsis model in baboons [64]. Furthermore, in models of endotoxin-induced shock and lung and kidney injury in rats, administration of activated protein C resulted in a significant improvement in organ function, associated with lower levels of inflammatory cytokines and less leukocyte infiltration [96]. Mice with a one-allele targeted disruption of the protein C gene (resulting in heterozygous protein C deficiency) have not only a more severe coagulation response to endotoxin but also demonstrate significant differences in inflammatory responses, as shown by higher levels of circulating pro-inflammatory cytokines [65]. It is likely that the effects of activated protein C on inflammation are mediated by the endothelial protein C receptor (EPCR), that may mediate downstream inflammatory processes [55]. Binding of activated protein C to the endothelial protein C receptor was shown to affect gene expression profiles of cells by inhibiting endotoxin-induced calcium fluxes in the cell and by blocking NFκB nuclear translocation, which is a prerequisite for increases in pro-inflammatory cytokines and adhesion molecules [102, 103]. Recent experiments also suggest that EPCR binding of activated protein C can result in activation of PAR-1 [104]. Like activated protein C, EPCR itself may have anti-inflammatory properties. Lastly, activated protein C is capable of inhibiting endothelial cell apoptosis, which also seems to be mediated by binding of activated protein C to the endothelial protein C receptor and seems to require PAR-1 [105, 106].

11.5
Diagnostic Approach to Coagulation Abnormalities in Sepsis

It is important to realize that apart from DIC there are several other reasons for coagulation abnormalities in patients with sepsis (Table 11.1). Although thrombocytopenia is common in patients with severe sepsis, this may also be caused by other (sometimes simultaneously occurring) diseases, such as immune thrombocytopenia, medication-induced bone marrow depression, heparin-induced thrombocytopenia, or thrombotic microangiopathies [29, 107]. It is very important to properly diagnose these causes of thrombocytopenia, since they may require distinct treatment strategies [33]. Laboratory tests can be helpful in differentiating the coagulopathy in sepsis from various other hemostatic disorders, such as vitamin K deficiency or liver failure. Since

Table 11.1 Routine laboratory value abnormalities in patients with sepsis, due to disseminated intravascular coagulation (DIC) or other causes.

Test	Abnormality	Causes other than DIC contributing to test result
Platelet count	Decreased	Sepsis, impaired production, major blood loss, hypersplenism hematophagocytosis, immune thrombocytopenia, microangiopathy, heparin-induced thrombocytopenia
Prothrombin time	Prolonged	Vitamin K deficiency, liver failure, major blood loss
aPTT	Prolonged	Liver failure, heparin treatment, major blood loss
Fibrin degradation products	Elevated	Surgery, trauma, infection, hematoma
Protease inhibitors	Decreased	Liver failure, capillary leakage

such conditions, however, may also occur simultaneously with for example DIC, this differentiation is not always simple [108].

11.5.1
Tests for Intravascular Fibrin Formation and Fibrin Degradation Products

According to the current understanding of sepsis-associated coagulation abnormalities the determination of soluble fibrin in plasma appears to be crucial [109–112]. In general, the sensitivity of these assays for severe coagulation activation or DIC is relatively higher than the specificity. Indeed, initial clinical studies indicate that if the concentration of soluble fibrin has increased above a defined threshold, a diagnosis of DIC can be made [34, 109, 111, 113]. Most of the clinical studies show a sensitivity of 90–100% for the diagnosis of DIC but a rather low specificity [114]. Another problem so far is that a reliable test is not available for accurately quantitating soluble fibrin in plasma. A recent study showed a wide discordance among various assays for soluble fibrin [115]. The commercial semiquantative assays can be used to support the data obtained by screening tests. Similarly, fibrin monomers can be measured and may confirm the presence of intravascular fibrin formation. Since soluble fibrin in plasma can only be generated intravascularly, this test will not be influenced by extravascular fibrin formation, which for example may occur during local inflammation or trauma.

Fibrin degradation products (FDPs) may be detected by specific ELISAs or by latex agglutination assays, allowing rapid and bed-side determination in emergency cases [116]. None of the available assays for fibrin degradation products discriminates between degradation products of cross-linked fibrin and fibrinogen degradation, which may cause spuriously high results [117, 118]. The specificity of high levels of fibrin degradation products is therefore limited and many other conditions, such as trauma, recent surgery, inflammation or venous thrombo-embolism, are associated with elevated FDPs. More recently developed tests are specifically aimed at the detection of neo-antigens on degraded cross-linked fibrin. One such test detects an epitope related to plasmin-degraded cross-linked γ-chain, resulting in fragment D-dimer. These tests better differentiate degradation of cross-linked fibrin from fibrinogen or fibrinogen degradation products [119]. D-dimer levels are high in patients with DIC, but also poorly distinguish patients with DIC from patients with venous thrombo-embolism, recent surgery or inflammatory conditions [116, 120].

11.5.2
Markers for Thrombin Generation and Coagulation Activation

Activation peptides that are released upon the conversion of a coagulation factor zymogen to an active protease are sensitive markers for coagulation activation. Examples of such markers are prothrombin activation fragment F1 + 2 (F1 + 2), and the activation peptides of factors IX and X [121–123]. Indeed, these markers are markedly elevated in most patients with sepsis. Elevated plasma concentrations of thrombin–antithrombin complexes may well reflect the increased generation of thrombin, and thrombin-mediated fibrinogen to fibrin conversion can be monitored by increased levels of fibrinogen activation peptide fibrinopeptide-A (FPA) [124, 125]. All these markers are increased in most patients with sepsis and their high sensitivity may be helpful in detecting even low-grade activation of coagulation. The specificity of high levels of markers for coagulation factor activation is probably limited, since many other conditions may lead to elevated plasma levels. Another drawback may be that these assays are very much dependent on optimal venous puncture, which may be difficult in sick patients and during routine (intensive) care. The most important disadvantage of these tests may be that their use is limited to specialized coagulation laboratories and that they are not available for routine use in most clinical centers. Thus, although these tests are very relevant for research on the pathogenesis of coagulation disturbances in sepsis and the effect of specific interventions in the coagulation cascade of patients with sepsis, their practical use in clinical medicine is so far limited.

11.5.3
Platelet Count and Coagulation Factors in Patients with Sepsis

The platelet count in sepsis is correlated with markers of thrombin generation, since thrombin-induced platelet aggregation is for a large part responsible for platelet consumption [107]. Since the normal platelet count may vary considerably a single determination is often not very helpful but a continuous drop in platelet count, determined in septic patients at intervals of about 4 h may indicate the generation of thrombin, causing intravascular platelet aggregation. As mentioned before, however, a low or decreasing platelet count is not very specific for DIC.

Consumption of coagulation factors leads to low levels of coagulation factors in patients with sepsis. In addition, impaired synthesis, for example due to impaired liver function or a vitamin K deficiency, and loss of coagulation proteins, due to massive bleeding, may play a role as well [126]. Although the accuracy of the measurement of one-stage clotting assays in DIC has been contested (due to the presence of activated coagulation factors in plasma), the level of coagulation factors appears to correlate well with the severity of DIC [126]. The low level of coagulation factors is reflected by prolonged coagulation screening tests, such as the prothrombin time or the activated partial thromboplastin time (aPTT). Plasma levels of factor VIII are paradoxically increased in most patients with DIC, probably due to massive release of von Willebrand factor from the endothelium in combination with acute phase behavior of factor VIII [127]. Measurement of fibrinogen has been widely advocated as a useful tool for the diagnosis of DIC but in fact is not very helpful to diagnose DIC in most cases [7]. Fibrinogen acts as an acute-phase reactant and despite ongoing consumption plasma levels can remain well within the normal range for a long period of time. In a consecutive series of patients the sensitivity of a low fibrinogen level for the diagnosis of DIC was only 28% and hypofibrinogenemia was detected in very severe cases of DIC only. Sequential measurements of fibrinogen might be more useful and provide diagnostic clues.

Plasma levels of physiological coagulation inhibitors, such as antithrombin III or protein C, are useful indicators of ongoing coagulation activation [18, 54]. Antithrombin is the principal inhibitor of thrombin and may be readily exhausted during continuous thrombin generation. Plasma levels of antithrombin III has been shown to be a potent predictor of survival in patients with sepsis and DIC [54]. Levels of protein C may also indicate the severity of the DIC. In patients with meningococcal septicemia, very low plasma levels of protein C are observed and this may play a pivotal role in the occurrence of purpura fulminans in these patients [128, 129]. In fact, the plasma level of protein C may also be regarded as a strong predictor of the outcome in DIC patients [130].

11.5.4
Diagnostic Management in Clinical Practice

Most of the newer, more sensitive, tests described in the previous section are presently available in specialized laboratories only and although these tests may be very helpful in clinical trials or other research, they can often not be used in a routine setting. A scoring system, utilizing simple laboratory tests that are available in almost all hospital laboratories, has been presented by the subcommittee on DIC of the International Society on Thrombosis and Haemostasis [131]. Following these objectives the committee proposes a 5-step diagnostic algorithm to calculate a DIC score, as summarized in Figure 11.4. The score can be calculated based on routinely available laboratory tests. Tentatively, and awaiting a definitive prospective validation, a score equal to or more than 5 is compatible with DIC, whereas a score of less than 5 may be indicative (but is *not* affirmative) for non-overt DIC. Initial prospective studies show that the sensitivity of the DIC score is 93%, whereas the specificity is 98% [132]. Interestingly, the severity of DIC according to this scoring system is related to mortality in patients with sepsis [133].

11.6 Supportive Treatment of Coagulation Abnormalities in Sepsis

The keystone of the treatment of hemostatic abnormalities in patients with sepsis is the specific treatment of the sepsis with appropriate antibiotics and control of the infectious source. However, in many cases additional supportive

1. Risk assessment: Does the patient have a underlying disorder known to be associated with overt DIC?
   If yes, proceed. If no, do not use this algorithm;

2. Order global coagulation tests (platelet count, prothrombin time [PT], fibrinogen, soluble fibrin monomers, or fibrin degradation products).

3. Score global coagulation test results:

- platelet count (> 100 = 0, < 100 = 1, < 50 = 2)
- elevated fibrin-related marker (e.g. soluble fibrin monomers/fibrin degradation products) (no increase: 0, moderate increase: 2, strong increase: 3)
- prolonged prothrombin time
  (< 3 sec. = 0, > 3 but < 6 sec. = 1, > 6 sec. = 2)
- fibrinogen level
  (> 1.0 g/L = 0, < 1.0 g/L = 1)

4. Calculate score.

5. If ≥5: compatible with overt DIC; repeat scoring daily.
   If <5: suggestive (not affirmative) for non-overt DIC; repeat next 1-2 days.

Figure 11.4 Diagnostic algorithm for the diagnosis of overt disseminated intravascular coagulation (DIC).

treatment, aimed at circulatory and respiratory support and replacement of organ function, is required. Coagulation abnormalities may proceed, even after proper treatment has been initiated. In those cases, supportive measures to manage the coagulation disorder may be considered and may positively affect morbidity and mortality. The growing insight into the various mechanisms that play a role in the coagulation abnormalities associated with sepsis has indeed been accommodating in the development of such supportive management strategies.

11.6.1 Plasma and Platelet Substitution Therapy

Low levels of platelets and coagulation factors may increase the risk of bleeding. However, plasma or platelet substitution therapy should not be instituted on the basis of laboratory results alone; it is indicated only in patients with active bleeding and in those requiring an invasive procedure or otherwise at risk for bleeding complications [134]. The suggestion that administration of blood components might "add fuel to the fire" has in fact never been proven in clinical or experimental studies. The presumed efficacy of treatment with plasma, fibrinogen, cryoprecipitate, or platelets is not based on randomized controlled trials but appears to be rational therapy in bleeding patients or in patients at risk for bleeding with a significant depletion of these hemostatic factors [135]. It may be necessary to use large volumes of plasma to correct the coagulation defect. Coagulation factor concentrates, such as prothrombin complex concentrate, may overcome this obstacle, but these compounds may lack essential factors, such as factor V. Moreover, in the older literature caution is advocated with the use of prothrombin complex concentrates in DIC, since it may worsen the coagulopathy due to small traces of activated factors in the concentrate. It is, however, not clear whether this is still relevant for the concentrates that are currently in use. Specific deficiencies in coagulation factors, such as fibrinogen, may be corrected by administration of purified coagulation factor concentrates.

11.6.2 Anticoagulants

Experimental studies have shown that heparin can at least partly inhibit the activation of coagulation in sepsis [136]. Uncontrolled case series in patients with sepsis and DIC have claimed to be successful. However, a beneficial effect of heparin on clinically important outcome events in patients with DIC has never been demonstrated in controlled clinical trials [137]. Also, the safety of heparin treatment is debatable in DIC patients who are prone to bleeding. Therapeutic doses of heparin are indicated in patients with clinically overt thromboembolism or extensive fibrin deposition, such as purpura fulminans or acral ischemia. Patients with sepsis may benefit from prophylaxis to prevent venous thromboembolism, which may not be achieved with standard low-dose subcutaneous heparin [138]. Theoretically, the most logical anticoagulant agent to use in DIC is directed against tissue factor activity. Potential agents include recombinant tissue factor pathway inhibitor, inactivated factor VIIa, and recombinant NAPc2, a potent and specific inhibitor of the ternary complex between tissue factor–factor VIIa and factor Xa [139]. Phase II trials of recombinant TFPI in patients with sepsis showed promising results [140] but a recently completed phase III trial did not show an overall survival benefit in patients that were treated with TFPI [140, 141]. Interestingly, in an *a priori* defined subgroup of patients with an INR of <1.2, TFPI treatment was associated with a more favorable outcome.

11.6.3 Restoration of Anticoagulant Pathways

In view of the deficient state of physiological anticoagulant pathways in patients with sepsis, restoration of these inhibitors may be a rational approach [142]. Since antithrombin is one of the most important physiological inhibitors of coagulation and based on successful preclinical results, the use of antithrombin III concentrates in patients with DIC has been studied relatively intensively. Most of the randomized controlled trials concern patients with sepsis, septic shock, or both. All trials show some beneficial effect in terms of improvement of laboratory parameters, shortening of the duration of DIC, or even improvement in organ function [7]. In the more recent clinical trials, very high doses of antithrombin concentrate to attain supraphysiological plasma levels were used. A series of relatively small trials showed a modest reduction in mortality in antithrombin-treated patients [143–145]; however, the effect did not reach statistical significance in any of the trials. A large-scale, multicenter, randomized controlled trial to directly address this issue showed no significant reduction in mortality of patients with sepsis who were treated with antithrombin concentrate [146]. Interestingly, *post-hoc* subgroup analyses indicated some benefit in patients who did not receive concomitant heparin, but this observation needs prospective validation.

Based on the notion that depression of the protein C system may significantly contribute to the pathophysiology of DIC, supplementation of (activated) protein C might be beneficial [142]. A beneficial effect of recombinant human activated protein C was demonstrated in two randomized controlled trials. Firstly, in a dose-ranging clinical trial, 131 patients with sepsis were enrolled [147]. The included patients received activated protein C by continuous infusion at doses ranging from 12 μg/kg/h to 30 μg/kg/h, or placebo. Based on D-dimer plasma levels the optimal dose of recombinant human activated protein C was determined to be 24 μg/kg/h. A subsequent phase III trial of activated protein C concentrate in patients with sepsis was prematurely stopped because of efficacy in reducing mortality in these patients [36]. All-cause mortality at 28 days after inclusion was 24.7% in the activated protein C group versus 30.8% in the control group (19.4% relative risk reduction). The administration of activated protein C was demonstrated to cause an amelioration of coagulation abnormalities and activated protein C-treated patients had reduced organ failure [148]. In view of the above-described effects that activated protein C has on inflammation, part of the success may have been a result of the beneficial effect on inflammatory pathways. Interestingly, a recent analysis of this trial demonstrated that patients who were classified as having DIC according to the DIC scoring system of the ISTH, derived relatively greater benefit from activated protein C treatment than patients that did not have overt DIC [19]. The relative risk reduction in mortality of patients with sepsis and DIC who received activated protein C was 38%, in comparison with a relative risk reduction of 18% in patients with sepsis who did not have DIC. This seems to underscore the importance of the coagulation derangement in the pathogenesis of sepsis and the point of impact that restoration of microvascular anticoagulant pathways may provide in the treatment of sepsis. Recombinant human-activated protein C has been licensed in most countries for treatment of patients with severe sepsis and two or more organ failures. The most frequently encountered adverse effect of activated protein C is bleeding. In the phase III study in patients with severe sepsis the incidence of major bleeding (i.e. bleeding reported as a serious adverse event) during the infusion period was 2.4% in the activated protein C group as compared with 1.0% in the control group ($p = 0.02$) [36]. During the 28-day study period the incidence of major bleeding was 3.5% in the activated protein C group and 2.0% in the placebo group ($p = 0.06$). Gastro-intestinal bleeding was the most frequently occurring bleeding complication in both groups. Most of the bleeding episodes were procedure-related or occurred in patients with a severely deranged coagulation system (aPTT > 120 s or PT (INR) > 3.0), whereas spontaneous bleeding was rare. Of note, severe thrombocytopenia (i.e. platelet count $<50 \times 10^9$/l) was an exclusion criterion for the trial but patients with lower platelet counts appeared to derive relatively more benefit from the administration of activated protein C than patients with higher platelet counts. Ongoing studies are focusing on the concomitant use of heparin in patients who receive activated protein C and the efficacy of activated protein C in patients with less severe sepsis.

References

1. Levi, M., ten Cate, H., van der Poll, T. and van Deventer, S.J. (**1993**) Pathogenesis of disseminated intravascular coagulation in sepsis. *JAMA*, **270**, 975–9.
2. Levi, M. and ten Cate, H. (**1999**) Disseminated intravascular coagulation. *N. Engl. J. Med.*, **341**, 586–92.
3. Colman, R.W., Robboy, S.J. and Minna, J.D. (**1979**) Disseminated intravascular coagulation: a reappraisal. *Annu. Rev. Med.*, **30**, 359–74.

4. Levi, M., van Gorp, E. and ten Cate, H. (**2002**) Disseminated Intravascular Coagulation, in *Blood: Pinciples and Practice of Hematology* (eds R.I. Handin, S.E. Lux and T.P. Stossel), J.B.Lippincott, Philadelphia.
5. Levi, M. and Marder, V.J. (**2006**) Coagulation abnormalities in sepsis, in *Hemostasis and Thrombosis: Basic Principles and Clinical Practice* (eds R.W. Colman, V.J. Marder, A.W. Clowes, J.N. George and S.Z. Goldhaber), Lippincott William and Wilkins, Philadelphia, pp. 1601–13.
6. Miller, D.L., Welty-Wolf, K., Carraway, M.S., Ezban, M., Ghio, A., Suliman, H. and Piantadosi, C.A. (**2002**) Extrinsic coagulation blockade attenuates lung injury and proinflammatory cytokine release after intratracheal lipopolysaccharide. *Am. J. Respir. Cell. Mol. Biol.*, **26**, 650–8.
7. Levi, M., ten Cate, H. and van der Poll, T. (**1999**) Disseminated intravascular coagulation: State of the art. *Thromb. Haemost.*, **82**, 695–705.
8. Ratnoff, O.D. and Nebehay, W.G. (**1962**) Multiple coagulative defects in a patient with the Waterhouse–Friderichsen syndrome. *Ann. Intern. Med.*, **56**, 627.
9. Levi, M., Keller, T.T., van Gorp, E. and ten Cate, H. (**2003**) Infection and inflammation and the coagulation system. *Cardiovasc. Res.*, **60**, 26–39.
10. Robboy, S.J., Major, M.C., Colman, R.W. and Minna, J.D. (**1972**) Pathology of disseminated intravascular coagulation (DIC). Analysis of 26 cases. *Hum. Pathol.*, **3**, 327–43.
11. Shimamura, K., Oka, K., Nakazawa, M. and Kojima, M. (**1983**) Distribution patterns of microthrombi in disseminated intravascular coagulation. *Arch. Pathol. Lab. Med.*, **107**, 543–7.
12. Coalson, J.J. *Pathology of Sepsis, Septic Shock, and Multiple Organ Failure. Perspective on Sepsis and Septic Shock,* Society of Critical Care Medicine, Fullerton, CA, (**1986**) pp. 27–59.
13. Creasey, A.A., Chang, A.C., Feigen, L., Wun, T.C., Taylor, F.B.J. and Hinshaw, L.B. (**1993**) Tissue factor pathway inhibitor reduces mortality from *Escherichia coli* septic shock. *J. Clin. Invest.*, **91**, 2850–6.
14. Kessler, C.M., Tang, Z., Jacobs, H.M. and Szymanski, L.M. (**1997**) The suprapharmacologic dosing of antithrombin concentrate for *Staphylococcus aureus*-induced disseminated intravascular coagulation in guinea pigs: substantial reduction in mortality and morbidity. *Blood*, **89**, 4393–401.
15. Taylor, F.B.J., Chang, A., Ruf, W., Morrissey, J.H., Hinshaw, L., Catlett, R., *et al.* (**1991**) Lethal E. coli septic shock is prevented by blocking tissue factor with monoclonal antibody. *Circ. Shock*, **33**, 127–34.
16. Taylor, F.B. J., Chang, A., Esmon, C.T., D'Angelo, A., Vigano-D'Angelo, S. and Blick, K.E. (**1987**) Protein C prevents the coagulopathic and lethal effects of *Escherichia coli* infusion in the baboon. *J. Clin. Invest.*, **79**, 918–25.
17. Welty-Wolf, K.E., Carraway, M.S., Miller, D.L., Ortel, T.L., Ezban, M., Ghio, A.J., *et al.* (**2001**) Coagulation blockade prevents sepsis-induced respiratory and renal failure in baboons. *Am. J. Respir. Crit. Care Med.*, **164**, 1988–96.
18. Fourrier, F., Chopin, C., Goudemand, J., Hendrycx, S., Caron, C., Rime, A., *et al.* (**1992**) Septic shock, multiple organ failure, and disseminated intravascular coagulation. Compared patterns of antithrombin III, protein C, and protein S deficiencies see comments]. *Chest*, **101**, 816–23.
19. Dhainaut, J.F., Yan, S.B., Joyce, D.E., Pettila, V., Basson, B.R., Brandt, J.T., *et al.* (**2004**) Treatment effects of drotrecogin alfa (activated) in patients with severe sepsis with or without overt disseminated intravascular coagulation. *J. Thromb. Haemost.*, **2**, 1924–33.

20. Vanderschueren, S., De Weerdt, A., Malbrain, M., Vankersschaever, D., Frans, E., Wilmer, A. and Bobbaers, H. (**2000**) Thrombocytopenia and prognosis in intensive care. *Crit. Care Med.*, **28**, 1871–6.
21. Strauss, R., Wehler, M., Mehler, K., Kreutzer, D., Koebnick, C. and Hahn, E.G. (**2002**) Thrombocytopenia in patients in the medical intensive care unit: bleeding prevalence, transfusion requirements, and outcome. *Crit. Care Med.*, **30**, 1765–71.
22. Oppenheim-Eden, A., Glantz, L., Eidelman, L.A. and Sprung, C.L. (**1999**) Spontaneous intracerebral hemorrhage in critically ill patients: incidence over six years and associated factors. *Intensive Care Med.*, **25**, 63–7.
23. Stephan, F., Hollande, J., Richard, O., Cheffi, A., Maier-Redelsperger, M. and Flahault, A. (**1999**) Thrombocytopenia in a surgical ICU. *Chest*, **115**, 1363–70.
24. Akca, S., Haji Michael, P., de Medonca, A., Suter, P.M., Levi, M. and Vincent, J.L. (**2002**) The time course of platelet counts in critically ill patients. *Crit. Care Med.*, **30**, 753–6.
25. Chakraverty, R., Davidson, S., Peggs, K., Stross, P., Garrard, C. and Littlewood, T.J. (**1996**) The incidence and cause of coagulopathies in an intensive care population. *Br. J. Haematol.*, **93**, 460–3.
26. MacLeod, J.B., Lynn, M., McKenney, M.G., Cohn, S.M. and Murtha, M. (**2003**) Early coagulopathy predicts mortality in trauma. *J. Trauma.*, **55**, 39–44.
27. Wheeler, A.P. and Bernard, G.R. (**1999**) Treating patients with severe sepsis. *N. Engl. J. Med.*, **340**, 207–14.
28. Levi, M., de Jonge, E. and van Der, P.T. (**2003**) Sepsis and disseminated intravascular coagulation. *J. Thromb. Thrombolysis.*, **16**, 43–7.
29. Baughman, R.P., Lower, E.E., Flessa, H.C. and Tollerud, D.J. (**1993**) Thrombocytopenia in the intensive care unit. *Chest*, **104**, 1243–7.
30. Mavrommatis, A.C., Theodoridis, T., Orfanidou, A., Roussos, C., Christopoulou-Kokkinou, V. and Zakynthinos, S. (**2000**) Coagulation system and platelets are fully activated in uncomplicated sepsis. *Crit. Care Med.*, **28**, 451–7.
31. Folman, C.C., Linthorst, G.E., van Mourik, J., van Willigen, G., de Jonge, E., Levi, M., *et al.* (**2000**) Platelets release thrombopoietin (Tpo) upon activation: another regulatory loop in thrombocytopoiesis? *Thromb. Haemost.*, **83**, 923–30.
32. Francois, B., Trimoreau, F., Vignon, P., Fixe, P., Praloran, V. and Gastinne, H. (**1997**) Thrombocytopenia in the sepsis syndrome: role of hemophagocytosis and macrophage colony-stimulating factor. *Am. J. Med.*, **103**, 114–20.
33. Warkentin, T.E., Aird, W.C. and Rand, J.H. (**2003**) Platelet-Endothelial Interactions: Sepsis, HIT, and Antiphospholipid Syndrome. *Hematology 2003. (Am. Soc. Hematol. Educ. Program)*, 497–519.
34. Shorr, A.F., Thomas, S.J., Alkins, S.A., Fitzpatrick, T.M. and Ling, G.S. (**2002**) D-dimer correlates with proinflammatory cytokine levels and outcomes in critically ill patients. *Chest*, **121**, 1262–8.
35. Owings, J.T., Gosselin, R.C., Anderson, J.T., Battistella, F.D., Bagley, M. and Larkin, E.C. (**2001**) Practical utility of the D-dimer assay for excluding thromboembolism in severely injured trauma patients. *J. Trauma.*, **51**, 425–9.
36. Bernard, G.R., Vincent, J.L., Laterre, P.F., LaRosa, S.P., Dhainaut, J.F., Lopez-Rodriguez, A., *et al.* (**2001**) Efficacy and safety of recombinant human activated protein C for severe sepsis. *N. Engl. J. Med.*, Mar 8, **344**(10), **344**, 699–709.
37. Gando, S., Nanzaki, S., Sasaki, S. and Kemmotsu, O. (**1998**) Significant

correlations between tissue factor and thrombin markers in trauma and septic patients with disseminated intravascular coagulation. *Thromb. Haemost.*, **79**, 1111–5.

38. Levi, M., van der Poll, T., ten Cate, H. and van Deventer, S.J. (**1997**) The cytokine-mediated imbalance between coagulant and anticoagulant mechanisms in sepsis and endotoxaemia. *Eur. J. Clin. Invest.*, **27**, 3–9.
39. Levi, M., van der Poll, T. and Buller, H.R. (**2004**) The bidirectional relationshiop between coagulation and inflammation. *Circulation*, **109**, 2698–704.
40. Aird, W.C. (**2001**) Vascular bed-specific hemostasis: role of endothelium in sepsis pathogenesis. *Crit. Care Med.*, **29**, S28–S34.
41. Levi, M., ten Cate, H., Bauer, K.A., van der Poll, T., Edgington, T.S., Buller, H.R., *et al.* (**1994**) Inhibition of endotoxin-induced activation of coagulation and fibrinolysis by pentoxifylline or by a monoclonal anti-tissue factor antibody in chimpanzees. *J. Clin. Invest.*, **93**, 114–20.
42. Biemond, B.J., Levi, M., ten Cate, H., Soule, H.R., Morris, L.D., Foster, D.L., *et al.* (**1995**) Complete inhibition of endotoxin-induced coagulation activation in chimpanzees with a monoclonal Fab fragment against factor VII/VIIa. *Thromb. Haemost.*, **73**, 223–30.
43. van Deventer, S.J., Buller, H.R., ten Cate, J.W., Aarden, L.A., Hack, C.E. and Sturk, A. (**1990**) Experimental endotoxemia in humans: analysis of cytokine release and coagulation, fibrinolytic, and complement pathways. *Blood*, **76**, 2520–6.
44. van der Poll, T., Buller, H.R., ten Cate, H., Wortel, C.H., Bauer, K.A., van Deventer, S.J., *et al.* (**1990**) Activation of coagulation after administration of tumor necrosis factor to normal subjects. *N. Engl. J. Med.*, **322**, 1622–7.
45. Ruf, W. and Edgington, T.S. (**1994**) Structural biology of tissue factor, the initiator of thrombogenesis *in vivo*. *FASEB. J.*, **8**, 385–90.
46. Osterud, B. and Flaegstad, T. (**1983**) Increased tissue thromboplastin activity in monocytes of patients with meningococcal infection: related to an unfavourable prognosis. *Thromb. Haemost.*, **49**, 5–7.
47. Franco, R.F., de Jonge, E., Dekkers, P.E., Timmerman, J.J., Spek, C.A., *et al.* (**2000**) The *in vivo* kinetics of tissue factor messenger RNA expression during human endotoxemia: relationship with activation of coagulation. *Blood*, **96**, 554–9.
48. Giesen, P.L., Rauch, U., Bohrmann, B., Kling, D., Roque, M., Fallon, J.T., *et al.* (**1999**) Blood-borne tissue factor: another view of thrombosis. *Proc. Natl. Acad. Sci. USA.*, **96**, 2311–5.
49. Osterud, B., Rao, L.V. and Olsen, J.O. (**2000**) Induction of tissue factor expression in whole blood - lack of evidence for the presence of tissue factor expression on granulocytes. *Thromb. Haemost.*, **83**, 861–7.
50. Rauch, U., Bonderman, D., Bohrmann, B., Badimon, J.J., Himber, J., Riederer, M.A. and Nemerson, Y. (**2000**) Transfer of tissue factor from leukocytes to platelets is mediated by CD15 and tissue factor. *Blood*, **96**, 170–5.
51. Vary, T.C. and Kimball, S.R. (**1992**) Regulation of hepatic protein synthesis in chronic inflammation and sepsis. *Am. J. Physiol.*, **262**, C445–C452.
52. Seitz, R., Wolf, M., Egbring, R. and Havemann, K. (**1989**) The disturbance of hemostasis in septic shock: role of neutrophil elastase and thrombin, effects of antithrombin III and plasma substitution. *Eur. J. Haematol.*, **43**, 22–8.
53. Opal, S.M., Kessler, C.M., Roemisch, J. and Knaub, S. (**2002**) Antithrombin, heparin, and heparan sulfate. *Crit. Care Med.*, **30**, S325–S331.
54. Mesters, R.M., Mannucci, P.M., Coppola, R., Keller, T.,

Ostermann, H. and Kienast, J. (**1996**) Factor VIIa and antithrombin III activity during severe sepsis and septic shock in neutropenic patients (see comments). *Blood*, **88**, 881–6.

55. Esmon, C.T. (**2002**) New mechanisms for vascular control of inflammation mediated by natural anticoagulant proteins. *J. Exp. Med.*, **196**, 561–4.
56. Zushi, M., Gomi, K., Yamamoto, S., Maruyama, I., Hayashi, T. and Suzuki, K. (**1989**) The last three consecutive epidermal growth factor-like structures of human thrombomodulin comprise the minimum functional domain for protein C-activating cofactor activity and anticoagulant activity. *J. Biol. Chem.*, **264**, 10351–3.
57. Bajzar, L., Morser, J. and Nesheim, M. (**1996**) TAFI, or plasma procarboxypeptidase B, couples the coagulation and fibrinolytic cascades through the thrombin-thrombomodulin complex. *J. Biol. Chem.*, **271**, 16603–8.
58. Taylor, F.B. Jr., Peer G.T., Lockhart M.S., Ferrell G. and Esmon C.T. (**2001**) Endothelial cell protein C receptor plays an important role in protein C activation *in vivo*. *Blood*, **97**, 1685–8.
59. Eckle, I., Seitz, R., Egbring, R., Kolb, G. and Havemann, K. (**1991**) Protein C degradation *in vitro* by neutrophil elastase. *Biol. Chem. Hoppe. Seyler.*, **372**, 1007–13.
60. Nawroth, P.P. and Stern, D.M. (**1986**) Modulation of endothelial cell hemostatic properties by tumor necrosis factor. *J. Exp. Med.*, **163**, 740–5.
61. Faust, S.N., Levin, M., Harrison, O.B., Goldin, R.D., Lockhart, M.S., Kondaveeti, S., *et al.* (**2001**) Dysfunction of endothelial protein C activation in severe meningococcal sepsis. *N. Engl. J. Med.*, **345**(6), 408–16.
62. Garcia de Frutos, P., Alim, R.I., Hardig, Y., Zoller, B. and Dahlback, B. (**1994**) Differential regulation of alpha and beta chains of C4b–binding protein during acute-phase response resulting in stable plasma levels of free anticoagulant protein S. *Blood*, **84**, 815–22.
63. Taylor, F., Chang, A., Ferrell, G., Mather, T., Catlett, R., Blick, K., *et al.* (**1991**) C4b-binding protein exacerbates the host response to *Escherichia coli*. *Blood*, **78**, 357–63.
64. Taylor, F.BJ., Dahlback, B., Chang, A.C., Lockhart, M.S., Hatanaka, K., Peer, G. and Esmon, C.T. (**1995**) Role of free protein S and C4b binding protein in regulating the coagulant response to *Escherichia coli*. *Blood*, **86**, 2642–52.
65. Levi, M., Dorffler-Melly, J., Reitsma, P.H., Buller, H.R., Florquin, S., van der Poll, T. and Carmeliet, P. (**2003**) Aggravation of endotoxin-induced disseminated intravascular coagulation and cytokine activation in heterozygous protein C deficient mice. *Blood*, **101**, 4823–7.
66. Broze, G.J., Jr., Girard T.J. and Novotny W.F. (**1990**) Regulation of coagulation by a multivalent Kunitz-type inhibitor. *Biochemistry*, **29**, 7539–46.
67. de Jonge, E., Dekkers, P.E., Creasey, A.A., Hack, C.E., Paulson, S.K., Karim, A., *et al.* (**2000**) Tissue factor pathway inhibitor (TFPI) dose-dependently inhibits coagulation activtion without influencing the fibrinolytic and cytokine response during human endotoxemia. *Blood*, **95**, 1124–9.
68. Gando, S., Kameue, T., Morimoto, Y., Matsuda, M., Hayakawa, M. and Kemmotsu, O. (**2002**) Tissue factor production not balanced by tissue factor pathway inhibitor in sepsis promotes poor prognosis. *Crit. Care Med.*, **30**, 1729–34.
69. Levi, M. (**2002**) The imbalance between tissue factor and tissue factor pathway inhibitor in sepsis. *Crit. Care Med.*, **30**, 1914–5.
70. van der Poll, T., Levi, M., Buller, H.R., van Deventer, S., de

Boer, J., Hack, C.E., *et al.* (**1991**) Fibrinolytic response to tumor necrosis factor in healthy subjects. *J. Exp. Med.*, **174**, 729–32.
71. Biemond, B.J., Levi, M., ten Cate, H., van der Poll, T., Buller, H.R., Hack, C.E. and ten Cate, J.W. (**1995**) Plasminogen activator and plasminogen activator inhibitor I release during experimental endotoxaemia in chimpanzees: effect of interventions in the cytokine and coagulation cascades. *Clin. Sci. (Colch.)*, **88**, 587–94.
72. Yamamoto, K. and Loskutoff, D.J. (**1996**) Fibrin deposition in tissues from endotoxin-treated mice correlates with decreases in the expression of urokinase-type but not tissue-type plasminogen activator. *J. Clin. Invest.*, **97**, 2440–51.
73. Pinsky, D.J., Liao, H., Lawson, C.A., Yan, S.F., Chen, J., Carmeliet, P., *et al.* (**1998**) Coordinated induction of plasminogen activator inhibitor-1 (PAI-1) and inhibition of plasminogen activator gene expression by hypoxia promotes pulmonary vascular fibrin deposition. *J. Clin. Invest.*, **102**, 919–28.
74. Hermans, P.W., Hibberd, M.L., Booy, R., Daramola, O., Hazelzet, J.A., de, G.R. and Levin, M. (**1999**) 4G/5G promoter polymorphism in the plasminogen-activator-inhibitor-1 gene and outcome of meningococcal disease. Meningococcal Research Group. *Lancet.*, **354**, 556–60.
75. Westendorp, R.G., Hottenga, J.J. and Slagboom, P.E. (**1999**) Variation in plasminogen-activator-inhibitor-1 gene and risk of meningococcal septic shock. *Lancet.*, **354**, 561–3.
76. Hack, C.E., Aarden, L.A. and Thijs, L.G. (**1997**) Role of cytokines in sepsis. *Adv. Immunol.*, **66**, 101–95.
77. Bauer, K.A., ten Cate, H., Barzegar, S., Spriggs, D.R., Sherman, M.L. and Rosenberg, R.D. (**1989**) Tumor necrosis factor infusions have a procoagulant effect on the hemostatic mechanism of humans. *Blood*, **74**, 165–72.
78. van der Poll, T., Levi, M., van Deventer, S.J., ten Cate, H., Haagmans, B.L., Biemond, B.J., *et al.* (**1994**) Differential effects of anti-tumor necrosis factor monoclonal antibodies on systemic inflammatory responses in experimental endotoxemia in chimpanzees. *Blood*, **83**, 446–51.
79. van der Poll, T., Coyle, S.M., Levi, M., Jansen, P.M., Dentener, M., Barbosa, K., *et al.* (**1997**) Effect of a recombinant dimeric tumor necrosis factor receptor on inflammatory responses to intravenous endotoxin in normal humans. *Blood*, **89**, 3727–34.
80. Hinshaw, L.B., Tekamp-Olson, P., Chang, A.C., Lee, P.A., Taylor, F.B., Jr., Murray, C.K., *et al.* (**1990**) Survival of primates in LD100 septic shock following therapy with antibody to tumor necrosis factor (TNF alpha). *Circ. Shock*, **30**, 279–92.
81. Abraham, E., Wunderink, R., Silverman, H., Perl, T.M., Nasraway, S., Levy, H., *et al.* (**1995**) Efficacy and safety of monoclonal antibody to human tumor necrosis factor alpha in patients with sepsis syndrome. A randomized, controlled, double-blind, multicenter clinical trial. TNF-alpha MAb Sepsis Study Group. *JAMA*, **273**, 934–41.
82. van der Poll, T., Levi, M., Hack, C.E., ten Cate, H., van Deventer, S.J., Eerenberg, A.J., *et al.* (**1994**) Elimination of interleukin 6 attenuates coagulation activation in experimental endotoxemia in chimpanzees. *J. Exp. Med.*, **179**, 1253–9.
83. Stouthard, J.M., Levi, M., Hack, C.E., Veenhof, C.H., Romijn, H.A., Sauerwein, H.P. and van der Poll, T. (**1996**) Interleukin-6 stimulates coagulation, not fibrinolysis, in humans. *Thromb. Haemost.*, **76**, 738–42.
84. Fischer, E., Marano, M.A., van Zee, K.J., Rock, C.S., Hawes, A.S., Thompson, W.A., *et al.* (**1992**) Interleukin-1 receptor blockade improves survival and hemodynamic

performance in Escherichia coli septic shock, but fails to alter host responses to sublethal endotoxemia. *J. Clin. Invest.*, **89**, 1551–7.

85. Jansen, P.M., Boermeester, M.A., Fischer, E., de, J.I., van, dP., Moldawer, L.L., *et al.* (**1995**) Contribution of interleukin-1 to activation of coagulation and fibrinolysis, neutrophil degranulation, and the release of secretory-type phospholipase A2 in sepsis: studies in nonhuman primates after interleukin-1 alpha administration and during lethal bacteremia. *Blood*, **86**, 1027–34.
86. Boermeester, M.A., van, L.P., Coyle, S.M., Wolbink, G.J., Hack, C.E. and Lowry, S.F. (**1995**) Interleukin-1 blockade attenuates mediator release and dysregulation of the hemostatic mechanism during human sepsis. *Arch. Surg.*, **130**, 739–48.
87. Pajkrt, D., van, dP., Levi, M., Cutler, D.L., Affrime, M.B., van, dE., *et al.* (**1997**) Interleukin-10 inhibits activation of coagulation and fibrinolysis during human endotoxemia. *Blood*, **89**, 2701–5.
88. Mileno, M.D., Margolis, N.H., Clark, B.D., Dinarello, C.A., Burke, J.F. and Gelfand, J.A. (**1995**) Coagulation of whole blood stimulates interleukin-1 beta gene expression. *J. Infect. Dis.*, **172**, 308–11.
89. Jones, A. and Geczy, C.L. (**1990**) Thrombin and factor Xa enhance the production of interleukin-1. *Immunology.*, **71**, 236–41.
90. Johnson, K., Choi, Y., DeGroot, E., Samuels, I., Creasey, A. and Aarden, L. (**1998**) Potential mechanisms for a proinflammatory vascular cytokine response to coagulation activation. *J. Immunol.*, **160**, 5130–5.
91. Sower, L.E., Froelich, C.J., Carney, D.H., Fenton, J.W. and Klimpel, G.R. (**1995**) Thrombin induces IL-6 production in fibroblasts and epithelial cells. Evidence for the involvement of the seven-transmembrane domain (STD) receptor for alpha-thrombin. *J. Immunol.*, **155**, 895–901.
92. van der Poll, T., de Jonge, E. and Levi, M. (**2001**) Regulatory role of cytokines in disseminated intravascular coagulation. *Semin. Thromb. Hemost.*, **27**, 639–51.
93. Coughlin, S.R. (**2000**) Thrombin signalling and protease-activated receptors. *Nature*, **407**, 258–64.
94. de Jonge, E., Friederich, P.W., Levi, M. and van der Poll, T. (**2003**) Activation of coagulation by administration of recombinant factor VIIa elcicits interleukin-6 and interleukin-8 release in healthy humen subjects. *Clin. Diagn. Lab. Immunol.*, **10**, 495–7.
95. Kaneider, N.C., Forster, E., Mosheimer, B., Sturn, D.H. and Wiedermann, C.J. (**2003**) Syndecan-4-dependent signaling in the inhibition of endotoxin-induced endothelial adherence of neutrophils by antithrombin. *Thromb. Haemost.*, **90**, 1150–7.
96. Okajima, K. (**2001**) Regulation of inflammatory responses by natural anticoagulants. *Immunol. Rev.*, **184**, 258–74.
97. Grey, S.T., Tsuchida, A., Hau, H., Orthner, C.L., Salem, H.H. and Hancock, W.W. (**1994**) Selective inhibitory effects of the anticoagulant activated protein C on the responses of human mononuclear phagocytes to LPS, IFN-gamma, or phorbol ester. *J. Immunol.*, **153**, 3664–72.
98. Yuksel, M., Okajima, K., Uchiba, M., Horiuchi, S. and Okabe, H. (**2002**) Activated protein C inhibits lipopolysaccharide-induced tumor necrosis factor-alpha production by inhibiting activation of both nuclear factor-kappa B and activator protein-1 in human monocytes. *Thromb. Haemost.*, **88**, 267–73.
99. Murakami, K., Okajima, K., Uchiba, M., Johno, M., Nakagaki, T., Okabe, H. and Takatsuki, K. (**1996**) Activated protein C attenuates endotoxin-induced pulmonary

vascular injury by inhibiting activated leukocytes in rats. *Blood*, **87**, 642–7.
100. Taylor, F.B., Jr. (**1994**) Studies on the inflammatory-coagulant axis in the baboon response to E. coli: regulatory roles of proteins C, S, C4bBP and of inhibitors of tissue factor. (Review; 13 refs). *Prog. Clin. Biol. Res.*, **388**, 175–94.
101. Taylor, F.BJ., Stearns-Kurosawa, D.J., Kurosawa, S., Ferrell, G., Chang, A.C., Laszik, Z., *et al.* (**2000**) The endothelial cell protein C receptor aids in host defense against Escherichia coli sepsis. *Blood*, **95**, 1680–6.
102. White, B., Schmidt, M., Murphy, C., Livingstone, W., O'Toole, D., Lawler, M., *et al.* (**2000**) Activated protein C inhibits lipopolysaccharide-induced nuclear translocation of nuclear factor kappaB (NF-kappaB) and tumour necrosis factor alpha (TNF-alpha) production in the THP-1 monocytic cell line. *Br. J. Haematol.*, **110**(1), 130–4.
103. Hancock, W.W., Grey, S.T., Hau, L., Akalin, E., Orthner, C., Sayegh, M.H. and Salem, H.H. (**1995**) Binding of activated protein C to a specific receptor on human mononuclear phagocytes inhibits intracellular calcium signaling and monocyte-dependent proliferative responses. *Transplantation*, **60**, 1525–32.
104. Riewald, M., Petrovan, R.J., Donner, A., Mueller, B.M. and Ruf, W. (**2002**) Activation of endothelial cell protease activated receptor 1 by the protein C pathway. *Science*, **296**, 1880–2.
105. Cheng, T., Liu, D., Griffin, J.H., Fernandez, J.A., Castellino, F., Rosen, E.D., *et al.* (**2003**) Activated protein C blocks p53-mediated apoptosis in ischemic human brain endothelium and is neuroprotective. *Nat. Med.*
106. Mosnier, L.O. and Griffin, J.H. (**2003**) Inhibition of staurosporine-induced apoptosis of endothelial cells by activated protein C requires protease activated receptor-1 and endothelial cell protein C receptor. *Biochem. J.*, **373**, 65–70.
107. Neame, P.B., Kelton, J.G., Walker, I.R., Stewart, I.O., Nossel, H.L. and Hirsh, J. (**1980**) Thrombocytopenia in septicemia: the role of disseminated intravascular coagulation. *Blood*, **56**, 88–92.
108. Levi, M., de Jonge, E. and Meijers, J. (**2002**) The diagnosis of disseminated intravascular coagulation. *Blood Rev.*, **16**, 217–23.
109. Dempfle, C.E., Pfitzner, S.A., Dollman, M., Huck, K., Stehle, G. and Heene, D.L. (**1995**) Comparison of immunological and functional assays for measurement of soluble fibrin. *Thromb. Haemost.*, **74**, 673–9.
110. Bredbacka, S., Blomback, M. and Wiman, B. (**1994**) Soluble fibrin: a predictor for the development and outcome of multiple organ failure. *Am. J. Hematol.*, **46**, 289–94.
111. Okajima, K., Uchiba, M., Murakami, K., Okabe, H. and Takatsuki, K. (**1996**) Determination of plasma soluble fibrin using a new ELISA method in patients with disseminated intravascular coagulation. *Am. J. Hematol.*, **51**, 186–91.
112. McCarron, B.I., Marder, V.J., Kanouse, J.J. and Francis, C.W. (**1999**) A soluble fibrin standard: comparable dose-response with immunologic and functional assays. *Thromb. Haemost.*, **82**, 145–8.
113. Dempfle, C.E. (**1999**) The use of soluble fibrin in evaluating the acute and chronic hypercoagulable state. (Review; 147 refs). *Thromb. Haemost.*, **82**, 673–83.
114. Horan, J.T. and Francis, C.W. (**2001**) Fibrin degradation products, fibrin monomer and soluble fibrin in disseminated intravascular coagulation. *Semin. Thromb. Hemost.*, **27**(6), 657–66.
115. McCarron, B.I., Marder, V.J. and Francis, C.W. (**1999**) Reactivity of soluble fibrin assays with plasmic degradation products of fibrin and in patients receiving fibrinolytic

therapy. *Thromb. Haemost.*, **82**, 1722–9.

116. Carr, J.M., McKinney, M. and McDonagh, J. (**1989**) Diagnosis of disseminated intravascular coagulation. Role of D-dimer. *Am. J. Clin. Pathol.*, **91**, 280–7.

117. Boisclair, M.D., Ireland, H. and Lane, D.A. (**1990**) Assessment of hypercoagulable states by measurement of activation fragments and peptides. (Review; 113 refs). *Blood Rev.*, **4**, 25–40.

118. Prisco, D., Paniccia, R., Bonechi, F., Francalanci, I., Abbate, R. and Gensini, G.F. (**1989**) Evaluation of new methods for the selective measurement of fibrin and fibrinogen degradation products. *Thromb. Res.*, **56**, 547–51.

119. Shorr, A.F., Trotta, R.F., Alkins, S.A., Hanzel, G.S. and Diehl, L.F. (**1999**) D-dimer assay predicts mortality in critically ill patients without disseminated intravascular coagulation or venous thromboembolic disease. *Intensive Care Med.*, **25**, 207–10.

120. Greenberg, C.S., Devine, D.V. and McCrae, K.M. (**1987**) Measurement of plasma fibrin D-dimer levels with the use of a monoclonal antibody coupled to latex beads. *Am. J. Clin. Pathol.*, **87**, 94–100.

121. Teitel, J.M., Bauer, K.A., Lau, H.K. and Rosenberg, R.D. (**1982**) Studies of the prothrombin activation pathway utilizing radioimmunoassays for the F2/F1 + 2 fragment and thrombin–antithrombin complex. *Blood*, **59**, 1086–97.

122. Bauer, K.A., Kass, B.L., ten Cate, H., Bednarek, M.A., Hawiger, J.J. and Rosenberg, R.D. (**1989**) Detection of factor X activation in humans. *Blood*, **74**, 2007–15.

123. ten Cate, H., Bauer, K.A., Levi, M., Edgington, T.S., Sublett, R.D., Barzegar, S., *et al.* (**1993**) The activation of factor X and prothrombin by recombinant factor VIIa *in vivo* is mediated by tissue factor. *J. Clin. Invest.*, **92**, 1207–12.

124. Takahashi, H., Wada, K., Niwano, H. and Shibata, A. (**1992**) Comparison of prothrombin fragment 1 + 2 with thrombin-antithrombin III complex in plasma of patients with disseminated intravascular coagulation. *Blood Coagul. Fibrinolysis*, **3**, 813–8.

125. Kario, K., Matsuo, T., Kodama, K., Matsuo, M., Yamamoto, K. and Kobayashi, H. (**1992**) Imbalance between thrombin and plasmin activity in disseminated intravascular coagulation. Assessment by the thrombin-antithrombin-III complex/plasmin-alpha-2-antiplasmin complex ratio. *Haemostasis*, **22**, 179–86.

126. Bick, R.L. (**1996**) Disseminated intravascular coagulation: objective clinical and laboratory diagnosis, treatment, and assessment of therapeutic response. *Semin. Thromb. Hemost.*, **22**, 69–88.

127. Hesselvik, J.F., Blomback, M., Brodin, B. and Maller, R. (**1989**) Coagulation, fibrinolysis, and kallikrein systems in sepsis: relation to outcome. *Crit. Care Med.*, **17**, 724–33.

128. Fijnvandraat, K., Derkx, B., Peters, M., Bijlmer, R., Sturk, A., Prins, M.H., *et al.* (**1995**) Coagulation activation and tissue necrosis in meningococcal septic shock: severely reduced protein C levels predict a high mortality. *Thromb. Haemost.*, **73**, 15–20.

129. Wuillemin, W.A., Fijnvandraat, K., Derkx, B.H., Peters, M., Vreede, W., ten Cate, H. and Hack, C.E. (**1995**) Activation of the intrinsic pathway of coagulation in children with meningococcal septic shock. *Thromb. Haemost.*, **74**, 1436–41.

130. Mesters, R.M., Helterbrand, J., Utterback, B.G., Yan, B., Chao, Y.B., Fernandez, *et al.* (**2000**) Prognostic value of protein C concentrations in neutropenic patients at high risk of severe septic complications. *Crit. Care Med.*, **28**, 2209–16.

131. Taylor, F.BJ., Toh, C.H., Hoots, W.K., Wada, H. and Levi, M.

(2001) Towards definition, clinical and laboratory criteria, and a scoring system for disseminated intravascular coagulation. *Thromb. Haemost.*, **86**, 1327–30.

132. Bakhtiari, K., Meijers, J.C., de Jonge, E. and Levi, M. (**2004**) Prospective validation of the international society of thrombosis and maemostasis scoring system for disseminated intravascular coagulation. *Crit. Care Med.*, **32**, 2416–21.
133. Dhainaut, J.F., Yan, S.B., Joyce, D.E., Pettila, V., Basson, B.R., Brandt, J.T., *et al.* (**2004**) Treatment effects of drotrecogin alfa (activated) in patients with severe sepsis with or without overt disseminated intravascular coagulation. *J. Thromb. Haemost.*, **2**, 1924–33.
134. Alving, B.M., Spivak, J.L. and DeLoughery, T.G. (**1998**) Consultative hematology: Hemostasis and transfusion issues in surgery and critical care medicine. *Hematology 1998* (Am. Soc. Hematol. Educ. Program), 320–41.
135. de Jonge, E., Levi, M., Stoutenbeek, C.P. and van Deventer, S.J. (**1998**) Current drug treatment strategies for disseminated intravascular coagulation. *Drugs*, **55**, 767–77.
136. du Toit, H., Coetzee, A.R. and Chalton, D.O. (**1991**) Heparin treatment in thrombin-induced disseminated intravascular coagulation in the baboon. *Crit. Care Med.*, **19**, 1195–200.
137. Feinstein, D.I. (**1982**) Diagnosis and management of disseminated intravascular coagulation: the role of heparin therapy. (Review; 34 refs). *Blood*, **60**, 284–7.
138. Dorffler-Melly, J., de Jonge, E., Pont, A.C., Meijers, J., Vroom, M.B., Buller, H.R. and Levi, M. (**2002**) Bioavailability of subcutaneous low-molecular-weight heparin to patients on vasopressors. *Lancet*, **359**, 849–50.
139. Vlasuk, G.P., Bergum, P.W., Bradbury, A.E., Loynds, P., Mant, T.G. and Rote, W.E. (**1997**) Clinical evaluation of rNAPc2, an inhibitor of the fVIIa/tissue factor coagulation complex. *Am. J. Cardiol.*, **80**, 66S.
140. Abraham, E., Reinhart, K., Svoboda, P., Seibert, A., Olthoff, D., Dal Nogare, A., *et al.* (**2001**) Assessment of the safety of recombinant tissue factor pathway inhibitor in patients with severe sepsis: a multicenter, randomized, placebo–controlled, single-blind, dose escalation study. *Crit. Care Med.*, **29**, 2081–9.
141. Abraham, E., Reinhart, K., Opal, S., Demeyer, I., Doig, C., Rodriguez, A.L., *et al.* (**2003**) Efficacy and safety of tifacogin (recombinant tissue factor pathway inhibitor) in severe sepsis: a randomized controlled trial. *JAMA*, **290**, 238–47.
142. Levi, M., de Jonge, J.E. and van der Poll, T. (**2001**) Rationale for restoration of physiological anticoagulant pathways in patients with sepsis and disseminated intravascular coagulation. *Crit. Care Med.*, **29**(7 Suppl), S90–S94.
143. Fourrier, F., Chopin, C., Huart, J.J., Runge, I., Caron, C. and Goudemand, J. (**1993**) Double-blind, placebo-controlled trial of antithrombin III concentrates in septic shock with disseminated intravascular coagulation. *Chest*, **104**, 882–8.
144. Eisele, B., Lamy, M., Thijs, L.G., Keinecke, H.O., Schuster, H.P., Matthias, F.R., *et al.* (**1998**) Antithrombin III in patients with severe sepsis. A randomized, placebo-controlled, double-blind multicenter trial plus a meta-analysis on all randomized, placebo-controlled, double-blind trials with antithrombin III in severe sepsis (see comments). *Intensive Care Med.*, **24**, 663–72.
145. Baudo, F., Caimi, T.M., de, C.F., Ravizza, A., Arlati, S., Casella, G., *et al.* (**1998**) Antithrombin III (ATIII) replacement therapy in patients with sepsis and/or postsurgical

complications: a controlled double-blind, randomized, multicenter study [see comments]. *Intensive Care Med.*, **24**, 336–42.

146. Warren, B.L., Eid, A., Singer, P., Pillay, S.S., Carl, P., Novak, I., *et al.* (**2001**) Caring for the critically ill patient. High-dose antithrombin III in severe sepsis: a randomized controlled trial. *JAMA*, **286**, 1869–78.
147. Bernard, G.R., Ely, E.W., Wright, T.J., Fraiz, J., Stasek, J.E., Jr., Russell J.A., *et al.* (**2001**) Safety and dose relationship of recombinant human activated protein C for coagulopathy in severe sepsis. *Crit. Care. Med.*, **29**, 2051–9.
148. Vincent, J.L., Angus, D.C., Artigas, A., Kalil, A., Basson, B.R., Jamal, H.H., *et al.* (**2003**) Effects of drotrecogin alfa (activated) on organ dysfunction in the PROWESS trial. *Crit. Care Med.*, **31**, 834–40.

12
Epithelial Barrier Dysfunction as a Mechanism in the Pathogenesis of Multiple Organ Dysfunction

Mitchell P. Fink, Antonella Cotoia, Donna Beer-Stolz, Shiguang Liu, Penny L. Sappington and Russell L. Delude

12.1
Introduction

The multiple organ dysfunction syndrome (MODS) is the most common cause of death among patients requiring care in an intensive care unit [1]. There is widespread agreement that MODS is the clinical manifestation of a dysregulated inflammatory response. Indeed, most of the published research regarding the pathogenesis of MODS has focused on the various signaling pathways that lead to the activation of the innate immune system and the elaboration of cytokines, oxidants, tissue-destructive enzymes and other pro-inflammatory mediators. Interestingly, however, the biochemical and cell biological basis for organ dysfunction *per se* remains very poorly understood.

12.2
Is "Cytopathic Hypoxia" the underlying Mechanism Responsible for MODS?

One possible explanation for cellular dysfunction in sepsis and MODS is an acquired intrinsic derangement in mitochondrial function leading to inadequate production of adenosine triphosphate (ATP), a pathological state, which has been termed "cytopathic hypoxia" [2, 3]. Data from numerous studies provide strong support for the notion that mitochondrial structure and function are impaired in septic animals [4–7]. Similarly, results from a number of clinical studies have confirmed that cellular respiration is deranged in patients with sepsis [8–10]. Furthermore, data obtained by Mervyn Singer's group in London, England provide strong, albeit still circumstantial, evidence that mitochondrial dysfunction contributes to the development of MODS in experimental animals with sepsis [5]. According to this hypothesis, mitochondrial dysfunction, and impaired ATP synthesis on this basis, leads to a re-prioritization of cellular activities in order to consume energy. In

Sepsis and Non-infectious Systemic Inflammation. Edited by J.-M. Cavaillon and C. Adrie

ISBN: 978-3-527-31935-0

other words, MODS is the clinical manifestation of cells throughout the body (or at least in key tissues) entering a state of partially suspended animation.

Data obtained by Berg *et al.*, however, raise the possibility that mitochondrial dysfunction is an epiphenomenon rather than a fundamental factor contributing to organ dysfunction in MODS [11]. Berg *et al.* carried out a study using cytomix-stimulated IEC-6 (nontransformed rat) enterocyte-like cells as a reductionist *in vitro* model of sepsis, and showed that the rate of ATP turnover actually increases in immunostimulated as compared to control monolayers [11]. The increase in ATP utilization (and production) is maintained by a marked increase in the rate of glycolysis. In related studies, Scharte *et al.* [12] and Bertges *et al.* [13] showed that incubating Caco-2 (human enterocyte-like) or IEC-6 cells with cytomix increases DNA binding by hypoxia-inducible factor (HIF)-1, the key transcription factor responsible for acute cellular adaptation to hypoxic conditions. Moreover, Scharte *et al.* showed that incubating IEC-6 cells with cytomix leads to the transcriptional activation of two key HIF-1 responsive genes, aldolase A and enolase-1, that encode proteins important in the glycolytic pathway [12]. Other results from both animal [14, 15] and clinical studies [16, 17] also support the view that sepsis is associated with a shift toward increased ATP production via glycolysis and away from oxidative phosphorylation.

12.3 What is the Role of Cellular Apoptosis in the Pathogenesis of MODS?

Massive apoptosis among lymphoid cells is a prominent feature of sepsis in both human patients and mice [18–21]. Moreover, using pharmacological or genetic approaches to limit lymphoid cell apoptosis improves survival in mice with bacterial peritonitis [18, 20], a finding that supports the view that increased programmed cell death of lymphocytes is significant in the pathogenesis of sepsis. Presumably, blocking lymphocytic apoptosis improves survival in experimental sepsis by improving host defenses against infection [22]. Alternatively, however, phagocytosis of apoptotic lymphocytes by macrophages may trigger the secretion of a pro-inflammatory cytokine-like protein, high mobility group box (HMGB)-1 (see below), exacerbating the deleterious systemic inflammatory response to infection and further promoting organ system injury [23].

Intestinal epithelial apoptosis also occurs in both patients and animals with sepsis [19, 24, 25], although most cells in the epithelial sheet are not affected. Despite these observations, neither apoptosis nor necrosis are prominent features in other organs, notably the lungs, liver or kidneys, that are commonly involved in cases of MODS [19, 26]. Thus, it is highly improbable that loss of cell mass *per se* can account for the development of lung, liver, gut or kidney dysfunction in patients with MODS.

12.4
Tight Junctions maintain Epithelial Polarity and Barrier Function

The normal functioning of the lungs, liver, kidneys and intestine, among other organs, depends on the establishment and maintenance of compositionally distinct compartments that are lined by sheets of epithelial cells. An essential element in this process is the formation of tight junctions (TJs) between adjacent cells making up the epithelial sheet. The TJ serves as a fence that differentiates the cytosolic membrane into apical and basolateral domains. This fence function preserves cellular polarity and, in combination with transcellular vectorial transport processes, generates distinct internal environments in the opposing compartments that are formed by the epithelial sheet. In addition, the TJ acts as a regulated semi-permeable barrier that limits the passive diffusion of solutes across the paracellular pathway between adjacent cells. Thus, the barrier function of the TJ is necessary to prevent dissipation of the concentration gradients that exist between the two compartments defined by the epithelium. In some organs, notably the gut and the lung, this barrier function is also important to prevent systemic contamination by microbes and toxins that are present in the external environment [27].

Situated immediately basal to the TJ is the adherens junction (AJ), which also is a continuous, circumferential belt of cell–cell interaction. The molecular architecture of the AJ is similar to that of the TJ. Importantly, the TJ is evolved from and stabilized by the AJ and is structurally and functionally interrelated to the AJ [28].

The formation of TJs involves the assembly of multiple different peripheral membrane proteins and at least three different integral membrane proteins [29]. Among the peripheral membrane proteins associated with TJs are the membrane-associated guanylate kinase-like proteins, ZO-1, ZO-2, and ZO-3. The integral membrane proteins involved in TJ formation include occludin and members of a large class of proteins called claudins. Both occludin and the claudins contain four transmembrane domains and are thought to be the actual points of cell–cell contact within the TJ [30]. ZO-1 has been shown to interact with the cytoplasmic tails of occludin and the claudins [31].

12.5
Epithelial Dysfunction as a Pathogenic Mechanism underlying MODS

Many of the organs commonly affected in MODS (e.g. the lungs, liver, kidneys and gut) depend on the proper functioning of an epithelial component. Therefore, it is reasonable to hypothesize that epithelial cell dysfunction is important in this syndrome. Of course, the proper functioning of epithelia depends not only on the formation of TJs, but also on the appropriate expression, localization and activity of many other cellular constituents (e.g. membrane pumps, cytoskeletal proteins, and cell-surface integrins and receptors). Indeed, data have accumulated from studies using experimental animals to support the view that sepsis is associated with altered expression of key pumps and transport proteins in epithelial cells. However, in view of the fundamental importance of TJs for the maintenance of epithelial polarization, vectorial transport and barrier function, there has been considerable interest in studying changes in the expression and localization of TJ proteins in sepsis or MODS.

The proper functioning of the lungs, kidneys, liver and gut depends on the generation and maintenance of compositionally distinct compartments. In the lungs, failure to maintain normal TJ formation would be expected to promote alveolar flooding, and hence pulmonary edema, on the basis of back-leakage of salt and water that is pumped from the apical side of the alveolar epithelium to the basolateral side. Of course, TJ dysfunction is not the only potential cause of impaired alveolar fluid clearance. Other causes might be decreased expression or function of Na^+, K^+-ATPase, the pump responsible for the vectorial transport of sodium ions [32–34], or decreased expression or function of the epithelial sodium channel [33].

Cholestatic jaundice is a fairly common clinical manifestation of MODS and is associated with an increased risk of mortality [35, 36]. Efforts to understand the pathophysiological mechanisms responsible for cholestatic jaundice due to sepsis have largely focused on alterations in the function and expression of various bile acid transporters [37–41] or water pores [42]. Nevertheless, another factor that could contribute to the development of intrahepatic cholestasis is back-leakage of bile from the canalicular spaces into the sinusoids [43, 44].

Alterations in the barrier function of the intestinal epithelium could permit the leakage of bacteria or microbial products, such as lipopolysaccharide (LPS), bacterial (CpG rich) DNA or flagellin, from the lumen of the gut into the systemic compartment, leading to the initiation or amplification of a deleterious inflammatory response. Additionally, and perhaps even more importantly, increased gut epithelial permeability might permit systemic absorption of cytotoxic proteins, such as exotoxin A, produced by luminal bacteria in the gut, particularly under conditions of stress to the host [45]. The notion that this process actually occurs in patients with MODS is supported by results from a number of clinical studies, which have documented increases in intestinal epithelial permeability in a variety of acute conditions that are associated with systemic inflammation [46–49]. Moreover, in several recent studies, increased intestinal permeability in critically ill patients has been shown to be associated with an increased risk of complications, MODS, or even mortality [48–52].

Acute renal failure in patients with sepsis or MODS is manifested by the development of azotemia and often, also oliguria. The mechanisms underlying development of acute renal failure due to sepsis are likely multiple and remain poorly understood. Important factors are thought to be excessive renal vasoconstriction [53], glomerular thrombin deposition [54] and decreased expression of urea transporters in renal epithelial cells [55]. Without minimizing the importance of these and other mechanisms, it is reasonable to postulate that another factor might be back-leakage of tubular fluid as a result of TJ dysfunction in the tubular epithelium. Certainly, evidence from both animal [56] and human studies [57–59] supports the concept of back-leakage of glomerular filtrate during acute renal failure.

Additionally, loss of "fence function" (i.e. disruption of renal epithelial polarization) as a consequence of TJ and AJ dysfunction can lead to aberrant apical localization of membrane proteins, such as the integrins, which are normally localized to the basolateral aspect of the cells [28]. The basolateral localization of integrins is important for the proper anchoring of tubular cells to the basement membrane. ATP depletion results in the redistribution of β_1-integrin subunits to the apical membrane, and the loss of anchorage allows the exfoliation of viable cells into the tubular lumen [60], potentially promoting back-leakage of tubular fluid via two mechanisms. First, denudation of the basement membrane on the basis of the detachment epithelial cells would be expected to expose a greater surface area for back-leakage of tubular fluid. Second, the presence of exfoliated cells within the tubular lumen would be expected to increase back pressure, further exacerbating pathological back-leakage of glomerular filtrate. Tubular cell detachment [61, 62] and obstruction [56] have been reported in studies of acute renal failure.

12.6 The Role of Nitric Oxide (NO) and Peroxynitrite ($ONOO^-$) in the Regulation of TJ Protein Expression and Function

The permeability of cultured epithelial monolayers increases when the cells are incubated with various pro-inflammatory cytokines [63–68]. The mechanisms responsible for cytokine-induced epithelial hyperpermeability are incompletely understood. It is known, however, that compounds that spontaneously release $NO^•$ increase the permeability of cultured intestinal epithelial cell monolayers [69, 70]. This observation is pertinent, since incubating Caco-2 human enterocyte-like cells with the pro-inflammatory cytokine, interferon (IFN)-γ, or a mixture of the pro-inflammatory cytokines, IFN-γ, tumor necrosis factor (TNF) and interleukin (IL)-1β, leads to increased expression of inducible $NO^•$ synthase (iNOS) and increased production of $NO^•$ [64, 71, 72]. Moreover, compounds that inhibit the enzymatic activity of iNOS have been shown to ameliorate the development of hyperpermeability induced by exposing Caco-2 cells to IFN-γ [64] or "cytomix" (IFN-γ plus TNF plus IL-1β) [72]. Similarly, L-N(6)-(1-iminoethyl)lysine (L-NIL), an isoform-selective iNOS inhibitor, blocks the development of hyperpermeability when Calu-3 (human alveolar epithelial) monolayers are incubated with cytomix [73]. Thus, IFN-γ or cytomix appear to increase intestinal epithelial permeability, at least in part, by increasing the production of $NO^•$ by enterocytes.

NO˙ reacts rapidly with superoxide radical anion ($O_2^{\cdot-}$) to form the potent oxidizing and nitrating species, peroxynitrite ($ONOO^-$) [74, 75]. Several lines of evidence support the view that $ONOO^-$ (or some related species) rather than NO˙ *per se* is responsible for the deleterious effects of NO˙ on intestinal epithelial barrier function. Thus, when Caco-2 monolayers are incubated with the NO˙ donor, SNAP, permeability is significantly increased, but the magnitude of the effect is small [70]. Furthermore, the permeability of Caco-2 monolayers is not affected when the cells are incubated with pyrogallol, a compound that spontaneously generates $O_2^{\cdot-}$ in aqueous solutions [70]. However, if Caco-2 cells are co-incubated with both SNAP and pyrogallol, then epithelial permeability is dramatically increased [70]. SNAP-induced hyperpermeability is also markedly enhanced by co-incubating the cells with diethyldithiocarbamate, a compound that is known to inactivate Cu-Zn superoxide dismutase and would thereby be expected to increase the concentration of endogenously generated $O_2^{\cdot-}$ [70]. Taken together, these findings support the view that $NO^{\cdot-}$ induced hyperpermeability is enhanced by the simultaneous availability of $O_2^{\cdot-}$; i.e. conditions favoring the formation of $ONOO^-$. Since ONOOH is a weak acid (pKa ~ 6.8) and many of the effects of $ONOO^-$ are thought to be mediated by an unstable form of the protonated species, studies from our group showing that $NO^{\cdot-}$ induced hyperpermeability is enhanced under mildly acidic conditions further support the notion that $ONOO^-$/ONOOH is the responsible moiety [63, 76].

The mechanism(s) responsible for $NO^{\cdot-}$ or $ONOO^-$-mediated intestinal epithelial hyperpermeability remain to be elucidated. However, it is known that NO˙ generated endogenously as the result of iNOS expression induced by incubating Caco-2 cells with cytomix or exogenously from the NO˙ donor DETA-NONOate decreases the expression and impaired proper localization of the TJ proteins, ZO-1, ZO-3, and occludin [65]. Furthermore, incubating Caco-2 cells with either DETA-NONOate or cytomix increases the expression of claudin-1 and promotes the accumulation of this protein in what appear to be vesicles within the cells. These findings support the view that NO˙ (or a related reactive species) increases epithelial permeability by causing derangements in the expression and/or localization of several key TJ proteins.

Sugi *et al.* proposed that one way that NO˙ might alter the expression or localization of various TJ proteins is by modulating the activity of the membrane pump, Na^+, K^+-ATPase [77]. In a series of studies using monolayers of T84 enterocyte-like cells, these investigators reported that intracellular sodium concentration and cell volume increase following exposure to the pro-inflammatory cytokine, IFN-γ. Additionally, Sugi *et al.* showed that incubating T84 cells with either NO˙ or IFN-γ decreases the expression and activity of Na^+, K^+-ATPase. Remarkably, growing the monolayers in medium with low sodium concentration inhibits the development of hyperpermeability following exposure to IFN-γ and also prevents IFN-γ-induced alterations in occludin expression. These findings suggest a pathway that involves the following steps. IFN-γ (and/or other pro-inflammatory cytokines) $\rightarrow$ iNOS induction → NO• production → inhibition of Na^+, K^+-ATPase expression and function → cell swelling → altered expression and/or targeting of TJ proteins (e.g. occludin) → hyperpermeability.

12.7 Functional iNOS Expression is essential for LPS-induced Alterations in Intestinal Barrier Function in Mice

When Han *et al.* injected C57Bl/6J mice with a small sublethal dose of *E. coli* LPS, intestinal mucosal permeability to the permeability probe, fluorescein isothiocyanate-labeled dextran (molecular mass, 4 kDa; FD4), increased significantly. Treatment of endotoxemic mice with L-NIL ameliorated LPS-induced ileal mucosal hyperpermeability. Basal ileal mucosal permeability in control iNOS knockout (iNOS$^{-/-}$) mice on a C57Bl/6J background was greater than that measured in control (wild-type) iNOS$^{+/+}$ mice, a finding that is consistent with reports that basal levels of NO• are required for normal gut homeostasis [78, 79]. Despite a basal defect in intestinal barrier function in iNOS$^{-/-}$ mice, permeability to FD4 failed to increase further when these mice were challenged with LPS.

Bacterial translocation from the gut lumen to mesenteric lymph nodes (MLN) is another measure of *in vivo* intestinal mucosal barrier function. Thus, induction of endotoxemia increased the number of bacteria that were recovered from MLN from wild-type (iNOS$^{+/+}$) mice [80]. Treatment of endotoxemic iNOS$^{+/+}$ mice with L-NIL to block iNOS-dependent NO• production decreased LPS-induced bacterial translocation. Similarly, LPS failed to induce bacterial translocation in iNOS$^{-/-}$ mice.

Immunohistochemical studies of ileal tissue from endotoxemic mice were performed using samples harvested 12 h after injection of LPS [80]. ZO-1 formed a continuous staining pattern around the enterocyte layer near the apical region of the lateral membrane of crypt and villous cells of the epithelium and the endothelium of the lamina propria from normal mice. Following injection of mice with LPS, ZO-1 staining was maintained in the crypts, but staining progressively decreased over the tips of the villi. In sections from endotoxemic mice, the staining patterns for ZO-1 were disrupted only in focal regions of the ileum; approximately 60% of the villi in a given section stained normally. If the endotoxemic mice were treated with L-NIL to pharmacologically block iNOS-dependent NO• production, then the correct targeting of ZO-1 in the ileal mucosa was preserved. Similar findings were obtained when staining was carried out for occludin instead of ZO-1.

Parallel experiments were performed using iNOS$^{-/-}$ mice [80]. The levels of occludin and ZO-1 in ileal mucosa from control iNOS$^{-/-}$ mice (i.e. those not challenged with LPS) were reproducibly lower than the levels of these proteins in control iNOS$^{+/+}$ mice. To some extent, these basal differences in occludin and ZO-1 expression confounded interpretation of the results obtained in

Injecting mice with LPS significantly increased the concentration of the NO• breakdown products, NO_2^- and NO_3^-, in bronchoalveolar lavage fluid (BALF) and serum. Treatment with L-NIL only partially inhibited the accumulation of NO_2^- and NO_3^- in serum, whereas treatment with the iNOS inhibitor almost completely blocked accumulation of these NO• metabolites in BALF.

NP-40 insoluble occludin and ZO-1 levels decreased within 6 h of injecting mice with a sublethal dose of LPS, and were maximally decreased at 12 h. By 18 h, NP-40 insoluble occludin and ZO-1 levels were starting to normalize. In lung tissue specimens from normal mice, ZO-1 was localized as a continuous line along the boundaries between neighboring bronchial and alveolar epithelial cells. The intensity of this staining was markedly reduced in lung tissue harvested from mice that had been challenged with LPS 12 h earlier.

12.10 HMGB1 as a Mediator of Epithelial Dysfunction

The HMGB family of DNA-binding proteins, namely HMGB1 (previously known as HMG1) and HMGB2 (previously known as HMG2), have molecular masses of about 28 kDa and share greater than 80% amino acid sequence identity [84, 85]. The HMGB proteins bend DNA by virtue of a conserved DNA binding domain, the so-called HMG1 box [86]. Each HMG1 box contains a string of 70–80 amino acid residues, which is folded into a characteristic, twisted, L-shaped structure [86, 87]. HMGB1 facilitates the binding of several regulatory protein complexes to DNA, particularly members of the nuclear hormone-receptor family [88, 89], V(D)J recombinases [90] and the tumor suppressor proteins, p53 and p73 [91].

In 1999, HMGB1 was identified as a cytokine-like mediator of LPS-induced mortality in mice [92]. Subsequently, these findings were extended by Yang *et al.*, who showed that HMGB1 is also a mediator of lethality in mice rendered septic by the induction of polymicrobial bacterial peritonitis [93]. Additional studies have documented that extracellular HMGB1 can promote TNF release from mononuclear cells [94] and increase the permeability of Caco-2 monolayers, via an NO•-dependent mechanism [95]. Interestingly, HMGB1 released by immunostimulated Caco-2 cells (see below) seems to be capable of further amplifying derangements in TJ function initially triggered by other pro-inflammatory cytokines [96]. Furthermore, treatment with a polyclonal anti-HMGB1 neutralizing antibody has been shown to ameliorate gut mucosal hyperpermeability and improve survival in mice subjected to hemorrhagic shock [97].

HMGB1 also has been implicated in the pathogenesis of human diseases. In the original report describing HMGB1 as a mediator of LPS-induced lethality, Wang *et al.* reported that circulating levels of this protein are increased in patients with severe sepsis [92]. Shortly thereafter, Ombrellino *et al.* described a patient with high circulating levels of HMGB1 following an episode of hemorrhagic shock [98]. More recently, increased levels of HMGB1 mRNA have been detected in whole blood samples from patients with septic shock, particularly among non-survivors [99]. Similarly, persistently high serum levels of HMGB1 protein have been detected in patients with septic shock and sepsis [100]. Elevated circulating levels of HMGB1 have also been documented and shown to correlate with severity of injury in victims of multiple trauma [97, 101].

HMGB1 is actively secreted by immunostimulated macrophages [92, 102–104], natural killer cells [105], plasmacytoid dendritic cells [106] and pituicytes [107]. Epithelial cells, including enterocytes, also secrete HMGB1 following immune stimulation. Kuniyasu *et al.* recently reported that WiDr human colon cancer cells constitutively release HMGB1 into culture supernatants [108]. In contrast, Liu *et al.* observed only very low levels of HMGB1 in the media of unstimulated Caco-2 cells [96]. However, following stimulation of the cells with cytomix, Liu and colleagues observed a large increase in the amount of HMGB1 released into the culture media. These investigators also showed that incubating Caco-2 cells with the synthetic Toll-like receptor (TLR) 2 ligand, FSL-1, or the TLR5 ligand, flagellin, caused a large increase in the amount of HMGB1 released into the media. Interestingly, the TLR4 agonist, LPS, failed to stimulate HGMB1 secretion by Caco-2 cells.

Since it is known that HMGB1 is released by necrotic cells [109], it is important that Liu *et al.* were able to document that incubating Caco-2 cells with cytomix for 48 h was neither associated with an increase in the number of cells taking up the vital dye, trypan blue, nor increased release of the intracellular enzyme, lactate dehydrogenase (LDH). These observations confirm findings previously reported by our laboratory, wherein we showed that incubation of Caco-2 cells with cytomix fails to increase staining with the fluorescent dye, ethidium homodimer-1, which only penetrates into dead cells [110]. Because Kuniyasu *et al.* showed that colon cancer cells release HMGB1 [108] and Caco-2 cells are cancer cells, it is noteworthy that we showed that cytomix-stimulated (but not resting) *primary* murine enterocyte cultures release HMGB1. Thus, the findings obtained by Liu *et al.* support the view that immunostimulated enterocytes (and not just colon cancer cells) secrete HMGB1, and the release of this protein by these cells is the result of an active process rather than being secondary to cell death.

12.11
Cross-talk between Gut Epithelial Cells and Intestinal Bacteria

When certain bacteria reach high densities of organisms per gram of tissue or milliliter of growth medium, the microorganisms increase their expression of virulence genes; this phenomenon is called "quarum sensing" [111]. Work carried out by the Alverdy laboratory at the University of Chicago during the past few years has shown that soluble compounds released by the intestinal epithelium during ischemia, hypoxia, or tissue injury, are potent activators of quarum sensing molecular circuitry in the opportunistic pathogen *Pseudomonas aeruginosa*. These soluble compounds include adenosine [112] and the endogenous opioid, dynorphin [113]. Activation of the quarum sensing virulence circuitry enhances expression of lectin protein, PA-I, which, in turn, promotes disruption of the intestinal epithelial barrier [111]. Thus, during the sorts of stress that occur during critical illness, cross-talk between the gut epithelium and luminal microbes can promote the expression of virulence genes in the microbes as well derangements in gut barrier function.

References

1. Mayr, V.D., Dünser, M.W., Greil, V., Jochberger, S., Luckner, G., Ulmer, H., Friesenecker, B.E., Takala, J. and Hasibeder, W.R. (**2006**) Causes of death and determinants of outcome in critically ill patients. *Crit. Care*, **10**, R154.
2. Fink, M.P. (**2001**) Cytopathic hypoxia: mitochondrial dysfunction as a mechanism contributing to organ dysfunction in sepsis. *Crit. Care Clin. North Am.*, **17**, 219–237.
3. Fink, M.P. (**2002**) Bench-to-bedside review: Cytopathic hypoxia. *Crit. Care*, **6**, 491–499.
4. King, C.J., Tytgat, S., Delude, R.L. and Fink, M.P. (**1999**) Ileal mucosal oxygen consumption is decreased in endotoxemic rats but is restored toward normal by treatment with aminoguanidine. *Crit. Care Med.*, **27**, 2518–2524.
5. Brealey, D., Karyampudi, S., Jacques, T.S., Novelli, M., Stidwill, R., Taylor, V., Smolenski, R.T. and Singer, M. (**2004**) Mitochondrial dysfunction in a long-term rodent model of sepsis and organ failure. *Am. J. Physiol. Regul. Integr. Comp. Physiol.*, **286**, R491–R497.
6. Crouser, E.D., Julian, M.W., Huff, J.E., Struck, J. and Cook, C.H. (**2006**) Carbamoyl phosphate synthase-1: a marker of mitochondrial damage and depletion in the liver during sepsis. *Crit. Care Med.*, **34**, 2439–2446.
7. Welty-Wolf, K.E., Simonson, S.G., Huang, Y.-C.T., Fracica, P.J., Patterson, J.W. and Piantadosi, C.A. (**1996**) Ultrastructural changes in skeletal muscle mitochondria in gram-negative sepsis. *Shock*, **5**, 378–384.
8. Boulos, M., Astiz, M.E., Barua, R.S. and Osman, M. (**2003**) Impaired mitochondrial function induced by serum from septic shock patients is attenuated by inhibition of nitric oxide synthase and poly(ADP-ribose) synthase. *Crit. Care Med.*, **31**, 353–358.
9. Brealey, D., Brand, M., Hargreaves, I., Heales, S., Land, J., Smolenski, R., Davies, N.A., Cooper, C.E. and Singer, M. (**2002**) Association between mitochondrial dysfunction and severity and outcome of septic shock. *Lancet*, **360**, 219–223.
10. Fredriksson, K., Hammarqvist, F., Strigard, K., Hultenby, K., Ljungqvist, O., Wernerman, J. and Rooyackers, O. (**2006**) Derangements in mitochondrial metabolism in intercostal and leg muscle of critically ill patients with sepsis-induced multiple organ failure. *Am. J. Physiol. Endocrinol. Metab.*, **291**, E1044–E1050.
11. Berg, S., Sappington, P.L., Guzik, L.J., Delude, R.L. and

Fink, M.P. (**2003**) Pro-inflammatory cytokines increase the rate of glycolysis and ATP turnover in cultured rat enterocytes. *Crit. Care Med.*, **31**, 1203–1212.

12. Scharte, M., Han, X., Bertges, D.J., Fink, M.P. and Delude, R.L. (**2003**) Cytokines induce HIF-1 DNA binding and the expression of HIF-1-dependent genes in cultured rat enterocytes. *Am. J. Physiol. Gastrointest. Liver Physiol.*, **284**, G373–G384.
13. Bertges, D.J., Berg, S., Fink, M.P. and Delude, R.L. (**2002**) Regulation of hypoxia-inducible factor 1 in enterocytic cells. *J. Surg. Res.*, **106**, 157–165.
14. L'Her, E. and Sebert, P. (**2001**) A global approach to energy metabolism in an experimental model of sepsis. *Am. J. Resp. Crit. Care Med.*, **164**, 1444–1447.
15. Bundgaard, H., Kjeldsen, K., Suarez Krabbe, K., van Hall, G., Simonsen, L., Qvist, J., Hansen, C.M., Moller, K., Fonsmark, L., Lav Madsen, Pand Klarlund Pedersen, B. (**2003**) Endotoxemia stimulates skeletal muscle Na+-K+-ATPase and raises blood lactate under aerobic conditions in humans. *Am. J. Physiol. Heart Circ. Physiol.*, **284**, H1028–H1034.
16. Levy, B. (**2006**) Lactate and shock state. the metabolic view. *Curr. Opin. Crit. Care*, **12**, 315–321.
17. Levy, B., Gibot, S., Franck, P., Cravoisy, A. and Bollaert, P.-E. (**2005**) Relation between muscle Na+K+ ATPase activity and raised lactate concentrations in septic shock: a prospective study. *Lancet*, **365**, 871–875.
18. Hotchkiss, R.S., Chang, K.C., Swanson, P.E., Tinsley, K.W., Hui, K.P., Klender, P., Xanthoudakis, S., Roy, S., Black, C., Grimm, E., Aspiotis, R., Han, Y., Nicholson, D.W. and Karl, I.E. (**2000**) Caspase inhibitors improve survival in sepsis. a critical role of the lymphocyte. *Nat. Immunol.*, **1**, 496–501.
19. Hotchkiss, R.S., Swanson, P.E., Freeman, B.D., Tinsley, K.W., Cobb, J.P., Matuschak, G.M., Buchman, T.G. and Karl, I.E. (**1999**) Apoptotic cell death in patients with sepsis, shock, and multiple organ dysfunction. *Crit. Care Med.*, **27**, 1230–1251.
20. Hotchkiss, R.S., Tinsley, K.W., Swanson, P.E., Chang, K.C., Cobb, J.P., Buchman, T.G., Korsmeyer, S.J. and Karl, I.E. (**1999**) Prevention of lymphocyte cell death in sepsis improves survival in mice. *Proc. Natl. Acad. Sci. USA*, **96**, 14541–14546.
21. Le Tulzo, Y., Pangault, C., Gacouin, A., Guilloux, V., Tribut, O., Amiot, L., Tattevin, P., Thomas, R., Fauchet, R. and Drenou, B. (**2002**) Early circulating lymphocyte apoptosis in human septic shock is associated with poor outcome. *Shock*, **18**, 487–494.
22. Weaver, J.G., Rouse, M.S., Steckelberg, J.M. and Badley, A.D. (**2004**) Improved survival in experimental sepsis with an orally administered inhibitor of apoptosis. *FASEB J.*, **18**, 1185–1191.
23. Qin, S., Wang, H., Yuan, R., Li, H., Ochani, M., Ochani, K., Rosas-Ballina, M., Czura, C.J., Huston, J.M., Miller, E., Lin, X., Sherry, B., Kumar, A., Larosa, G., Newman, W., Tracey, K.J. and Yang, H. (**2006**) Role of HMGB1 in apoptosis-mediated sepsis lethality. *J. Exp. Med.*, **203**, 1637–1642.
24. Coopersmith, C.M., Stromberg, P.E., Dunne, W.M., Davis, C.G., Amiot, D.M2., Buchman, T.G., Karl, I.E. and Hotchkiss, R.S. (**2002**) Inhibition of intestinal epithelial apoptosis and survival in a murine model of pneumonia-induced sepsis. *J. Amer. Med. Assoc.*, **287**, 1716–1721.
25. Clark, J.A., Clark, A.T., Hotchkiss, R.S., Buchman, T.G. and Coopersmith, C.M. (**2008**) Epidermal growth factor treatment decreases mortality and is associated with

improved gut integrity in sepsis. *Shock*.

26. Dear, J.W., Yasuda, H., Hu, X., Hieny, S., Yuen, P.S., Hewitt, S.M., Sher, A. and Star, R.A. (**2006**) Sepsis-induced organ failure is mediated by different pathways in the kidney and liver: acute renal failure is dependent on MyD88 but not renal cell apoptosis. *Kidney Int.*, **69**, 832–836.
27. Stevenson, B.R. (**1999**) Understanding tight junction clinical physiology at the molecular level. *J. Clin. Invest.*, **104**, 3–4.
28. Lee, D.B., Huang, E. and Ward, H.J. (**2006**) Tight junction biology and kidney dysfunction. *Am. J. Phyiol. Renal Physiol.*, **290**, F20–F34.
29. Tsukita, S., Furuse, M. and Itoh, M. (**2001**) Multifunctional strands in tight junctions. *Nat. Rev. Mol. Cell Biol.*, **2**, 285–293.
30. (**2000**) The structure and function of claudins, cell adhesion molecules at tight junctions. *Ann. NY Acad. Sci.*, **915**, 129–135.
31. Furuse, M., Itoh, M., Hirase, T., Nagafuchi, A., Yonemura, S. and Tsukita, S. (**1994**) Direct association of occludin with ZO-1 and its possible involvement in the localization of occludin in tight junctions. *J Cell Biol.*, **127**, 1617–1626.
32. Mairbaurl, H., Wodopia, R., Eckes, S., Schulz, S. and Bartsch, P. (**1997**) Impairment of cation transport in A549 cells and rat alveolar epithelial cells by hypoxia. *Am. J. Physiol.*, **273**, L797–L806.
33. Planes, C., Escoubet, B., Blot-Chabaud, M., Friedlander, G., Farman, N. and Clerici, C. (**1997**) Hypoxia downregulates expression and activity of epithelial sodium channels in rat alveolar epithelial cells. *Am. J. Respir. Cell Mol. Biol.*, **17**, 508–518.
34. Vadász, I., Morty, R.E., Kohstall, M.G., Olschewski, H., Grimminger, F., Seeger, W. and Ghofrani, H.A. (**2005**) Oleic acid inhibits alveolar fluid reabsorption: a role in acute respiratory distress syndrome? *Am. J. Resp. Crit. Care Med.*, **171**, 469–479.
35. Franson, T.R., LaBrecque, D.R., Buggy, B.P., Harris, G.J. and Hoffmann, R.G. (**1989**) Serial bilirubin determinations as a prognostic marker in clinical infections. *Am. J. Med. Sci.*, **297**, 149–152.
36. Vincent, J.L., Angus, D.C., Artigas, A., Kalil, A., Basson, B.R., Jamal, H.H., Johnson, G.3. and Bernard, G.R, Recombinant Human Activated Protein C Worldwide Evaluation in Severe Sepsis (PROWESS) Study Group. (**2003**) Effects of drotrecogin alfa (activated) on organ dysfunction in the PROWESS trial. *Crit. Care Med.*, **31**, 834–840.
37. Green, R.M., Beier, D. and Gollan, J.L. (**1996**) Regulation of hepatocyte bile salt transporters by endotoxin and inflammatory cytokines in rodents. *Gastro*, **111**, 193–198.
38. Moseley, R.H., Wang, W., Takeda, H., Lown, K., Schick, L., Ananthanarayanan, M. and Suchy, F.J. (**1996**) Effect of endotoxin on bile acid transport in rat liver, a potential model for sepsis-associated cholestasis. *Am. J. Physiol.*, **271**, G137–G146.
39. Bolder, U., Ton-Nu, H.T., Schteingart, C.D., Frick, E. and Hofmann, A.F. (**1997**) Hepatocyte transport of bile acids and organic anions in endotoxemic rats: impaired uptake and secretion. *Gastro*, **112**, 214–225.
40. Trauner, M., Arrese, M., Lee, H., Boyer, J.L. and Karpen, S.J. (**1998**) Endotoxin downregulates rat hepatic ntcp gene expression via decreased activity of critical transcription factors. *J. Clin. Invest.*, **101**, 2092–2100.
41. Andrejko, K.M., Raj, N.R., Kim, P.K., Cereda, M. and Deutschman, C.S. (**2008**) IL-6 modulates sepsis-induced decreases in transcription of hepatic

organic anion and bile acid transporters. *Shock*.

42. Lehmann, G.L., Carreras, F.I., Soria, L.R., Gradilone, S.A. and Marinelli, R.A. (**2008**) LPS induces the TNF{alpha}-mediated down-regulation of rat liver aquaporin-8: role in sepsis-associated cholestasis. *Am. J. Physiol. Gastrointest. Liver Physiol.*, **294**, G567–G575.
43. Trauner, M., Meier, P.J. and Boyer, J.L. (**1998**) Molecular pathogenesis of cholestasis. *N. Engl. J. Med.*, **339**, 1217–1227.
44. Anderson, J.M. (**1996**) Leaky junctions and cholestasis: a tight correlation. *Gastro.*, **110**, 1662–1665.
45. Alverdy, J., Holbrook, C., Rocha, F., Seiden, L., Licheng, R., Musch, M., Chang, E., Ohman, D. and Suh, S. (**2000**) Gut-derived sepsis occurs when the right pathogen with the right virulence genes meets the right host: evidence for *in vivo* virulence expression in *Pseudomonas aeruginosa*. *Ann. Surg.*, **232**, 480–489.
46. Langkamp-Henken, B., Donovan, T.B., Pate, L.M., Maull, C.D. and Kudsk, K.A. (**1995**) Increased intestinal permeability following blunt and penetrating trauma. *Crit. Care Med.*, **23**, 660–664.
47. Pape, H.-C., Dwenger, A., Regel, G., Auf'm'Kolck, M., Gollub, F., Wisner, D., Sturm, J.A. and Tscherne, H. (**1994**) Increased gut permeability after multiple trauma. *Br. J. Surg.*, **81**, 850–852.
48. Kompan, L., Kremzar, B., Gadzijev, E. and Prosek, M. (**1999**) Effects of early enteral nutrition on intestinal permeability and the development of multiple organ failure after multiple injury. *Intensive Care Med.*, **25**, 157–161.
49. Faries, P.L., Simon, R.J., Martella, A.T., Lee, M.J. and Machiedo, G.W. (**1998**) Intestinal permeability correlates with severity of injury in trauma patients. *J. Trauma*, **44**, 1031–1036.
50. Ammori, B.J., Leeder, P.C., King, R.F.G.J., Barclay, G.R., Martin, I.G., Larvin, M. and McMahon, M.J. (**1999**) Early increase in intestinal permeability in patients with severe acute pancreatitis: correlation with endotoxemia, organ failure, and mortality. *J. Gastrointest. Surg.*, **3**, 252–262.
51. Doig, C.J., Sutherland, L.R., Sandham, J.D., Fick, G.H., Verhoef, M. and Meddings, J.B. (**1998**) Increased intestinal permeability is associated with the development of multiple organ dysfunction syndrome in critically ill ICU patients. *Am. J. Respir. Crit. Care Med.*, **158**, 444–451.
52. Oudemans-van Straaten, H.M., Jansen, P.G., Hoek, F.J., van Deventer, S.J., Sturk, A., Stoutenbeek, C.P., Tytgat, G.N., Wildevuur, C.R. and Eysman, L. (**1996**) Intestinal permeability, circulating endotoxin, and postoperative systemic responses in cardiac surgery patients. *J. Cardiothorac. Vasc. Anesth.*, **10**, 187–194.
53. Wang, W., Jittikanont, S., Falk, S.A., Li, P., Feng, L., Gengaro, P.E., Poole, B.D., Bowler, R.P., Day, B.J., Crapo, J.D. and Schrier, R.W. (**2003**) Interaction among nitric oxide, reactive oxygen species, and antioxidants during endotoxemia-related acute renal failure. *Am. J. Phyiol. Renal Physiol.*, **284**, F532–F537.
54. Yamamoto, K., Shimokawa, T., Yi, H., Isobe, K., Kojima, T., Loskutoff, D.J. and Saito, H. (**2004**) Aging accelerates endotoxin-induced thrombosis : increased responses of plasminogen activator inhibitor-1 and lipopolysaccharide signaling with aging. *Am. J. Pathol.*, **161**, 1805–1814.
55. Schmidt, C., Hocherl, K. and Bucher, M. (**2007**) Cytokine-mediated regulation of urea transporters during experimental endotoxemia. *Am. J. Phyiol. Renal Physiol.*, **292**, F1479–F1489.
56. Donohoe, J.F., Venkatachalam, M.A., Bernard, D.B. and Levinsky, N.G.

(**1978**) Tubular leakage and obstruction after renal ischemia: structural-functional correlations. *Kidney Int.*, **13**, 208–222.

57. Kwon, O., Nelson, W.J., Sibley, R., Huie, P., Scandling, J.D., Dafoe, D., Alfrey, E. and Myers, B.D. (**1998**) Backleak, tight junctions, and cell-cell adhesion in postischemic injury to the renal allograft. *J. Clin. Invest.*, **101**, 2054–2064.
58. Moran, S.M. and Myers, B.D. (**1985**) Pathophysiology of protracted acute renal failure in man. *J. Clin. Invest*, **76**, 1440–1448.
59. Myers, B.D., Chui, F., Hilberman, M. and Michaels, A.S. (**1979**) Transtubular leakage of glomerular filtrate in human acute renal failure. *Am. J. Physiol.*, **237**, F319–F325.
60. Lieberthal, W., McKenney, J.B., Kiefer, C.R., Snyder, L.M., Kroshian, V.M. and Sjaastad, M.D. (**1997**) Integrin-mediated adhesion between renal tubular cells after anoxic injury. *J. Am. Soc. Nephrol.*, **8**, 175–183.
61. Graber, M., Lane, B., Lamia, R. and Pastoriza-Munoz, E. (**1991**) Bubble cells: renal tubular cells in the urinary sediment with characteristics of viability. *J. Am. Soc. Nephrol.*, **1**, 999–1004.
62. Racusen, L.C., Fivush, B.A., Li, Y.L., Slatnik, I. and Solez, K. (**1991**) Dissociation of tubular cell detachment and tubular cell death in clinical and experimental "acute tubular necrosis." *Lab. Invest.*, **64**, 546–556.
63. Unno, N., Hodin, R.A. and Fink, M.P. (**1999**) Acidic conditions exacerbate interferon-gamma-induced intestinal epithelial hyperpermeability: role of peroxynitrous acid. *Crit. Care Med.*, **27**, 1429–1436.
64. Unno, N., Menconi, M.J., Smith, M. and Fink, M.P. (**1995**) Nitric oxide mediates interferon-gamma-induced hyperpermeability in cultured human intestinal epithelial monolayers. *Crit. Care Med.*, **23**, 1170–1176.
65. Han, X., Fink, M.P. and Delude, R.L. (**2003**) Proinflammatory cytokines cause NO•-dependent and independent changes in expression and localization of tight junction proteins in intestinal epithelial cells. *Shock*, **19**, 229–237.
66. Adams, R.B., Planchon, S.M. and Roche, J.K. (**1993**) IFN-γ modulation of epithelial barrier function: time course, reversibility, and site of cytokine binding. *J. Immunol.*, **150**, 2356–2363.
67. Planchon, S.M., Martins, C.AP., Guerrant, R.L. and Roche, J.K. (**1994**) Regulation of intestinal epithelial barrier function by TGF-b1. Evidence for its role in abrogating the effects of T cell cytokine. *J. Immunol.*, **153**, 5730–5739.
68. Madara, J.L. and Stafford, J. (**1989**) Interferon-γ directly affects barrier function of cultured intestinal epithelial monolayers. *J. Clin. Invest.*, **83**, 724–727.
69. Salzman, A.L., Menconi, M.J., Unno, N., Ezzell, R.M., Casey, D.M., Gonzalez, P.K. and Fink, M.P. (**1995**) Nitric oxide dilates tight junctions and depletes ATP in cultured Caco-2BBe intestinal epithelial monolayers. *Am. J. Physiol.*, **268**, G361–G373.
70. Menconi, M.J., Unno, N., Smith, M., Aguirre, D.E. and Fink, M.P. (**1998**) The effect of nitric oxide donors on the permeability of cultured intestinal epithelial monolayers: role of superoxide, hydroxyl radical, and peroxynitrite. *Biochem. Biophys. Acta*, **1425**, 189–203.
71. Chavez, A., Morin, M.J., Unno, N., Fink, M.P. and Hodin, R.A. (**1999**) Acquired interferon-γ responsiveness during Caco-2 cell differentiation: effects on iNOS gene expression. *Gut*, **44**, 659–665.
72. Chavez, A., Menconi, M.J., Hodin, R.A. and Fink, M.P. (**1999**) Cytokine-induced epithelial hyperpermeability: role of nitric oxide. *Crit. Care Med.*, **27**, 2246–2251.
73. Han, X., Fink, M.P., Uchiyama, T. and Delude, R.L. (**2004**) Increased

iNOS activity is essential for the development of pulmonary epithelial tight junction dysfunction in endotoxemic mice. *Am. J. Physiol. Lung Cell Mol. Physiol.*, **286**, L259–L267.
74. Wink, D.A. and Mitchell, J.B. (**1998**) Chemical biology of nitric oxide: insights into regulatory, cytotoxic, and cytoprotective mechanisms of nitric oxide. *Free Rad. Biol. Med.*, **25**, 434–456.
75. Pryor, W.A. and Squadrito, G.L. (**1995**) The chemistry of peroxynitrite: a product from the reaction of nitric oxide with superoxide. *Am. J. Physiol.*, **268**, L699–L722.
76. Unno, N., Menconi, M.J., Smith, M., Aguirre, D.G. and Fink, M.P. (**1997**) Hyperpermeability of intestinal epithelial monolayers is induced by NO: effect of low extracellular pH. *Am. J. Physiol.*, **272**, G923–G934.
77. Sugi, K., Musch, M.W., Field, M. and Chang, E.B. (**2001**) Inhibition of Na^+, K^+-ATPase by interferon gamma down-regulates intestinal epithelial transport and barrier function. *Gastro.*, **120**, 1393–1403.
78. Kubes, P. (**1993**) Ischemia–reperfusion in the feline small intestine: a role for nitric oxide. *Am. J. Physiol.*, **264**, G143–G149.
79. Kubes, P. (**1992**) Nitric oxide modulates epithelial permeability in the feline small intestine. *Am. J. Physiol.*, **262**, G1138–G1142.
80. Han, X., Fink, M.P., Yang, R. and Delude, R.L. (**2004**) Increased iNOS activity is essential for intestinal epithelial tight junction dysfunction in endotoxemic mice. *Shock*, **21**, 261–270.
81. Lora, L., Mazzon, E., Martines, D., Fries, W., Muraca, M., Martin, A., d'Odorico, A., Naccarato, R. and Citi, S. (**1997**) Hepatocyte tight-junctional permeability is increased in rat experimental colitis. *Gastro.*, **113**, 1347–1354.
82. Han, X., Fink, M.P., Uchiyama, T., Yang, R. and Delude, R.L. (**2004**) Increased iNOS activity is essential for the development of hepatic epithelial tight junction dysfunction in endotoxemic mice. *Am. J. Physiol. Gastrointest. Liver Physiol.*, **286**, G126–G136.
83. Notterpek, L., Roux, K.J., Amici, S.A., Yazdanpour, A., Rahner, C. and Fletcher, B.S. (**2001**) Peripheral myelin protein 22 is a constituent of intercellular junctions in epithelia. *Proc. Natl Acad. Sci. USA*, **98**, 14404–14409.
84. Goodwin, G.H., Sanders, C. and Johns, E.W. (**1973**) A new group of chromatin-associated proteins with a high content of acidic and basic amino acids. *Eur. J. Biochem.*, **38**, 14–19.
85. Sanders, C. (**1977**) A method for the fractionation of the high-mobility-group non-histome chromosomal proteins. *Biochem. Biophys. Res. Comm.*, **78**, 1034–1042.
86. Thomas, J.O. and Travers, A.A. (**2001**) HMG1 and 2, and related 'architectural' DNA-binding proteins. *Trends Biochem. Sci.*, **26**, 167–174.
87. Bianchi, M.E. (**1994**) Prokaryotic HU and eukaryotic HMG1: a kinked relationship. *Mol. Microbiol.*, **14**, 1–5.
88. Prendergast, P., Onate, S.A., Christensen, K. and Edwards, D.P. (**1994**) Nuclear accessory factors enhance the binding of progesterone receptor to specific target DNA. *J. Steroid Biochem. Mol. Biol.*, **48**, 1–13.
89. Zhang., C.C.., Krieg, S. and Shapiro, D.J. (**1999**) HMG-1 stimulates estrogen response element binding by estrogen receptor from stably transfected HeLa cells. *Mol. Endocrinol.*, **13**, 632–643.
90. Ciubotaru, M. and Schatz, D.G. (**2004**) Synapsis of recombination signal sequences located in *cis* and DNA underwinding in V(D)J recombination. *Mol. Cell Biol.*, **24**, 8727–8744.
91. Stros, M., Muselikova-Polanska, E., Pospisilova, S. and Strauss, F. (**2004**) High-affinity binding of tumor-suppressor protein p53 and HMGB1 to hemicatenated DNA loops. *Biochemistry*, **43**, 7215–7225.

92. Wang, H., Bloom, O., Zhang, M., Vishnubhakat, J.M., Ombrellino, M., Che, J., Frazier, A., Yang, H., Ivanova, S., Borovikova, L., Manogue, K.R., Faist, E., Abraham, E., Andersson, J., Andersson, U., Molina, P.E., Abumrad, N.N., Sama, A. and Tracey, K.J. (**1999**) HMG-1 as a late mediator of endotoxin lethality in mice. *Science*, **285**, 248–251.
93. Yang, H., Ochani, M., Li, J., Tanovic, M., Harris, H.E., Susarla, S., Ulloa, L., Wang, H., DiRaimo, R., Czura, C.J., Wang, H., Warren, H.S., Fink, M.P., Fenton, M.J., Andersson, U. and Tracey, K.J. (**2004**) Reversing established sepsis with antagonists of endogenous HMGB1. *Proc. Natl Acad. Sci. USA*, **101**, 296–301.
94. Andersson, U., Wang, H., Palmblad, K., Aveberger, A.C., Bloom, O., Erlandsson-Harris, H., Janson, A., Kokkola, R., Zhang, M., Yang, H. and Tracey, K.J. (**2001**) High mobility group 1 protein (HMG-1) stimulates proinflammatory cytokine synthesis in human monocytes. *J. Exp. Med.*, **192**, 565–570.
95. Sappington, P.L., Yang, R., Yang, H., Tracey, K.J., Delude, R.L. and Fink, M.P. (**2002**) HMGB1 B box increases the permeability of Caco-2 enterocytic monolayers and impairs intestinal barrier function in mice. *Gastro*, **123**, 790–802.
96. Liu, S., Stolz, D.B., Sappington, P.L., Macias, C.A., Killeen, M.E., Tenhunen, J.J., Delude, R.L. and Fink, M.P. (**2006**) HMGB1 is secreted by immunostimulated enterocytes and contributes to cytomix-induced hyperpermeability of Caco-2 monolayers. *Am. J. Physiol. Cell Physiol.*, **290**, C990–C999.
97. Yang, R., Harada, N., Mollen, K.P., Prince, J.M., Levy, R.M., Englert, J.A., Gallowitsch-Puerta, M., Yang, L., Yang, H., Tracey, K.J., Harbrecht, B.G., Billiar, T.R. and Fink, M.P. (**2006**) Anti-HMGB1 neutralizing antibody ameliorates gut barrier dysfunction and improves survival after hemorrhagic shock. *Mol. Med.*, **12**, 104–114.
98. Ombrellino, M., Wang, H., Ajemian, M.S., Talhouk, A., Scher, L.A., Friedman, S.G. and Tracey, K.J. (**1999**) Increased serum concentrations of high-mobility-group protein 1 in haemorrhagic shock. *Lancet*, **354**, 1446–1447.
99. Pachot, A., Monneret, G., Voirin, N., Leissner, P., Venet, F., Bohe, J., Payen, D., Bienvenu, J., Mougin, B. and Lepape, A. (**2005**) Longitudinal study of cytokine and immune transcription factor mRNA expression in septic shock. *Clin. Immunol.*, **114**, 61–69.
100. Sunden-Cullberg, J., Norrby-Teglund, A., Rouhianen, A., Rauvala, H., Herman, G., Tracey, K.J., Lee, M.L., Andersson, J., Tokics, L. and Treutiger, C.J. (**2005**) Persistent elevation of high mobility group box-1 protein (HMGB1) in patients with severe sepsis and septic shock. *Crit. Care Med.*, **33**, 564–573.
101. Fei, J., Yu, H.J., Zhou, J., Huang, X.K., Liang, H.P. and Jiang, Y.G. (**2005**) Study on high mobility group-1 protein in patients with multiple trauma. *Zhongguo Wei Zhong Bing Ji Jiu Yi Xue*, **17**, 273–275.
102. Rendon-Mitchell, B., Ochani, M., Li, J., Han, J., Wang, H., Susarla, S., Czura, C., Mitchell, R.A., Chen, G., Sama, A.E., Tracey, K.J. and Wang, H. (**2003**) IFN-gamma induces high mobility group box 1 protein release partly through a TNF-dependent mechanism. *J. Immunol.*, **170**, 3890–3897.
103. Gardella, S., Andrei, C., Ferrera, D., Lotti, L.V., Torrisi, M.R., Bianchi, M.E. and Rubartelli, A. (**2002**) The nuclear protein HMGB1 is secreted by monocytes via a non-classical, vesicle-mediated secretory pathway. *EMBO Rep.*, **3**, 995–1001.
104. Bonaldi, T., Talamo, F., Scaffidi, P., Perrera, D., Porto, A., Bachi, A., Rubartelli, A., Agresti, A. and

Bianchi, M.E. (**2003**) Monocytic cells hyperacetylate chromatin protein HMGB1 to redirect it towards secretion. *EMBO J.*, **22**, 5551–5560.

105. Semino, C., Angelini, G., Poggi, A. and Rubartelli, A. (**2005**) NK/iDC interaction results in IL-18 secretion by DCs at the synaptic cleft followed by NK cell activation and release of the DC maturation factor HMGB1. *Blood*.
106. Dumitriu, I.E., Baruah, P., Bianchi, M.E., Manfredi, A.A. and Rovere-Querini, P. (**2005**) Requirement of HMGB1 and RAGE for the maturation of human plasmacytoid dendritic cells. *Eur. J. Immunol.*, **35**, 2184–2190.
107. Wang, H., Vishnubhakat, J.M., Bloom, O., Zhang, M., Ombrellino, M., Sama, A. and Tracey, K.J. (**2002**) Proinflammatory cytokines (tumor necrosis factor and interleukin 1) stimulate release of high mobility group protein-1 by pituicytes. *Surgery*, **126**, 389–392.
108. Kuniyasu, H., Yano, S., Sasaki, T., Sasahira, T., Sone, S. and Ohmori, H. (**2005**) Colon cancer cell-derived high mobility group 1/amphoterin induces growth inhibition and apoptosis in macrophages. *Am. J. Pathol.*, **166**, 751–760.
109. Scaffidi, P., Misteli, T. and Bianchi, M.E. (**2002**) Release of chromatin protein HMGB1 by necrotic cells triggers inflammation. *Nature*, **418**, 191–195.
110. Sappington, P.L., Fink, M.E., Yang, R., Delude, R.L. and Fink, M.P. (**2003**) Ethyl pyruvate provides durable protection against inflammation-induced intestinal epithelial barrier dysfunction. *Shock*, **20**, 521–528.
111. Alverdy, J.C. (**2007**) The re-emerging role of the intestinal microflora in critical illness and inflammation: why the gut hypothesis of sepsis syndrome will not go away. *J. Leukocyte Biol.*
112. Patel, N.J., Zaborina, O., Wu, L., Wang, Y., Wolfgeher, D.J., Valuckaite, V., Ciancio, M.J., Kohler, J.E., Shevchenko, O., Colgan, S.P., Chang, E.B., Turner, J.R. and Alverdy, J.C. (**2007**) Recognition of intestinal epithelial HIF-1alpha activation by *Pseudomonas aeruginosa*. *Am. J. Physiol. Gastrointest. Liver Physiol.*, **292**, G134–G142.
113. Zaborina, O., Lepine, F., Xiao, G., Valuckaite, V., Chen, Y., Li, T., Ciancio, M., Zaborin, A., Petroff, E., Turner, J.R., Rahme, L.G., Chang, E. and Alverdy, J.C. (**2007**) Dynorphin activates quorum sensing quinolone signaling in *Pseudomonas aeruginosa*. *PLoS Pathog.*, **3**, e35.

13
Neural Regulation of Cytokines

Jared M. Huston and Kevin J. Tracey

13.1
Introduction

The immune system, like other physiological systems, is regulated though feedback loops that maintain homeostasis. Positive and negative feedback loops maintain an "appropriate" or healthy level of immune activation during infection or injury. Controlling systems may occasionally fail, which invariably results in significant morbidity and mortality. While excessive activation of inhibitory or counter-regulatory pathways may be associated with immunodeficiency or damage due to secondary infections, inadequate restraint of the immune system can precipitate an excessive release of toxic inflammatory mediators, resulting in tissue damage or organ dysfunction. Cytokines are a class of immune-derived proteins that have been implicated as mediators of inflammatory disease pathogenesis. The principal dogma of the "cytokine theory of disease" states that immune system-derived cytokines are both necessary and sufficient to cause the clinical signs and symptoms of inflammatory diseases, ranging from arthritis and colitis to sepsis [1].

Some of the first studies of cytokine functions revealed mediator roles in physiological responses via interaction with nerves. For example, neural cells are reported to express cytokine receptors that transduce intracellular responses implicated in the neuro-immune responses underlying the development of fever and anorexia. Cytokines also enhance hypothalamic–pituitary outflow, which stimulates the release of glucocorticoids from the adrenal glands. This later response is an example of cytokines activating a counter-regulatory feedback loop as glucocorticoids are potent inhibitors of cytokine release. In fact, there are many examples of anti-inflammatory humoral pathways that protect against protracted, excessive cytokine release such as the release of soluble receptors that neutralize cytokine damage, and the production of "anti-inflammatory" cytokines that suppress the synthesis and activity of potentially damaging "pro-inflammatory cytokines". Recent work has revealed that in addition to these well-known humoral anti-inflammatory pathways, there

Sepsis and Non-infectious Systemic Inflammation. Edited by J.-M. Cavaillon and C. Adrie

ISBN: 978-3-527-31935-0

are direct, neural control systems that inhibit cytokine release and protect against tissue and organ damage. Here we review some of the anatomical and physiological connections between the nervous and immune systems, with an emphasis on how efferent signals in the vagus nerve specifically regulate cytokine release to prevent damage during injury, ischemia, or infection.

13.2
Cytokine Stimulation of the Nervous System

The immune system relays information to the nervous system, and the nervous system to the immune system, via many mediators, including cytokines [2]. Cytokines can also activate neural signals directly, as in the case of interleukin-1, which activates an afferent signal via the vagus nerve to mediate fever. Cytokines may also interact with the nervous system by crossing the blood–brain barrier via saturable transport systems [3]. Transporters have been identified for a number of cytokines, including interleukin (IL)-1 [4] and TNFα [5]. Inflammatory mediators produced in the periphery may also activate the endothelial cells of the blood–brain barrier to synthesize transcription factors and inflammatory cytokines in addition to IL-1 and TNFα [6]. In a series of recent studies, some cytokines, most notably IL-6, were found to communicate with the central nervous system by binding to the endothelial cells comprising the blood–brain barrier, which subsequently released second messengers including nitric oxide and prostaglandins, to alter the activity of neurons controlling body temperature, hypothalamic–pituitary–adrenal (HPA) axis activation, fever, and other behavioral signs of illness [7–10].

In addition to transport systems and the release of second messengers, brain endothelial cells themselves are also capable of secreting cytokines [11]. Verma and colleagues [11] reported that the secretion of cytokines by brain endothelial cells is polarized in that the lipid composition and the populations of transporters and receptors differ between the luminal (blood-facing) and abluminal (brain-facing) surfaces [11–17]. Using an *in vitro* mouse blood endothelial cell monolayer culture system, Verma *et al.* [11] identified that IL-1α, IL-10, granulocyte-macrophage colony stimulating factor, IL-6 and TNFα are all secreted from brain endothelial cells, and that a greater secretion of these cytokines occurs on the luminal membrane in both constitutive and lipopolysaccharide (LPS)-stimulated conditions.

Once activated by infection or injury, potentially through one or more of the above-described pathways, the nervous system has multiple means of responding to the inflammatory signals to regulate inflammation via neuroendocrine pathways, the synthesis and release of neuropeptides, and neuronal mechanisms such as the cholinergic anti-inflammatory pathway, a direct route for inhibiting cytokines via signals in the vagus nerve.

13.3
Neuro-Endocrine Mediation of Cytokines

13.3.1
Hypothalamic–Pituitary–Adrenal (HPA) Axis

The HPA is composed of the paraventricular nucleus in the hypothalamus, the anterior pituitary gland, and the adrenal glands. Corticotropin-releasing hormone is secreted from the paraventricular nucleus in response to physical and emotional stresses or trauma. Corticotropin releasing hormone then enters the hypophyseal portal blood supply to stimulate the secretion of adrenocorticotropin hormone (ACTH) from the anterior pituitary gland. ACTH enters the circulation and stimulates the expression and secretion of glucocorticoids, which inhibit cytokine production and suppress inflammation. Secretion of corticotrophin releasing hormone is self-regulated, and is down regulated by both ACTH and cortisol, which also inhibits ACTH release. The HPA axis can be negatively modulated by other regulatory pathways, including the sympathetic nervous system and the neuropeptide arginine vasopressin [18] and positively modulated by serotonergic, cholinergic, and histaminergic pathways [19, 20].

Cortisol, the primary effector molecule of the HPA axis, exerts a diverse array of effects on multiple body systems, including the immune system. The effects of glucocorticoids are dose-dependent and physiological concentrations of circulating glucocorticoids are typically lower than levels measured following pharmacologic use of these drugs. Whereas high pharmacologic doses are known to result in immunosuppression, physiologic levels can augment certain aspects of the inflammatory response [20, 21]. Glucocorticoids enter immune cells via passive diffusion then bind to the inactive glucocorticoid receptor in the cytoplasm, which dissociates from a multiprotein complex and translocates to the nucleus. The intranuclear glucocorticoid receptor complex binds to specific glucocorticoid response elements (GREs) that regulate gene transcription, and also interact with various transcription factors including activator protein 1 (AP-1) and nuclear factor (NF)-κB [20].

Glucocorticoids modulate cytokines, adhesion molecules, and chemoattractant expression to regulate immune cell trafficking, maturation, and differentiation. The anti-inflammatory actions of glucocorticoids reduce leukocyte trafficking to sites of inflammation via inhibition of chemoattractant and adhesion molecule expression, including intracellular cell adhesion molecule 1 (ICAM-1), endothelial-leukocyte adhesion molecule 1 (ELAM-1), vascular cell adhesion molecule 1 (VCAM-1), and E- and L-selectin. Glucocorticoids suppress immune system function by reducing prostaglandin and nitric oxide production, inducing apoptosis of macrophages and monocytes, inactivating neutrophils, and reducing the activities of antigen presenting dendritic cells.

An intact and fully functional HPA axis is critical to host survival during systemic inflammatory and infectious conditions [22]. Numerous animal and human studies have shown that impairment of the HPA axis via adrenalectomy, hypophysectomy, or glucocorticoid receptor blockade, leads to increased infection severity, exaggerated pro-inflammatory cytokine production and tissue injury, and in some cases, increased mortality [22–27]. As summarized by Webster and Sternberg [22], administration of synthetic glucocorticoids reverses these observed effects in that the severity of infection is reduced and protection against lethality is observed.

Initial studies in the 1950s suggested that administration of supplemental steroids to critically ill patients with adrenal insufficiency improved overall outcome. Since subsequent studies did not substantiate this result, steroid replacement therapy was abandoned. Recently, however, researchers re-evaluated the clinical utility of glucocorticoid administration in critically ill patients. These studies resulted in a modification of our current standards of practice in that most ICUs now routinely analyze critically ill patients for adrenal insufficiency, and will supplement glucocorticoids and mineralocorticoids to those meeting specific inclusion and exclusion criteria.

13.3.2 Hypothalamic–Pituitary–Gonadal (HPG) Axis

The hypothalamic–pituitary connection also may regulate immune function via modulation of sex hormone production through activation of the HPG axis. In general, physiologic concentrations of estrogen enhance immune function, whereas testosterone and dehydroepiandrosterone (DHEA) exert immunosuppressive effects [21, 28, 29]. Clinical observations indicate that women are at 2- to 10-fold higher risk for developing inflammatory and autoimmune diseases, including rheumatoid arthritis, multiple sclerosis, and systemic lupus erythematosus. Conversely, sepsis inhibits HPG axis function (and therefore gonadal function), either through a direct cytokine effect on hypothalamic neurons and/or the gonads or indirectly through corticotrophin releasing hormone or endogenous opioid concentrations. Several experimental studies proposed that female sex hormones maintain beneficial cell-mediated immune responses and improve outcome in sepsis, which suggests that sepsis-induced tissue injury and mortality may result from immunosuppression, rather than an exaggerated pro-inflammatory response.

Differences in immune regulation during sepsis between men and women [30] have also been reported. Cytokine production was prospectively evaluated in 106 male and 37 female polytraumatized patients, and premenopausal (<50 years of age) women were found to have significantly less MODS and sepsis and significantly lower plasma cytokine levels as compared with age-matched males. The levels of sex steroids were not measured in this study; however, these results suggested that women were protected from MODS and sepsis due to the effects of estrogen on plasma cytokines. This conclusion is in contrast to previously published work by Eachempati *et al.*

who studied gender-based differences in clinical outcome in patients with sepsis [31]. Eachempati reported that the female gender was an independent predictor of increased mortality in critically ill surgical patients with documented infection [31]. Clearly, further research in this area is warranted.

13.3.3 Hypothalamic–Growth Hormone Axis

The effects of growth hormone (GH), another modulator of the immune system, are mediated primarily through insulin-like growth factor-1 (IGF-1) [21]. Both GH and IGF-1 can increase the survival and proliferation of lymphoid cells, which led to the suggestion that GH itself functions as a cytokine [32, 33]. Critically ill patients suffer from increased protein turnover and negative nitrogen balance, which may compromise tissue repair, wound healing, and immune function [34]. This negative nitrogen balance is due, at least in part, to an increased resistance to GH. The administration of high doses of GH to patients with sepsis has been found to improve nitrogen balance [34]. While GH appears therefore to be beneficial in abrogating the catabolic response to injury, surgery, and sepsis, equally optimistic findings in GH-treated critically patients have not been reported. Takala *et al.* reported the results from two prospective, multi-center, double-blind, randomized, placebo-controlled studies examining the effect of the exogenous administration of high doses of GH to critically ill patients. In both studies, treatment with GH significantly increased mortality compared to the placebo-treated patients. In surviving patients, length of stay in the intensive care unit and duration of mechanical ventilation were also prolonged in GH-treated patients [34]. To date, the role of GH in the regulation of immunity remains unclear.

13.3.4 Hypothalamic–Pituitary–Thyroid (HPT) Axis

Thyroxine (T_4) and triiodothyronine (T_3) secretion by the thyroid gland is controlled by thyroid-stimulating hormone (TSH) from the pituitary gland and thyrotropin-releasing hormone (TRH) from the hypothalamus. Thyroid hormones can potentiate immune cells and immune responses [35–37], although the role of thyroid hormones in the regulation of immunity remains controversial. TSH is inhibited during inflammation [21, 35], and specifically, IL-1 can suppress TSH whereas the anti-inflammatory cytokine IL-2 stimulates this axis [38, 39]. Further, the HPT axis also can alter the function of the HPA axis and vice versa. Experimental studies have found that administration of T_4 activates the HPA axis and protects rats from an inflammatory challenge [40]. In turn, chronic activation of the HPA suppresses the production of TSH and therefore inhibits the conversion of T_4 to T_3.

13.4 Direct Neural Control of Cytokines

13.4.1 The Peripheral Nervous System

Neuropeptides released from peripheral nerves regulate pain, touch, and temperature perception; these molecules can also regulate inflammation [41]. Some examples include corticotropin releasing hormone, substance P, and calcitonin gene-related peptide. These peptides exert important control over the pro-inflammatory response during pathogen clearance, and contribute to the mediation of the hallmark signs of pain, edema, heat and erythema.

Locally-acting neuropeptides can also mediate anti-inflammatory effects. Alpha-melanocyte-stimulating hormone (α-MSH) blocks Toll-like receptor 4 (TLR4) signaling in macrophages, inhibits NK-κB activation, and suppresses TNF, IL-1, IL-2, and IFNγ production. α-MSH also induces IL-10 production and limits pro-inflammatory cytokine-mediated organ damage. Exogenous and endogenous opioids, including endorphins and enkephalins, exert anti-inflammatory effects by inhibiting chemotaxis of neutrophils and monocytes. Opioids also may limit inflammation indirectly via stimulation of the HPA axis [42]. Vasoactive intestinal peptide inhibits inflammation by reducing pro-inflammatory cytokine production, limiting NK-cell activity, preventing macrophage activation, stimulating the production of IL-10, and decreasing TLR2 and TLR4 mRNA expression by intestinal epithelial cells [41].

13.4.2 The Sympathetic Nervous System

The autonomic nervous system, comprised of the sympathetic, parasympathetic, and enteric nervous systems, regulates visceral body functions. The sympathetic and parasympathetic divisions typically function in opposition to maintain homeostatic levels of vital organ function including heart rate, blood pressure, cardiac contractility, respiratory rate, gastrointestinal secretion and motility, and body temperature [24]. The autonomic nervous system receives and sends neural signals to other regions of the brain, including the brain stem and hypothalamus. The autonomic nervous system is modulated by higher brain regions, as evidenced by individuals who learn through biofeedback to decrease their heart rate by increasing parasympathetic activity [24].

Beginning in the early 1980s, evidence regarding the interaction between the autonomic nervous system and other vital organ systems, including the immune system, began to emerge. Direct neural input to the highly specialized chromaffin cells of the adrenal medulla via the sympathetic trunks results in the release of epinephrine, or adrenaline, directly into the general circulation following stimulation of these nervous connections. While originally characterized as a potent and critical vasomodulatory hormone, recent studies have provided considerable insight into the varied immunomodulatory effects of epinephrine. The sympathetic nervous system is also capable of controlling immune function via neuropeptide Y released from the adrenal medulla. The sympathetic nervous system also influences immune responses via regional modulation of immune cells. Effector cells of the immune system are situated to maximize exposure to potential inflammatory stimuli in primary and secondary lymphoid organs, including the thymus, bone marrow, spleen, and lymph nodes, and these sites receive noradrenergic innervation [43]. Nerve terminals originate in nuclei located within the brain stem, which then give rise to preganglionic fibers. In turn, these fibers exit the central nervous system via the thoracic and lumbar spinal nerves to synapse in the paravertebral sympathetic chains or prevertebral ganglia. Postganglionic fibers then innervate other ganglia, organs and tissues where norepinephrine is released at nerve endings.

Cathecholaminergic nerve terminals innervate both the vascular smooth muscle and specific parenchymal compartments of the various lymphoid organs, and participate in the control of blood flow and leukocyte trafficking to organs [43]. Catecholaminergic fibers are rich in zones of T cells, macrophages, and plasma cells, but regions containing immature B cells are poorly innervated [43]. Interestingly, the majority of catecholaminergic fibers in lymphoid organs do not appear to form "classical" synapses with target cells. Instead, norepinephrine appears to be released into an extra-neuronal space from where it often has to diffuse a considerable distance prior to contacting the target cell [43]. Catecholamines, such as norepinephrine, exert their effects by binding to one of two receptors: alpha (α) or beta (β) adrenergic receptors. The majority of lymphoid cells express β-adrenergic receptors; however, different lymphoid cells express variable densities and receptor subtypes. The β_2-adrenergic receptor is the most prevalent and physiologically important of the receptor subtypes identified to date.

Catecholaminergic fibers can either inhibit or amplify inflammation. For example, norepinephrine inhibits the production of pro-inflammatory cytokines and upregulates the production of anti-inflammatory cytokines, such as IL-10, from monocytes and dendritic cells [41]. Similar to the effect of glucocorticoids, the upregulation or anti-inflammatory cytokines results in a shift from a Th1 to a Th2 immune response. In addition, catecholaminergic fibers also decrease natural killer (NK)-cell activity and dendritic cell chemotaxis. The immunosupessive activity of the sympathetic nervous system, however, is not universal: in some cases, activation of the sympathetic nervous system can augment inflammatory responses. For example, stimulation of the $\alpha 2$ adrenoreceptor by norepinephrine can increase TNF production in response to LPS [43, 44].

Major trauma, burns, and surgery often lead to severe immunosuppression that can increase a patient's susceptibility to develop infection and/or sepsis. In a 1998 study by Woiciechowsky *et al.*, elevated catecholamine concentrations contributed to this post-injury immunocompromised state via an increased production of anti-inflammatory Th2 cytokines and a concomitant reduction in Th1 cytokine levels [45]. Further, high levels of circulating IL-10 correlated with an increased incidence of infection and inhibition of adrenergic signaling via β-adrenergic receptor blockade prevented the increased rate of infection [45]. Despite the important role of catecholaminergic signaling on regulation of immune responses, the clinical utility of these hormones or related receptor antagonists to individuals with sepsis remains to be established.

13.5 Neural Control of Cytokines via the Inflammatory Reflex

The inflammatory reflex, illustrated in Figure 13.1, is a vagus nerved-based reflex arc which maps the concept that a neural mechanism can coordinate and modulate cytokine responses [46]. The inflammatory reflex includes both an afferent (sensory) and efferent (motor) branch which function in concert to form a discrete and rapid mechanism capable of modulating the activity of cytokine-producing cells [24]. Thus, the brain is able to rapidly regulate the cytokine response in a localized, highly controlled, organ-specific manner [47].

13.5.1 The Afferent Arm of the Inflammatory Reflex

Inflammatory mediators activate afferent vagus nerve signals, which are transmitted to the medullary reticular formation, locus ceruleus, hypothalamus, and dorsal vagal complex ultimately resulting in an increased release of adrenocorticotropin hormone (ACTH) from the anterior pituitary gland [24, 46]. Further, afferent neural signals stimulate an increase in systemic glucocorticoid levels capable of inhibiting pro-inflammatory cytokines [48–50]. Melanocyte-stimulating hormone, a potent anti-inflammatory protein that inhibits the synthesis of cytokines, is also released in inflammatory conditions [51]. Ascending sensory fibers of the vagus nerve that synapse in the nucleus tractus solitarius of the upper medulla can also activate nerve signals to inhibit cytokine release [52].

13.5.2 The Cholinergic Anti-inflammatory Pathway

The motor neural arc of the inflammatory reflex, referred to as the cholinergic anti-inflammatory pathway, inhibits inflammation by suppressing cytokine synthesis [24, 46, 47] via release of acetylcholine in organs of the reticuloendothelial system, including spleen, liver, and gastrointestinal tract [24, 53, 54]. Acetylcholine binds to the α7 nicotinic acetylcholine receptor

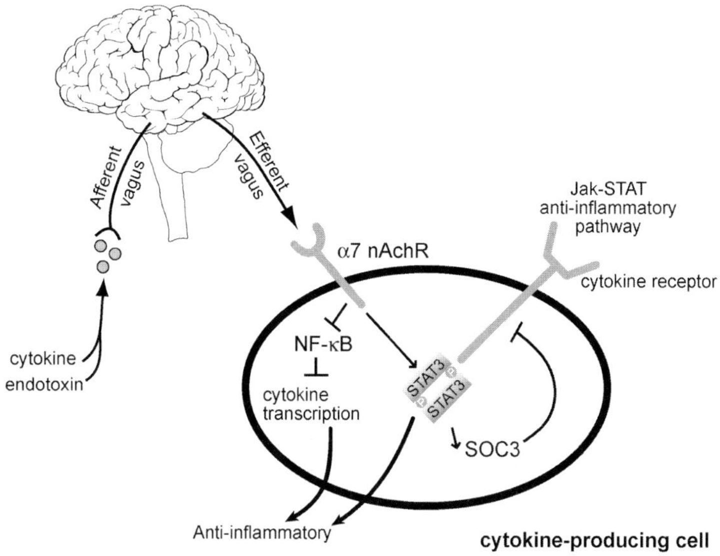

Figure 13.1 The inflammatory reflex. Vagal stimulation results in the release of acetylcholine which, inhibits NF-κB and stimulates the STAT3-SOC3 anti-inflammatory pathway via α7 nicotinic receptors (nAChR) on activated macrophages and other cytokine-producing cells. Overall, the release of TNF, HMGB1 and other pro-inflammatory cytokines implicated in inflammatory conditions is inhibited. (From [46]; reproduced with permission of Blackwell Publishing.)

(α7nAChR) expressed on the surface of activated macrophages (and other cytokine-producing cells) whereby cytokine synthesis and release is prevented by inhibiting nuclear factor (NF)-κB (i.e. decreasing transcription activity of the transcription factor NF-κB subunit p65) and by stimulating the JAK-STAT anti-inflammatory pathway [47, 55–58]. This suppresses pro-inflammatory cytokine release and prevents injury to tissues and organs [47].

The molecular basis for signaling via the cholinergic anti-inflammatory pathway was elucidated by a series of cell culture and animal studies. Borovikova *et al.* demonstrated that exposure of macrophages to acetylcholine resulted in a significant reduction in the synthesis of pro-inflammatory cytokines following stimulation of the macrophages with endotoxin [53]. Further, animal models revealed that activation of the cholinergic anti-inflammatory pathway via electrical stimulation of the vagus nerve significantly reduced the synthesis of TNF in the liver, spleen, and heart, and suppressed circulating concentrations of TNF during lethal endotoxemia [52, 53]. Finally, interruption of the cholinergic anti-inflammatory pathway via surgical vagotomy significantly exacerbated pro-inflammatory cytokine production. Activation of the cholinergic anti-inflammatory pathway via vagus nerve stimulation reduced pro-inflammatory cytokine production and prevented tissue injury in multiple models of systemic inflammation, including hemorrhagic shock, ischemia–reperfusion injury, myocardial ischemia, ileus, experimental arthritis, and pancreatitis [51, 52, 57, 59–66]. Pharmacologic manipulation of the cholinergic anti-inflammatory pathway via administration of nicotine, a potent α7 acetylcholine receptor agonist, inhibited pro-inflammatory cytokine production and rescued animals from lethal polymicrobial sepsis in a cecal ligation and puncture (CLP) model [67]. Most recently, Huston and colleagues reported that non-electrical, mechanical vagus nerve stimulation significantly inhibited pro-inflammatory cytokine production and rescued animals from CLP sepsis, even when treatment was initiated 24 h after the onset of sepsis [68]. Interestingly, the protective effects of the activated cholinergic anti-inflammatory pathway do not appear to be limited to suppression of cytokine synthesis. Nicotine agonists inhibit the expression of endothelial cell adhesion molecules [66]. Further, nicotinic agonists directly inhibited endothelial cell activation in human microvascular endothelial cells (HuMVECs) whereas mecamylamine, a nAChR antagonist blocked this inhibition. Based on mechanistic studies in this same report, nicotine appears to inhibit TNF-induced endothelial cell activation by blocking entry of the TNF-induced nuclear factor-κB. This study also reported that in the carrageenan air pouch model, electrical stimulation of the vagus nerve and administration of cholinergic agonists both resulted in a significant blockage of leukocyte migration in mice [66].

The receptor that is required for the activity of the cholinergic anti-inflammatory pathway is the α7 subunit of the nicotinic acetylcholine receptor (α7nAChR). In α7nAChR-knockout animals, vagus nerve stimulation failed to suppress pro-inflammatory cytokine production whereas vagus nerve stimulation significantly reduced cytokine levels in wild-type littermates [55]. Moreover, α7nAChR-knockout animals mounted a significantly elevated pro-inflammatory cytokine response following an endotoxin challenge compared with wild-type control animals [55]. A recent study by Huston *et al.*, designed to further characterize the anatomic and physiologic components of the cholinergic anti-inflammatory pathway, revealed that the spleen is an important component of the reflex for the vagus nerve to exert its protective, anti-inflammatory effects [69]. Organs of the reticuloendothelial system, including the spleen, liver, and lungs, house macrophages and monocytes are responsible for the immediate, early response to circulating pathogens. These pathogens release toxic mediators, such as LPS, which then stimulate tissue macrophages to secrete pro-inflammatory cytokines. The spleen is a major source of systemic TNF production during murine endotoxemia, and splenectomy significantly reduces circulating TNF levels in endotoxin-challenged mice [69]. Further, vagus nerve stimulation significantly reduced TNF synthesis in the spleens of wild-type mice, but was ineffective in α7nAChR-knockout animals. Vagus nerve stimulation also failed to inhibit systemic TNF production in splenectomized mice or in mice following interruption of the common celiac branch of the abdominal vagus nerve. Together, this series of experiments suggest that anti-inflammatory vagus nerve signaling traverses this branch of the vagus nerve, which synapses in the spleen. Finally, pharmacological exploitation of the cholinergic anti-inflammatory pathway via administration of nicotine, a potent α7nAChR agonist, not only failed to inhibit pro-inflammatory cytokine production or protect against sepsis lethality in splenectomized mice, but also exacerbated the inflammatory cytokine response and accelerated lethality. Together, these findings indicate that the spleen is critical for the physiological interface of cholinergic control of immune responses via anti-inflammatory signaling of the vagus nerve [69].

13.5.3 Clinical Implications of the Cholinergic Anti-inflammatory Pathway and Future Directions

From an epidemiological perspective, it is interesting to consider individuals who have undergone surgical interruption of the abdominal vagus nerve for the treatment of peptic ulcer disease. To date, there have not been comprehensive studies of this unique patient population with respect to clinical outcomes in sepsis. Confounding this analysis is that little is known about compensatory mechanisms that may be upregulated during chronically denervated subjects, a phenomenon that has been observed in vagotomized mice which exhibited a transient enhanced sensitivity to inflammation that lessened over time [70].

Another group includes subjects who have electrical cervical vagus nerve stimulators for the treatment of medically refractory epilepsy and depression. The increase in efferent vagus nerve signaling in these patients could theoretically activate the cholinergic anti-inflammatory pathway. Indeed, a comparison of the electrical stimulation parameters used clinically to control epilepsy and those used experimentally to inhibit systemic inflammation revealed very little difference between the two groups in terms of basal cytokine expression [71]. Unfortunately, these subjects have not been studied under conditions of inflammation and elevated cytokine production, and there is no evidence in animals that vagus nerve stimulation can suppress unstimulated cytokine levels, so no effect would be expected in healthy humans.

An evolving area of research that may provide some insight into the neurological basis of inflammatory conditions is the study of heart rate variability. Increases in efferent vagus nerve signaling lead to heart rate slowing and increases in heart rate variability (i.e. changes in beat-to-beat variability of heart rhythm). Clinical studies of patients with sepsis and other inflammatory disorders indicate that a decrease in heart rate variability is a negative prognostic factor that correlates to elevated cytokine levels. It is possible that decreases in heart rate variability reflect diminished cardiac vagus nerve activity and reduced cholinergic anti-inflammatory pathway function. It is plausible that a reduction in cholinergic anti-inflammatory pathway signaling leads to exaggerated and deleterious inflammatory responses from excessive cytokine release.

References

1. Tracey, K.J. (**2005**) *Fatal Sequence: The Killer Within*, Dana Press.
2. Quan, N. (**2006**) Brain's firewall: Blood-brain barrier activity regulates neuroimmune information flow. *Brain Behav. Immun.*, **20**, 447–8.
3. Banks, W.A., Kastin, A.J. and Durham, D.A. (**1989**) Bidirectional transport of interleukin-1 alpha across the blood-brain barrier. *Brain Res. Bull.*, **23**, 433–7.
4. Plotkin, S.R., Banks, W.A., Maness, L.M. and Kastin, A.J. (**2000**) Differential transport of rat and human interleukin-1alpha across the blood–brain barrier and blood–testis barrier in rats. *Brain Res.*, **881**, 57–61.
5. Pan, W., Csernus, B. and Kastin, A.J. (**2003**) Upregulation of p55 and p75 receptors mediating TNF-alpha transport across the injured blood-spinal cord barrier. *J. Mol. Neurosci.*, **21**, 173–84.
6. Ohtsuki, S. (**2004**) New aspects of the blood–brain barrier in transporters; its physiological roles in the central nervous system. *Biol. Pharm. Bull.*, **27**, 1489–96.
7. Elmquist, J.K., Scammell, T.E. and Saper, C.B. (**1997**) Mechanisms of CNS response to systemic immune challenge: the febrile response. *Trends Neurosci.*, **20**, 565–70.
8. Gaillard, R.C. (**1998**) Cytokines in the neuroendocrine system. Cytokines in the neuroendocrine system. *Int. Rev. Immunol.*, **17**, 181–216.
9. Hopkins, S.J. (**2007**) Central nervous system recognition of peripheral inflammation: a neural, hormonal collaboration. *Acta. Biomed.*, **78**(Suppl 1), 231–47.
10. Inoue, W., Matsumura, K., Yamagata, K., Takemiya, T., Shiraki, T. and Kobayashi, S. (**2002**) Brain-specific endothelial induction of prostaglandin E(2) synthesis enzymes and its temporal relation to fever. *Neurosci. Res.*, **44**, 51–61.
11. Verma, S., Nakaoke, R., Dohgu, S. and Banks, W.A. (**2006**) Release of cytokines by brain endothelial cells: A polarized response to lipopolysaccharide. *Brain Behav. Immun.*, **20**, 449–55.
12. Banks, W.A. and Broadwell, R.D. (**1994**) Blood to brain and brain to blood passage of native horseradish peroxidase, wheat germ agglutinin, and albumin: pharmacokinetic and morphological assessments. *J. Neurochem.*, **62**, 2404–19.
13. Betz, A.L. and Goldstein, G.W. (**1978**) Polarity of the blood–brain barrier: neutral amino acid transport into isolated brain capillaries. *Science*, **202**, 225–7.
14. Davson, H. and Segal, M.B. (**1996**) Special aspects of the blood-brain barrier, in *Physiology of the CNS and Blood-Brain Barriers*, CRC Press, Boca Raton, FL, pp. 303–485.
15. Taylor, E.M. (**2002**) The impact of the efflux transporters in the brain on the development of drugs for CNS disorders. *Clin. Pharmacokinet.*, **41**, 81–92.
16. Vorbrodt, A.W. (**1994**) Glycoconjugates and anionic sites in the blood-brain barrier, in (eds M. Nicolini and P.F. Zatta), *Glycobiology and the Brain*, Pergamon Press, Oxford, pp. 37–62.
17. Zambenedettiet, P., Giordano, R. and Zatta, P. (**1996**) Identification of lectin binding sites in the rat brain. *Glycoconj.*, **13**, 341–6.
18. Scott, L.V. and Dinan, T.G. (**1998**) Vasopressin and the regulation of hypothalamic-pituitary-adrenal axis function: implications for the pathophysiology of depression. *Life Sci.*, **62**, 1985–98.
19. Calogero, A.E., Gallucci, W.T., Chrousos, G.P. and Gold, P.W. (**1988**) Catecholamine effects upon rat hypothalamic corticotropin-releasing hormone secretion *in vitro*. *J. Clin. Invest.*, **82**, 839–46.
20. Webster, J.I., Tonelli, L. and Sternberg, E.M. (**2002**)

Neuroendocrine regulation of immunity. *Annu. Rev. Immunol.*, **20**, 125–63.
21. Eskandari, F., Webster, J.I. and Sternberg, E.M. (**2003**) Neural immune pathways and their connection to inflammatory diseases. *Arthritis. Res. Ther.*, **5**, 251–65.
22. Webster, J.I. and Sternberg, E.M. (**2004**) Role of the hypothalamic-pituitary-adrenal axis, glucocorticoids and glucocorticoid receptors in toxic sequelae of exposure to bacterial and viral products. *J. Endocrinol.*, **181**, 207–21.
23. Ledingham, I.M. and Watt, I. (**1983**) Influence of sedation on mortality in critically ill multiple trauma patients. *Lancet*, **1**, 1270.
24. Tracey, K.J. (**2002**) The inflammatory reflex. *Nature*, **420**, 853–9.
25. Bertini, R., Bianchi, M. and Ghezzi, P. (**1988**) Adrenalectomy sensitizes mice to the lethal effects of interleukin 1 and tumor necrosis factor. *J. Exp. Med.*, **167**, 1708–12.
26. Butler, L.D., Layman, N.K., Riedl, P.E., Cain, R.L., Shellhaas, J., Evans, G.F. and Zuckerman, S.H. (**1989**) Neuroendocrine regulation of *in vivo* cytokine production and effects: I. *In vivo* regulatory networks involving the neuroendocrine system, interleukin-1 and tumor necrosis factor-alpha. *J. Neuroimmunol.*, **24**, 143–53.
27. Bloom, O., Wang, H., Ivanova, S., Vishnubhakat, J.M., Ombrellino, M. and Tracey, K.J. (**1998**) Hypophysectomy, high tumor necrosis factor levels, and hemoglobinemia in lethal endotoxemic shock. *Shock*, **10**, 395–400.
28. Cutolo, M. (**1998**) The roles of steroid hormones in arthritis. *Br. J. Rheumatol.*, **37**, 597–9.
29. Cutolo, M. and Wilder, R.L. (**2000**) Different roles for androgens and estrogens in the susceptibility to autoimmune rheumatic diseases. *Rheum. Dis. Clin. North Am.*, **26**, 825–39.
30. Frink, M., Pape, H.C., van Griensven, M., Krettek, C., Chaudry, I.H. and Hildebrand, F. (**2007**) Influence of sex and age on MODS and cytokines after multiple injuries. *Shock*, **27**, 151–6.
31. Eachempati, S.R., Hydo, L. and Barie, P.S. (**1999**) Gender-based differences in outcome in patients with sepsis. *Arch. Surg.*, **134**, 1342–7.
32. van Buul-Offers, S.C. and Kooijman, R. (**1998**) The role of growth hormone and insulin-like growth factors in the immune system. *Cell. Mol. Life Sci.*, **54**, 1083–94.
33. Jeay, S., Sonenshein, G.E., Postel-Vinay, M.C., Kelly, P.A. and Baixeras, E. (**2002**) Growth hormone can act as a cytokine controlling survival and proliferation of immune cells: new insights into signaling pathways. *Mol. Cell. Endocrinol.*, **188**, 1–7.
34. Takala, J., Ruokonen, E., Webster, N.R., Nielsen, M.S., Zandstra, D.F., Vundelinckx, G. and Hinds, C.J. (**1999**) Increased mortality associated with growth hormone treatment in critically ill adults. *N. Engl. J. Med.*, **341**, 785–92.
35. Pawlikowski, M., Stepien, H. and Komorowski, J. (**1994**) Hypothalamic–pituitary–thyroid axis and the immune system. *Neuroimmunomodulation*, **1**, 149–52.
36. Kruger, T.E. (**1996**) Immunomodulation of peripheral lymphocytes by hormones of the hypothalamus–pituitary–thyroid axis. *Adv. Neuroimmunol.*, **6**, 387–95.
37. Wang, H.C. and Klein, J.R. (**2001**) Immune function of thyroid stimulating hormone and receptor. *Crit. Rev. Immunol.*, **21**, 323–37.
38. Rettori, V., Jurcovicova, J. and McCann, S.M. (**1987**) Central action of interleukin-1 in altering the release of TSH, growth hormone, and prolactin in the male rat. *J. Neurosci. Res.*, **18**, 179–83.
39. Witzke, O., Winterhagen, T., Saller, B., Roggenbuck, U., Lehr, I., Philipp, T., Mann, K. and Reinhardt, W. (**2001**) Transient stimulatory effects on pituitary–thyroid axis in patients

treated with interleukin-2. *Thyroid*, **11**, 665–70.
40. Rittenhouse, P.A. and Redei, E. (**1997**) Thyroxine administration prevents streptococcal cell wall-induced inflammatory responses. *Endocrinology*, **138**, 1434–9.
41. Sternberg, E.M. (**2006**) Neural regulation of innate immunity: a coordinated nonspecific host response to pathogens. *Nat. Rev. Immunol.*, **6**, 318–28.
42. Mellon, R.D. and Bayer, B.M. (**1999**) The effects of morphine, nicotine and epibatidine on lymphocyte activity and hypothalamic-pituitary-adrenal axis responses. *J. Pharmacol. Exp. Ther.*, **288**, 635–42.
43. Elenkov, I.J., Wilder, R.L., Chrousos, G.P. and Vizi, E.S. (**2000**) The sympathetic nerve–an integrative interface between two supersystems: the brain and the immune system. *Pharmacol. Rev.*, **52**, 595–638.
44. Pavlov, V.A., Wang, H., Czura, C.J., Friedman, S.G. and Tracey, K.J. (**2003**) The cholinergic anti-inflammatory pathway: a missing link in neuroimmunomodulation. *Mol. Med.*, **9**, 125–34.
45. Woiciechowsky, C., Asadullah, K., Nestler, D., Eberhardt, B., Platzer, C., Schöning, B., Glöckner, F., Lanksch, W.R., Volk, H.D. and Döcke, W.D. (**1998**) Sympathetic activation triggers systemic interleukin-10 release in immunodepression induced by brain injury. *Nat. Med.*, **4**, 808–13.
46. Oke, S.L. and Tracey, K.J. (**2007**) The inflammatory reflex and the role of complementary and alternative medical therapies. *Ann. NY. Acad. Sci.*, Sep 28 [Epub ahead of print].
47. Tracey, K.J. (**2005**) Fat meets the cholinergic antiinflammatory pathway. *J. Exp. Med.*, **202**, 1017–21.
48. Sternberg, E.M., *et al.* (**1989**) Inflammatory mediator-induced hypothalamic-pituitary-adrenal axis activation is defective in streptococcal cell wall arthritis-susceptible Lewis rats. *Proc. Natl. Acad. Sci. USA.*, **86**, 2374–78.
49. Watkins, L.R. and Maier, S.F. (**1999**) Implications of the immune-to-brain communication for sickness and pain. *Proc. Natl. Acad. Sci. USA.*, **96**, 7710–13.
50. Sternberg, E.M. (**1997**) Neural-immune interactions in health and disease. *J. Clin. Invest.*, **100**, 2641–7.
51. Zhang, M. *et al.* (**1999**) Spermine inhibition of monocyte activation and inflammation. *Mol. Med.*, **5**, 595–605.
52. Bernik, T.R., Friedman, S.G., Ochani, M., DiRaimo, R., Ulloa, L., Yang, H., Sudan, S., Czura, C.J., Ivanova, S.M. and Tracey, K.J. (**2002**) Pharmacological stimulation of the cholinergic antiinflammatory pathway. *J. Exp. Med.*, **195**, 781–8.
53. Borovikova, L.V., Ivanova, S., Zhang, M., Yang, H., Cotchkina, G.I., Watkins, L.R., Wang, H., Abumrad, N., Eaton, J.W. and Tracey, K.J. (**2000**) Vagus nerve stimulation attenuates the systemic inflammatory response to endotoxin. *Nature*, **405**, 458–62.
54. Fleshner, M. *et al.* (**1998**) Thermogenic and corticosterone responses to intravenous cytokines (IL-1 beta and THF-alpha) are attenuated by a subdiaphragmatic vagotomy. *J. Neuroimmunol.*, **86**, 134–41.
55. Wang, H., Yu, M., Ochani, M., Amella, C.A., Tanoic, M., Susaria, S., Li, J.H., Wang, H., Yang, H., Ulloa, L., Al-Abed, Y., Czura, C.J. and Tracey, K.J. (**2003**) Nicotinic acetylcholine receptor α7 subunit is an essential regulator of inflammation. *Nature*, **421**, 384–8.
56. Wang, H. *et al.* (**2004**) Cholinergic agonists inhibit HMGB1 release and improve survival in experimental sepsis. *Nat. Med.*, **10**, 1216–21.
57. de Jong, W.J. *et al.* (**2005**) Stimulation of the vagus nerve attenuates macrophage activation by activating the Jak2-STAT3 signaling pathway. *Nat. Immunol.*, **6**, 844–51.
58. Metz, C.N. and Tracey, K.J. (**2005**) It takes nerve to dampen inflammation. *Nat. Immunol.*, **6**, 756–7.
59. Bernik, T.R., Friedman, S.G., Ochani, M., DiRaimo, R., Susarla, S.,

Czura, C.J. and Tracey, K.J. (**2002**) Cholinergic antiinflammatory pathway inhibition of tumor necrosis factor during ischemia reperfusion. *J. Vasc. Surg.*, **36**, 1231–6.

60. Tracey, K.J. (**2007**) Physiology and immunology of the cholinergic antiinflammatory pathway. *J. Clin. Invest.*, **117**, 289–96.
61. Mioni, C. *et al.* (**2005**) Activation of an efferent cholinergic pathway produces strong protection against myocardial ischemia/reperfusion injury in rats. *Crit. Care. Med.*, **33**, 2621–28.
62. Altavilla, D. *et al.* (**2006**) Activation of the cholinergic antiinflammatory pathway reduces NF-kappab activation, blunts TNF-alpha production, and protects against splanchic artery occlusion shock. *Shock*, **25**, 500–6.
63. Guarini, S. *et al.* (**2003**) Efferent vagal fibre stimulation blunts nuclear factor–kappaB activation and protects against hypovolemic hemorrhagic shock. *Circulation*, **107**, 1189–94.
64. Borovikova, L.V. *et al.* (**2000**) Role of vagus nerve signaling in CNI-1493-mediated suppression of acute inflammation. *Auton. Neurosci.*, **85**, 141–7.
65. van Westerloo, D.J. *et al.* (**2006**) The vagus nerve and nicotinic receptors modulate experimental pancreatitis severity in mice. *Gastroenterology*, **130**, 1822–30.
66. Saeed, R.W. *et al.* (**2005**) Cholinergic stimulation blocks endothelial cell activation and leukocyte recruitment during inflammation. *J. Exp. Med.*, **201**, 1113–23.
67. Wang, H., Liao, H., Ochani, M., Justiniani, M., Lin, X., Yang, L., Al-Abed, Y., Wang, H., Metz, C., Miller, E.J., Tracey, K.J. and Ulloa, L. (**2004**) Cholinergic agonists inhibit HMGB1 release and improve survival in experimental sepsis. *Nat. Med.*, **10**, 1216–21.
68. Huston, J.M., Puerta, M.G., Ochani, M., Ochani, K., Yuan, R., Rosas-Ballina, M., Ashok, M., Goldstein, R.S., Chavan, S., Pavlov, V.A., Metz, C.N., Yang, H., Czura, C.J., Wang, H. and Tracey, K.J. (**2007**) Transcutaneous vagus nerve stimulation reduces serum high mobility group box1 levels and improves survival in murine sepsis. *Crit. Care Med.*, Sep 25 [Epub ahead of print].
69. Huston, J.M., Ochani, M., Rosas-Ballina, M., Liao, H., Ochani, K., Pavlov, V.A., Gallowitsch-Puerta, M., Ashok, M., Czura, C.J., Foxwell, B., Tracey, K.J. and Ulloa, L. (**2006**) Splenectomy inactivates the cholinergic antiinflammatory pathway during lethal endotoxemia and polymicrobial sepsis. *J. Exp. Med.*, **203**, 1623–28.
70. Ghia, J.E., Blennerhassett, P. and Collins, S.M. (**2007**) Vagus nerve integrity and experimental colitis. *Am. J. Physiol. Gastrointest. Liver. Physiol.*, **293**, G560–7.
71. Corcoran, C., Connor, T.J., O'Keane, V. and Garland, M.R. (**2005**) The effects of vagus nerve stimulation on pro- and anti-inflammatory cytokines in humans: a preliminary report. *Neuroimmunomodulation*, **12**, 307–9.

14
Genetic Polymorphisms and Other Parameters that Affect Predisposal to Infection and Outcome

Jean-Marc Cavaillon and Christophe Adrie

14.1
Non-genetic Parameters that Affect the Occurrence of Infection, Organ Failure and Outcome

The key study by Sorensen *et al.* [1] established that susceptibility and premature death due to infection was the most heritable human disease. Since then, many investigations have demonstrated genetic predisposal to susceptibility to infection and death due to sepsis. However, as illustrated on Figure 14.1, many other events are associated with susceptibility to infection, the severity of the infectious process, the occurrence of organ failure and eventually death. These parameters are either linked to the infectious agent itself, or to the quality of treatment in the intensive care unit, or finally to the patient him/herself. As reviewed below, numerous parameters linked to the patient will affect the natural history of the infectious disease. In this context with such a myriad of co-factors, it is fascinating that a single nucleotide polymorphism (SNP) could be pointed out as a key element in susceptibility and outcome.

14.1.1
The Infectious Agent and Site of Infection

It is becoming increasingly clear that the nature of the infection underlying sepsis is a major determinant of outcome. Both the site of infection, and may be to a lesser extent, the nature of the organism have a significant impact on survival from sepsis, and there is a significant interaction between them. It has been amply demonstrated that the risk of death from sepsis is highly dependent on the site of infection [2]. Infections emanating from the urinary tract and intravascular catheter sites are less likely to be lethal than sepsis associated with soft tissue, intraabdominal, or pulmonary sources of infection. Pneumonia remains associated with the highest mortality with both Gram-negative bacteria (such as *Pseudomonas aeruginosa, Escherichia coli, Klebsiella pneumoniae* but not *Bacillus fragilis*) and

Sepsis and Non-infectious Systemic Inflammation. Edited by J.-M. Cavaillon and C. Adrie

ISBN: 978-3-527-31935-0

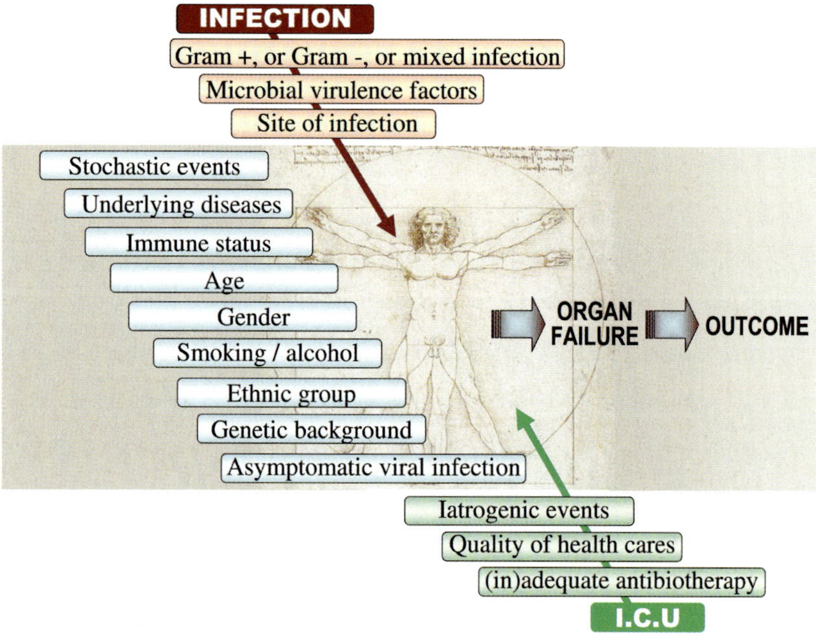

Figure 14.1 The natural evolution of a systemic inflammatory disorder or sepsis is greatly influenced by parameters linked to the pathogen(s), the ICU treatments and many other elements associated with the patient.

Gram-positive bacteria (such as *Staphylococcus aureus, Streptococcus pyogenes* but not *Streptococcus pneumoniae*) [3], and with Candida spp. infection. A high mortality is also observed for central nervous system infection by Candida spp. and *P. aeruginosa*. Gram-positive bacteria have been the main source of sepsis since the late 1980s [4], but Gram-negative bacteremia remains more frequent in developing countries [5]. Not surprisingly, antibiotic-resistant strains, particularly in the case of Gram-negative nosocomial infections, are associated with higher mortality [6]. Indeed, mortality after hospital-acquired sepsis is higher than that in patients with community-acquired infection [7, 8]. This is most probably the consequence of the occurrence of another insult among hospitalized patients (diseases, surgery) that concomitantly enhances the severity of the infectious disease. In a survey of 510 papers including 55 854 clinical infections, Cohen *et al.* summarized the present situation in industrialized countries [3]. Gram-negative bacteremia was clearly associated with a higher mortality rate than infection with Gram-positive organisms, the only exception being infection with *Streptococcus pyogenes*. For other infections, the identity of the infecting organism is of little consequence for most patients provided that appropriate, prompt antimicrobial therapy is administered [9]. Indeed the authors refer to studies that failed to find any differences in terms of mortality between infection with Gram-positive and Gram-negative bacteria [7, 10]. However, when instead of a global analysis in terms of Gram staining, the outcome is studied as a function of a given microorganism, it appears that the mortality ranges from 20% (*Escherichia coli*), to 46% (Pseudomonas species), with intermediate values (30% *Staphylococcus aureus*; 35% *Streptococcus pneumoniae*; 42% Klebsiella species) [10].

14.1.2 Parameters Linked to the Patients

14.1.2.1 Stochastic Events

Stochastic (from Greek *stokhastikos*, "aim at, guess") is an adjective used to characterize processes that are randomly determined. The concept of stochastic events was considered early on by geneticists studying inbred mice to explain differences between genetically identical individuals. For example, highly inbred mice (MRL/Ipr or B6-lpr) reproducibly develop an autoimmune syndrome resembling human systemic lupus erythematosus, and produce high titers of autoantibodies to the nuclear ribonucleoprotein Sm. However, an approximately 25% incidence occurs. It was reported that, although the synthesis of anti-Sm antibodies is under genetic control, the expression of this capability in an individual animal is governed by stochastic events [11]. One of the most striking examples that illustrates a stochastic event is the concept of lethal dose 50 (LD50) in inbred mice. How can it be explained that among strictly genetically identical mice given the same number of bacteria, half will survive when half will die? In other words, there are many other parameters that affect the innate immune response during an infectious process that are not genetically determined. This is illustrated by a recent study on gene expression in liver of inbred mice. A significant discrepancy exists between individuals: for example, in Balb/c mice, among the 2382 genes studied, a difference in expression greater than 1.5-fold was found for 19% of the genes, and a difference in expression higher than 2-fold was found for 3% of the genes [12]. Individual variability among genetically identical individuals correlates with trait anxiety [13]. C57Bl/6 Mice were submitted to a free choice open field test and classified as "long emergence latency" (LEL) or "short emergence latency" (SEL). Under unstressed control conditions, LEL mice expressed higher levels of hippocampal mRNA for the glucocorticoid receptor than SEL mice. Following stress exposure, LEL mice displayed higher plasma corticosterone than SEL mice. Altogether these examples illustrate that the level of gene expression cannot be fully predicted by the genetic background. Furthermore, other stochastic events occur at the cell level, as illustrated in the case of the association of the glucocorticoid receptor to the promoter of genes. It was reported that stochastic molecular events were critical determinants in the extent of transcriptional activity induced by hormones within individual members of a cell population [14]. In humans the best examples of stochastic events are given by the phenotypic discordance between monozygotic twins such as their susceptibility to diseases. To some extent these discrepancies are due to epigenetic factors that change over the lifetime of a multicellular organism. It has been proposed that epigenetic drift during development can be stochastic or determined by environmental factors [15]. We will not review all the environmental events that can alter/modify/change the immune status: exercise, hygiene, dietary lipids, probiotics, exposure to chemicals or pollutants, smoking, alcohol, and stress. Psychoneuroimmunology is an important new area of investigation to further understand the consequence of depression or life stress on inflammatory and infectious disease risk [16]. Aberg *et al.* [17] reported that in mice psychological stress increases the production of endogenous glucocorticoids which downregulated epidermal expression of murine antimicrobial peptides. As a consequence, a cutaneous skin infection induced by group A *Streptococcus pyogenes* (GAS) was more severe. When Coe *et al.* [18] showed that stress occurring *in utero* can affect the immune system, it became easier to appreciate that there will always be numerous unknown events in the history of a given patient that will finally influence their body reaction to a given insult.

14.1.2.2 **Underlying Diseases**

About 50% of ICU patients with severe sepsis present chronic comorbidities according to Knaus' definitions [19, 20] and many others present other underlying risk factors for sepsis such as age, nutritional factors, diabetes or alcoholism [21, 22] often in association. Overall the comorbidities appear to be a risk factor for the development of infection and severe sepsis, and influence poor outcome in the short as well as the long term. Interestingly, Quartin *et al.* [23] showed that survival of sepsis patients was lower compared to controls from the general population 8 years after the initial hospitalization, even when comorbidities were accounted for. Sepsis is an acute process, and its major manifestation, acute organ dysfunction, is strongly associated with short-term mortality. However acute organ dysfunction was not associated with long-term mortality in those who survived the original insult as shown in the SUPPORT study cohort, a heterogeneous cohort of critically ill subjects [24] and in subjects hospitalized with pneumonia [25]. Rather, factors such as advancing age and underlying health status appeared to be more important [26]. The overall increased burden of chronic health conditions augments the risk of infection and sepsis. However, survivors of severe sepsis may develop a higher burden of chronic disease, which in turn may represent a risk factor for subsequent illnesses, thereby initiating a spiral of events that eventually leads to death [27].

14.1.2.3 **Age**

The occurrence of sepsis is increased in patients older than 55 years, and the incidence increases on a logarithmic scale [28]. Of course, underlying disorders, especially those that chronically impair immune host response are more frequent in older patients and contribute to enhanced sensitivity and death [29]. Even more interestingly, Angus' study revealed in the oldest patients with no comorbidity, that mortality increased linearly as a function of age. This observation reflects data obtained in animal models. In one of the most relevant approaches to human sepsis, the cecal ligature and puncture (CLP) model, it was shown that mortality was far higher in the oldest mice [30]. Furthermore, when compared with young mice, aged animals had higher plasma levels of IL-6 and TNF 24 h after CLP. In addition, aged mice derived no benefit from antibiotic therapy initiated 12 h after CLP. Indeed, the presence of TNF mRNA in the lungs and heart was more pronounced in aged mice after CLP [31]. In aged mice, CLP led to increased splenic apoptosis and gut epithelial cell death as compared with young septic animals [32], and caused functional and histological changes consistent with acute renal failure in humans [33]. Aged mice are also known to be more sensitive to bacterial LPS. This increased sensitivity in senescent mice was associated with increased expression of NF-κB in the tissues [34], and correlated with elevated plasma levels of nitric oxide and TNF [35]. Visceral adipocyte tissues from old mice expressed higher mRNA levels of IL-1β, IL-6, TNF, and cyclooxygenase-2 than those of young mice. Adipocytes of old mice favored the production of TNF and IL-6 by macrophages, probably through a ceramide-dependent mechanism [36]. It is possible that it may not only be ageing that is associated with the enhanced capacity to produce inflammatory cytokines but it may also be linked to an enhanced responsiveness to signals delivered by inflammatory cytokines, such as a hepatic hyperresponsiveness to IL-1β [37], or an enhanced sensitivity to TNF-induced lethality [38]. In humans, monocyte function seems to be increased in the elderly. Leukocytes of elderly persons produce higher amounts of IL-1, IL-6, IL-8 and TNF after induction with LPS than those from young donors [39]. However, this topic remains controversial. Nevertheless, in the NORASEPT II study in which 930 patients with septic shock were enrolled, TNF levels were significantly higher in the oldest group of patients [40]. Furthermore, in human volunteers receiving an intravenous bolus of LPS, the elderly subjects showed a more prolonged fever response as compared to the young controls [41]. The elderly group also showed larger initial increases in TNF and sTNFR-I levels and prolonged increased levels of sTNFR-I. Finally the elderly group showed a more rapid increase in C-reactive protein levels than did the young group. Altogether, these observations tend to suggest that age is a parameter on its own that enhances inflammatory response as well as sensitivity to infection.

14.1.2.4 Gender

Sex hormones [42] or sex-related gene polymorphisms [43, 44] may protect women against sepsis and death from sepsis. Differences in hormone profiles have been widely suggested as the cause of gender-based differences in the incidence and outcome of sepsis. In mice, proestrus females tolerated polymicrobial sepsis better than males [45], and survival improved in males after testosterone receptor blockade [46]. However this gender effect may depend on the bacteria; in a mouse model, the pronounced sensitivity of females to Listeria and of males to *Streptococcus pyogenes* was revealed by significantly higher lethality rates [47]. Interestingly, in IL-10 knockout (KO) mice there was no longer a difference between genders in terms of sensitivity to Listeria. However, as one moves away from cell culture and animal models, the clear dichotomy of response influenced by gender becomes even more blurred. Epidemiological studies produced even more conflicting results, perhaps reflecting effects of age, case-mix differences, nature of the injury preceding development of sepsis (e.g. trauma or burns), infection source, co-morbidities, and menopausal status [42, 48–54]. Another possible source of gender-based differences may be the reported lower use of invasive procedures in critically ill women compared to men, despite greater severity of illness in women, and even after adjustment for age [55]. Recently, in a very large cohort of patients, mortality from severe sepsis was higher in men than in women after adjustment for confounding factors [56]. This difference was due to higher mortality in men older than 50 years compared to (postmenopausal) women of the same age; mortality was not significantly different between younger men and women. In this study, the level of care and rate of invasive procedures were similar in women and men. Further clinical studies that assess the diverse effects of gender and sex hormones on host inflammatory responses are of great importance to open new therapeutic opportunities, and to improve the care given to men and women who develop critical illnesses.

14.1.2.5 **Lifestyle**

The presence of certain factors linked to lifestyle leads to a higher incidence of complications and a poorer prognosis in ICU patients. By multivariate analysis, urban residence was found to be one of five factors predictive of increased intensive care mortality of patients after cardiac surgery [57]. Of course, a history of smoking that alters lung function in healthy subjects, increases the risk factor for pneumococcal pneumonia [58], and identifies high-risk patients for surgical complications [59]. Interestingly, passive smoking in children under 12 years of age was significantly more frequent in meningococcal patients than among the control population [60]. Not only smoking, but also alcohol abuse is a prominent risk factor for case fatality in bacteremic patients [61]. Indeed chronic alcohol abuse is associated with a persistent fever, delayed resolution of symptoms, increased rates of bacteremia, increased use of intensive care, prolonged duration of hospital stay. For example, chronic alcohol abuse was associated with an increased incidence of acute respiratory distress syndrome and severity of multiple organ dysfunction in patients with septic shock [62]. If smoking and alcohol abuse identify at risk patients, other parameters may identify lower risk patients. This is the case of vaccination, and as expected prior vaccination against pneumococcus is associated with improved survival, decreased chance of respiratory failure or other complications, and decreased length of stay among hospitalized patients with community-acquired pneumonia [63].

14.1.2.6 Asymptomatic Viral Infection

Viruses can be considered as a co-morbidity factor. Asymptomatic viruses (e.g. *Herpes simplex* virus, hepatitis B virus, cytomegalovirus, Human T lymphotropic virus type 1) can enhance susceptibility and sensitivity of patients to bacterial infection. Such a synergy has been well established in animal models. For example, infection with a non-cytopathogenic virus, lymphocytic choriomeningitis virus (LCMV), was found to sensitize mice to low doses of LPS. This hypersensitivity correlated with the hyperproduction of TNF, as a reflection of virus-induced production of interferon-gamma (IFNγ) [64]. Similarly, a remarkable synergism of adenoviruses and LPS was reported in mice, leading to a dramatic increase in LPS-induced TNF production and major sensitization to LPS-induced lethal shock [65]. In addition, the severity of a bacterial infection can be enhanced following a viral infection. This has been illustrated in mouse models of synergy: influenza virus infection preceding *Streptococcus pneumoniae* challenge primed the animals for pneumonia infection and led to 100% mortality [66]. Similarly, a model of sequential influenza A virus and *Neisseria meningitidis* infection in mice led to fatal meningococcal pneumonia in an otherwise non-lethal bacterial infection. The susceptibility of mice to bacterial superinfection was correlated with the peak interferon-gamma production in the lungs [67]. Although, proof is still inconclusive in the human setting, it is quite probable that viral infections may accelerate poor outcome in infectious or non-infectious SIRS patients.

14.1.2.7 Ethnic Group

Disparities in health outcomes among various ethnic groups living in developed countries are becoming more evident. For example, the annual incidences of meningitis and bacteremic pneumonia, in Alaskan native children <2 years old were 8–10 times higher than those for other US groups [68]. Along the same lines, neonatal sepsis was the strongest predictor of mortality among African infants, even greater than birth at 28 weeks or less [69]. In adults, among renal transplant recipients, the evaluation of episodes of infection demonstrated an increased mortality rate for white compared to black recipients despite similar severity of illness [70]. Similarly, *Staphylococcus aureus* bacteremia is more likely to occur in certain ethnic groups [71]. Adjusting for differences in poverty in their region of residence, blacks had a higher population-based incidence of severe sepsis than whites, and Hispanics had a lower incidence [72]. However, age-adjusted case fatality rates for hospitalized white and black patients with sepsis were similar [73].

14.2 Parameters Linked to Health Care

14.2.1 Antibiotherapy

The initial empiric selection of antibiotics may need to be changed after identification of the causative microbial agents and the results of the antibiogram. Numerous studies have revealed that initial inadequate antibiotic coverage of microorganisms was associated with a significantly higher mortality than adequate initial therapy, even though the change in therapy led to clinical resolution. The critical importance of an appropriate early antibiotic therapy has been emphasized for critically ill patients with bloodstream infections [74], and pneumonia acquired in the intensive care unit [75]. However, the early mortality rate was unaffected by the appropriateness of empirical antibiotic therapy [76], and mortality attributable to inappropriate antibiotic treatment increased with the severity of illness at ICU admission [77]. In a study that addressed both appropriate antibiotherapy and three genetic polymorphisms (TNFα 308 promoter; IL-10-1082 promoter; and in the first intron of the Lymphotoxin-α gene), in septic shock patients, the delay in initiation of adequate antibiotic therapy was the only independent predictor of mortality [78].

14.2.2 Iatrogenic Events

Critically ill patients require a higher density of interventions, invasive procedures, and management decisions than other patients. They receive multiple medications, which increases the risk of mistakes in drug dosages, dosing intervals, and target patients [79]. Finally, the large number of invasive procedures used in the ICU carries a high risk of nosocomial infection. Adverse events are common and often occurred in combination in individual patients [80]. Several preventable or non-preventable adverse events independently contribute to death underlining the need to create a safe ICU environment in order to prevent those which are preventable.

Some of the iatrogenic events can be demonstrated in animal models. Among them, mechanical ventilation with conventional tidal volume function is a cofactor in the initiation of acute lung injury, as illustrated by an enhanced transcriptional response to LPS [81]. A major complication of mechanical ventilation is the development of ventilator-associated pneumonia, and experiments have associated mechanical ventilation and *Staphylococcus aureus* infection. It has been reported that mechanical ventilation exacerbated both pulmonary and systemic inflammation in response to bacteria and contributed to the pathogenesis of both acute lung injury and multiple organ dysfunction syndrome [82].

14.2.3 Quality of Health Care

Burnout has been described as a condition in which professionals "lose all concern, all emotional feeling for the people they work with, and come to treat them in a detached or even dehumanized way". Professional burnout is a psychological syndrome arising in response to chronic interpersonal stressors on the job. Burnout is a problem that is specific to the work context, in contrast to depression, which tends to pervade every domain of an individual's life. Physical illness, emotional problems, increased turnover, absenteeism, and poor job performance and negative attitudes in general are some of the problems in a long list of difficulties that have been associated with burnout. Moreover, high level of burnout is frequent amongst the intensivists or medical staff in ICUs [83, 84], and is more related to organizational factors than factors related to the patients but could affect the quality of patient care [85]. Patients admitted on weekends had a higher risk-adjusted mortality than did patients admitted on weekdays. Disparities in resources, expertise, and healthcare providers working during weekends may explain the observed differences in weekend mortality [86]. Although women, overall, received better care than men, they received a higher quality of preventive and chronic care and a lower quality of acute care [87]. However the differences in the observed quality of care between sociodemographic subgroups are small compared with those between observed and desirable quality of health care in all subgroups [87].

14.3 Genetic Polymorphisms

After the above review of all parameters that affect the clinical course and outcome of SIRS patients, it is fascinating that just a single nucleotide polymorphism (SNP), (i.e. a single base change in the DNA sequence), can discriminate between patients that will or will not be susceptible to infection, will or will not develop multiple organ failure, and will or will not ultimately die. There is no doubt that a genetic background underlines individual susceptibility to infection. In 1988, Sorensen *et al.* [1] published the most convincing study that demonstrated the genetic influence for risk of death from an infectious cause. The risk of death from infection appeared to be more influenced by genetic background than the risk of death from cancer or cardiovascular or cerebrovascular causes. In parallel, many deficiencies of innate immunity (e.g. complement factors, cytokines, cytokine receptors, signaling molecules etc.) were associated with an increased susceptibility to infection. However, there

growth factor (VEGF) has recently been identified among the factors that contribute to sepsis-induced lethality [105], and since numerous SNPs have been reported for the VEGF gene, studies attempting to identify whether some SNPs correlate with susceptibility to infection and adverse effects should soon appear in the literature.

14.3.2
Innate Immune Receptors

Polymorphisms of genes coding for PAMPs receptors have been associated with increased susceptibility to infection and sepsis. This is particularly true for TLR4 (Table 14.2). While one SNP (Asp299Gly) was found to be associated with increased susceptibility to infection, it was not the case for other SNPs (Thr399Ile and Ala896Gly) [101, 106, 107]. However, other studies failed to find such an association for the Asp299Gly SNP of TLR4 [108]. In a study of 197 white patients with systemic meningococcal infection, no SNP was significantly over-represented, but rather rare TLR4 mutations were markedly frequent [109]. The Asp299Gly SNP of TLR4 occurs at a frequency of 6 to 10% in Caucasian populations and was shown to be associated with reduced airway responsiveness to inhaled LPS in human volunteers [110]. Despite this, the authors showed in this study that THP-1 cells transfected with the TLR4 expressing the Asp299Gly mutation had an impaired response to LPS, it was then clearly established that monocytes expressing the heterozygous Asp299Gly mutation did not exhibit any functional defect in cytokine release after LPS stimulation [111]. Ethnic differences should also to be taken into consideration, and no association between TLR4 and TLR2 polymorphisms and bacteremia has been observed among Korean patients [112]. An association of CD14 polymorphism and the occurrence of sepsis with Gram-negative bacteria was reported among Caucasian critically ill patients but was not found among Japanese patients [113, 114]. As mentioned for the SNPs of genes coding for cytokines, controversial studies can be found in the literature regarding innate immune receptors. For example, the association between TLR2 SNP (Arg753Gln) and susceptibility to *S. aureus* infection was first reported in a study that included 110 controls and 91 patients with septic shock [115], whereas a later study which included 696 controls and 420 patients, found no associations between this polymorphism and the occurrence of invasive *S. aureus* infection, serious morbidity and mortality [116].

14.3.3
Coagulation Factors

The coagulation process is closely linked to inflammation, and the induction of tissue factor (TF) by endotoxin, TNF, IL-1 or other inflammatory mediators,

Table 14.2 Influence of polymorphisms of genes coding for receptors of innate immunity on the occurrence of infection and outcome in patients with sepsis or non-infectious systemic inflammatory response syndrome.

Genes	Studied populations	Polymorphism associated with		References
		Occurrence of infection	Mortality or survival	
TLR4 (Asp299Gly; Thr399Ile)	Septic shock	Gram-negative infections	No	[106]
	ICU patients	Gram-negative infections	No	[148]
	Healthy population	severe acute infection	–	[107]
	Major visceral surgery	No	No	[108]
	Burn injury	Severe sepsis	–	[124]
	Sepsis, severe sepsis & septic shock	–	No	[101]
	Meningococcal disease	No	–	[149]
	Meningococcal infection	More frequent rare mutations	–	[109]
MD2 (-1625G)	Trauma	Sepsis	–	[150]
TLR2 (Arg753Gln)	Septic shock	*Staphylococcus aureus* infection	–	[115]
	Severe *S. aureus* infection	No	No	[116]
	Critically ill	Sepsis with Gram-negative bacteria	No	[113]
	Trauma	Sepsis in African Americans	–	[151]
CD14 (C-159T)	Severe sepsis	No	No	[152]
	Septic shock	Septic shock	Yes	[153]
	ICU patients	No	No	[148]
	Caucasian critically ill	sepsis with Gram-negative bacteria	No	[113]
	Japanese critically ill	No	No	[114]
	Burn injury	Severe sepsis	–	[124]
	Sepsis, severe sepsis & septic shock	–	No	[101]
Nod2 (Leu1007fsinsC)	Sepsis	–	Yes	[154]

(*continued overleaf*)

Table 14.2 *(continued)*.

Genes	Studied populations	Polymorphism associated with		References
		Occurrence of infection	Mortality or survival	
LPS Binding Protein (291C)	Trauma	No	No	[155]
LPS Binding Protein (Gly98)	Sepsis	sepsis only in male patients	Yes	[43]
Bactericidal Permeability Increasing Protein	Sepsis	No	No	[43]
Mannose Binding Lectin	SIRS	Sepsis, severe sepsis, septic shock	Yes	[156]
	Respiratory infections	Bacterial but not fungal infection	–	[157]
	Critically ill	Sepsis	No	[113]
	Severe sepsis, septic shock	Severe sepsis, septic shock	No	[158]
	Acute pyelonephritis	Septic shock	–	[159]
	ARDS	Septic shock	Yes	[160]

is the key element that initiates the coagulation cascade in response to inflammation. Various SNPs are known for the TF gene. Among them, the -1208 D/I polymorphism is functional in that it regulates basal TF-mRNA in circulating monocytes and circulating microparticle-associated TF-procoagulant activity *in vivo*. However, as assessed in human volunteers receiving endotoxin, it does not influence the relative increase in TF-mRNA or coagulation activation [117]. Certain treatments that aimed to downregulate the coagulation process, improved the outcome of animals in different septic models. These experimentally successful approaches and the potential beneficial effect of activated protein C in human septic shock, illustrate that coagulation is important during sepsis. Accordingly, numerous investigations have undertaken the analysis of the putative association of SNP of coagulation or fibrinolysis factors with infection, adverse effects and outcome (Table 14.3). Genotypes associated with increased expression of PAI-1 were associated with increased susceptibility to community-acquired pneumonia in elderly whites [118]. In addition, SNPs of other molecules that are not directly involved in coagulation may still be influential. This was reported in human volunteers injected with LPS and genotyped for the Ser128Arg SNP of E-selectin. It was shown that the

S128R allele was associated with a 50 to 80% enhancement in thrombin generation [119].

The observation that one given treatment may or may not be effective depending upon SNPs opens a new area of investigation in the field of sepsis and SIRS patients. Such an observation has been reported for the use of activated protein C in sepsis (Drs K.R. Walley and J. Russell, unpublished data).

Table 14.3 Influence of polymorphisms of genes coding for coagulation factors on the occurrence of infection and outcome in patients with sepsis or non-infectious systemic inflammatory response syndrome.

Genes	Studied populations	Polymorphism associated with			References
		Occurrence of infection	Adverse events	Mortality or survival	
Factor V	Severe sepsis	–	No	Yes	[161]
	Severe sepsis	–	Severe sepsis	No	[162]
Plasminogen Activator Inhibitor-1	Meningococcal disease	–	/	Yes	[163]
	Meningococcal disease	–	Vascular complications	Yes	[164]
	Trauma	Sepsis	Multiple organ failure	Yes	[165]
	Sepsis, severe sepsis & septic shock	–	No	No	[101]
	Community-acquired pneumonia	Yes	–	–	[118]
Urokinase plasminogen activator	Sepsis, severe sepsis & septic shock	–	No	No	[101]
Thrombin activatable thrombolysis inhibitor	Meningococcal disease	–	–	Yes	[166]
Fibrinogen	Critically ill	Yes	Severity of organ dysfunction	Yes	[167]
Urokinase	Infection-associated acute lung injury	–	Mechanical ventilation	Yes	[168]
Protein C	Severe sepsis	–	Organ dysfunction	Yes	[169]
	Meningococcal disease (children)	Yes	Sepis	–	[170]

14.3.4
Others

Different SNPs for various molecules which play a role in inflammation and innate immunity have been reported to be associated with either the occurrence of sepsis, ARDS, or organ failure (Table 14.4). Although so far, very few reports have addressed the SNPs of receptors for factors involved in innate immunity, many more studies are likely to be published within the next few years. In contrast, numerous reports have established a link between the Fcγ receptor IIa (CD32) and susceptibility to meningococcal infection. Most interestingly, neutrophils expressing the FcγRIIa with the allotype most frequently observed among patients, were less efficient at phagocytosing *N. meningitidis* opsonized with polyclonal IgG2 antibodies [120].

Not surprisingly, reports of SNPs of signaling molecules have now begun to be published. Of course, mutations leading to stop codons or to non-functional proteins are associated with very severe deficiencies and increased sensitivity to infection. This was reported for IRAK-4, a deficiency of which leads to hyporesponsiveness to LPS and recurrent bacterial infection [121]. Interestingly, the SNP Ser180Leu of Mal/TIRAP, an adapter molecule essential for MyD88-dependent signaling downstream of TLR2 and TLR4, was found to be protective for heterozygote patients against infectious diseases, including invasive pneumococcal disease among European patients, and bacteremia, malaria and tuberculosis among African patients [122].

A very interesting study has revealed the role played by mitochondrial DNA. In this report the haplogroups of mitochondrial DNA among 150 patients hopitalized with severe sepsis and 542 age-matched controls from the Northeast of England were compared. The frequency of the mitochondrial DNA haplogroup H did not differ between patients and controls, but it was a strong independent predictor of outcome during severe sepsis, conferring a 2.12-fold (95% CI 1.02–4.43) increased chance of survival at 180 days compared with individuals without the haplogroup H [123]. Furthermore, the authors noted that the most extreme core temperature in the first 24 h was significantly higher among the patients with the H haplogroup. It is worth recalling that the H haplogroup is associated with enhanced respiratory chain activity.

In conclusion, in this chapter we have reviewed all the parameters that influence the occurrence of infection, organ failure and eventually death. More than 8 millions SNPs are known in the human genome and still only a few have been studied with regard to their influence in infectious or non-infectious SIRS patients. The study by Sorensen *et al.* [1] that established the substantial heritability of susceptibility to death due to infection, is far from being convincingly illustrated by the numerous SNP studies. Most probably, a combination of many SNPs should be investigated in a study, rather than only one. Finally, it should not be forgotten that other factors or events greatly influence the occurrence of infection and patients' outcome, as illustrated in Figure 14.1.

Table 14.4 Influence of polymorphisms of genes coding for modulators of Inflammation and innate immunity, receptors and signaling molecules, on the occurrence of infection and outcome in patients with sepsis or non-infectious systemic inflammatory response syndrome.

Genes	Studied populations	Polymorphism associated with			References
		Occurrence of infection	Adverse events	Mortality or survival	
Modulators of inflammation and innate immunity					
Surfactant protein-B	Community-acquired pneumonia	–	ARDS/Septic shock	No	[171]
Angiotensin Converting Enzyme	Coronary artery bypass grafting	–	ARDS	–	[172]
	Community-acquired pneumonia	–	No	No	[173]
Caspase-12	African Americans	Sepsis	–	–	[174]
HSP 70	Severe multiple trauma	–	Liver failure	No	[175]
	Community-acquired pneumonia	–	Septic shock	Trend ($p = 0.06$)	[176]
β-defensin 1	Severe sepsis	Sepsis	–	Yes	[177]
Myosin light chain kinase	Sepsis	Sepsis	Acute lung injury	–	[178]
C-reactive protein (CRP)	Bacteremia	Streptococcus pneumonia	–	Yes	[179]

(continued overleaf)

Table 14.4 *(continued).*

Genes	Studied populations	Polymorphism associated with			References
		Occurrence of infection	Adverse events	Mortality or survival	
Receptors					
FcγRIIa	Meningococcal septic shock	Yes	–	–	[120]
	Meningococcal infection	Yes	Yes	–	[180]
	Streptococcus pneumoniae infection	Yes	–	–	[181]
	Severe acute respiratory syndrome (SARS)	–	Severe course	No	[182]
TNF R	Severe sepsis & septic shock	No	No	No	[100]
Signaling Molecules					
IRAK 1	Sepsis	–	Shock Ventilator free days	Yes	[183]
Transcription factor NRF2	Major trauma	–	Acute lung injury	–	[184]
Mal/TIRAP	Infectious diseases	Yes	–	–	[122]
Mitochondrial function	Severe sepsis	No	–	Yes	[123]

References

1. Sorensen, T.I., Nielsen, G.G., Andersen, P.K. and Teasdale, T.W. (**1988**) Genetic and environmental influences on premature death in adult adoptees. *N. Engl. J. Med.*, **318**, 727–732.
2. Pedersen, G., Schonheyder, H.C. and Sorensen, H.T. (**1997**) Antibiotic therapy and outcome of monomicrobial gram-negative bacteraemia: a 3-year population-based study. *Scand. J. Infect. Dis.*, **29**, 601–606.
3. Cohen, J., Cristofaro, P., Carlet, J. and Opal, S. (**2004**) New method of classifying infections in critically ill patients. *Crit. Care Med.*, **32**, 1510–1526.
4. Martin, G.S., Mannino, D.M., Eaton, S. and Moss, M. (**2003**) The epidemiology of sepsis in the United States from 1979 through 2000. *N. Engl. J. Med.*, **348**, 1546–1554.
5. Tanriover, M.D., Guven, G.S., Sen, D., Unal, S. and Uzun, O. (**2006**) Epidemiology and outcome of sepsis in a tertiary-care hospital in a developing country. *Epidemiol. Infect.*, **134**, 315–322.
6. Guillemot, D., Gasquet, I., Vallet, O., David, M.F., Laurent, C. and Mathieu, D. (**2005**) Thirty-day mortality of nosocomial systemic bacterial infections according to antibiotic susceptibility in an 800-bed teaching hospital in France. *Clin. Microbiol. Infect.*, **11**, 502–504.
7. Alberti, C., Brun-Buisson, C., Goodman, S.V., *et al.* (**2003**) Influence of systemic inflammatory response syndrome and sepsis on outcome of critically ill infected patients. *Am. J. Respir. Crit. Care Med.*, **168**, 77–84.
8. Annane, D., Aegerter, P., Jars-Guincestre, M.C. and Guidet, B. (**2003**) Current epidemiology of septic shock: the CUB-Rea Network. *Am. J. Respir. Crit. Care Med.*, **168**, 165–172.
9. Wheeler, A.P. (**2007**) Recent developments in the diagnosis and management of severe sepsis. *Chest*, **132**, 1967–1976.
10. Opal, S.M., Garber, G.E., LaRosa, S.P., *et al.* (**2003**) Systemic host responses in severe sepsis analyzed by causative microorganism and treatment effects of drotrecogin alfa (activated). *Clin. Infect. Dis.*, **37**, 50–58.
11. Eisenberg, R.A., Craven, S.Y., Warren, R.W. and Cohen, P.L. (**1987**) Stochastic control of anti-Sm autoantibodies in MRL/Mp-lpr/lpr mice. *J. Clin. Invest.*, **80**, 691–697.
12. Pritchard, C., Coil, D., Hawley, S., Hsu, L. and Nelson, P.S. (**2006**) The contributions of normal variation and genetic background to mammalian gene expression. *Genome. Biol.*, **7**, R26.
13. Jakovcevski, M., Schachner, M. and Morellini, F. (**2008**) Individual variability in the stress response of C57BL/6J male mice correlates with trait anxiety. *Genes. Brain. Behav.*, **7**, 235–243.
14. Voss, T.C., John, S. and Hager, G.L. (**2006**) Single–cell analysis of glucocorticoid receptor action reveals that stochastic post-chromatin association mechanisms regulate ligand-specific transcription. *Mol. Endocrinol.*, **20**, 2641–2655.
15. Poulsen, P., Esteller, M., Vaag, A. and Fraga, M.F. (**2007**) The epigenetic basis of twin discordance in age-related diseases. *Pediatr. Res.*, **61**, 38R–42R.
16. Irwin, M.R. (**2007**) Human psychoneuroimmunology: 20 Years of discovery. *Brain. Behav. Immun.*.
17. Aberg, K.M., Radek, K.A., Choi, E.H., *et al.* (**2007**) Psychological stress downregulates epidermal antimicrobial peptide expression and increases severity of cutaneous infections in mice. *J. Clin. Invest.*, **117**, 3339–3349.
18. Coe, C.L., Lubach, G.R. and Shirtcliff, E.A. (**2007**) Maternal stress during pregnancy predisposes for iron deficiency in infant monkeys

impacting innate immunity. *Pediatr. Res.*, **61**, 520–524.
19. Adrie, C., Alberti, C., Chaix-Couturier, C., *et al.* (**2005**) Epidemiology and economic evaluation of severe sepsis in France: age, severity, infection site, and place of acquisition (community, hospital, or intensive care unit) as determinants of workload and cost. *J. Crit. Care*, **20**, 46–58.
20. Knaus, W.A., Draper, E.A., Wagner, D.P. and Zimmerman, J.E. (**1985**) APACHE II: a severity of disease classification system. *Crit. Care Med.*, **13**, 818–829.
21. Esper, A., Burnham, E.L. and Moss, M. (**2006**) The effect of alcohol abuse on ARDS and multiple organ dysfunction. *Minerva. Anestesiol.*, **72**, 375–381.
22. LaCroix, A.Z., Lipson, S., Miles, T.P. and White, L. (**1989**) Prospective study of pneumonia hospitalizations and mortality of U.S. older people: the role of chronic conditions, health behaviors, and nutritional status. *Public Health Rep.*, **104**, 350–360.
23. Quartin, A.A., Schein, R.M., Kett, D.H. and Peduzzi, P.N. (**1997**) Magnitude and duration of the effect of sepsis on survival: Department of Veterans Affairs Systemic Sepsis Cooperative Studies Group. *JAMA*, **277**, 1058–1063.
24. Jaagosild, P., Dawson, N.V., Thomas, C., *et al.* (**1998**) Outcomes of acute exacerbation of severe congestive heart failure: quality of life, resource use, and survival. SUPPORT Investigators: The Study to Understand Prognosis and Preferences for Outcomes and Risks of Treatments. *Arch. Intern. Med.*, **158**, 1081–1089.
25. Clermont, G., Angus, D.C., Linde-Zwirble, W.T., Griffin, M.F., Fine, M.J. and Pinsky, M.R. (**2002**) Does acute organ dysfunction predict patient-centered outcomes? *Chest*, **121**, 1963–1971.
26. Angus, D.C., Laterre, P.F., Helterbrand, J., *et al.* (**2004**) The effect of drotrecogin alfa (activated) on long-term survival after severe sepsis. *Crit. Care Med.*, **32**, 2199–2206.
27. Yende, S. and Angus, D.C. (**2007**) Long-term outcomes from sepsis. *Curr. Infect. Dis. Rep.*, **9**, 382–386.
28. Angus, D.C., Linde-Zwirble, W.T., Lidicker, J., Clermont, G., Carcillo, J. and Pinsky, M.R. (**2001**) Epidemiology of severe sepsis in the United States: analysis of incidence, outcome, and associated costs of care. *Crit. Care Med.*, **29**, 1303–1310.
29. Dhainaut, J.F., Claessens, Y.E., Janes, J. and Nelson, D.R. (**2005**) Underlying disorders and their impact on the host response to infection. *Clin. Infect. Dis.*, **41**(Suppl 7), S481–489.
30. Turnbull, I.R., Wlzorek, J.J., Osborne, D., Hotchkiss, R.S., Coopersmith, C.M. and Buchman, T.G. (**2003**) Effects of age on mortality and antibiotic efficacy in cecal ligation and puncture. *Shock*, **19**, 310–313.
31. Saito, H., Sherwood, E.R., Varma, T.K. and Evers, B.M. (**2003**) Effects of aging on mortality, hypothermia, and cytokine induction in mice with endotoxemia or sepsis. *Mech. Ageing Dev.*, **124**, 1047–1058.
32. Turnbull, I.R., Buchman, T.G., Javadi, P., Woolsey, C.A., Hotchkiss, R.S., Karl, I.E. and Coopersmith, C.M. (**2004**) Age disproportionately increases sepsis-induced apoptosis in the spleen and gut epithelium. *Shock*, **22**, 364–368.
33. Miyaji, T., Hu, X., Yuen, P.S., Muramatsu, Y., Iyer, S., Hewitt, S.M., Star, R.A. (**2003**) Ethyl pyruvate decreases sepsis-induced acute renal failure and multiple organ damage in aged mice. *Kidney Int.*, **64**, 1620–1631.
34. Chang, C.K. and LoCicero, J., 3rd. (**2004**) Overexpressed nuclear factor kappaB correlates with enhanced expression of interleukin-1beta and inducible nitric oxide synthase in aged murine lungs to endotoxic

stress. *Ann. Thorac. Surg.*, **77**, 1222–1227; discussion 1227.

35. Chorinchath, B.B., Kong, L.Y., Mao, L. and McCallum, R.E. (**1996**) Age-associated differences in TNF-alpha and nitric oxide production in endotoxic mice. *J. Immunol.*, **156**, 1525–1530.
36. Wu, D., Ren, Z., Pae, M., Guo, W., Cui, X., Merrill, A.H. and Meydani, S.N. (**2007**) Aging up-regulates expression of inflammatory mediators in mouse adipose tissue. *J. Immunol.*, **179**, 4829–4839.
37. Rutkute, K., Karakashian, A.A., Giltiay, N.V., Dobierzewska, A. and Nikolova-Karakashian, M.N. (**2007**) Aging in rat causes hepatic hyperresposiveness to interleukin-1beta which is mediated by neutral sphingomyelinase-2. *Hepatology*, **46**, 1166–1176.
38. Tateda, K., Matsumoto, T., Miyazaki, S. and Yamaguchi, K. (**1996**) Lipopolysaccharide-induced lethality and cytokine production in aged mice. *Infect. Immun.*, **64**, 769–774.
39. Rink, L., Cakman, I. and Kirchner, H. (**1998**) Altered cytokine production in the elderly. *Mech. Ageing Dev.*, **102**, 199–209.
40. Marik, P.E. and Zaloga, G.P. (**2001**) The effect of aging on circulating levels of proinflammatory cytokines during septic Shock Norasept II Study Investigators. *J. Am. Geriatr. Soc.*, **49**, 5–9.
41. Krabbe, K.S., Bruunsgaard, H., Hansen, C.M., *et al.* (**2001**) Ageing is associated with a prolonged fever response in human endotoxemia. *Clin. Diagn. Lab. Immunol.*, **8**, 333–338.
42. Schroder, J., Kahlke, V., Staubach, K.H., Zabel, P. and Stuber, F. (**1998**) Gender differences in human sepsis. *Arch. Surg.*, **133**, 1200–1205.
43. Hubacek, J.A., Stuber, F., Frohlich, D., *et al.* (**2001**) Gene variants of the bactericidal/permeability increasing protein and lipopolysaccharide binding protein in sepsis patients, gender-specific genetic predisposition to sepsis. *Crit. Care Med.*, **29**, 557–561.
44. Schroder, J., Kahlke, V., Book, M. and Stuber, F. (**2000**) Gender differences in sepsis: genetically determined? *Shock*, **14**, 307–310; discussion 310–303.
45. Zellweger, R., Wichmann, M.W., Ayala, A., Stein, S., DeMaso, C.M. and Chaudry, I.H. (**1997**) Females in proestrus state maintain splenic immune functions and tolerate sepsis better than males. *Crit. Care Med.*, **25**, 106–110.
46. Angele, M.K., Wichmann, M.W., Ayala, A., Cioffi, W.G. and Chaudry, I.H. (**1997**) Testosterone receptor blockade after hemorrhage in males. Restoration of the depressed immune functions and improved survival following subsequent sepsis. *Arch. Surg.*, **132**, 1207–1214.
47. Pasche, B., Kalaydjiev, S., Franz, T.J., *et al.* (**2005**) Sex-dependent susceptibility to Listeria monocytogenes infection is mediated by differential interleukin-10 production. *Infect. Immun.*, **73**, 5952–5960.
48. Wichmann, M.W., Inthorn, D., Andress, H.J. and Schildberg, F.W. (**2000**) Incidence and mortality of severe sepsis in surgical intensive care patients: the influence of patient gender on disease process and outcome. *Intensive Care Med.*, **26**, 167–172.
49. Offner, P.J., Moore, E.E. and Biffl, W.L. (**1999**) Male gender is a risk factor for major infections after surgery. *Arch. Surg.*, **134**, 935–938; discussion 938–940.
50. Crabtree, T.D., Pelletier, S.J., Gleason, T.G., Pruett, T.L. and Sawyer, R.G. (**1999**) Gender-dependent differences in outcome after the treatment of infection in hospitalized patients. *JAMA*, **282**, 2143–2148.
51. Bindl, L., Buderus, S., Dahlem, P., *et al.* (**2003**) Gender-based differences

in children with sepsis and ARDS: the ESPNIC ARDS Database Group. *Intensive Care Med.*, **29**, 1770–1773.

52. Mostafa, G., Huynh, T., Sing, R.F., Miles, W.S., Norton, H.J. and Thomason, M.H. (**2002**) Gender-related outcomes in trauma. *J. Trauma*, **53**, 430–434; discussion 434–435.
53. George, R.L., McGwin, G., Jr., Schwacha M.G., Metzger J., Cross J.M., Chaudry I.H. and Rue L.W., 3rd. (**2005**) The association between sex and mortality among burn patients as modified by age. *J. Burn Care Rehabil.*, **26**, 416–421.
54. Eachempati, S.R., Hydo, L. and Barie, P.S. (**1999**) Gender-based differences in outcome in patients with sepsis. *Arch. Surg.*, **134**, 1342–1347.
55. Valentin, A., Jordan, B., Lang, T., Hiesmayr, M. and Metnitz, P.G. (**2003**) Gender-related differences in intensive care: a multiple-center cohort study of therapeutic interventions and outcome in critically ill patients. *Crit. Care Med.*, **31**, 1901–1907.
56. Adrie, C., Azoulay, E., Francais, A., *et al.* (**2007**) Influence of gender on the outcome of severe sepsis: a reappraisal. *Chest.*, **132**, 1786–1793.
57. Thompson, M.J., Elton, R.A., Mankad, P.A., Campanella, C., Walker, W.S., Sang, C.T. and Cameron, E.W. (**1997**) Prediction of requirement for, and outcome of, prolonged mechanical ventilation following cardiac surgery. *Cardiovasc. Surg.*, **5**, 376–381.
58. Marrie, T.J. (**1999**) Pneumococcal pneumonia: epidemiology and clinical features. *Semin. Respir. Infect.*, **14**, 227–236.
59. Peivandi, A.A., Kasper-Konig, W., Quinkenstein, E., Loos, A.H. and Dahm, M. (**2003**) Risk factors influencing the outcome after surgical treatment of complicated deep sternal wound complications. *Cardiovasc. Surg.*, **11**, 207–212.
60. Haneberg, B., Tonjum, T., Rodahl, K. and Gedde-Dahl, T.W. (**1983**) Factors preceding the onset of meningococcal disease, with special emphasis on passive smoking: symptoms of ill health. *NIPH Ann.*, **6**, 169–173.
61. Huttunen, R., Laine, J., Lumio, J., Vuento, R. and Syrjanen, J. (**2007**) Obesity and smoking are factors associated with poor prognosis in patients with bacteraemia. *BMC Infect. Dis.*, **7**, 13.
62. Moss, M., Parsons, P.E., Steinberg, K.P., *et al.* (**2003**) Chronic alcohol abuse is associated with an increased incidence of acute respiratory distress syndrome and severity of multiple organ dysfunction in patients with septic shock. *Crit. Care Med.*, **31**, 869–877.
63. Fisman, D.N., Abrutyn, E., Spaude, K.A., Kim, A., Kirchner, C. and Daley, J. (**2006**) Prior pneumococcal vaccination is associated with reduced death, complications, and length of stay among hospitalized adults with community-acquired pneumonia. *Clin. Infect. Dis.*, **42**, 1093–1101.
64. Nansen, A., Christensen, J.P., Marker, O. and Thomsen, A.R. (**1997**) Sensitization to lipopolysaccharide in mice with asymptomatic viral infection: role of T cell-dependent production of interferon-gamma. *J. Infect. Dis.*, **176**, 151–157.
65. Fejer, G., Szalay, K., Gyory, I., *et al.* (**2005**) Adenovirus infection dramatically augments lipopolysaccharide-induced TNF production and sensitizes to lethal shock. *J. Immunol.*, **175**, 1498–1506.
66. McCullers, J.A. and Rehg, J.E. (**2002**) Lethal synergism between influenza virus and Streptococcus pneumoniae: characterization of a mouse model and the role of platelet-activating factor receptor. *J. Infect. Dis.*, **186**, 341–350.
67. Alonso, J.M., Guiyoule, A., Zarantonelli, M.L., *et al.* (**2003**) A model of meningococcal bacteremia after respiratory superinfection in

influenza A virus-infected mice. *FEMS Microbiol. Lett.*, **222**, 99–106.

68. Davidson, M., Parkinson, A.J., Bulkow, L.R., Fitzgerald, M.A., Peters, H.V. and Parks, D.J. (**1994**) The epidemiology of invasive pneumococcal disease in Alaska, 1986–1990 – ethnic differences and opportunities for prevention. *J. Infect. Dis.*, **170**, 368–376.
69. Claydon, J.E., Mitton, C., Sankaran, K. and Lee, S.K. (**2007**) Ethnic differences in risk factors for neonatal mortality and morbidity in the neonatal intensive care unit. *J. Perinatol.*, **27**, 448–452.
70. Pelletier, S.J., Isaacs, R.B., Raymond, D.P., *et al.* (**2004**) Ethnic disparities in outcome from posttransplant infections. *Shock*, **22**, 197–203.
71. Hill, P.C., Birch, M., Chambers, S., *et al.* (**2001**) Prospective study of 424 cases of Staphylococcus aureus bacteraemia: determination of factors affecting incidence and mortality. *Intern. Med. J.*, **31**, 97–103.
72. Barnato, A.E., Alexander, S., Linde-Zwirble, W.T. and Angus, D.C. (**2007**) Racial variation in the incidence, care, and outcomes of severe sepsis. *Am. J. Respir. Crit. Care Med.*, **177**, 279–284.
73. Dombrovskiy, V.Y., Martin, A.A., Sunderram, J. and Paz, H.L. (**2007**) Occurrence and outcomes of sepsis: influence of race. *Crit. Care Med.*, **35**, 763–768.
74. Ibrahim, E.H., Sherman, G., Ward, S., Fraser, V.J. and Kollef, M.H. (**2000**) The influence of inadequate antimicrobial treatment of bloodstream infections on patient outcomes in the ICU setting. *Chest*, **118**, 146–155.
75. Alvarez-Lerma, F. (**1996**) Modification of empiric antibiotic treatment in patients with pneumonia acquired in the intensive care unit. ICU-Acquired Pneumonia Study Group. *Intensive Care Med.*, **22**, 387–394.
76. Garnacho-Montero, J., Garcia-Garmendia, J.L., Barrero-Almodovar, A., Jimenez-Jimenez, F.J., Perez-Paredes, C. and Ortiz-Leyba, C. (**2003**) Impact of adequate empirical antibiotic therapy on the outcome of patients admitted to the intensive care unit with sepsis. *Crit. Care Med.*, **31**, 2742–2751.
77. Valles, J., Rello, J., Ochagavia, A., Garnacho, J. and Alcala, M.A. (**2003**) Community-acquired bloodstream infection in critically ill adult patients: impact of shock and inappropriate antibiotic therapy on survival. *Chest*, **123**, 1615–1624.
78. Garnacho-Montero, J., Aldabo-Pallas, T., Garnacho-Montero, C., Cayuela, A., Jimenez, R., Barroso, S. and Ortiz-Leyba, C. (**2006**) Timing of adequate antibiotic therapy is a greater determinant of outcome than are TNF and IL-10 polymorphisms in patients with sepsis. *Crit. Care*, **10**, R111.
79. Bates, D.W., Cullen, D.J., Laird, N., *et al.* (**1995**) Incidence of adverse drug events and potential adverse drug events: Implications for prevention. ADE Prevention Study Group. *JAMA*, **274**, 29–34.
80. Garrouste-Orgeas, M., Timsit, J., Soufir, L., *et al.* (**2008**) Impact of adverse events on the outcomes in ICU patients *Crit. Care Med.*, **36**, 2041–2047.
81. Altemeier, W.A., Matute-Bello, G., Gharib, S.A., Glenny, R.W., Martin, T.R. and Liles, W.C. (**2005**) Modulation of lipopolysaccharide-induced gene transcription and promotion of lung injury by mechanical ventilation. *J. Immunol.*, **175**, 3369–3376.
82. Dhanireddy, S., Altemeier, W.A., Matute-Bello, G., O'Mahony, D.S., Glenny, R.W., Martin, T.R. and Liles, W.C. (**2006**) Mechanical ventilation induces inflammation, lung injury, and extra-pulmonary organ dysfunction in experimental pneumonia. *Lab. Invest.*, **86**, 790–799.
83. Embriaco, N., Azoulay, E., Barrau, K., Kentish, N., Pochard, F., Loundou, A.

and Papazian, L. (**2007**) High level of burnout in intensivists: prevalence and associated factors. *Am. J. Respir. Crit. Care Med.*, **175**, 686–692.
84. Poncet, M.C., Toullic, P., Papazian, L., *et al.* (**2007**) Burnout syndrome in critical care nursing staff. *Am. J. Respir. Crit. Care Med.*, **175**, 698–704.
85. McCue, J.D. (**1982**) The effects of stress on physicians and their medical practice. *N. Engl. J. Med.*, **306**, 458–463.
86. Saposnik, G., Baibergenova, A., Bayer, N. and Hachinski, V. (**2007**) Weekends: a dangerous time for having a stroke? *Stroke*, **38**, 1211–1215.
87. Asch, S.M., Kerr, E.A., Keesey, J., Adams, J.L., Setodji, C.M., Malik, S. and McGlynn, E.A. (**2006**) Who is at greatest risk for receiving poor-quality health care? *N. Engl. J. Med.*, **354**, 1147–1156.
88. Denlinger, L.C., Angelini, G., Schell, K., *et al.* (**2005**) Detection of human P2X7 nucleotide receptor polymorphisms by a novel monocyte pore assay predictive of alterations in lipopolysaccharide-induced cytokine production. *J. Immunol.*, **174**, 4424–4431.
89. Stuber, F., Klaschik, S., Lehmann, L.E., Schewe, J.C., Weber, S. and Book, M. (**2005**) Cytokine promoter polymorphisms in severe sepsis. *Clin. Infect. Dis.*, **41**(Suppl 7), S416–420.
90. de Craen, A.J., Posthuma, D., Remarque, E.J., van den Biggelaar, A.H., Westendorp, R.G. and Boomsma, D.I. (**2005**) Heritability estimates of innate immunity: an extended twin study. *Genes Immun.*, **6**, 167–170.
91. Kellum, J.A., Kong, L., Fink, M.P., *et al.* (**2007**) Understanding the inflammatory cytokine response in pneumonia and sepsis: results of the Genetic and Inflammatory Markers of Sepsis (GenIMS) Study. *Arch. Intern. Med.*, **167**, 1655–1663.
92. Terry, C.F., Loukaci, V. and Green, F.R. (**2000**) Cooperative influence of genetic polymorphisms on interleukin 6 transcriptional regulation. *J. Biol. Chem.*, **275**, 18138–18144.
93. Eskdale, J., Gallagher, G., Verweij, C.L., Keijsers, V., Westendorp, R.G. and Huizinga, T.W. (**1998**) Interleukin 10 secretion in relation to human IL-10 locus haplotypes. *Proc. Natl. Acad. Sci. USA*, **95**, 9465–9470.
94. Jacob, C.O., Fronek, Z., Lewis, G.D., Koo, M., Hansen, J.A. and McDevitt, H.O. (**1990**) Heritable major histocompatibility complex class II-associated differences in production of tumor necrosis factor alpha: relevance to genetic predisposition to systemic lupus erythematosus. *Proc. Natl. Acad. Sci. USA*, **87**, 1233–1237.
95. Westendorp, R.G., Langermans, J.A., de Bel, C.E., Meinders, A.E., Vandenbroucke, J.P., van Furth, R. and van Dissel, J.T. (**1995**) Release of tumor necrosis factor: an innate host characteristic that may contribute to the outcome of meningococcal disease. *J. Infect. Dis.*, **171**, 1057–1060.
96. Nadel, S., Newport, M.J., Booy, R. and Levin, M. (**1996**) Variation in the tumor necrosis factor-alpha gene promoter region may be associated with death from meningococcal disease. *J. Infect. Dis.*, **174**, 878–880.
97. Mira, J.P., Cariou, A., Grall, F., *et al.* (**1999**) Association of TNF2, a TNF-alpha promoter polymorphism, with septic shock susceptibility and mortality: a multicenter study. *JAMA*, **282**, 561–568.
98. O'Keefe, G.E., Hybki, D.L. and Munford, R.S. (**2002**) The G->A single nucleotide polymorphism at the -308 position in the tumor necrosis factor-alpha promoter increases the risk for severe sepsis after trauma. *J. Trauma*, **52**, 817–825; discussion 825–816.
99. Waterer, G.W., Quasney, M.W., Cantor, R.M. and Wunderink, R.G. (**2001**) Septic shock and respiratory failure in community-acquired

pneumonia have different TNF polymorphism associations. *Am. J. Respir. Crit. Care Med.*, **163**, 1599–1604.

100. Gordon, A.C., Lagan, A.L., Aganna, E., *et al.* (**2004**) TNF and TNFR polymorphisms in severe sepsis and septic shock: a prospective multicentre study. *Genes Immun.*, **5**, 631–640.

101. Jessen, K.M., Lindboe, S.B., Petersen, A.L., Eugen-Olsen, J. and Benfield, T. (**2007**) Common TNF-alpha, IL-1 beta, PAI-1, uPA, CD14 and TLR4 polymorphisms are not associated with disease severity or outcome from Gram negative sepsis. *BMC Infect. Dis.*, **7**, 108.

102. Bajwa, E.K., Yu, C.L., Gong, M.N., Thompson, B.T. and Christiani, D.C. (**2007**) Pre–B-cell colony-enhancing factor gene polymorphisms and risk of acute respiratory distress syndrome. *Crit. Care Med.*, **35**, 1290–1295.

103. Wattanathum, A., Manocha, S., Groshaus, H., Russell, J.A. and Walley, K.R. (**2005**) Interleukin-10 haplotype associated with increased mortality in critically ill patients with sepsis from pneumonia but not in patients with extrapulmonary sepsis. *Chest*, **128**, 1690–1698.

104. Jaber, B.L., Rao, M., Guo, D., Balakrishnan, V.S., Perianayagam, M.C., Freeman, R.B., Pereira, B.J. (**2004**) Cytokine gene promoter polymorphisms and mortality in acute renal failure. *Cytokine*, **25**, 212–219.

105. Yano, K., Liaw, P.C., Mullington, J.M., *et al.* (**2006**) Vascular endothelial growth factor is an important determinant of sepsis morbidity and mortality. *J. Exp. Med.*, **203**, 1447–1458.

106. Lorenz, E., Mira, J.P., Frees, K.L. and Schwartz, D.A. (**2002**) Relevance of mutations in the TLR4 receptor in patients with gram-negative septic shock. *Arch. Intern. Med.*, **162**, 1028–1032.

107. Kiechl, S., Lorenz, E., Reindl, M., *et al.* (**2002**) Toll-like receptor 4 polymorphisms and atherogenesis. *N. Engl. J. Med.*, **347**, 185–192.

108. Feterowski, C., Emmanuilidis, K., Miethke, T., *et al.* (**2003**) Effects of functional Toll-like receptor-4 mutations on the immune response to human and experimental sepsis. *Immunology*, **109**, 426–431.

109. Smirnova, I., Mann, N., Dols, A., Derkx, H.H., Hibberd, M.L., Levin, M. and Beutler, B. (**2003**) Assay of locus-specific genetic load implicates rare Toll-like receptor 4 mutations in meningococcal susceptibility. *Proc. Natl. Acad. Sci. USA*, **100**, 6075–6080.

110. Arbour, N.C., Lorenz, E., Schutte, B.C., *et al.* (**2000**) TLR4 mutations are associated with endotoxin hyporesponsiveness in humans. *Nat. Genet.*, **25**, 187–191.

111. von Aulock, S., Schroder, N.W., Gueinzius, K., *et al.* (**2003**) Heterozygous toll-like receptor 4 polymorphism does not influence lipopolysaccharide-induced cytokine release in human whole blood. *J. Infect. Dis.*, **188**, 938–943.

112. Yoon, H.J., Choi, J.Y., Kim, C.O., *et al.* (**2006**) Lack of Toll-like receptor 4 and 2 polymorphisms in Korean patients with bacteremia. *J. Korean. Med. Sci.*, **21**, 979–982.

113. Sutherland, A.M., Walley, K.R. and Russell, J.A. (**2005**) Polymorphisms in CD14, mannose-binding lectin, and Toll-like receptor-2 are associated with increased prevalence of infection in critically ill adults. *Crit. Care Med.*, **33**, 638–644.

114. Nakada, T.A., Hirasawa, H., Oda, S., *et al.* (**2005**) Influence of toll-like receptor 4, CD14, tumor necrosis factor, and interleukine-10 gene polymorphisms on clinical outcome in Japanese critically ill patients. *J. Surg. Res.*, **129**, 322–328.

115. Lorenz, E., Mira, J.P., Cornish, K.L., Arbour, N.C. and Schwartz, D.A. (**2000**) A novel polymorphism in the toll-like receptor 2 gene and its potential association with staphylococcal infection. *Infect. Immun.*, **68**, 6398–6401.

116. Moore, C.E., Segal, S., Berendt, A.R., Hill, A.V. and Day, N.P. (**2004**) Lack of association between Toll-like receptor 2 polymorphisms and susceptibility to severe disease caused by *Staphylococcus aureus*. *Clin. Diagn. Lab. Immunol.*, **11**, 1194–1197.
117. Marsik, C., Endler, G., Halama, T., *et al.* (**2006**) Polymorphism in the tissue factor region is associated with basal but not endotoxin-induced tissue factor-mRNA levels in leukocytes. *J. Thromb. Haemost.*, **4**, 745–749.
118. Yende, S., Angus, D.C., Ding, J., *et al.* (**2007**) 4G/5G PAI-1 Polymorphisms and haplotypes are associated with pneumonia. *Am. J. Respir. Crit. Care Med.*, **176**, 1129–1137.
119. Jilma, B., Marsik, C., Kovar, F., Wagner, O.F., Jilma-Stohlawetz, P. and Endler, G. (**2005**) The single nucleotide polymorphism Ser128Arg in the E-selectin gene is associated with enhanced coagulation during human endotoxemia. *Blood*, **105**, 2380–2383.
120. Bredius, R.G., Derkx, B.H., Fijen, C.A., *et al.* (**1994**) Fc gamma receptor IIa (CD32) polymorphism in fulminant meningococcal septic shock in children. *J. Infect. Dis.*, **170**, 848–853.
121. Medvedev, A.E., Lentschat, A., Kuhns, D.B., *et al.* (**2003**) Distinct mutations in IRAK-4 confer hyporesponsiveness to lipopolysaccharide and interleukin-1 in a patient with recurrent bacterial infections. *J. Exp. Med.*, **198**, 521–531.
122. Khor, C.C., Chapman, S.J., Vannberg, F.O., *et al.* (**2007**) A Mal functional variant is associated with protection against invasive pneumococcal disease, bacteremia, malaria and tuberculosis. *Nat. Genet.*, **39**, 523–528.
123. Baudouin, S.V., Saunders, D., Tiangyou, W., *et al.* (**2005**) Mitochondrial DNA and survival after sepsis: a prospective study. *Lancet*, **366**, 2118–2121.
124. Barber, R.C., Chang, L.Y., Arnoldo, B.D., Purdue, G.F., Hunt, J.L., Horton, J.W. and Aragaki, C.C. (**2006**) Innate immunity SNPs are associated with risk for severe sepsis after burn injury. *Clin. Med. Res.*, **4**, 250–255.
125. Stuber, F., Petersen, M., Bokelmann, F. and Schade, U. (**1996**) A genomic polymorphism within the tumor necrosis factor locus influences plasma tumor necrosis factor-alpha concentrations and outcome of patients with severe sepsis. *Crit. Care Med.*, **24**, 381–384.
126. Majetschak, M., Flohe, S., Obertacke, U., *et al.* (**1999**) Relation of a TNF gene polymorphism to severe sepsis in trauma patients. *Ann. Surg.*, **230**, 207–214.
127. Fang, X.M., Schroder, S., Hoeft, A. and Stuber, F. (**1999**) Comparison of two polymorphisms of the interleukin-1 gene family: interleukin-1 receptor antagonist polymorphism contributes to susceptibility to severe sepsis. *Crit. Care Med.*, **27**, 1330–1334.
128. Read, R.C., Camp, N.J., di Giovine, F.S., *et al.* (**2000**) An interleukin-1 genotype is associated with fatal outcome of meningococcal disease. *J. Infect. Dis.*, **182**, 1557–1560.
129. Ma, P., Chen, D., Pan, J. and Du, B. (**2002**) Genomic polymorphism within interleukin-1 family cytokines influences the outcome of septic patients. *Crit. Care Med.*, **30**, 1046–1050.
130. Watanabe, E., Hirasawa, H., Oda, S., *et al.* (**2005**) Cytokine-related genotypic differences in peak interleukin-6 blood levels of patients with SIRS and septic complications. *J. Trauma*, **59**, 1181–1189; discussion 1189–1190.
131. Stassen, N.A., Breit, C.M., Norfleet, L.A. and Polk, H.C., Jr. (**2003**) IL-18 promoter polymorphisms correlate with the development of post-injury sepsis. *Surgery*, **134**, 351–356.

132. Marshall, R.P., Webb, S., Hill, M.R., Humphries, S.E. and Laurent, G.J. (**2002**) Genetic polymorphisms associated with susceptibility and outcome in ARDS. *Chest*, **121**, 68S–69S.
133. Stafford-Smith, M., Podgoreanu, M., Swaminathan, M., *et al.* (**2005**) Association of genetic polymorphisms with risk of renal injury after coronary bypass graft surgery. *Am. J. Kidney Dis.*, **45**, 519–530.
134. Sutherland, A.M., Walley, K.R., Manocha, S. and Russell, J.A. (**2005**) The association of interleukin 6 haplotype clades with mortality in critically ill adults. *Arch. Intern. Med.*, **165**, 75–82.
135. Schluter, B., Raufhake, C., Erren, M., *et al.* (**2002**) Effect of the interleukin-6 promoter polymorphism (-174 G/C) on the incidence and outcome of sepsis. *Crit. Care Med.*, **30**, 32–37.
136. Stassen, N.A., Leslie-Norfleet, L.A., Robertson, A.M., Eichenberger, M.R. and Polk, H.C., Jr. (**2002**) Interferon-gamma gene polymorphisms and the development of sepsis in patients with trauma. *Surgery*, **132**, 289–292.
137. Gao, L., Flores, C., Fan-Ma, S., *et al.* (**2007**) Macrophage migration inhibitory factor in acute lung injury: expression, biomarker, and associations. *Transl. Res.*, **150**, 18–29.
138. Villar, J., Perez-Mendez, L., Flores, C., *et al.* (**2007**) A CXCL2 polymorphism is associated with better outcomes in patients with severe sepsis. *Crit. Care Med.*, **35**, 2292–2297.
139. Hildebrand, F., Stuhrmann, M., van Griensven, M., Meier, S., Hasenkamp, S., Krettek, C. and Pape, H.C. (**2007**) Association of IL-8-251A/T polymorphism with incidence of Acute Respiratory Distress Syndrome (ARDS) and IL-8 synthesis after multiple trauma. *Cytokine*, **37**, 192–199.
140. Gallagher, P.M., Lowe, G., Fitzgerald, T., Bella, A., Greene, C.M., McElvaney, N.G. and O'Neill, S.J. (**2003**) Association of IL-10 polymorphism with severity of illness in community acquired pneumonia. *Thorax*, **58**, 154–156.
141. Schaaf, B.M., Boehmke, F., Esnaashari, H., *et al.* (**2003**) Pneumococcal septic shock is associated with the interleukin-10-1082 gene promoter polymorphism. *Am. J. Respir. Crit. Care Med.*, **168**, 476–480.
142. Lowe, P.R., Galley, H.F., Abdel-Fattah, A. and Webster, N.R. (**2003**) Influence of interleukin-10 polymorphisms on interleukin-10 expression and survival in critically ill patients. *Crit. Care Med.*, **31**, 34–38.
143. Schroder, O., Laun, R.A., Held, B., Ekkernkamp, A. and Schulte, K.M. (**2004**) Association of interleukin-10 promoter polymorphism with the incidence of multiple organ dysfunction following major trauma: results of a prospective pilot study. *Shock*, **21**, 306–310.
144. Gong, M.N., Thompson, B.T., Williams, P.L., Zhou, W., Wang, M.Z., Pothier, L. and Christiani, D.C. (**2006**) Interleukin-10 polymorphism in position -1082 and acute respiratory distress syndrome. *Eur. Respir. J.*, **27**, 674–681.
145. Stanilova, S.A., Miteva, L.D., Karakolev, Z.T. and Stefanov, C.S. (**2006**) Interleukin-10-1082 promoter polymorphism in association with cytokine production and sepsis susceptibility. *Intensive Care Med.*, **32**, 260–266.
146. Arnalich, F., Lopez-Maderuelo, D., Codoceo, R., *et al.* (**2002**) Interleukin-1 receptor antagonist gene polymorphism and mortality in patients with severe sepsis. *Clin. Exp. Immunol.*, **127**, 331–336.
147. Endler, G., Marculescu, R., Starkl, P., *et al.* (**2006**) Polymorphisms in the interleukin-1 gene cluster in children and young adults with systemic meningococcemia. *Clin. Chem.*, **52**, 511–514.
148. Agnese, D.M., Calvano, J.E., Hahm, S.J., Coyle, S.M., Corbett, S.A., Calvano, S.E. and

Lowry, S.F. (**2002**) Human toll-like receptor 4 mutations but not CD14 polymorphisms are associated with an increased risk of gram-negative infections. *J. Infect. Dis.*, **186**, 1522–1525.

149. Read, R.C., Pullin, J., Gregory, S., *et al.* (**2001**) A functional polymorphism of toll-like receptor 4 is not associated with likelihood or severity of meningococcal disease. *J. Infect. Dis.*, **184**, 640–642.

150. Gu, W., Shan, Y.A., Zhou, J., *et al.* (**2007**) Functional significance of gene polymorphisms in the promoter of myeloid differentiation-2. *Ann. Surg.*, **246**, 151–158.

151. McDaniel, D.O., Hamilton, J., Brock, M., *et al.* (**2007**) Molecular analysis of inflammatory markers in trauma patients at risk of postinjury complications. *J. Trauma*, **63**, 147–157; discussion 157–148.

152. Hubacek, J.A., Stuber, F., Frohlich, D., Book, M., Wetegrove, S., Rothe, G. and Schmitz, G. (**2000**) The common functional C(-159)T polymorphism within the promoter region of the lipopolysaccharide receptor CD14 is not associated with sepsis development or mortality. *Genes Immun.*, **1**, 405–407.

153. Gibot, S., Cariou, A., Drouet, L., Rossignol, M. and Ripoll, L. (**2002**) Association between a genomic polymorphism within the CD14 locus and septic shock susceptibility and mortality rate. *Crit. Care Med.*, **30**, 969–973.

154. Brenmoehl, J., Herfarth, H., Gluck, T., *et al.* (**2007**) Genetic variants in the NOD2/CARD15 gene are associated with early mortality in sepsis patients. *Intensive Care Med.*, **33**, 1541–1548.

155. Barber, R.C. and O'Keefe, G.E. (**2003**) Characterization of a single nucleotide polymorphism in the lipopolysaccharide binding protein and its association with sepsis. *Am. J. Respir. Crit. Care Med.*, **167**, 1316–1320.

156. Garred, P., J, J.S., Quist, L., Taaning, E. and Madsen, H.O. (**2003**) Association of mannose-binding lectin polymorphisms with sepsis and fatal outcome, in patients with systemic inflammatory response syndrome. *J. Infect. Dis.*, **188**, 1394–1403.

157. Gomi, K., Tokue, Y., Kobayashi, T., Takahashi, H., Watanabe, A., Fujita, T. and Nukiwa, T. (**2004**) Mannose-binding lectin gene polymorphism is a modulating factor in repeated respiratory infections. *Chest*, **126**, 95–99.

158. Gordon, A.C., Waheed, U., Hansen, T.K., *et al.* (**2006**) Mannose-binding lectin polymorphisms in severe sepsis: relationship to levels, incidence, and outcome. *Shock*, **25**, 88–93.

159. Smithson, A., Munoz, A., Suarez, B., *et al.* (**2007**) Association between mannose-binding lectin deficiency and septic shock following acute pyelonephritis due to *Escherichia coli*. *Clin. Vaccine Immunol.*, **14**, 256–261.

160. Gong, M.N., Zhou, W., Williams, P.L., Thompson, B.T., Pothier, L. and Christiani, D.C. (**2007**) Polymorphisms in the mannose binding lectin-2 gene and acute respiratory distress syndrome. *Crit. Care Med.*, **35**, 48–56.

161. Kerlin, B.A., Yan, S.B., Isermann, B.H., *et al.* (**2003**) Survival advantage associated with heterozygous factor V Leiden mutation in patients with severe sepsis and in mouse endotoxemia. *Blood*, **102**, 3085–3092.

162. Yan, S.B. and Nelson, D.R. (**2004**) Effect of factor V Leiden polymorphism in severe sepsis and on treatment with recombinant human activated protein C. *Crit. Care Med.*, **32**, S239–246.

163. Hermans, P.W., Hibberd, M.L., Booy, R., Daramola, O., Hazelzet, J.A., de Groot, R. and Levin, M. (**1999**) 4G/5G promoter polymorphism in the plasminogen-activator-inhibitor-1 gene and outcome of meningococcal

disease: Meningococcal Research Group. *Lancet*, **354**, 556–560.
164. Haralambous, E., Hibberd, M.L., Hermans, P.W., Ninis, N., Nadel, S. and Levin, M. (**2003**) Role of functional plasminogen-activator-inhibitor-1 4G/5G promoter polymorphism in susceptibility, severity, and outcome of meningococcal disease in Caucasian children. *Crit. Care Med.*, **31**, 2788–2793.
165. Menges, T., Hermans, P.W., Little, S.G., *et al.* (**2001**) Plasminogen-activator-inhibitor-1 4G/5G promoter polymorphism and prognosis of severely injured patients. *Lancet*, **357**, 1096–1097.
166. Kremer Hovinga, J.A., Franco, R.F., Zago, M.A., ten Cate, H., Westendorp, R.G. and Reitsma, P.H. (**2004**) A functional single nucleotide polymorphism in the thrombin-activatable fibrinolysis inhibitor (TAFI) gene associates with outcome of meningococcal disease. *J. Thromb. Haemost.*, **2**, 54–57.
167. Manocha, S., Russell, J.A., Sutherland, A.M., Wattanathum, A. and Walley, K.R. (**2007**) Fibrinogen-beta gene haplotype is associated with mortality in sepsis. *J. Infect.*, **54**, 572–577.
168. Arcaroli, J., Sankoff, J., Liu, N., Allison, D.B., Maloney, J. and Abraham, E. (**2008**) Association between urokinase haplotypes and outcome from infection-associated acute lung injury. *Intensive Care Med.*, **34**, 300–307.
169. Walley, K.R. and Russell, J.A. (**2007**) Protein C -1641 AA is associated with decreased survival and more organ dysfunction in severe sepsis. *Crit. Care Med.*, **35**, 12–17.
170. Binder, A., Endler, G., Rieger, S., Geishofer, G., Resch, B., Mannhalter, C. and Zenz, W. (**2007**) Protein C promoter polymorphisms associate with sepsis in children with systemic meningococcemia. *Hum. Genet.*, **122**, 183–190.
171. Quasney, M.W., Waterer, G.W., Dahmer, M.K., Kron, G.K., Zhang, Q., Kessler, L.A. and Wunderink, R.G. (**2004**) Association between surfactant protein B + 1580 polymorphism and the risk of respiratory failure in adults with community-acquired pneumonia. *Crit. Care Med.*, **32**, 1115–1119.
172. Marshall, R.P., Webb, S., Bellingan, G.J., *et al.* (**2002**) Angiotensin converting enzyme insertion/deletion polymorphism is associated with susceptibility and outcome in acute respiratory distress syndrome. *Am. J. Respir. Crit. Care Med.*, **166**, 646–650.
173. van de Garde, E.M., Endeman, H., Deneer, V.H., *et al.* (**2008**) Angiotensin-converting enzyme insertion/deletion polymorphism and risk and outcome of pneumonia. *Chest.*, **133**, 220–225.
174. Saleh, M., Vaillancourt, J.P., Graham, R.K., *et al.* (**2004**) Differential modulation of endotoxin responsiveness by human caspase-12 polymorphisms. *Nature*, **429**, 75–79.
175. Schroder, O., Schulte, K.M., Ostermann, P., Roher, H.D., Ekkernkamp, A. and Laun, R.A. (**2003**) Heat shock protein 70 genotypes HSPA1B and HSPA1L influence cytokine concentrations and interfere with outcome after major injury. *Crit. Care Med.*, **31**, 73–79.
176. Waterer, G.W., ElBahlawan, L., Quasney, M.W., Zhang, Q., Kessler, L.A., Wunderink, R.G. (**2003**) Heat shock protein 70-2+1267 AA homozygotes have an increased risk of septic shock in adults with community-acquired pneumonia. *Crit. Care Med.*, **31**, 1367–1372.
177. Chen, Q.X., Lv, C., Huang, L.X., Cheng, B.L., Xie, G.H., Wu, S.J. and Fang, X.M. (**2007**) Genomic variations within DEFB1 are associated with the susceptibility to and the fatal outcome of severe sepsis in Chinese Han population. *Genes Immun.*, **8**, 439–443.
178. Gao, L., Grant, A., Halder, I., *et al.* (**2006**) Novel polymorphisms in the myosin light chain kinase gene

confer risk for acute lung injury. *Am. J. Respir. Cell Mol. Biol.*, **34**, 487–495.

179. Eklund, C., Huttunen, R., Syrjanen, J., Laine, J., Vuento, R. and Hurme, M. (**2006**) Polymorphism of the C-reactive protein gene is associated with mortality in bacteraemia. *Scand. J. Infect. Dis.*, **38**, 1069–1073.
180. Platonov, A.E., Shipulin, G.A., Vershinina, I.V., Dankert, J., van de Winkel, J.G. and Kuijper, E.J. (**1998**) Association of human Fc gamma RIIa (CD32) polymorphism with susceptibility to and severity of meningococcal disease. *Clin. Infect. Dis.*, **27**, 746–750.
181. Yuan, F.F., Wong, M., Pererva, N., Keating, J., Davis, A.R., Bryant, J.A. and Sullivan, J.S. (**2003**) FcgammaRIIA polymorphisms in *Streptococcus pneumoniae* infection. *Immunol. Cell Biol.*, **81**, 192–195.
182. Yuan, F.F., Tanner, J., Chan, P.K., *et al.* (**2005**) Influence of FcgammaRIIA and MBL polymorphisms on severe acute respiratory syndrome. *Tissue Antigens*, **66**, 291–296.
183. Arcaroli, J., Silva, E., Maloney, J.P., He, Q., Svetkauskaite, D., Murphy, J.R. and Abraham, E. (**2006**) Variant IRAK-1 haplotype is associated with increased nuclear factor-kappaB activation and worse outcomes in sepsis. *Am. J. Respir. Crit. Care Med.*, **173**, 1335–1341.
184. Marzec, J.M., Christie, J.D., Reddy, S.P., *et al.* (**2007**) Functional polymorphisms in the transcription factor NRF2 in humans increase the risk of acute lung injury. *FASEB J.*, **21**, 2237–2246.

15
Altered Immune Status and Leukocytes Reprogramming

Jean-Marc Cavaillon, Minou Adib-Conquy and Christophe Adrie

15.1
Introduction

Sepsis and conditions leading to non-infectious systemic inflammatory response syndrome (SIRS) such as burns, trauma, major surgery, hemorrhage, and resuscitated cardiac arrest are associated with alterations in immune status. This was first established *in vivo* by demonstrating a reduced delayed type hypersensitivity (DTH) reaction of skin test to recall antigens. The altered DTH response in intensive care unit (ICU) patients was often linked with a higher risk for the development of sepsis and death [1]. Failure to develop DTH was associated with a reduced recruitment of T-lymphocytes in the skin [2]. In addition, ICU patients with reduced DTH made significantly less antibody to tetanus toxoid, a T-cell-dependent antigen, whereas a normal antibody response was noted with T-cell-independent antigens. Following these *in vivo* observations, numerous studies addressed the reactivity of circulating leukocytes in *in vitro* analysis. Lymphocytes display a reduced proliferation to antigens, mitogens, and in a mixed lymphocyte reaction [3, 4], enhanced apoptosis [5], and an impaired capacity to release cytokines upon *in vitro* activation [6, 7]. The altered capacity to release cytokines is not specific to lymphocytes, since monocytes, dendritic cells and neutrophils exhibit similar alterations. Furthermore, altered properties have also been reported for circulating NK cells (Figure 15.1). It is worth mentioning that most studies with monocytes have used endotoxin from Gram-negative bacteria (lipopolysaccharide, LPS) to perform these investigations. However, hyporesponsiveness of monocytes to LPS is not a generalized phenomenon, and some of the terms often used in literature such as anergy, immunodepression or immunoparalysis seem inaccurate. As a matter of fact, hyporesponsiveness was first described with LPS but is not similar with all stimuli, and is not observed for all cytokines [8]. Actually, all signaling pathways are not dysregulated. Accordingly, the term "cellular reprogramming", previously proposed by Zhang and Morrison [9] to qualify endotoxin tolerance, appears to

Sepsis and Non-infectious Systemic Inflammation. Edited by J.-M. Cavaillon and C. Adrie

ISBN: 978-3-527-31935-0

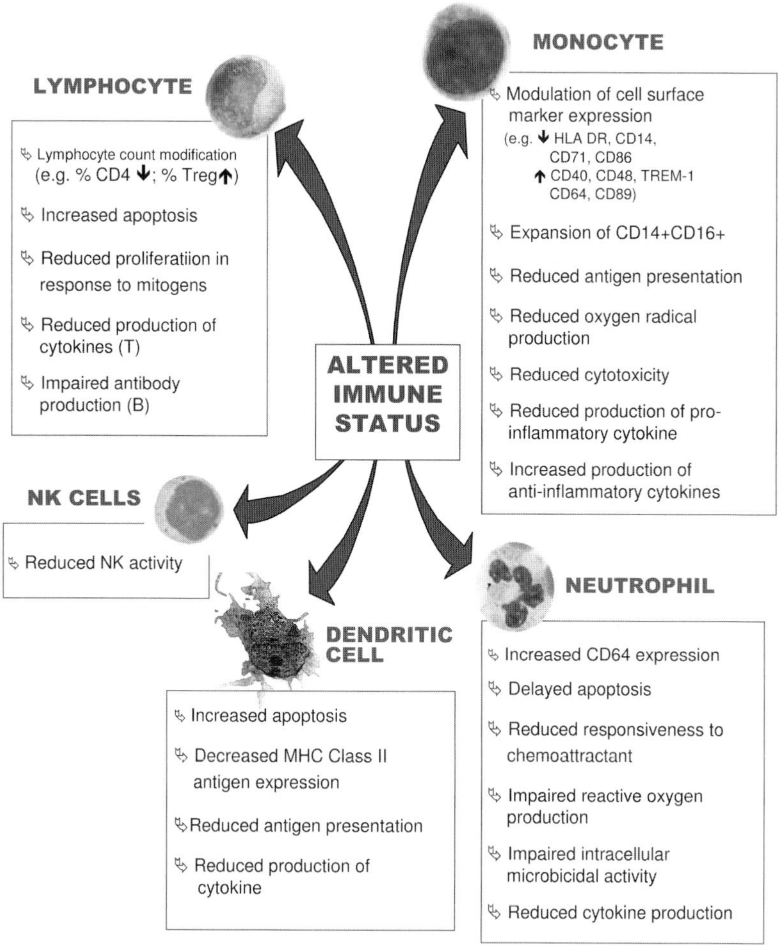

Figure 15.1 Circulating leukocytes show modified immune status in patients with sepsis or non-infectious SIRS.

be the most appropriate description of the profound alteration in the immune response observed during severe inflammation. Endotoxin tolerance was first introduced to describe the unresponsiveness of an organism or cell to a second exposure to endotoxin after a first encounter [10]. Indeed, this phenomenon shares many similarities with the events reported in sepsis and SIRS [11].

15.2
Leukocytes Surface Markers

Rapid mobilization and subsequent redistribution of leukocytes occurs during sepsis and SIRS. The nature of the leukocytes found within the bloodstream reflects: (1) the disappearance of activated cells that bind to endothelium and migrate towards inflammatory tissues, (2) the enhanced apoptosis of lymphocytes and the delayed apoptosis of neutrophils, and (3) the boost of hematopoiesis that leads to the release of freshly produced leukocytes from the bone marrow. These events lead to markedly different circulating leukocytes subpopulations as compared to healthy controls. B and T-lymphopenia, and neutrophilia are hallmarks of sepsis that can be mimicked in human volunteers receiving a bolus of LPS. Lymphopenia affects both CD4+ and CD8+ populations, and among CD4+ cells, the increased percentage of circulating regulatory T-cells (Treg, CD4+ CD25+) was indeed due to a decrease in the CD4+ CD25− lymphocyte population [12]. A decrease in circulating myeloid and plasmacytoid dendritic cells has been observed and shown to correlate with fatal outcome [13]. Among monocytes, an expansion of CD14+ CD16+ blood monocytes in human sepsis has been reported [14].

Not only does the percentage of the subpopulations differ but also the expression of cell surface markers markedly changes. For instance, as summarized in Table 15.1, numerous cell surface markers are either up- or down-regulated on monocytes. It is probable that some of the alterations in monocyte functions are the consequences of these modulations, because many of these surface markers contribute to monocyte activation. The reduced expression of HLA-DR molecules on monocytes is a hallmark of sepsis and SIRS [15, 16], and is also observed following LPS injection into human healthy volunteers [17]. Down-regulation was more pronounced in super-infected trauma patients and associated with poor outcome [15]. Decreased MHC Class II expression has been universally described in sepsis but only the evolution through the time discriminated survivors from non-survivors [16, 18]. As expected, the decreased expression of MHC Class II antigen results in an altered antigen presentation capacity [19]. Interestingly, IL-10, present in septic

Table 15.1 Modulation of cell surface markers on monocytes during sepsis and SIRS.

Down-regulated	Up-regulated
HLA-DR	TNF receptor p50
TNF receptor p75	CD40
CD14	FcγR1 (CD64), FcγR2 (CD32), FcγR3 (CD16)
Transferrin receptor (CD71)	FcαRI (CD89)
Co-stimulatory molecule B7 (CD86)	TLR2 and TLR4[a]
GM-CSF receptor	Mac-1 (CD11b)
	Tissue factor (CD142)
	Hemoglobin scavenger receptor (CD163)
	Triggering receptor expressed on myeloid cells-1 (TREM-1)

[a] Up-regulation has been reported in sepsis and after surgery; unchanged or reduced levels have also been reported in other non-infectious SIRS.

identically altered Th1 and Th2 populations was reported in patients after successful resuscitation following cardiac arrest [48] and in trauma patients [49].

15.5.2 Neutrophils

Ex vivo cytokine (IL-1β, IL-1 receptor antagonist [IL-1Ra], and IL-8) production is also reduced in neutrophils obtained from septic patients after LPS exposure compared to healthy controls [50–52]. This has been further confirmed after an i.v. administration of LPS to healthy human volunteers whose neutrophils displayed a reduced capacity to produce chemokines upon *in vitro* stimulation [53]. Furthermore, the reduced *ex vivo* cytokine production by neutrophils from septic and non-infectious SIRS patients was observed with some (e.g. LPS) but not all stimuli (e.g. heat-killed *Staphylococcus aureus*). It is worth mentioning that, in contrast to monocytes, neutrophils could not be rendered tolerant to endotoxin by a previous *in vitro* exposure to LPS [51]. On the contrary, *in vitro* LPS-pretreatment of neutrophils from healthy subjects led to enhanced IL-8 production after re-exposure to LPS. Similarly, neutrophils isolated from sepsis and SIRS patients have been shown to be primed for elastase release, and for superoxide production. Not only LPS, but also TNF, platelet activating factor (PAF) and sera from patients may contribute to the priming effect.

15.5.3 Monocytes

Monocyte reactivity to LPS has been under particular scrutiny. Upon activation with LPS, monocytes from septic and non-septic SIRS patients display a diminished capacity to release TNF, IL-1α, IL-1β, IL-6 [54], and IL-12 [55]. Not only is the cell machinery affected but there is also a reduced number of cytokine-producing cells as assessed by flow-cytometry analysis [56]. Once again similar findings were observed in healthy volunteers after LPS exposure [57]. Most interestingly, as summarized in Table 15.2, the impaired capacity of monocytes to produce inflammatory cytokines in response to LPS has been described in very numerous clinical settings, including different types of bacterial and parasitic infections, different types of non-infectious SIRS (trauma, surgery, resuscitation after cardiac arrest (RCA)) or in patients with a severe organ dysfunction (pancreatitis, heart or liver failure). However, as discussed below, the capacity of monocytes from sepsis or SIRS patients to produce cytokines, can be unchanged or even enhanced when other activators are used instead of LPS, or when other

Table 15.2 Impaired production of cytokines by human monocytes in response to LPS.

Clinical setting	Cytokines	References
Sepsis	IL-1α, IL-1β, IL-6, TNF	[54]
	IL-12	[119]
LPS administered to humans volunteers	IL-1β, IL-6, IL-8, TNF	[57]
Infections:		
Haemophilus influenzae	IL-1	[120]
Leishmania	IL-1β	[121]
Borrelia burgdorferi	TNF	[122]
Salmonella typhi	TNF, IL-1	[123]
Dengue virus	TNF, IL-1β	[124]
Schistosoma haematobium	TNF, IL-8, IL-10	[125]
Wuchereria bancrofti	IL-1β	[126]
Plasmodium falciparum	TNF	[127]
HIV	TNF	[128]
Periodontitis	IL-1β, IL-12p70	[129]
Surgery	IL-1, TNF	[130]
Cardio-pulmonary bypass	TNF	[131]
Pancreatitis	TNF	[132]
Trauma	IL-1	[133]
	TNF, IL-6, IL-8	[134]
	IL-12	[73]
Severe heart failure	IL-1, TNF	[135]
Resuscitated cardiac arrest	TNF, IL-6	[48]
Acute coronary syndrome	TNF, IL-6	[136]
Chronic liver failure	TNF	[137]
Liver cirrhosis	TNF	[138]

cytokines are studied (IL-1Ra, IL-10, macrophage migration inhibitory factor (MIF)).

15.5.4 Dendritic Cells

More recently, investigations on splenic dendritic cells have established in animal models of sepsis [58] or after trauma-hemorrhage [59] that these cells also display a reduced capacity to produce TNF, IL-6, IL-12 and IFNγ in response to TLR agonists, while the production of IL-10 was unchanged or even enhanced. In a model of lung infection post-cecal ligature and puncture (CLP), it was suggested that the defective IL-12 production by lung dendritic cells was responsible for the altered lung cell-mediated immune response [60].

15.6
Altered *Ex Vivo* Cytokine Production is not a Generalized Phenomenon

15.6.1
Nature of the Insult

The *ex vivo* altered capacity of circulating leukocytes isolated from patients with infectious or non-infectious SIRS to produce TNF after LPS exposure best characterizes reprogrammation. However this subtle modification may differ depending upon the nature of the stress. For example, the *ex vivo* production of TNF in response to LPS is not reduced beyond 2 days after surgery [61], while trauma patients display a long-lasting hyporeactivity for several days after their admission [62]. Similarly, the *ex vivo* production of IL-8 upon LPS activation in whole blood samples was shown to be lower among patients with sepsis as compared to healthy controls, whereas it was unchanged in patients who underwent surgery and cardiopulmonary by-pass [63]. The use of anesthetic drugs before the insult may limit cellular reprogramming following surgical injury as opposed to trauma, or burn. If this holds true, it would imply that neuromediators generated during the insult contribute to cellular reprogramming.

15.6.2
Nature of the Cell Culture

Experiments carried out using whole-blood assays are probably the most difficult to interpret. First, certain cytokines can be produced by different cell types present in the blood sample (e.g. IL-8, IL-1Ra), and the responsiveness of each individual population may be differently affected by the insult. For example, in sepsis, LPS-activated whole blood samples display an enhanced release of IL-1Ra as compared to healthy controls [64], whereas IL-1Ra produced by isolated neutrophils is reduced [52]. Second, the presence of immunosuppressive factor(s) within the plasma may play an inhibitory role during *ex vivo* culture that may mask or influence the individual behavior of leukocytes. For example, in sepsis, the *ex vivo* induction of TNF, IL-6 and IL-10 by heat killed *Escherichia coli* was shown to be reduced in whole blood as compared to healthy controls [65], whereas no difference was observed for isolated monocytes [66].

15.6.3
Nature of the Cytokines Produced

Most studies have reported a decreased *ex vivo* production of pro-inflammatory cytokines (i.e. TNF, IL-1α, IL-1β, IL-6, and IL-12) by leukocytes from sepsis or SIRS patients. However, the production of granulocyte-colony stimulating factor (G-CSF) was shown to be enhanced in a longitudinal analysis of LPS-activated whole blood samples from ICU patients [67]. More surprisingly, the release of MIF, which is considered to be a pro-inflammatory cytokine, was enhanced upon stimulation with different activators in patients who were not treated with glucocorticoids, whereas in most cases the *ex vivo* production was similar to controls in patients treated with glucocorticoids [68]. When anti-inflammatory cytokines (i.e. IL-1Ra, IL-10) were investigated, no modification or even an enhanced production was reported. We recently observed an enhanced production of IL-10 by monocytes from septic patients in response to both LPS and Pam_3CysSK_4 (a synthetic lipopeptide used as a specific toll-like receptor-2 (TLR2) ligand) [66]. A Similar enhanced IL-10 production was observed with circulating leukocytes after surgery or trauma [69]. In RCA patients, we observed an unaltered production of IL-10 [48]. The fact that after LPS stimulation, monocytes can display a reduced production of TNF and an unaltered or even enhanced production of IL-10, further illustrates that the sensing of LPS by monocytes is accompanied by a modification of the intracellular signaling pathways that limits the production of pro-inflammatory cytokines and maintains or favors that of anti-inflammatory ones.

15.6.4
Nature of the Activating Agent: Response to Whole Bacteria

Although the use of highly specific TLR agonists is useful to further understand the alteration of specific signaling pathways within cells from SIRS patients, the response to whole bacteria may represent a more relevant and physiological approach to monitor immune status. For instance, in contrast to LPS and Pam_3CysSK_4, the production of TNF by isolated monocytes of septic or RCA patients in response to heat-killed *Escherichia coli* or *Staphylococcus aureus* was not diminished when compared to that obtained with cells from healthy donors [66]. Similarly, isolated neutrophils from sepsis patients produced normal levels of IL-1β in response to *S. aureus*. In whole blood assays, TNF production induced by *S. aureus* was unaltered in trauma and RCA patients while contradictory results were reported for sepsis. In contrast, TNF production in whole blood assays was diminished in sepsis and trauma patients in response to *E. coli* [65, 69]. Bacteria can activate monocytes following their interaction with various receptors on the cell surface, and also after phagocytosis. In healthy volunteers and sepsis patients, TNF and IL-10 production in response to *S. aureus* was reduced when phagocytosis was prevented by cytochalasin D. These results suggest that both surface receptors and internal sensors are involved in cytokine production [70]. In addition to the surface sensors (TLR2 and TLR4), TLR9 is an intracellular receptor that recognizes bacterial DNA. Nod1 is an intracellular sensor of fragments of Gram negative-derived peptidoglycan and Nod2 detects fragments of any bacterial peptidoglycan by their basic structure, the muramyl dipeptide (MDP). We showed that Nod1 and Nod2 mRNA expression was similar in monocytes of healthy controls and patients. This may explain the maintained responsiveness to MDP and whole bacteria we observed in septic patients [66].

Cross-tolerance has been regularly reported in experimental models of endotoxin tolerance between LPS and other TLR ligands, LPS and whole bacteria or LPS and cytokines. For instance, LPS-sensitized human monocytes were also hyporeactive, in terms of TNF production, to heat-killed *Streptococcus pyogenes, S. aureus* and Zymosan [71]. Revisiting the concept of cross-tolerance with Gram-positive bacteria in a mouse model of endotoxin tolerance, we showed that the phenomenon was only transient in the blood compartment [72]: 24 h after the first injection of LPS, blood cells were tolerant to LPS but displayed a normal response to *S. aureus*, similar to observations in the blood of septic patients.

15.7 Does Altered *Ex Vivo* Cytokine Production Correlate with Severity?

It would be worth knowing whether the degree of immunodysregulation is associated with the occurrence of infection and/or the outcome. In our initial study, we showed that the recovery of a normal level of *ex vivo* cytokine production, observed among survivors during a survey of septic patients, was not found in patients who ultimately died [54]. Among trauma patients, the number of monocytes producing TNF or IL-12 in response to LPS or IFNγ+ LPS, respectively, was markedly depressed among the patients who developed adverse clinical conditions (i.e. acute respiratory distress syndrome (ARDS), sepsis, body fluid infection) [73]. A significant inverse correlation ($r = 0.68$, $p = 0.03$) was found between the number of IL-12-producing monocytes at 2 days post-injury and the duration of SIRS. Another study showed that the first quartile of ICU patients who had the lowest *ex vivo* LPS-induced TNF production required mechanical ventilation for a significantly longer period of time, and there was a trend for a longer stay in ICU and a higher frequency of infection [74]. In contrast, among patients who underwent major elective surgery, *ex vivo* TNF production was not a discriminating factor between those who developed sepsis and those who did not [75]. However, when LPS-induced IL-12p70 was investigated, the authors reported that the patients who developed post-operative sepsis had the lowest capacity to produce this cytokine on days 1 and 4 post operation, but more surprisingly on the pre-operative day as well. Furthermore, those with the lowest production developed sepsis with organ injury. Altogether, these studies suggest that the lowest *ex vivo* cytokine production is associated with the highest risk to develop infection. By analysis of the nuclear factor-κB within monocytes it was found that this was also a discriminating factor between survivors and non-survivors (see below).

15.8 Molecular Mechanisms of Altered TLR-Dependent TNF Production

Few attempts have been undertaken to decipher the intracellular and molecular mechanisms responsible for the altered responsiveness of monocytes, particularly to LPS. The negative regulation of the LPS-induced TLR4 signaling pathways has been investigated. NF-κB, is the main transcription factor required for the expression of the genes coding for inflammatory molecules. NF-κB exists as an active p65p50 heterodimer whereas its p50p50 homodimer behaves as an inhibitory form. A significant decrease in the ratio between the p65p50 heterodimer and the p50p50 homodimer was reported for monocytes of septic and trauma patients as compared to healthy volunteers [76]. The ratio was even lower in non-surviving patients. This observation resembles the situation which had been described to occur within tolerant monocytic cell lines [77]. Several other molecules have recently been described that negatively regulate LPS-activated signaling pathways and to contribute to endotoxin tolerance. Interleukin (IL)-1 receptor associated kinase (IRAK)-M prevents the dissociation of IRAK-1 and IRAK-4 from myeloid differentiation 88 (MyD88) and the formation of IRAK–TRAF6 complex, and is a negative regulator of TLR signaling. This so-called "endotoxin tolerance" is significantly reduced in IRAK-M deficient mice [78] which concurs with a recent report that *ex vivo* LPS-stimulated monocytes from septic patients express IRAK-M mRNA more rapidly than cells from healthy donors [79]. Other inhibitory molecules in the TLR pathway have been under scrutiny and may play an important role in the adaptive mechanism to inflammatory processes. Toll interacting protein (Tollip) is an adaptor protein that potently suppresses the activity of IRAKs after TLR activation [80]. Suppressor of cytokine signaling-1 (SOCS-1) is one of eight members of a family involved in the negative regulation of cytokine signal transduction pathways, particularly the JAK/STAT pathway [81]. An LPS-inducible splicing variant of MyD88, termed "MyD88 short" (MyD88s), is defective in its ability to induce IRAK phosporylation and behaves as a dominant-negative inhibitor of LPS-induced NF-κB activation [82]. Single immunoglobulin IL-1 receptor-related molecule (SIGIRR), a member of the TLR/IL-1R superfamily, is a negative modulator of the signaling induced by IL-1 or LPS [83]. The contribution of these molecules has recently been studied in sepsis and resuscitated cardiac arrest patients, a model of non-infectious SIRS. We did not observe any modification of the expression of mRNA coding for Tollip and SOCS-1 in monocytes from sepsis and RCA patients [66]. In contrast, we found a significantly enhanced expression of mRNA coding for MyD88s and SIGIRR molecules in monocytes from septic patients that was associated with a reduction of TNF in response to LPS. In RCA patients, only SIGIRR mRNA expression was enhanced, while MyD88s mRNA expression was similar to that found in healthy controls. It can be hypothesized that the enhanced expression of SIGIRR mRNA is associated with reduced responsiveness to LPS while normal levels of MyD88s may be associated with the normal responsiveness to Pam3CysSK4 observed specifically in these patients. Indeed, so far SIGIRR has been shown to inhibit the TLR4 signaling pathway but not TLR2, whereas MyD88 is shared by both receptors. Thus, despite similarities in the alteration of the immune status of sepsis and RCA patients we observed a few differences [66]. These observations may well reflect the fact that analysis was carried out at a very early stage after cardiac arrest, since sepsis develops insidiously over a long period of time without a definite starting point. Similarly, impaired activation of extracellular signal-regulated kinase (ERK) was reported in LPS-activated monocytes from septic patients but not in those from non-infectious SIRS patients [84].

15.9 Desensitizing Agents in Plasma

The presence of deactivating or immunosuppressive agents within the bloodstream most probably contributes to the hyporeactivity of circulating leukocytes. In the late 1970s it was reported that sera of burns patients were able to suppress the proliferative response of normal cells [85]. Prins *et al.* [86] showed that sera from septic patients had the capacity to down-regulate TNF production by activated monocytes from healthy donors. The fact that "septic plasma" behaves as an immunosuppressive milieu [87] is illustrated in human volunteers by the capacity of endotoxin to induce plasma inhibitors [88]. Furthermore the plasma obtained from successfully resuscitated cardiac arrest patients was able to blunt TNF production by leukocytes from healthy controls after LPS exposure [11, 48]. The effects of septic or SIRS plasma are not limited to leukocytes and their capacity to induce cardiac myocyte apoptosis and to impair mitochondrial function have also been reported.

15.10 Mediators and Cells that are Responsible for Leukocyte Reprogramming and Altered Immune Status

As summarized in Table 15.3, numerous mediators can dampen the inflammatory response and limit the capacity of leukocytes to produce pro-inflammatory cytokines in response to LPS. The fact that the response of whole blood samples of septic shock patients is enhanced in samples collected after plasma filtration and adsorption strongly suggests that blood leukocytes

Table 15.3 Immunosuppressive mediators that can act during sepsis and SIRS to alter the immune status.

Type of mediator	Mediators
Cytokines	Interleukin-10 (IL-10) Interleukin-13 (IL-13) Transforming growth factor-β (TGFβ
Released intracellular molecules	Heat Shock Proteins (HSP) Ubiquitin
Plasma factor	Ligand of TREM-2
Lipid mediators	Prostaglandins
Hormones	Glucocorticoids α-Melanocyte Stimulating Hormone (αMSH)
Neuromediators	Adrenalin Acetylcholine Vasoactive Intestinal Peptide (VIP) Pituitary Adenylate Cyclase-Activating Polypeptide (PACAP) Urocortin Adrenomodulin Corticostatin
Purine nucleoside	Adenosine

are surrounded by an inhibitory milieu [89]. Some of these plasma factors are able to neutralize endotoxin. These include soluble CD14 [90] and LPS-binding protein (LBP) [91] which favors the transfer of LPS to lipoproteins that are known to neutralize endotoxin [92]. Among the mediators that negatively modulate the response of circulating leukocytes, IL-10 and TGFβ are the main identified anti-inflammatory cytokines [93, 94]. It is worth mentioning that TGFβ can be released by apoptotic T-lymphocytes [95]. Of interest is the work of Fumeaux and Pugin [20] who showed that IL-10 was also in part responsible for the reduced expression of HLA-DR molecules on monocytes. IL-10 release can be favored by the action of catecholamines which are known to contribute to the altered responsiveness of circulating leukocytes [96]. Many other neuromediators could be responsible for the reduced reactivity of circulating leukocytes [97]; this is particularly the case for acetylcholine [98]. Other deactivating agents such as heat shock proteins, ubiquitin, ligand of TREM-2, glucocorticoids or prostaglandins may possibly contribute to alterations in immune status. However, direct evidence to links these mediators with the observed reduced *ex vivo* cytokine release are still rare in sepsis or SIRS [99].

Animal models have allowed the identification of cell subsets that can contribute to alterations in immune status. Among IL-10-producing cells,

Treg contributes to some of the aspects of sepsis-induced lymphoid immune suppression since depletion of CD4+ CD25+ cells *in vivo* before CLP, markedly restored CD4+ CD25− proliferative capacity and Th1 cytokine release without altering plasma pro-inflammatory cytokine levels. Depletion of CD25+ cells before induction of sepsis did not alter septic mortality [100]. In contrast, in the same CLP model, adoptive transfer of *in vitro*-stimulated Treg in both prevention and therapeutic modes significantly increased peritoneal TNF-alpha production, improved bacterial clearance, and survival [101]. Interleukin 10 is also expressed by heterogeneous, immature, predominantly myeloid progenitor cells such as GR-1+ CD11b+. This population is dramatically increased and remains elevated in the spleen, lymph nodes, and bone marrow during polymicrobial sepsis [102]. GR-1+ cell depletion *in vivo* prevents both the sepsis-induced augmentation of Th2 cell-dependent antibody production and depression of Th1 cell-dependent antibody production. Thus, GR-1+ CD11b+ cells contribute to sepsis-induced T cell suppression and favor Th2 polarization [102].

15.11 Compartmentalization

During sepsis and SIRS, cells derived from tissues are fully responsive to or even primed by *ex vivo* stimuli, in contrast to cells derived from hematopoietic compartments (blood, spleen, etc.), which are hyporeactive [103]. This dichotomy illustrates the concept of compartmentalization, which occurs within the whole body during sepsis. A nice example is given by the *ex vivo* production of IL-12 by dendritic cells which is reduced after sepsis in cells derived from the spleen and enhanced in cells derived from the peritoneal cavity [104]. The nature of the insult (e.g. burn, hemorrhage, trauma, peritonitis), the cellular composition of each compartment (e.g. nature of phagocytes, nature of endothelial cells), and its micro-environment (e.g. local presence of granulocyte-macrophage colony stimulating factor (GM-CSF) in the lungs, low levels of arginine in the liver, release of endotoxin from the gut), and leukocyte recruitment, have a great influence on local inflammation and tissue injury. High levels of pro-inflammatory mediators (e.g. IL-1, TNF, IFNγ, HMGB-1, MIF) produced locally and released into the bloodstream initiate remote organ injury as a consequence of organ cross-talk. Anti-inflammatory mediators predominate within the bloodstream to avoid igniting new inflammatory foci [105]. However, their presence within tissues may not always be sufficient to prevent the initiation of a deleterious inflammatory response in the different compartments [106]. From a very simplistic point of view it was suggested that during sepsis, SIRS predominates within the inflamed tissues, while in the blood the leukocytes rather exhibit a hyporeactivity as a reflection of a compensatory anti-inflammatory response syndrome (CARS); both SIRS and CARS being present concomitantly.

15.12
Reversal of Reprogramming

Reversal of endotoxin tolerance can be achieved *in vitro* with the use of IFNγ or GM-CSF [107, 108]. These cytokines prevent endotoxin tolerance induced by low but not by high doses of LPS, by inhibiting IRAK degradation and by promoting its association with MyD88 after a second LPS stimulation, which in turn leads to NF-κB activation and TNF production [109]. Similarly, *in vivo*, both IFNγ and GM-CSF restore the systemic response to LPS in endotoxin-tolerant mice [110].

Numerous experiments have demonstrated that pre-treatment with a given inflammatory cytokine primes the immune system and renders the animal more resistant to infection. These experiments have to be distinguished from those showing that this pre-treatment reverses the immune alteration due to infection or SIRS. For example, in a rat model, it was shown that IFNγ restored host defenses against *S. aureus* after hemorrhagic shock [111]. In cancer patients treated with LPS, repeated injections with IFNγ resulted in a transient attenuation of circulatory cytokines. Subcutaneous injection of IFNγ reversed the reduction of serum levels of TNF, IL-6 and GM-CSF in response to a second challenge with LPS, but did not affect IL-8 levels [112]. *In vitro* pretreatment with IFNγ of peripheral blood mononuclear cells from septic patients, led to an enhanced production of TNF in response to LPS [113]. *In vivo* treatment of patients with IFNγ after admission to the intensive care unit, led to a recovery of the capacity of the cells to respond to *ex vivo* stimulation by LPS, and the expression of HLA-DR on the surface of monocytes was also restored [113].

GM-CSF was similarly successfully used to restore immune status. Hyporesponsiveness of whole blood to LPS induced by trauma, sepsis, or cardiac surgery as assessed by TNF production, respiratory burst activity or HLA-DR expression, can be overridden *in vitro* by pre-incubation with GM-CSF [114, 115]. Administration of a daily dose of 5 μg/kg recombinant human GM-CSF over a period of 3 days was achieved in nine consecutive patients with severe sepsis and multi-organ dysfunction. It was reported that GM-CSF upregulated HLA-DR expression on monocytes and concomitantly increased the *ex vivo* whole blood TNF-alpha response to LPS [116].

Most probably other means and other cytokines will be described in the near future to restore the immune status. For example, a Siberian team [117] showed in patients with surgical sepsis that extracorporeal immunotherapy with autologous mononuclear cells treated with IL-2 resulted in a significantly improved level of HLA-DR expression on monocytes and an enhanced *ex vivo* TNF production. Cytokine-based immunotherapy notably decreased the mortality of patients with generalized surgical infection to 14.6%, which was lower than the mortality in the control group (34.5%). In a mouse model of lung infection with *Aspergillus fumigatus* after CLP, the intrapulmonary transfer of bone marrow-derived dendritic cells prevented fatal infection. This therapy reduced the overall inflammatory response and fungal growth in the lung, and promoted the balance between pro-inflammatory and suppressive cytokines in the lung [118].

15.13 Clinical Meaning

As mentioned above there are some data that suggest a correlation between altered immune status and increased probability of developing sepsis among ICU patients. But, there is less evidence to suggest that altered immune status is associated with increased mortality. Therefore, despite similarities in endotoxin tolerance, this observation appears to constitute a major difference. Indeed, endotoxin tolerance has been associated with increased resistance to infection and LPS-induced lethality, and protection against injuries following various insults such as ischemia–reperfusion. Of course, the human clinical setting does not correspond to a single exposure to LPS. Thus, overall, the leukocyte reprogramming observed in ICU patients is a far more complex phenomenon that endotoxin tolerance. It combines concomitantly the inhibition of some signaling pathways while others are maintained or even enhanced. What has been observed with purified LPS does not seem to parallel the observations from whole bacteria. What is true for circulating leukocytes appears quite different for leukocytes present within tissues. The results of these fascinating varieties of adaptive modifications to stress remain to be further deciphered, since the aim is to manipulate them in order to improve patient outcome.

References

1. Meakins, J..L., Pietsch, J.B., Bubenick, O., Kelly, R., Rode, H., Gordon, J.. and MacLean, L.D. (**1977**) Delayed hypersensitivity: indicator of acquired failure of host defenses in sepsis and trauma. *Ann. Surg.*, **186**, 241–250.
2. Christou, N.V., Meakins, J..L., Gordon, J., *et al.* (**1995**) The delayed hypersensitivity response and host resistance in surgical patients. 20 years later. *Ann. Surg.*, **222**, 534–546; discussion 546–538.
3. Miller, C.L. and Baker, C.C. (**1979**) Changes in lymphocyte activity after thermal injury. The role of suppressor cells. *J. Clin. Invest.*, **63**, 202–210.
4. Keane, R.M., Birmingham, W., Shatney, C.M., Winchurch, R.A and Munster, A.M. (**1983**) Prediction of sepsis in the multitraumatic patient by assays of lymphocyte responsiveness. *Surg. Gynecol Obstet.*, **156**, 163–167.
5. Le Tulzo, Y., Pangault, C., Gacouin, A., *et al.* (**2002**) Early circulating lymphocyte apoptosis in human septic shock is associated with poor outcome. *Shock*, **18**, 487–494.
6. Wood, J.J., Rodrick, M..L., O'Mahony, J.B., Palder, S.B., Saporoschetz, I., D'Eon, P. and Mannick, J.A. (**1984**) Inadequate interleukin 2 production. A fundamental immunological deficiency in patients with major burns. *Ann. Surg.*, **200**, 311–320.

7. Abraham, E. and Regan, R.F. (**1985**) The effects of hemorrhage and trauma on interleukin 2 production. *Arch. Surg.*, **120**, 1341–1344.
8. Cavaillon, J.M., Adrie, C., Fitting, C. and Adib-Conquy, M. (**2005**) Reprogramming of circulatory cells in sepsis and SIRS. *J. Endotoxin Res.*, **11**, 311–320.
9. Zhang, X. and Morrison, D.C. (**1993**) Lipopolysaccharide structure-function relationship in activation versus reprogramming of mouse peritoneal macrophages. *J. Leukoc. Biol.*, **54**, 444–450.
10. Cross, A.S. (**2002**) Endotoxin tolerance-current concepts in historical perspective. *J. Endotoxin Res.*, **8**, 83–98.
11. Cavaillon, J.M., Adrie, C., Fitting, C. and Adib-Conquy, M. (**2003**) Endotoxin tolerance: is there a clinical relevance? *J. Endotoxin Res.*, **9**, 101–107.
12. Venet, F., Pachot, A., Debard, A..L., Bohe, J., Bienvenu, J., Lepape, A. and Monneret, G. (**2004**) Increased percentage of CD4+ CD25+ regulatory T cells during septic shock is due to the decrease of CD4+CD25- lymphocytes. *Crit. Care Med.*, **32**, 2329–2331.
13. Guisset, O., Dilhuydy, M.S., Thiebaut, R., *et al.* (**2007**) Decrease in circulating dendritic cells predicts fatal outcome in septic shock. *Intensive Care Med.*, **33**, 148–152.
14. Fingerle, G., Pforte, A., Passlick, B., Blumenstein, M., Strobel, M. and Ziegler-Heitbrock, H.WL. (**1993**) The novel subset of CD14+/CD16+ blood monocytes is expanded in sepsis patients. *Blood*, **82**, 3170–3176.
15. Hershman, M.J., Cheadle, W.G., Wellhausen, S.R., Davidson, P.F. and Polk, H.C, Jr. (**1990**) Monocyte HLA-DR antigen expression characterizes clinical outcome in the trauma patient. *Br. J. Surg.*, **77**, 204–207.
16. Tschaikowsky, K., Hedwig-Geissing, M., Schiele, A., Bremer, F., Schywalsky, M. and Schuttler, J. (**2002**) Coincidence of pro- and anti-inflammatory responses in the early phase of severe sepsis: Longitudinal study of mononuclear histocompatibility leukocyte antigen-DR expression, procalcitonin, C-reactive protein, and changes in T-cell subsets in septic and postoperative patients. *Crit. Care Med.*, **30**, 1015–1023.
17. Weijer, S., Lauw, F.N., Branger, J., van den Blink, B. and van der Poll, T. (**2002**) Diminished interferon-gamma production and responsiveness after endotoxin administration to healthy humans. *J. Infect. Dis.*, **186**, 1748–1753.
18. Monneret, G., Lepape, A., Voirin, N., *et al.* (**2006**) Persisting low monocyte human leukocyte antigen-DR expression predicts mortality in septic shock. *Intensive Care Med,*, **32**, 1175–1183.
19. Ayala, A., Ertel, W. and Chaudry, I.H. (**1996**) Trauma-induced suppression of antigen presentation and expression of major histocompatibility class II antigen complex in leukocytes. *Shock*, **5**, 79–90.
20. Fumeaux, T. and Pugin, J. (**2002**) Role of interleukin-10 in the intracellular sequestration of human leukocyte antigen-DR in monocytes during septic shock. *Am. J. Respir. Crit. Care Med.*, **166**, 1475–1482.
21. Morlion, B.J., Torwesten, E., Kuhn, K.S., Puchstein, C. and Furst, P. (**2000**) Cysteinyl-leukotriene generation as a biomarker for survival in the critically ill. *Crit. Care Med.*, **28**, 3655–3658.
22. von Knethen, A., Tautenhahn, A., Link, H., Lindemann, D. and Brune, B. (**2005**) Activation-induced depletion of protein kinase C alpha provokes desensitization of monocytes/macrophages in sepsis. *J. Immunol.*, **174**, 4960–4965.
23. Stephan, F., Yang, K., Tankovic, J., *et al.* (**2002**) Impairment of polymorphonuclear neutrophil functions precedes nosocomial infections in critically ill patients. *Crit. Care Med.*, **30**, 315–322.

24. Cummings, C.J., Martin, T.R., Frevert, C.W., *et al.* (**1999**) Expression and function of the chemokine receptors CXCR1 and CXCR2 in sepsis. *J. Immunol.*, **162**, 2341–2346.
25. Solomkin, J.S., Bass, R.C., Bjornson, H.S., Tindal, C.J. and Babcock, G.F. (**1994**) Alterations of neutrophil responses to tumor necrosis factor alpha and interleukin-8 following human endotoxemia. *Infect. Immun.*, **62**, 943–947.
26. Hechtman, D.H., Cybulsky, M.I., Fuchs, H.J., Baker, J.B. and Gimbrone, M.A, Jr. (**1991**) Intravascular IL-8. Inhibitor of polymorphonuclear leukocyte accumulation at sites of acute inflammation. *J. Immunol.*, **147**, 883–892.
27. Keel, M., Ungethum, U., Steckholzer, U., Niederer, E., Hartung, T., Trentz, O. and Ertel, W. (**1997**) Interleukin-10 counterregulates proinflammatory cytokine-induced inhibition of neutrophil apoptosis during severe sepsis. *Blood*, **90**, 3356–3363.
28. Liacos, C., Katsaragakis, S., Konstadoulakis, M.M., *et al.* (**2001**) Apoptosis in cells of bronchoalveolar lavage: a cellular reaction in patients who die with sepsis and respiratory failure. *Crit. Care Med.*, **29**, 2310–2317.
29. Hotchkiss, R.S., Tinsley, K.W., Swanson, P.E., *et al.* (**2001**) Sepsis-induced apoptosis causes progressive profound depletion of B and CD4+ T lymphocytes in humans. *J. Immunol.*, **166**, 6952–6963.
30. Adrie, C., Bachelet, M., Vayssier-Taussat, M., *et al.* (**2001**) Mitochondrial membrane potential and apoptosis peripheral blood monocytes in severe human sepsis. *Am. J. Respir. Crit. Care Med.*, **164**, 389–395.
31. Hotchkiss, R.S., Tinsley, K.W., Swanson, P.E., *et al.* (**2002**) Depletion of dendritic cells, but not macrophages, in patients with sepsis. *J. Immunol.*, **168**, 2493–2500.
32. Ayala, A., Herdon, C.D., Lehman, D..L., DeMaso, C.M., Ayala, C.A. and Chaudry, I.H. (**1995**) The induction of accelerated thymic programmed cell death during polymicrobial sepsis: control by corticosteroids but not tumor necrosis factor. *Shock*, **3**, 259–267.
33. Coopersmith, C.M., Stromberg, P.E., Dunne, W.M., *et al.* (**2002**) Inhibition of intestinal epithelial apoptosis and survival in a murine model of pneumonia-induced sepsis. *JAMA*, **287**, 1716–1721.
34. Coopersmith, C.M., Chang, K.C., Swanson, P.E., *et al.* (**2002**) Overexpression of Bcl-2 in the intestinal epithelium improves survival in septic mice. *Crit. Care Med.*, **30**, 195–201.
35. Grassme, H., Kirschnek, S., Riethmueller, J., *et al.* (**2000**) CD95/CD95 ligand interactions on epithelial cells in host defense to *Pseudomonas aeruginosa*. *Science*, **290**, 527–530.
36. Hotchkiss, R.S., Tinsley, K.W., Swanson, P.E., *et al.* (**1999**) Prevention of lymphocyte cell death in sepsis improves survival in mice. *Proc. Natl. Acad. Sci. USA*, **96**, 14541–14546.
37. Braun, J.S., Novak, R., Herzog, K.H., Bodner, S.M., Cleveland, J.L. and Tuomanen, E.I. (**1999**) Neuroprotection by a caspase inhibitor in acute bacterial meningitis. *Nat. Med.*, **5**, 298–302.
38. Hotchkiss, R.S., Swanson, P.E., Knudson, C.M., *et al.* (**1999**) Overexpression of Bcl-2 in transgenic mice decreases apoptosis and improves survival in sepsis. *J. Immunol.*, **162**, 4148–4156.
39. Hotchkiss, R.S. and Nicholson, D.W. (**2006**) Apoptosis and caspases regulate death and inflammation in sepsis. *Nat. Rev. Immunol.*, **6**, 813–822.
40. Byrne, A. and Reen, D.J. (**2002**) Lipopolysaccharide induces rapid production of IL-10 by monocytes in

the presence of apoptotic neutrophils. *J. Immunol.*, **168**, 1968–1977.

41. Jia, S.H., Li, Y., Parodo, J., Kapus, A., Fan, L., Rotstein, O.D. and Marshall, J.C. (**2004**) Pre-B cell colony-enhancing factor inhibits neutrophil apoptosis in experimental inflammation and clinical sepsis. *J. Clin. Invest.*, **113**, 1318–1327.
42. Qin, S., Wang, H., Yuan, R., *et al.* (**2006**) Role of HMGB1 in apoptosis-mediated sepsis lethality. *J. Exp. Med.*, **203**, 1637–1642.
43. Wang, H., Bloom, O., Zhang, M., *et al.* (**1999**) HMG-1 as a late mediator of endotoxin lethality in mice. *Science*, **285**, 248–251.
44. Lehner, M.D., Ittner, J., Bundschuh, D.S., van Rooijen, N., Wendel, A. and Hartung, T. (**2001**) Improved innate immunity of endotoxin-tolerant mice increases resistance to *Salmonella enterica* serovar *typhimurium* infection despite attenuated cytokine response. *Infect. Immun.*, **69**, 463–471.
45. Rayhane, N., Fitting, C., Lortholary, O., Dromer, F. and Cavaillon, J.M. (**2000**) Administration of endotoxin associated with lipopolysaccharide tolerance protects mice against fungal infection. *Infect. Immun.*, **68**, 3748–3753.
46. Ayala, A., Deol, Z.K., Lehman, D..L., Herdon, C.D. and Chaudry, I.H. (**1994**) Polymicrobial sepsis but not low-dose endotoxin infusion causes decreased splenocyte IL-2/IFN-gamma release while increasing IL-4/IL-10 production. *J. Surg. Res.*, **56**, 579–585.
47. Muret, J., Marie, C., Fitting, C., Payen, D. and Cavaillon, J.M. (**2000**) *Ex vivo* T-lymphocyte derived cytokine production in SIRS patients is influenced by experimental procedures. *Shock*, **13**, 169–174.
48. Adrie, C., Adib-Conquy, M., Laurent, I., *et al.* (**2002**) Successful cardiopulmonary resuscitation after cardiac arrest as a "sepsis-like" syndrome. *Circulation*, **106**, 562–568.
49. Puyana, J.C., Pellegrini, J.D., De, A.K., Kodys, K., Silva, W.E and, Miller, C.L. (**1998**) Both T-helper-1- and T-helper-2-type lymphokines are depressed in posttrauma anergy. *J. Trauma*, **44**, 1037–1045; discussion 1045–1036.
50. McCall, C.E., Grosso-Wilmoth, L.M., LaRue, K., Guzman, R.N. and Cousart, S.L. (**1993**) Tolerance to endotoxin-induced expression of the interleukin-1 beta gene in blood neutrophils of humans with the sepsis syndrome. *J. Clin. Invest.*, **91**, 853–861.
51. Marie, C., Muret, J., Fitting, C., Losser, M.R., Payen, D. and Cavaillon, J.M. (**1998**) Reduced *ex vivo* interleukin-8 production by neutrophils in septic and nonseptic systemic inflammatory response syndrome. *Blood*, **91**, 3439–3446.
52. Marie, C., Muret, J., Fitting, C., Payen, D. and Cavaillon, J.M. (**2000**) Interleukin-1 receptor antagonist production during infectious and noninfectious systemic inflammatory response syndrome. *Crit. Care Med.*, **28**, 2277–2282.
53. Schultz, M.J., Olszyna, D.P., de Jonge, E., Verbon, A., van Deventer, S.J. and van der Poll, T. (**2000**) Reduced *ex vivo* chemokine production by polymorphonuclear cells after *in vivo* exposure of normal humans to endotoxin. *J. Infect. Dis.*, **182**, 1264–1267.
54. Munoz, C., Carlet, J., Fitting, C., Misset, B., Bleriot, J.P. and Cavaillon, J.M. (**1991**) Dysregulation of *in vitro* cytokine production by monocytes during sepsis. *J. Clin. Invest.*, **88**, 1747–1754.
55. Ertel, W., Kremer, J.P., Kenney, J., Steckholzer, U., Jarrar, D., Trentz, O. and Schildberg, F.W. (**1995**) Downregulation of proinflammatory cytokine release in whole blood from septic patients. *Blood*, **85**, 1341–1347.
56. Fumeaux, T., Dufour, J., Stern, S. and Pugin, J. (**2004**) Immune monitoring of patients with septic shock by measurement of intraleukocyte cytokines. *Intensive Care Med.*, **30**, 2028–2037.

57. Granowitz, E.V., Porat, R., Mier, J.W., *et al.* (**1993**) Intravenous endotoxin suppresses the cytokine response of peripheral blood mononuclear cells of healthy humans. *J. Immunol.*, **151**, 1637–1645.
58. Flohe, S.B., Agrawal, H., Schmitz, D., Gertz, M., Flohe, S. and Schade, F.U. (**2006**) Dendritic cells during polymicrobial sepsis rapidly mature but fail to initiate a protective Th1-type immune response. *J. Leukoc. Biol.*, **79**, 473–481.
59. Kawasaki, T., Fujimi, S., Lederer, J.A., *et al.* (**2006**) Trauma-hemorrhage induces depressed splenic dendritic cell functions in mice. *J. Immunol.*, **177**, 4514–4520.
60. Wen, H., Hogaboam, C.M., Gauldie, J. and Kunkel, S.L. (**2006**) Severe sepsis exacerbates cell-mediated immunity in the lung due to an altered dendritic cell cytokine profile. *Am. J. Pathol.*, **168**, 1940–1950.
61. Cabie, A., Fitting, C., Farkas, J.C., Laurian, C., Cormier, J.M., Carlet, J. and Cavaillon, J.M. (**1992**) Influence of surgery on in-vitro cytokine production by human monocytes. *Cytokine*, **4**, 576–580.
62. Adib-Conquy, M., Asehnoune, K., Moine, P. and Cavaillon, J.M. (**2001**) Long-term-impaired expression of nuclear factor-kappa B and I kappa B alpha in peripheral blood mononuclear cells of trauma patients. *J. Leukoc. Biol.*, **70**, 30–38.
63. Marie, C., Fitting, C., Muret, J., Payen, D. and Cavaillon, J.M. (**2000**) Interleukin 8 production in whole blood assays: is interleukin 10 responsible for the downregulation observed in sepsis? *Cytokine*, **12**, 55–61.
64. van Deuren, M., van der Ven-Jongekrijg, J., Demacker, P.N., *et al.* (**1994**) Differential expression of proinflammatory cytokines and their inhibitors during the course of meningococcal infections. *J. Infect. Dis.*, **169**, 157–161.
65. Haupt, W., Zirngibl, H., Riese, J., Stehr, A., Linde, H.J. and Hohenberger, W. (**1997**) Depression of tumor necrosis factor-alpha, interleukin-6, and interleukin-10 production: a reaction to the initial systemic hyperactivation in septic shock. *J. Invest. Surg.*, **10**, 349–355.
66. Adib-Conquy, M., Adrie, C., Fitting, C., Gattolliat, O., Beyaert, R. and Cavaillon, J.M. (**2006**) Up-regulation of MyD88s and SIGIRR, molecules inhibiting Toll-like receptor signaling, in monocytes from septic patients. *Crit. Care Med.*, **34**, 2377–2385.
67. Weiss, M., Fischer, G., Barth, E., Boneberg, E., Schneider, E.M., Georgieff, M. and Hartung, T. (**2001**) Dissociation of LPS-induced monocytic *ex vivo* production of granulocyte colony-stimulating factor (G-CSF) and TNF-alpha in patients with septic shock. *Cytokine*, **13**, 51–54.
68. Maxime, V., Fitting, C., Annane, D. and Cavaillon, J.M. (**2005**) Corticoids normalize leukocyte production of macrophage migration inhibitory factor in septic shock. *J. Infect. Dis.*, **191**, 138–144.
69. Adib-Conquy, M., Moine, P., Asehnoune, K., *et al.* (**2003**) Toll-like receptor-mediated tumor necrosis factor and interleukin-10 production differ during systemic inflammation. *Am. J. Respir. Crit. Care Med.*, **168**, 158–164.
70. Kapetanovic, R., Nahori, M.A., Balloy, V., Fitting, C., Philpott, D.J., Cavaillon, J.M. and Adib-Conquy, M. (**2007**) Contribution of phagocytosis and intracellular sensing for cytokine production by *Staphylococcus aureus*-activated macrophages. *Infect. Immun.*, **75**, 830–837.
71. Cavaillon, J.M., Pitton, C. and Fitting, C. (**1994**) Endotoxin tolerance is not a LPS-specific phenomenon: partial mimicry with IL-1, IL-10 and TGFbeta. *J. Endotoxin Res.*, **1**, 1038–1045.
72. Fitting, C., Dhawan, S. and Cavaillon, J.M. (**2004**)

Compartmentalization of tolerance to endotoxin. *J. Infect. Dis.*, **189**, 1295–1303.

73. Spolarics, Z., Siddiqi, M., Siegel, J.H., Garcia, Z.C., Stein, D.S., Denny, T. and Deitch, E.A. (**2003**) Depressed interleukin-12-producing activity by monocytes correlates with adverse clinical course and a shift toward Th2-type lymphocyte pattern in severely injured male trauma patients. *Crit. Care Med.*, **31**, 1722–1729.
74. Heagy, W., Hansen, C., Nieman, K., Cohen, M., Richardson, C., Rodriguez, J.L. and West, M.A. (**2000**) Impaired *ex vivo* lipopolysaccharide-stimulated whole blood tumor necrosis factor production may identify "septic" intensive care unit patients. *Shock*, **14**, 271–276; discussion 276–277.
75. Hensler, T., Heidecke, C.D., Hecker, H., *et al.* (**1998**) Increased susceptibility to postoperative sepsis in patients with impaired monocyte IL-12 production. *J. Immunol.*, **161**, 2655–2659.
76. Adib-Conquy, M., Adrie, C., Moine, P., *et al.* (**2000**) NF-kappaB expression in mononuclear cells of patients with sepsis resembles that observed in lipopolysaccharide tolerance. *Am. J. Respir. Crit. Care Med.*, **162**, 1877–1883.
77. Ziegler-Heitbrock, H.W., Wedel, A., Schraut, W., *et al.* (**1994**) Tolerance to lipopolysaccharide involves mobilization of nuclear factor kappa B with predominance of p50 homodimers. *J. Biol. Chem.*, **269**, 17001–17004.
78. Kobayashi, K., Hernandez, L.D., Galan, J.E., Janeway, C.A, Jr.., Medzhitov, R. and Flavell R.A. (**2002**) IRAK-M is a negative regulator of Toll-like receptor signaling. *Cell*, **110**, 191–202.
79. Escoll, P., del Fresno, C., Garcia, L., Valles, G., Lendinez, M.J., Arnalich, F. and Lopez-Collazo, E. (**2003**) Rapid up-regulation of IRAK-M expression following a second endotoxin challenge in human monocytes and in monocytes isolated from septic patients. *Biochem. Biophys. Res. Commun.*, **311**, 465–472.
80. Zhang, G Ghosh, S. (**2002**) Negative regulation of toll-like receptor-mediated signaling by Tollip. *J. Biol. Chem.*, **277**, 7059–7065.
81. Nakagawa, R., Naka, T., Tsutsui, H., *et al.* (**2002**) SOCS-1 participates in negative regulation of LPS responses. *Immunity*, **17**, 677–687.
82. Janssens, S., Burns, K., Tschopp, J. and Beyaert, R. (**2002**) Regulation of interleukin-1- and lipopolysaccharide-induced NF-kappaB activation by alternative splicing of MyD88. *Curr. Biol.*, **12**, 467–471.
83. Wald, D., Qin, J., Zhao, Z., *et al.* (**2003**) SIGIRR, a negative regulator of Toll-like receptor-interleukin 1 receptor signaling. *Nat. Immunol.*, **4**, 920–927
84. West, M.A., Koons, A., Crandall, M., Skinner, R., Worley, M. and Shapiro, M.B. (**2007**) Whole blood leukocyte mitogen activated protein kinases activation differentiates intensive care unit patients with systemic inflammatory response syndrome and sepsis. *J. Trauma*, **62**, 805–811.
85. Constantian, M.B. (**1978**) Association of sepsis with an immunosuppressive polypeptide in the serum of burn patients. *Ann. Surg.*, **188**, 209–215.
86. Prins, J.M., Kuijper, E.J., Mevissen, M..L., Speelman, P. and van Deventer, S.J. (**1995**) Release of tumor necrosis factor alpha and interleukin 6 during antibiotic killing of Escherichia coli in whole blood: influence of antibiotic class, antibiotic concentration, and presence of septic serum. *Infect. Immun.*, **63**, 2236–2242.
87. Cavaillon, J.M. (**2002**) "Septic plasma": an immunosuppressive milieu. *Am. J. Respir. Crit. Care Med.*, **166**, 1417–1418.

88. Spinas, G.A., Bloesch, D., Kaufmann, M.T., Keller, U. and Dayer, J.M. (**1990**) Induction of plasma inhibitors of interleukin 1 and TNF-alpha activity by endotoxin administration to normal humans. *Am. J. Physiol.*, **259**, R993–997.
89. Ronco, C., Brendolan, A., Lonnemann, G., *et al.* (**2002**) A pilot study of coupled plasma filtration with adsorption in septic shock. *Crit. Care Med.*, **30**, 1250–1255.
90. Kitchens, R..L., Thompson, P.A., Viriyakosol, S., O'Keefe, G.E. and Munford, R.S. (**2001**) Plasma CD14 decreases monocyte responses to LPS by transferring cell-bound LPS to plasma lipoproteins. *J. Clin. Invest.*, **108**, 485–493.
91. Vreugdenhil, A.C., Snoek, A.M., van't Veer, C., Greve, J.W. and Buurman, W.A. (**2001**) LPS-binding protein circulates in association with apoB-containing lipoproteins and enhances endotoxin-LDL/VLDL interaction. *J. Clin. Invest.*, **107**, 225–234.
92. Cavaillon, J.M., Fitting, C., Haeffner-Cavaillon, N., Kirsch, S.J. and Warren, H.S. (**1990**) Cytokine response by monocytes and macrophages to free and lipoprotein-bound lipopolysaccharide. *Infect. Immun.*, **58**, 2375–2382.
93. Brandtzaeg, P., Osnes, L., Ovstebo, R., Joo, G.B., Westvik, A.B. and Kierulf, P. (**1996**) Net inflammatory capacity of human septic shock plasma evaluated by a monocyte-based target cell assay: identification of interleukin-10 as a major functional deactivator of human monocytes. *J. Exp. Med.*, **184**, 51–60.
94. Ayala, A., Knotts, J.B., Ertel, W., Perrin, M.M., Morrison, M.H. and Chaudry, I.H. (**1993**) Role of interleukin 6 and transforming growth factor-beta in the induction of depressed splenocyte responses following sepsis. *Arch. Surg.*, **128**, 89–94; discussion 94-85.
95. Chen, W., Frank, M.E., Jin, W. and Wahl, S.M. (**2001**) TGF-beta released by apoptotic T cells contributes to an immunosuppressive milieu. *Immunity*, **14**, 715–725.
96. Asehnoune, K., Fitting, C., Edouard, A.R., Minville, V., Benhamou, D., Cavaillon, J.M. and Moine, P. (**2006**) beta2-Adrenoceptor blockade partially restores *ex vivo* TNF production following hemorrhagic shock. *Cytokine*, **34**, 212–218.
97. Gonzalez-Rey, E., Chorny, A. and Delgado, M. (**2007**) Regulation of immune tolerance by anti-inflammatory neuropeptides. *Nat. Rev. Immunol.*, **7**, 52–63.
98. Tracey, K.J. (**2007**) Physiology and immunology of the cholinergic antiinflammatory pathway. *J. Clin. Invest.*, **117**, 289–296.
99. Majetschak, M., Krehmeier, U., Bardenheuer, M., Denz, C., Quintel, M., Voggenreiter, G. and Obertacke, U. (**2003**) Extracellular ubiquitin inhibits the TNF-alpha response to endotoxin in peripheral blood mononuclear cells and regulates endotoxin hyporesponsiveness in critical illness. *Blood*, **101**, 1882–1890.
100. Wisnoski, N., Chung, C.S., Chen, Y., Huang, X. and Ayala, A. (**2007**) The contribution of CD4+ CD25+ T-regulatory-cells to immune suppression in sepsis. *Shock*, **27**, 251–257.
101. Heuer, J.G., Zhang, T., Zhao, J., *et al.* (**2005**) Adoptive transfer of *in vitro*-stimulated CD4+ CD25+ regulatory T cells increases bacterial clearance and improves survival in polymicrobial sepsis. *J. Immunol.*, **174**, 7141–7146.
102. Delano, M.J., Scumpia, P.O., Weinstein, J.S., *et al.* (**2007**) MyD88-dependent expansion of an immature GR-1(+)CD11b(+) population induces T cell suppression and Th2 polarization in sepsis. *J. Exp. Med.*, **204**, 1463–1474.
103. Cavaillon, J.M., Adib-Conquy, M., Cloez-Tayarani, I. and Fitting, C.

(**2001**) Immunodepression in sepsis and SIRS assessed by *ex vivo* cytokine production is not a generalized phenomenon: a review. *J. Endotoxin Res.*, **7**, 85–93.

104. Ding, Y., Chung, C.S., Newton, S., Chen, Y., Carlton, S., Albina, J.E. and Ayala, A. (**2004**) Polymicrobial sepsis induces divergent effects on splenic and peritoneal dendritic cell function in mice. *Shock*, **22**, 137–144.
105. Munford, R.S. and Pugin, J. (**2001**) Normal responses to injury prevent systemic inflammation and can be immunosuppressive. *Am. J. Respir. Crit. Care Med.*, **163**, 316–321.
106. Cavaillon, J.M. and Annane, D. (**2006**) Compartmentalization of the inflammatory response in sepsis and SIRS. *J. Endotoxin Res.*, **12**, 151–170.
107. Randow, F., Docke, W.D., Bundschuh, D.S., Hartung, T., Wendel, A. and Volk, H.D. (**1997**) *In vitro* prevention and reversal of lipopolysaccharide desensitization by IFN-gamma, IL-12, and granulocyte-macrophage colony-stimulating factor. *J. Immunol.*, **158**, 2911–2918.
108. Haas, J.G., Meyer, N., Riethmuller, G. and Ziegler-Heitbrock, H.W. (**1990**) Inhibition of lipopolysaccharide-induced *in vitro* desensitization by interferon-gamma. *Eur. J. Immunol.*, **20**, 1181–1184.
109. Adib-Conquy, M. and Cavaillon, J.M. (**2002**) Gamma interferon and granulocyte/monocyte colony-stimulating factor prevent endotoxin tolerance in human monocytes by promoting interleukin-1 receptor-associated kinase expression and its association to MyD88 and not by modulating TLR4 expression. *J. Biol. Chem.*, **277**, 27927–27934.
110. Bundschuh, D.S., Barsig, J., Hartung, T., Randow, F., Docke, W.D., Volk, H.D. and Wendel, A. (**1997**) Granulocyte-macrophage colony-stimulating factor and IFN-gamma restore the systemic TNF-alpha response to endotoxin in lipopolysaccharide-desensitized mice. *J. Immunol.*, **158**, 2862–2871.
111. Livingston, D.H. and Malangoni, M.A. (**1988**) Interferon-gamma restores immune competence after hemorrhagic shock. *J. Surg. Res.*, **45**, 37–43.
112. Mackensen, A., Galanos, C. and Engelhardt, R. (**1991**) Modulating activity of interferon-gamma on endotoxin-induced cytokine production in cancer patients. *Blood*, **78**, 3254–3258.
113. Docke, W.D., Randow, F., Syrbe, U., *et al.* (**1997**) Monocyte deactivation in septic patients: restoration by IFN-gamma treatment. *Nat. Med.*, **3**, 678–681.
114. Flohe, S., Borgermann, J., Dominguez, F.E., *et al.* (**1999**) Influence of granulocyte-macrophage colony-stimulating factor (GM-CSF) on whole blood endotoxin responsiveness following trauma, cardiopulmonary bypass, and severe sepsis. *Shock*, **12**, 17–24.
115. Williams, M.A., White, S.A., Miller, J.J., Toner, C., Withington, S., Newland, A.C. and Kelsey, S.M. (**1998**) Granulocyte-macrophage colony-stimulating factor induces activation and restores respiratory burst activity in monocytes from septic patients. *J. Infect. Dis.*, **177**, 107–115.
116. Nierhaus, A., Montag, B., Timmler, N., *et al.* (**2003**) Reversal of immunoparalysis by recombinant human granulocyte-macrophage colony-stimulating factor in patients with severe sepsis. *Intensive Care Med.*, **29**, 646–651.
117. Pal'tsev, A.V., Ovechkin, A.V., Zakharova, N.F., *et al.* (**2000**) [Cytokines in the treatment of a generalized surgical infection]. *Anesteziol Reanimatol*, 27–30.
118. Benjamim, C.F., Lundy, S.K., Lukacs, N.W., Hogaboam, C.M and Kunkel, S.L. (**2005**) Reversal of long-term sepsis-induced immunosuppression by dendritic cells. *Blood*, **105**, 3588–3595.

119. Ertel, W., Keel, M., Neidhardt, R., *et al.* (**1997**) Inhibition of the defense system stimulating interleukin-12 interferon-γ pathway during critical illness. *Blood*, **89**, 1612–1620.
120. Helminen, M., Vesikari, T. (**1990**) Interleukin-1 production in bacterial meningitis. *Scand. J. Infect. Dis.*, **22**, 105–108.
121. Ho, J.L., Badaro, R., Schwartz, A., *et al.* (**1992**) Diminished in vitro production of interleukin-1 and tumor necrosis factor-alpha during acute visceral leishmaniasis and recovery after therapy. *J. Infect. Dis.*, **165**, 1094–1102.
122. Diterich, I., Harter, L., Hassler, D., *et al.* (**2001**) Modulation of cytokine release in *ex vivo*-stimulated blood from borreliosis patients. *Infect. Immun.*, **69**, 687–694.
123. House, D., Chinh, N.T., Hien, T.T., *et al.* (**2002**) Cytokine release by lipopolysaccharide-stimulated whole blood from patients with typhoid fever. *J. Infect. Dis.*, **186**, 240–245.
124. Suharti, C., van Gorp, E.C., Dolmans, W.M., *et al.* (**2003**) Cytokine patterns during dengue shock syndrome. *Eur. Cytokine Netw.*, **14**, 172–177.
125. van der Kleij, D., van den Biggelaar, A.H., Kruize, Y.C., *et al.* (**2004**) Responses to Toll-like receptor ligands in children living in areas where schistosome infections are endemic. *J. Infect. Dis.*, **189**, 1044–1051.
126. Sasisekhar, B., Aparna, M., Augustin, D.J., *et al.* (**2005**) Diminished monocyte function in microfilaremic patients with lymphatic filariasis and its relationship to altered lymphoproliferative responses. *Infect. Immun.*, **73**, 3385–3393.
127. Berry, A., Chene, G., Benoit-Vical, F., *et al.* (**2005**) *Ex vivo* and *in vitro* impairment of CD36 expression and tumor necrosis factor-alpha production in human monocytes in response to Plasmodium falciparum-parasitized erythrocytes. *J. Parasitol.*, **91**, 316–322.
128. Brenchley, J.M., Price, D.A., Schacker, T.W., *et al.* (**2006**) Microbial translocation is a cause of systemic immune activation in chronic HIV infection. *Nat. Med.*, **12**, 1365–1371.
129. Fokkema, S.J., Loos, B.G., Slegte, C., *et al.* (**2002**) A type 2 response in lipopolysaccharide (LPS)-stimulated whole blood cell cultures from periodontitis patients. *Clin. Exp. Immunol.*, **127**, 374–378.
130. Cabié, A., Fitting, C., Farkas, J-C., *et al.* (**1992**) Influence of surgery on *in vitro* cytokine production by human monocytes. *Cytokine*, **4**, 576–580.
131. Wilhelm, W., Grundmann, U., Rensing, H., *et al.* (**2002**) Monocyte deactivation in severe human sepsis or following cardiopulmonary by pass. *Shock*, **17**, 354–360.
132. Kylanpaa, M.L., Mentula, P., Kemppainen, E., *et al.* (**2005**) Monocyte energy is present in patients with severe acute pancreatitis and is significantly alleviated by granulocyte-macrophage colony-stimulating factor and interferon-gamma in vitro. *Pancreas*, **31**, 23–27.
133 Faist, E, *Differential release and impact of antibiotic-induced endotoxin* (**1995**) Raven Press.
134. Keel, M., Schregenberger, N., Steckholzer, U., *et al.* (**1996**) Endotoxin tolerance after severe injury and its regulatory mechanisms. *Journal of Trauma-Injury Infection & Critical Care*, **41**, 430–437; discussion 437–438.
135. Shimokawa, H., Kuroiwa-Matsumoto, M., Takeshita, A. (**1998**) Cytokine generation capacities of monocytes are reduced in patients with severe heart failure. *Am. Heart J.*, **136**, 991–1002.
136. del Fresno, C., Soler-Rangel, L., Soares-Schanoski, A., *et al.* (**2007**) Inflammatory responses associated with acute coronary syndrome up-regulate IRAK-M and induce

endotoxin tolerance in circulating monocytes. *J. Endotoxin Res.*, **13**, 39–52.

137. Wasmuth, H.E., Kunz, D., Yagmur, E., *et al.* (**2005**) Patients with acute on chronic liver failure display "sepsis-like" immune paralysis. *J. Hepatol.*, **42**, 195–201.

138. Lin, C.Y., Tsai, I.F., Ho, Y.P., *et al.* (**2007**) Endotoxemia contributes to the immune paralysis in patients with cirrhosis. *J. Hepatol.*, **46**, 816–826.

PART IV
Experimental Models and Therapies

Sepsis and Non-infectious Systemic Inflammation. Edited by J.-M. Cavaillon and C. Adrie

ISBN: 978-3-527-31935-0

16
Experimental Models of Sepsis and Non-infectious SIRS

Raghavan Raju, William J. Hubbard and Irshad H. Chaudry

16.1
Introduction

In sepsis, microbes invade the circulatory system and can produce sequelae that are potentially debilitating or even life-threatening. Because the blood of vertebrates is not colonized by microbial organisms (as with skin or intestine), it is implicit that for sepsis to occur, mechanical barriers must be breached. In addition, if the sepsis is persistent, it also becomes evident that the immune system is not effective, and may even become dysfunctional or compromised.

There are multiple points and modes of entry for microbes, which can be endogenous or exogenous, making the picture of sepsis inescapably quite complex. The sites which are "pre-loaded" with endogenous flora generally would be anatomical sites that are able to communicate with the "outside world" (e.g. skin, gut, urogenital tract, liver via bile duct, and lungs). Exogenous microbes may be introduced by foreign objects, such as from a penetrating wound where the penetrating object is contaminated with environmental microbes. Thus, injury almost always has the capability to rapidly introduce substantial quantities of microbes, arising either from the environment or from the patient's own flora, or both. A fact that is confounding to medical care is that injury-related sepsis is also likely to take place in a setting where the patient is severely stressed and compromised, perhaps even in a state of septic shock. In light of this etiologic diversity, the picture of sepsis can at best be viewed as a matrix with no single unifying set of conditions, save for microbes in the blood.

Under this umbrella of intrinsic complexity, designing an experimental model of sepsis is daunting, and can at best be a compromise which is not all-encompassing. Thus, the experimental designer must at the outset determine the set of conditions that best define their test goals, selected from the "matrix". That notwithstanding, there are some conditions which involve polymicrobial organisms, breech mechanical and immunological barriers to entry, evoke inflammatory responses, produce systemic effects, generate evaluable, and

Sepsis and Non-infectious Systemic Inflammation. Edited by J.-M. Cavaillon and C. Adrie

ISBN: 978-3-527-31935-0

statistically useful data, such as an LD_{50}, that would be highly desirable to incorporate into a productive and useful model.

It is difficult to perform controlled studies in patients with sepsis because of the heterogeneity in genetic make-up of the patients, co-existence of one or more disease conditions and other variables. These complexities encourage the use of animal models, to study sepsis. Several different models such as peritonitis, endotoxicosis, live bacterial infusion and abscesses in extremities have been employed to study sepsis. In this chapter, we will attempt to provide an overview of some of the most utilized models for studying the pathophysiology of sepsis and non-infectious systemic inflammatory response syndrome (SIRS).

16.2 Definition of Sepsis, Endotoxemia and Bacteremia

The term "sepsis" has been used indiscriminately in the literature and it refers to different conditions in different studies. We have previously described sepsis as an acute infection wherein the host is toxic (febrile, anorexic, weak, lethargic, etc.) because of invasive infection, and "septic shock" occurs when the events of sepsis lead to circulatory failure [1]. Thus the sepsis syndrome is a manifestation of an acute bacterial infection. However, it should be recognized that serious viral and fungal infections may also be associated with signs and symptoms of "sepsis" [1]. Endotoxemia refers to a condition in which the host is injected with a low or high dose of endotoxin, and endotoxic shock occurs when the effects of endotoxin lead to circulatory collapse. Bacteremia accordingly refers to the presence of bacteria in the blood, whether or not it leads to, or results in the toxic state of sepsis.

16.2.1 Endotoxin Models

Gram-positive sepsis differs from Gram-negative sepsis in that the organisms often arise from skin, wounds, soft-tissue structures, and catheter sites rather than enteric or genitourinary sources [2]. Whereas Gram-positive bacteria depend on the production of powerful exotoxins (e.g. tetanus, botulism, diphtheria), with Gram-negative bacteria, it is principally the cell wall component, endotoxin that is implicated in pathogenesis [3]. Though detectable levels of the endotoxin, lipopolysaccharide (LPS) are found in up to 75% of patients with sepsis in the intensive care setting [4], treatment of Gram-negative bacteremia with antibodies to endotoxin have not yielded promising results [5, 6]. While a bolus infusion of LPS in human volunteers produces a rapid increase in serum levels of LPS followed by a precipitous fall, prolonged low-dose infusion induces influenza-like symptoms with fever, tachycardia, raised white cell count and a catabolic state ([7] and the references therein). Endotoxin stimulates the release of inflammatory mediators from various cell types. When sepsis is a systemic inflammatory response caused by infection, the systemic inflammatory response as seen in endotoxinemia due to exogenously administered LPS is referred to as non-infectious SIRS.

LPS exerts many of its biologic effects by stimulating host cells to produce bioactive inflammatory mediators. Monocytes and macrophages respond to LPS with the synthesis and release of proinflammatory cytokines, such as TNF-α, IL-1, IL-6, and IL-8. In the 1960s, it was found that C3H/HeN and C3H/OuJ substrains demonstrated sensitivity to LPS, mice of the C3H/HeJ strain were tolerant to the lethal effect of LPS [8, 9]. Detailed genetic studies revealed that C3H/HeJ mice had a point mutation on Toll-like receptor (TLR)-4, a component of the LPS receptor that prevented response to LPS [10]. Hence, the mutation on TLR-4 receptor rendered C3H/HeJ mice resistant to the lethal effect of LPS. It is known that LPS activates systemic inflammatory responses through TLR-4 [11]. Stimulation of TLRs leads to secretion of "early response" cytokines, such as TNF-α, and these in turn are thought to cause the systemic changes associated with the systemic inflammatory response syndrome, or SIRS [12].

Owing to variations in delivery of microbes (e.g., bolus or slow infusion) and the availability of numerous types of endotoxin, all obtainable in a relatively purified form, many endotoxin models have been described. There are mainly four categories of LPS-infusion models that reproduce SIRS or septic shock in patients [13]. They include models that utilize small sublethal doses of LPS, that provide aggressive resuscitation of intravascular volume, utilize continuous infusion of LPS and those that use LPS injected intraperitoneally [14–19].

While mice, cats and dogs are relatively endotoxin resistant, rabbits, sheep and non-human primates show an enhanced response. It has been observed that the sensitivity of endotoxin-resistant animals may be increased by prior sensitization with endotoxin or killed Gram-negative bacteria. In a rat model of endotoxemia produced by a bolus administration of endotoxin (40 mg/kg), cardiac index, mean blood pressure and central venous pressure decreased and were accompanied by compensatory increases in heart rate and total peripheral vascular resistance [20]. In contrast, lower doses using slow, continuous infusion produce a hyperdynamic response with an early increase in cardiac output [7, 15, 21]. Unfortunately, however, the slow endotoxin infusion models do not subsequently produce a hypodynamic circulatory state [22].

Endotoxin administration into experimental animals and humans elicited a strong TNF-α response [23] and passive immunization with neutralizing goat anti-TNF alpha IgG improved survival from 8 to 75% in rats administered LPS [24]. Kinetic studies in mice showed a rapid spike in TNF-α production followed by a peak of IL-1 activity at 6 h [25]. However, clinical trials that used either monoclonal antibodies to TNF-α or TNF-α soluble receptors (TNF-SR) were not successful [26–29], thus undermining the significance of this model. On the contrary, there was a dose-dependent increase in mortality with the TNF-SR treatment in a randomized, double-blind, placebo-controlled, multicenter trial [27]. Pooled data from randomized trials using neutralizing TNF antibodies in human sepsis show only a small benefit, though statistically significant, of 3% for treated patients [22, 23, 30]. Haudek and colleagues described acute and subacute endotoxemic models in baboons, the first evoked by bolus injection of LPS (1 mg, 0.1 mg, or 4 ng per kg of *Escherichia coli* (*E. coli*) LPS), and the second evoked by infusion of 1.5 mg/kg of *E. coli* LPS over 30 min [31]. They claimed that LPS-induced kinetics of cytokine release, as well as of hemodynamic and hematologic changes in baboons, were similar to those observed in humans, even though baboons required approximately 10^4-fold higher initial LPS dose to develop these manifestations.

While non-infectious inflammatory stimuli, such as that due to endotoxin, may mimic the systemic inflammatory response of generalized infection, the quantitative nature of the hematologic abnormalities, inflammatory cytokine activation and coagulation abnormalities are generally greater in true sepsis associated with an underlying infection [32].

16.2.2
Bacteremia Models

Several studies have investigated bolus or continuous intravenous administration of live bacteria. When rats were infused for 5 h with a high dose ($\sim 2 \times 10^{10}$) of *E. coli*, they developed early hypotension, reduced cardiac output (measured by thermal dilution technique) and did not survive [33]. However, rats infused with lower doses of the bacteria survived a 5-h infusion with hypotension and reduced cardiac output occurring later in the course of bacteremia. Heart rate was markedly elevated in both septic groups. The results indicated that severe Gram-negative bacteremia produces myocardial depression in the rat [33]. However injection of a bolus dose of bacteria in rats resulted in an initial increase in cardiac output that was subsequently followed by the normalization of cardiac output [34].

Hinshaw etal evaluated the effect of steroid/antibiotic treatments in an *E. coli*-induced bacteremia model in the baboons. When antibiotics were administered 30 min after the start of a 2-h intravenous administration of LD_{100} *E. coli*, all baboons survived, but delaying the steroid treatment until all *E. coli* were infused resulted in an 85% survival rate [35, 36]. Another study was undertaken to determine whether baboons would recover when initiation of treatment was delayed until they had sustained *E. coli*-induced systemic hypotension. Under those conditions, all eight untreated animals died within 42 h, five of the eight treated baboons survived when the steroid/antibiotic treatment was delayed until 2 h after the administration of *E. coli* [37]. These baboon models were designed to parallel the clinical situation by delaying initiation of treatment until severe sustained systemic hypotension had occurred.

Crocker *et al.* compared the pulmonary clearance and tissue retention of blood-borne *Pseudomonas aeruginosa* in dogs and pigs during continuous 6-h intravenous infusion of bacteria [38]. In contrast to controls, experimental pigs developed pulmonary artery (PA) hypertension and pulmonary failure manifested by hypoxemia, increased intrapulmonary shunting, non-cardiogenic pulmonary edema, and congestive atelectasis, a pattern of pulmonary failure similar to sepsis-induced ARDS in humans. In dogs, PA pressures were unchanged from baseline, no edema was detected, and comparable hyperventilation was associated with an increase in PaO_2 [38]. This study demonstrated host-dependent variation in different experimental models. However, in a porcine model using similar doses of three different live bacteria (*Pseudomonas aeruginosa, Staphylococcus aureus* and *E. coli*) it was found that the hemodynamic and pulmonary changes depended on the bacterial species used [39]. Some serotypes of the *Streptococcus pneumoniae* are more aggressive than others and may affect the severity of disease [40]. It has been observed that differences in bacterial serotype play a role in determining the disease outcome in animal models, regardless of the host's immunogenetic background [41, 42].

Shaw and Wolfe developed a conscious bacteremic dog model with hemodynamic and metabolic responses similar to those in humans who had been intra-arterially infused twice with *E. coli*, resuscitated and studied after 24 h [43]. The animals were hemodynamically stable, cardiac output and heart rate was increased and arterial pressure was not altered. Using radioactive and stable isotopes, they also assessed various metabolic parameters and found an increase in the metabolic rate, due to increased oxidation of free fatty acids and triglycerides. They also observed hyperinsulinemia and hyperglycemia as seen in septic patients. This model mimics many of the features of clinical bacteremia and avoids the confounding effects of anesthesia and surgical manipulation [44].

16.3 Exogenous Peritonitis Models

A variety of different models have been used to induce peritonitis in experimental animals and these include inoculation of bacterial cultures or fecal materials into the peritoneal cavity or perforation of bowel to allow contamination with gastrointestinal contents [45–47]. Fecal material or bacteria were encapsulated in fibrin clots or gelatin to ensure slow release and prolonged disease process. In one such model, gelatin capsules containing pooled colonic contents and barium sulfate were implanted into the pelvic region of Wistar rats; the animals developed generalized peritonitis and those that survived developed discrete intra-abdominal abscesses [46]. Using a similar approach, Fink and colleagues rendered chronically instrumented dogs bacteremic by implanting in the peritoneal cavity a fibrin clot containing viable *E. coli* [48]. One day later, there was an increase in cardiac output, heart rate and mean pulmonary artery pressure while mean systemic arterial pressure and systemic vascular resistance were decreased. Natanson *et al.* [49] developed a canine model that simulates the human cardiovascular response to septic shock in conscious, un-sedated dogs by implanting an *E. coli*-infected clot into the peritoneum, resulting in bacteremia, and severe systolic and diastolic cardiac dysfunction. Nakatani *et al.* [50] developed an intraperitoneal abscess model in the rat by inoculating with a fecal pellet made of sterile rat feces, agar, and a known number and strain of bacteria. A single, uniform abscess was formed in 100% of the animals that led to a peritonitis stage with leucopenia, hypoglycemia, body weight loss, and slight fever followed by the abscess stage with leukocytosis and a slight hyperglycemia [50].

In another study, implantation of 0.5% bovine fibrin clots containing 2×10^8 *E. coli* into the rat peritoneal cavity significantly reduced the 24-h mortality rate from 100 to 0% compared to bacteria in a similar volume of saline solution [47]. The mortality rate could be controlled by the encapsulation method and the number of bacteria inoculated. For example the same authors further observed that as few as 10^2 *E. coli* per fibrin clot produce abscesses, but 10^7 or more are required to produce death; without fibrin a dose of less than 10^7 *E. coli* neither killed nor produced intraperitoneal infections [47]. The defined bacterial as well as the fecal pellet models are generally associated with an early hyperdynamic cardiovascular response similar to that observed in patients. Additionally, studies using these models have shown that surviving rats may manifest metabolic and inflammatory changes for 5 or more days, especially when fluids and antibiotics are given [51, 52]. Models of bacteria-induced peritonitis without the use of adjuvants were also employed by some investigators. Ghiselli *et al.* [53] developed a model of *E. coli*-induced peritonitis to test the efficacy of antibiotics in a rat model of Gram-negative bacteremic shock. Rats were given an intraperitoneal injection of 2×10^{10} CFU of *E. coli* which produced a lethality rate of 100%. In guinea pigs fed *ad libitum*, controlled intraperitoneal infusion of bacteria by an implanted 7-day osmotic pump resulted in peritonitis and abscess formation with a 50% survival 14–18 days after pump implantation [54]. This model was used to study the relative influence of dietary composition on outcome, since survival could be extended to 2 weeks or more in the presence of continuing bacteremia.

16.3.1
Cecal Ligation and Perforation Models

In sepsis research, cecal ligation and perforation (CLP) developed by Chaudry and colleagues is a widely used model and has been successfully used in different species [1]. The models discussed thus far, i.e. endotoxemia, endotoxic shock or bacteremia may not simulate the progressive cardiovascular responses of polymicrobial sepsis, such as an existence of an early hyperdynamic phase followed by a late, hypodynamic phase. Sepsis in the CLP model is due to peritoneal contamination with mixed flora in the presence of devitalized tissue and thus bears an obvious resemblance to clinical problems such as perforated appendicitis and diverticulitis [1, 44]. Hence, the CLP model mimics many features of clinical peritonitis [1, 55–58].

16.3.1.1 The CLP Model

Rats are fasted overnight, anesthetized and a 2-cm ventral midline incision made. The cecum is isolated and ligated just below the ileocecal valve and the opposite ends of the cecum are punctured twice with an 18-gauge needle. The cecum is then placed back in the abdominal cavity which is closed in two layers. After the procedure, the animals are resuscitated with crystalloid fluid (5 ml/100 g BW) subcutaneously and returned to the cages. During the initial 10–12 h the rats that underwent the CLP procedure demonstrate hyperdynamic, hypermetabolic, hyperglycemic, hyperinsulinemic state with high blood lactates. At about 16–20 h after the procedure, they demonstrate a hypodynamic, hypometabolic, hypoglycemic, hypoinsulinemic condition. If fluid resuscitation is not provided following CLP, the animals do not demonstrate a hyperdynamic phase circulatory state [1] Within 1 h after the CLP, ascites fluid collected from the peritoneal cavity is positive for various microbes (*E. coli, S. bovis, P. mirabilis, B. fragilis,* etc.). The cecal bacteria begin to drain continuously into peritoneal cavity soon after performing the puncture. Studies have also shown that the percentage of cecum ligated was the principal determinant of mortality [57]. Although the model has been applied mostly to small rodents, a CLP model of neonatal sepsis has been developed for piglets, demonstrating its adaptability to large animals [58, 59]. In another study rats from 10 to 28 days of age were also used to study CLP-induced sepsis [60].

16.3.1.2 CLP as a Model of Sepsis

The CLP model has been successfully developed in several species including rat, mouse, sheep, goat and pig. CLP differs from other models such as endotoxemia as the former produces a biphasic hemodynamic response whereas administration of a large dose of endotoxin produces immediate hypotension and low tissue perfusion. Also, endotoxin-tolerant C3H/HeJ mice have been shown to have similar mortality rates after CLP as endotoxin-sensitive C3H/HeN mice [61, 62]. In the CLP model, it is possible to measure active hepatocellular function and various hemodynamic parameters such as cardiac output, circulating blood volume, and left ventricular performance, by using an *in vivo* hemoreflectometer and *in vivo* heart performance analyzer [56]. In one study, it was found that the circulating blood volume, as determined by indocyanine green clearance, decreased significantly at 20 h after CLP (late sepsis) though, systemic hematocrit increased significantly even prior to this observed decrease [63]. This model also mimics perforated appendicitis or diverticulitis as the punctured cecum can be excised at specific intervals and the peritoneum is irrigated with warm saline. Moreover, antibiotic therapy was found to significantly decrease mortality due to CLP, regardless of injury or severity, mimicking the clinical condition [64–66]. An age-dependent increase in mortality was also observed in several studies [65, 67]. Aged septic mice had increased splenic apoptosis and gut epithelial cell death compared with either young septic animals or aged sham animals [67].

In studies to test energy deficit, changes in cardiac mitochondrial content and caspase activation during sepsis, mitochondrial dysfunction and increased apoptosis were observed in the heart after CLP [68]. Sepsis often results in coagulopathy, and CLP has been used to assess the usefulness of anticoagulants in resuscitation therapies [59, 69]. The CLP model facilitated the identification of the efficacy of heparin in a fluid resuscitation regimen for sepsis by preserving endothelial cell, liver and heart function in sepsis, and is also found useful in hemorrhagic shock [59–61]. Since the intestinal flora is dependent on diet, one can modify the septic responses by changing cecal flora through the use of different diets [56]. One such study indicated that when the rats were fed a meat diet instead of a usual grain diet, the preparation was less lethal [1].

Besides analysis of survival, the cytokine expression profile in CLP has been investigated extensively. A detailed review can be found in Hubbard *et al.* [59]. The main proinflammatory cytokines, IL-6 and TNF-α, have been shown to increase following CLP. Furthermore, it has been reported that high levels of IL-6 strongly correlate with mortality following CLP, a phenomenon that also occurs in human sepsis [62, 70, 71]. Eskandari *et al.* [72] observed that CLP as well as LPS administration elevates TNF bioactivity in mice with peak plasma levels at 12 h after CLP and 90 min after LPS. Though anti-TNF treatment reduced the circulating TNF levels, it did not reduce mortality in either model [72]. In another study, passive immunization with neutralizing goat anti-TNF-α IgG improved survival from 8 to 75% in rats administered LPS intravenously [24]. Overall, while neutralization of TNF conferred protection against lethality after administration of endotoxin, passive immunization against TNF has been found to have no beneficial effect in lethal models of CLP [72–75]. The observation that the injection of recombinant TNF into animals [76, 77] and humans [78, 79] reproduced the signs and symptoms of sepsis combined with the observation that anti-TNF ameliorates the shock symptoms in endotoxin models, formed the compelling basis for clinical trials of TNF-based treatments [80–84]. Unfortunately, anti-TNF-α-based treatments were unsuccessful in the management of patients with sepsis [85]. When we look retrospectively at the preclinical studies that formed the basis of these clinical trials, it clearly shows that anti-TNF-α treatments did not improve sepsis symptoms in the CLP models.

The model of CLP in rodents is very simple, elegant, widely used, and well understood. It is also possible to adapt CLP to other appropriate animal species, although mice and rats have been used most frequently [59].

16.3.1.3 Potential Limitations of the CLP Model

As in any other experimental animal model, CLP models also used inbred strains of animals, mostly mice and rats. Most of the inbred strains are congenic and do not address the experimental variations associated with genetic diversity seen in outbred humans. Physiological measurements are more difficult in small animals such as rodents, the most preferred species for CLP used thus far, due to the size of these animals. This limitation is overcome by investigators using large animals for study. It is also suggested that the CLP model results in the formation of an intra-abdominal abscess and that the CLP mice more likely die from a fulminant bacterial release after necrosis of the ligated and ischemic cecum than from the persistent abscess with low-grade inflammation [86]. The single-hit model of sepsis may not be analogous to the patient with sepsis syndrome since the animals are initially healthy [56]. This has been addressed using a two-hit model of trauma–hemorrhage followed by CLP and an increased mortality rate was found in rats subjected to both procedures serially [63].

The limitations of the CLP model described above are outweighed by the advantages and the similarity of this model to human sepsis syndromes such as that which occurs due to perforation of the appendix or diverticulum and the polymicrobial origin. CLP is also characterized by focal infection, septicemia and release of bacterial products into the periphery. The CLP model remains one of the most used experimental animal models for sepsis and septic shock.

16.4 Animal Models versus Human Sepsis

While most experimental animal models utilize young and healthy animals, the results obtained may not reflect the vast majority of sepsis as seen in infants and aged humans. When humans with sepsis are treated in intensive care units, and because the animal studies lack both optimized ventilator support and blood pressure support, it is difficult to extrapolate the outcomes to septic human patients [87]. Patients with sepsis receive antibiotics, narcotic agents, dopamine etc., but there is no experimental sepsis model that includes the administration of these agents [22]. Most of the animals employed for the study of sepsis, mice, rats and baboons, are less sensitive to endotoxin than humans. Also, as mentioned with respect to the CLP model, animal models use inbred strains, whereas outbred humans are genetically diverse. Such diverse genetics of humans results in differential susceptibility to pathogens. Single nucleotide polymorphism (SNP) such as TLR-4 point mutation in C3H/HeJ mice imparting endotoxin tolerance have been described, but extensive SNP analyses in human patients have not yet yielded promising results [10, 88]. Identification of susceptibility genes in sepsis might be further complicated because this is a polygenic syndrome. Therefore, the pathogenesis of sepsis involves several factors that encompass a long chain of events from genetic interaction and pathogen recognition to factors that control host responses [88–91].

Notwithstanding such differences and deficiencies, animal models still remain the most promising preclinical model as well as the model of choice to understand disease processes and treatment modalities at least until appropriate *in silico* models are developed.

Acknowledgments

This work was supported by NIH grants R37 GM39519 and RO1 GM 37127.

References

1. Wichterman, K.A., Baue, A.E. and Chaudry, I.H. (**1980**) Sepsis and septic shock: a review of laboratory models and a proposal. *J. Surg. Res.*, **29**, 189–201.
2. Bone, R.C. (**1994**) Gram–positive organisms and sepsis. *Arch. Intern. Med.*, **54**, 26–34.
3. Steven, O. and Jonnathan, C. (**1999**) Clinical Gram-positive sepsis: Docs it fundamentally differ from Gram-negative bacterial sepsis? *Crit. Care Med.*, **27**, 1608–16.
4. Goldie, A.S., Fearon, K.C.H., Ross, J.A., Barclay, G.R., Jackson, R.E., Grant, I.S., *et al.* (**1995**) Natural cytokine antagonists and endogenous antiendotoxin core antibodies in sepsis syndrome. The Sepsis Interventional Group. *JAMA*, **274**, 172–7.
5. McCloskey, R.V., Straube, R.C., Sanders, C., Smith, S.M. and Smith, C.R (**1994**) The CHESS Trial Study Group. Treatment of septic shock with human monoclonal antibody HA-1A: a randomized, double-blind, placebo-controlled trial. *Ann. Intern. Med.*, **120**, 1–5.
6. Cross, A.S. (**1994**) Antiendotoxin antibodies: A dead end? *Ann. Int. Med.*, **121**, 58–60.
7. Parker, S.J. and Watkins, P.E. (**2001**) Experimental models of Gram-negative sepsis. *Brit. J. Surg.*, **88**, 22–30.
8. Heppner, G.. and Weiss, D.W. (**1965**) High susceptibility of strain of mice to endotoxin and endotoxin-red blood cell mixtures. *J. Bacteriol.*, **90**, 696–703.
9. Sultzer, B.M. (**1968**) Genetic control of leukocyte responses to endotoxin. *Nature (London)*, **219**, 1253–4.
10. Poltorak, A., He, X., Smirnova, I., Liu, M.-Y., van Huffel, C., Du, X., Birdwell, D., Alejos, E., Silva, M., Galanos, C., *et al.* (**1998**) Defective LPS signaling in C3H/HeJ and C57BL/10ScCr mice: mutations in *Tlr4* gene. *Science*, **282**, 2085–8.
11. Beutler, B. (**2001**) Sepsis begins at the interface of pathogen and host. *Biochem. Soc. Trans.*, **29**, 853–9.
12. Beutler, B. and Rietschel, E.T. (**2003**) "Innate immune sensing and its roots" the story of endotoxin. *Nature Rev. Immunol.*, **3**, 169–76.
13. Schultz, M.J. and van der Poll, T. (**2002**) Animal and human models for sepsis. *Ann. Med.*, **34**, 573–81.
14. Fink, M.P., Fiallo, V., Stein, K.L. and Gardiner, W.M. (**1987**) Systemic and regional hemodynamic changes after intraperitoneal endotoxin in rabbits: development of a new model of the clinical syndrome of hyperdynamic sepsis. *Circ. Shock*, **22**, 73–81.
15. Fink, M.P., Morrissey, P.E., Stein, K.L., Clement, R.E., Fiallo, V. and Gardiner, W.M. (**1988**) Systemic

and regional hemodynamic effects of cyclo-oxygenase and thromboxane synthetase inhibition in normal and hyperdynamic endotoxemic rabbits. *Circ. Shock*, **26**, 41–57.
16. Fink, M.P., Cohn, S.M., Lee, P.C., Rothschild, H.R., Deniz, Y.F., Wang, H, *et al.* (**1989**) Effect of lipopolysaccharide on intestinal intramucosal hydrogen ion concentration in pigs: evidence of gut ischemia in a normodynamic model of septic shock. *Crit. Care Med.*, **17**, 641–6.
17. Breslow, M.J., Miller, C.F., Parker, S.D Walman, A.T. and Traystman, R.J. (**1987**) Effect of vasopressors on organ blood flow during endotoxin shock in pigs. *Am. J. Physiol.*, **252**, H291–300.
18. Emerson, T.E and Jr., Gill C.C. (**1967**) Effects of slow intravenous endotoxin infusion on hemodynamics and survival in dogs. *J. Appl. Physiol.*, **22**, 874–7.
19. Fish, R.E., Lang, C.H. and Spitzer, J.A. (**1986**) Regional blood flow during continuous low-dose endotoxin infusion. *Circ. Shock*, **18**, 267–75.
20. Brackett, D.J., Schaefer, C.F., Tompkins, P., Fagraeus, L., Peters, L.J. and Wilson, M.F. (**1985**) Evaluation of cardiac output, total peripheral vascular resistance, and plasma concentrations of vasopressin in the conscious, unrestrained rat during endotoxemia. *Circ. Shock*, **17**, 273–84.
21. Johnston, T.D., Hampton, W.W. and Fry, D.E. (**1989**) Septic hemodynamics produced by infusion of endotoxin. *Curr. Surg.*, **46**, 101–3.
22. Chaudry, I.H. (**1999**) Lessons learned in the last century and future directions. *Arch. Surg.*, **134**, 922–9.
23. Remick, D.G. and Ward, P.A. (**2005**) Evaluation of endotoxin models for the study of sepsis. *Shock*, **24**(Suppl 1), 7–11.
24. Bagby, G.J., Plessala, K.J., Wilson, L.A., Thompson, J.J. and Nelson, S. (**1991**) Divergent efficacy of antibody to tumor necrosis factor-alpha in intravascular and peritonitis models of sepsis. *J. Infect. Dis.*, **163**, 83–8.
25. Remick, D.G., Strieter, R.M., Lynch, J.P 3rd, Nguyen D., Eskandari M. and Kunkel S.L. (**1989**) *In vivo* dynamics of murine tumor necrosis factor-alpha gene expression. Kinetics of dexamethasone-induced suppression. *Lab. Invest.*, **60**, 766–71.
26. Suffredini, A.F., Reda, D., Banks, S.M., Tropea, M., Agosti, J.M. and Miller, R. (**1995**) Effects of recombinant dimeric TNF receptor on human inflammatory responses following intravenous endotoxin administration. *J. Immunol.*, **155**, 5038–45.
27. Fisher, C.J Jr, Agosti J.M, Opal S.M, Lowry S.F, Balk R.A, Sadoff J.C, Abraham E,Schein R.M. and Benjamin E (**1996**) Treatment of septic shock with the tumor necrosis factor receptor: Fc fusion protein. The Soluble TNF Receptor Sepsis Study Group. *N. Engl. J. Med.*, **334**, 1697–1702.
28. Abraham, E., Anzueto, A., Gutierrez, G., Tessler, S., San Pedro, G., Wunderink, R., Dal Nogare, A., Nasraway, S., Berman, S., Cooney, R., Levy, H., Baughman, R., Rumbak, M., Light, R.B., Poole, L., Allred, R., Constant, J., Pennington, J. and Porter, S. (**1998**) Double-blind randomised controlled trial of monoclonal antibody to human tumour necrosis factor in treatment of septic shock. NORASEPT II Study Group. *Lancet*, **351**, 929–33.
29. Arndt, P and Abraham, E (**2001**) Immunological therapy of sepsis: experimental therapies. *Intensive Care Med.*, **27**, S104–15.
30. Marshall, J.C., Deitch, E., Moldawer, L.L., Opal, S., Redl, H. and van del Poll, T. (**2005**) Preclinical models of shock and sepsis: What can they tell us? *Shock*, **24**, 1–6.
31. Haudek, S.B., Natmessnig, B.E., Furst, W., Bahrami, S., Schlag, G. and Redl, H. (**2003**) Lipopolysaccharide dose response in baboons. *Shock*, **20**, 431–6.
32. Opal, M.S. and Horn, D.L. (**1999**) The microbial aspects of sepsis: Does the

organism and its treatment affect outcome? *Sepsis*, **3**, 51–5.
33. Pass, L.J., Schloerb, P.R., Pearce, F.J. and Drucker, W.R. (**1984**) Cardiopulmonary response of the rat to Gram-negative bacteremia. *Am. J. Physiol.*, **246**, H344–50.
34. Sato, T., Isoyama, T., Tnaka, J., Jones, R.T., Cowley, R.A. and Trump, B.F. (**1982**) The pathophysiology of septic shock: Changes in hemodynamics in rats following live *E. coli* injection. An application of the thermodilution method for measurement of cardiac output. *Adv. Shock Res.*, **7**, 25–42.
35. Hinshaw, L.B., Archer, L.T., Beller-Todd, B.K., Coalson, J.J., Flournoy, D.J., Passey, R., Benjamin, B. and White, G.L. (**1980**) Survival of primates in LD100 septic shock following steroid/antibiotic therapy. *J. Surg. Res.*, **28**, 151–70.
36. Hinshaw, L.B., Archer, L.T., Beller-Todd, B.K., Benjamin, B. and Flournoy, D.J. (**1981**) Survival of primates in lethal septic shock following delayed treatment with steroid. *Circ. Shock*, **8**, 291–300.
37. Hinshaw, L.B., Beller-Todd, B.K., Archer, L.T., Benjamin, B., Flournoy, D.J., Passey, R. and Wilson, M.F. (**1981**) Effectiveness of steroid/antibiotic treatment in primates administered LD100 *Escherichia coli*. *Ann. Surg.*, **194**, 51–6.
38. Crocker, S.H., Eddy, D.O., Obenauf, R.N., Wismar, B.L. and Lowery, B.D. (**1981**) Bacteremia: host-specific lung clearance and pulmonary failure. *J. Trauma*, **1**, 215–20.
39. Dehring, D.J., Crocker, S.H., Wismar, B.L., Steinberg, S.M., Lowery, B.D. and Cloutier, C.T. (**1983**) Comparison of live bacteria infusions in a porcine model of acute respiratory failure. *J. Surg. Res.*, **34**, 151–8.
40. Alanee, S.R., McGee, L., Jackson, D., Chiou, C.C., Feldman, C., Morris, A.J., Ortqvist, A., Rello, J., Luna, C.M., Baddour, L.M., Ip, M., Yu, V.L. and Klugman, K.P International Pneumococcal Study Group. (**2007**) Association of serotypes of *Streptococcus pneumoniae* with disease severity and outcome in adults: an international study. *Clin. Infect. Dis.*, **45**, 46–51.
41. Mizrachi-Nebenzahl, Y., Lifshitz, S., Teitelbaum, R., Novick, S., Levi, A., Benharroch, D., Ling, E. and Dagan, R. (**2003**) Differential activation of the immune system by virulent *Streptococcus pneumoniae* strains determines recovery or death of the host. *Clin. Exp. Immunol.*, **134**, 23–31.
42. Madsen, L.W., Aalbaek, B., Nielsen, O.L. and Jensen, H.E. (**2001**) Aerogenous infection of microbiologically defined minipigs with *Streptococcus suis* serotype 2. A new model. *APMIS.*, **109**, 412–8.
43. Shaw, J.H. and Wolfe, R.R. (**1984**) A conscious septic dog model with hemodynamic and metabolic responses similar to responses of humans. *Surgery*, **95**, 553–61.
44. Fink, M.P. and Heard, S.O. (**1990**) Laboratory models of Sepsis and septic shock. *J. Surg. Res.*, **49**, 186–96.
45. Nichols, R.L., Smith, J.W. and Balthazar, E.R. (**1978**) Peritonitis and intra-abdominal abscess: an experimental model for the evaluation of human disease. *J. Surg. Res.*, **5**, 129–34.
46. Weinstein, W.M., Onderdonk, A.B., Bartlett, J.G. and Gorbach, S.L. (**1974**) Experimental intra-abdominal abscesses in rats: development of an experimental model. *Infect. Immun.*, **10**, 1250–5.
47. Ahrenholz, D.H. and Simmons, R.L. (**1980**) Fibrin in peritonitis. I. Beneficial and adverse effects of fibrin in experimental E. coli peritonitis. *Surgery*, **88**, 41–7.
48. Fink, M.P., MacVittie, T.J, and Casey, L.C. (**1984**) Inhibition of prostaglandin synthesis restores normal hemodynamics in canine hyperdynamic sepsis. *Ann. Surg.*, **200**, 619–26.
49. Natanson, C., Fink, M.P., Ballantyne, H.K., MacVittie, T.J., Conklin, J.J. and Parrillo, J.E. (**1986**)

Gram–negative bacteremia produces both severe systolic and diastolic cardiac dysfunction in a canine model that simulates human septic shock. *J. Clin. Invest.*, **78**, 259–70.

50. Nakatani, T., Sato, T., Marzella, L., Hirai, F., Trump, B.F. and Siegel, J.H. (**1984**) Hepatic and systemic metabolic responses to aerobic and anaerobic intra-abdominal abscesses in a highly reproducible chronic rat model. *Circ. Shock*, **13**, 271–94.
51. Bryant, P., Shumate, M., Yumet, G., Lang, C.H., Vary, T.C. and Cooney, R.N. (**2003**) Capsaicin-sensitive nerves regulate the metabolic response to abdominal sepsis. *J. Surg. Res.*, **112**, 152–61.
52. Deitch, E.A. (**2005**) Rodent models of intra-abdominal infection. *Shock*, **24**(Suppl 1), 19–23.
53. Ghiselli, R., Giacometti, A., Cirioni, O., Mocchegiani, F., Orlando, F., D'Amato, G., Sisti, V., Scalise, G. and Saba, V (**2004**) Cecropin B enhances [beta]-lactams activities in experimental rat models of Gram-negative septic shock. *Ann. Surg.*, **239**, 251–56.
54. Alexander, J.W., Gonce, S.J., Miskell, P.W., Peck, M.D. and Sax, H. (**1989**) A new model for studying nutrition in peritonitis. The adverse effect of overfeeding. *Ann. Surg.*, **209**, 334–40.
55. Yang, S. and Hauptman, J.G. (**1994**) The efficacy of heparin and antithrombin III in fluid-resuscitated cecal ligation and puncture. *Shock*, **2**, 433–7.
56. Wang, P. and Chaudry, I.H. (**1998**) A single hit model of polymicrobial sepsis: Cecal ligation and puncture. *Sepsis*, **2**, 227–33.
57. Singleton, K.D. and Wischmeyer, P.E. (**2003**) Distance of cecum ligated influences mortality, tumor necrosis factor-alpha and interleukin-6 expression following cecal ligation and puncture in the rat. *Eur. Surg. Res.*, **35**, 486–91.
58. Kato, T., Hussein, M.H., Sugiura, T., Suzuki, S., Fukuda, S., Tanaka, T., Kato, I. and Togari, H. (**2004**) Development and characterization of a novel porcine model of neonatal sepsis. *Shock*, **21**, 329–35.
59. Hubbard, W.J., Choudhry, M., Schwacha, M.G., Kerby, J.D., Rue, L.W3rd., Bland K.I. and Chaudry, I.H. (**2005**) Cecal ligation and puncture. *Shock*, **24**(Suppl. 1), 52–7.
60. Witek-Janusek, L. and Ratmeyer, J.K. (**1991**) Sepsis in the young rat: maternal milk protects during cecal ligation and puncture sepsis but not during endotoxemia. *Circ. Shock*, **33**, 200–6.
61. McMasters, K.M., J.C. Peyton, D.J. Hadjiminas, and W.G. Cheadle. (**1994**) Endotoxin and tumour necrosis factor do not cause mortality from caecal ligation and puncture. *Cytokine*, **36**, 530–6.
62. Ayala, A. and Chaudry, I.H. (**1996**) Immune dysfunction in murine polymicrobial sepsis: mediators, macrophages, lymphocytes and apoptosis. *Shock*, **6**(Suppl. 1), S27–38.
63. Wang, P., Ba, Z.F., Reich, S.S., Zhou, M., Holme, K.R. and Chaudry, I.H. (**1996**) Effects of nonanticoagulant heparin on cardiovascular and hepatocellular function after hemorrhagic shock. *Am. J. Physiol.*, **270**, H1294–1302.
64. Coopersmith, C.M., Amiot, D.M 2nd., Stromberg P.E., Dunne W.M., Davis C.G., Osborne D.F., Husain K.D., Turnbull I.R., Karl I.E., Hotchkiss R.S. and Buchman T.G. (**2003**) Antibiotics improve survival and alter the inflammatory profile in a murine model of sepsis from Pseudomonas aeruginosa pneumonia. *Shock*, **19**, 408–14.
65. Turnbull, I.R., Wlzorek, J.J., Osborne, D., Hotchkiss, R.S., Coopersmith, C.M. and Buchman, T.G. (**2003**) Effects of age on mortality and antibiotic efficacy in cecal ligation and puncture. *Shock*, **19**, 310–3.
66. Turnbull, I.R., Javadi, P., Buchman, T.G., Hotchkiss, R.S., Karl, I.E. and Coopersmith, C.M. (**2004**) Antibiotics improve survival in

sepsis independent of injury severity but do not change mortality in mice with markedly elevated interleukin 6 levels. *Shock*, **21**, 121–5.
67. Turnbull, I.R., Buchman, T.G., Javadi, P., Woolsey, C.A., Hotchkiss, R.S., Karl, I.E. and Coopersmith, C.M. (**2004**) Age disproportionately increases sepsis-induced apoptosis in the spleen and gut epithelium. *Shock*, **22**, 364–8.
68. Watts, J.A., Kline, J.A., Thornton, L.R., Grattan, R.M. and Brar, S.S. (**2004**) Metabolic dysfunction and depletion of mitochondria in hearts of septic rats. *J. Mol. Cell. Cardiol.*, **36**, 141–50.
69. Morrison, A.M., Wang, P. and Chaudry, I.H (**1996**) A novel nonanticoagulant heparin prevents vascular endothelial cell dysfunction during hyperdynamic sepsis. *Shock*, **6**, 46–51.
70. Rittirsch, D., Hoesel, L.M, and Ward, P.A. (**2007**) The disconnection between animal models of sepsis and human sepsis. *J. Leukoc. Biol.*, **81**, 137–43.
71. Song, M, and Kellum, J.A. (**2005**) Interleukin-6. *Crit. Care Med.*, **33**, S463–5.
72. Eskandari, M.K., Bolgos, G., Miller, C., Nguyen, D.T., DeForge, L.E. and Remick, D.G. (**1992**) Anti-tumor necrosis factor antibody therapy fails to prevent lethality after cecal ligation and puncture or endotoxemia. *J. Immunol.*, **148**, 2724–30.
73. Evans, G.F., Snyder, Y.M., Butler, L.D. and Zuckerman, S.H. (**1989**) Differential expression of interleukin-1 and tumor necrosis factor in murine septic shock models. *Circ. Shock*, **29**, 279–90.
74. van der Poll, T. and Lowry, S.F. (**1995**) Tumor necrosis factor in sepsis: mediator of multiple organ failure or essential part of host defense? *Shock*, **3**, 1–12.
75. van der Poll, T., Marchant, A., Buurman, W.A., Berman, L., Keogh, C.V., Lazarus, D.D., Nguyen, L., Goldman, M., Moldawer, L.L. and Lowry, S.F. (**1995**) Endogenous IL-10 protects mice from death during septic peritonitis. *J. Immunol.*, **155**, 5397–401.
76. Tracey, K., Beutler, B., Lowry, S, *et al.* (**1986**) Shock and tissue injury induced by recombinant human cachectin. *Science*, **243**, 470–4.
77. Remick, D., Kunkel, R., Larrick, J, *et al.* (**1987**) Acute *in vivo* effects of human recombinant tumor necrosis factor. *Lab. Invest.*, **56**, 583–90.
78. Selby, P., Hobbs, S., Viner, C, *et al.* (**1987**) Tumour necrosis factor in man: clinical and biological observations. *Br. J. Cancer*, **56**, 803–8.
79. Sherman, M., Spriggs, D., Arthur, K, *et al.* (**1988**) Recombinant human tumor necrosis factor administered as a 5-day continuous infusion in cancer patients: phase I toxicity and effects on lipid metabolism. *J. Clin. Oncol.*, **6**, 344–50.
80. Hinshaw, L.B., Tekamp-Olson, P., Chang, A.C, *et al.* (**1990**) Survival of primates in LD100 septic shock following therapy with antibody to tumor necrosis factor (TNF alpha). *Circ. Shock*, **30**, 279–92.
81. Walsh, C.J., Sugerman, H.J., Mullen, P.G, *et al.* (**1992**) Monoclonal antibody to tumor necrosis factor alpha attenuates cardiopulmonary dysfunction in porcine Gram-negative sepsis. *Arch. Surg.*, **127**, 138–44.
82. Jesmok, G., Lindsey, C., Duerr, M, *et al.* (**1992**) Efficacy of monoclonal antibody against human recombinant tumor necrosis factor in *E coli*-challenged swine. *Am. J. Pathol.*, **141**, 1197–1207.
83. Zanotti, S., Kumar, A. and Kumar, A. (**2002**) Cytokine modulation in sepsis and septic shock. *Expert Opin. Investig. Drugs*, **11**, 1061–75.
84. Adrie, C. and Pinsky, M.R. (**2000**) The inflammatory balance in human sepsis. *Intensive Care Med.*, **26**, 364–75.
85. Khalil, A.A., Hall, J.C., Aziz, F.A. and Price, P. (**2006**) Tumour necrosis factor: implications for surgical patients. *ANZ. J. Surg.*, **76**, 1010–6.
86. Maier, S., Traeger, T., Entleutner, M., Westerholt, A., Kleist, B., Huser, N., Holzmann, B., Stier, A., Pfeffer, K.

and Heidecke, C.D. (**2004**) Cecal ligation and puncture versus colon ascendens stent peritonitis: two distinct animal models for polymicrobial sepsis. *Shock*, **21**, 505–11.

87. Esmon, C.T. (**2004**) Why do animal models (sometimes) fail to mimic human sepsis? *Crit. Care Med.*, **32**(Suppl), S219–22.
88. Arcaroli, J., Fessler, M.B. and Abraham, E. (**2005**) Genetic polymorphisms and sepsis. *Shock*, **24**, 300–12.
89. Adrie, C., Adib-Conquy, M., Laurent, I., Monchi, M., Vinsonneau, C., Fitting, C., Fraisse, F., Dinh-Xuan, A.T., Carli, P., Spaulding, C., Dhainaut, J.F. and Cavaillon, J.M. (**2002**) Successful cardiopulmonary resuscitation after cardiac arrest as a "sepsis-like" syndrome. *Circulation*, **106**, 562–8.
90. Annane, D., Bellissant, E. and Cavaillon, J.M. (**2005**) Septic shock. *Lancet*, **365**, 63–78.
91. Adib-Conquy, M., Monchi, M., Goulenok, C., Laurent, I., Thuong, M., Cavaillon, J.M. and Adrie, C. (**2007**) Increased plasma levels of soluble triggering receptor expressed on myeloid cells 1 and procalcitonin after cardiac surgery and cardiac arrest without infection. *Shock*, Jun 7; epub.

17
The Past, Present and Future of Therapies for Sepsis and Non-infectious Systemic Inflammation

Nitin Seam and Anthony F. Suffredini

17.1
Introduction

The therapies for sepsis and severe non-infectious systemic inflammation reflect an integration of achievements in understanding the pathogenesis of infection, the physiology of circulatory shock and the mechanisms of acute inflammation. Prior to the 20th century, infectious diseases were the leading causes of death in the developed world [1]. During the last century, the development of hygienic practices, sanitation, immunization and potent antibiotics decreased the severity and lethality of many infections. In parallel to the success in preventing and treating infections, the contribution of acute inflammation to shock and the morbid events associated with critical illness and injury was appreciated as the essential host factor that affected survival. In less than three decades, dramatic strides have been made in understanding the pathways and therapies for infectious and non-infectious acute inflammation that accompanies severe illness or injury. This chapter will review some of the important developments in our understanding of the treatment of severe inflammation associated with infection or injury.

17.2
Historical Elements in the Treatment of Infection

In the 16th century, the principles of Hippocrates (circa 480–370 BC) were still the standard of medical care. Health was defined by the proper proportion of four basic humors: blood, phlegm, black and yellow bile and treatments were designed to counteract humoral imbalances for a variety of illnesses [2]. Ambrose Paré (circa 1537), a French Renaissance surgeon, revolutionized the approach to controlling the source of infection with his treatment of battlefield injuries. In lieu of wound cauterization with scalding oil or branding irons, Paré introduced dressings of egg, oil of roses and turpentine as well as blood

Sepsis and Non-infectious Systemic Inflammation. Edited by J.-M. Cavaillon and C. Adrie

ISBN: 978-3-527-31935-0

vessel ligatures to prevent the spread of infection [3]. He further described source control of decubitus ulcers by employing drainage of abscesses and removal of loose bone chips [4].

Almost three centuries later, a Hungarian surgeon, Ignaz Phillip Semmelweiss (circa 1840), highlighted the role of the caregiver in the spread of puerperal fever. Infection spread from the hands of the examining physician through lacerations in the vagina, with subsequent spread to the bloodstream [5]. In Vienna, Austria (1847), he instituted a program of prophylactic hand washing with soap, warm water and chlorine as well as disinfection of instruments and linens [6]. This approach rapidly reduced mortality on the obstetric ward [7]. These concepts were extended by Joseph Lister at Glasgow University, who used antiseptic dressings of carbolic acid to control infection of contaminated wounds [8].

In 1798, Edward Jenner transformed the world of infectious diseases with his observations that inoculation with cowpox pus provided protection against smallpox, a major killer in Europe [9]. Nearly 75 years later, investigators including Louis Pasteur and Robert Koch provided the groundwork demonstrating that microorganisms are the causes of many diseases [10]. Pasteur helped to develop the burgeoning field of immunotherapy by showing that weakened strains of infections could serve as vaccines to cholera, anthrax and rabies. Emil von Behring and Shibasaburo Kitasato (1890) used the principles of immunotherapy to develop antitoxins to the scourges of diphtheria and tetanus [11, 12]. After immunizing animals with sublethal doses of bacterial toxins, in 1894 they used the resultant immune serum to treat children with diphtheria with dramatic results. Reduction of diphtheria mortality was demonstrated both with serum therapy and later with active immunization [13].

In the early 20th century, treatment of infection was limited to the most basic supportive care including the use of topical antimicrobial agents (e.g. antiseptic soaps) and drainage of infected foci. If available, immune serum from either animals or survivors of specific infections might be administered [14]. Paul Ehrlich (1909), a pioneer in the new field of chemotherapy, developed a novel parenteral arsenic derivative, Salvarsan, to treat syphilis [15]. This chemotherapeutic approach to treat infection provided the intellectual underpinnings for the seminal discovery of penicillin, which dramatically transformed the treatment of serious infections.

In 1928, Sir Alexander Fleming discovered the inhibitory effects of the *Penicillium* mold on a staphylococcal culture [16]. After demonstrating that penicillin was bacteriostatic and bactericidal, Fleming injected it into animals and noted no apparent toxicity. However, the publication of his results initially generated little enthusiasm, as many believed that immunizations, rather than chemotherapy, would cure bacterial infections. Ten years later, Howard Florey and co-workers demonstrated the efficacy of penicillin in a murine model of streptococcal infection and human clinical trials began in 1941 [17, 18]. In the immediate post World War II era, the use of penicillin and sulfonamides greatly diminished the severity of Gram-positive infections, but Gram-negative infections increased in frequency and severity [19, 20].

17.3 Historical Elements in the Treatment of Inflammation Associated with Infection or Injury

In the first century AD, Aulus Cornelius Celsus noted the four cardinal signs of inflammation: "rubor, et tumor cum calore et dolore" (redness and swelling with heat and pain) as a response of flesh to microbes after surgery [21]. However, a deeper understanding of inflammatory responses associated with injury and infection awaited the development of microscopic examination of infected and injured tissues. The early period of research in host immune response to microbial pathogens was recognized in 1908 when Paul Ehrlich and Ilya Metchnikov were awarded the Nobel Prize in Medicine for Ehrlich's observations on humoral immunity and Metchnikov's work on phagocytosis [22].

An early link between infection, inflammation and adrenal dysfunction was made by Waterhouse and Friderichsen who described severe infections associated with peripheral vascular collapse and adrenal hemorrhage [23, 24]. In the late 1920s and 1930s, investigators could sustain adrenalectomized animal using organic extracts of the adrenals [25]. By the 1930s, Rich described characteristic histologic changes in the adrenal cortex without hemorrhage in cases of severe meningococcal, streptococcal and diphtheria infections associated with peripheral vascular collapse [26]. Case series during this period describe the salutary effect of the use of animal adrenal extracts in humans with enteric fever [27] and pneumococcal pneumonia [28]. While corticosteroids had beneficial effects on the signs and symptoms of infections, their administration could be complicated by the recrudescence of chronic infections such as tuberculosis [29].

Wesley Spink and Max Harry Weil, two visionaries in septic shock research, examined the clinical course of patients with documented Gram-negative bacteremia at the University of Minnesota hospitals from 1950 to 1955 [30]. The effects of corticosteroids were reviewed in a small subgroup of patients in shock who required vasopressor therapy. Weighing the potential benefits and harm of using corticosteroids in patients with severe infection, Spink provided a prescient conclusion,

> *"When one is confronted with a critically ill patient it may be necessary to administer an agent or agents that can elicit harmful side-effects. Adrenocorticotropin and the corticosteroids have profound metabolic effects when administered to human subjects. . . . these hormonal agents, when given over a brief period of time to critically ill patients with infectious diseases, or to those with debilitating complications as a result of infections, will often produce dramatic improvement without provoking harmful effects. In some instances the judicious and prompt use of these steroids can prevent fatalities."* [31].

By the 1960s, several investigations provided further support for the use of corticosteroids in sepsis. J. Melby *et al.* induced lethal inflammation in dogs with endotoxin and found that the rise in liver transaminases, a marker of cell injury, was attenuated by pharmacologic doses of cortisol [32]. Richard Lillehei *et al.* showed that pre-treatment with high-dose corticosteroids increased survival in dogs given a lethal dose of endotoxin [33]. These and other experiments prompted clinical investigations of high-dose steroids in humans.

Hundreds of anecdotal reports and trials using historical controls to evaluate the effects of corticosteroids in sepsis were published in the literature [34]. During a 25-year period (1963–1988) only nine randomized controlled trials (seven blinded and two open label) were conducted to assess the effects of high doses of steroids in sepsis [34]. One of the nine trials had remarkably high survival rates but was a statistical outlier in comparison to other trials from that period. When this trial was removed from a meta-analysis, the remaining eight trials of short courses of high doses of steroids were found to be harmful and deleterious to survival [34] (Figure 17.1).

17.4 Historical Developments in the Treatment of Shock

In 1831, Laënnec suggested an association between infection and shock, noting the weakness of heart sounds in severe febrile conditions [35]. In 1899, Romberg and Pässler showed that rabbits inoculated with lethal doses of diphtheria and pneumococcus developed hypotension independent of injury to the heart [36]. By the early 20th century, physicians recognized that death from infectious diseases was not due to cardiac failure. Janeway wrote:

> *"We must in most cases abandon the idea of cardiac death at the height of acute infectious diseases, such as pneumonia, typhoid fever and the septic fevers... In place of heart failure, we must write vasomotor failure."* [37].

Dale and Laidlaw suggested a mechanism for vasomotor failure by showing that histamine produced secondary shock caused by pre-capillary arteriolar contraction along with capillary dilation [38].

The historic belief that "hemorrhage is shock and shock is hemorrhage" was repudiated by Brooks and Blalock (among others), who demonstrated in animals that hemorrhage first decreased cardiac output, then blood pressure as opposed to histamine-induced shock which first decreased blood pressure, then cardiac output [39]. By the 1930s, Atchley differentiated "medical shock" due to "anhydremia or toxin" causing vasomotor collapse from "surgical shock" due to hemorrhage or trauma [40]. Since Warfield also believed that shock from infectious disease was due to peripheral circulatory collapse rather

Study, Year (Reference)	Patients (Deaths), n (n) Steroid Group	Control Group	Relative Survival Benefit (95% CI)
Bennett et al., 1963 (35)	96 (54)	98 (32)	0.65 (0.50–0.84)
Klastersky et al., 1971 (17)	46 (24)	39 (22)	1.10 (0.69–1.75)
Schumer, 1976 (16)	86 (9)	86 (33)	1.45 (0.69–1.73)
Thompson et al., 1976 (36)	28 (22)	32 (25)	0.98 (0.37–2.59)
Lucas and Ledgerwood, 1984 (37)	23 (5)	25 (5)	0.98 (0.73–1.31)
Sprung et al., 1984 (18)	43 (33)	16 (11)	0.74 (0.29–1.89)
Bone et al., 1987 (19)	191 (65)	190 (48)	0.88 (0.78–1.01)
VA, 1987 (38)	112 (23)	111 (24)	1.01 (0.89–1.16)
Luce et al., 1988 (39)	38 (22)	37 (20)	0.92 (0.55–1.53)
Bollaert et al., 1998 (31)	22 (7)	19 (12)	1.85 (1.01–3.40)
Briegel et al., 1999 (32)	20 (3)	20 (4)	1.06 (0.80–1.42)
Yildiz et al., 2002 (34)	20 (8)	20 (12)	1.50 (0.79–2.83)
Annane et al., 2002 (30)	150 (82)	149 (91)	1.17 (0.89–1.52)
Summary			
All studies ($n = 13$)	875 (357)	842 (339)	1.01 (0.94–1.09)
Pre-1989 studies with reference 16 ($n = 9$)	663 (257)	634 (220)	0.97 (0.89–1.04)
Pre-1989 studies without reference 16 ($n = 8$)	577 (248)	548 (187)	0.89 (0.82–0.97)
Post-1997 studies ($n = 4$)	212 (100)	208 (119)	1.23 (1.01–1.50)

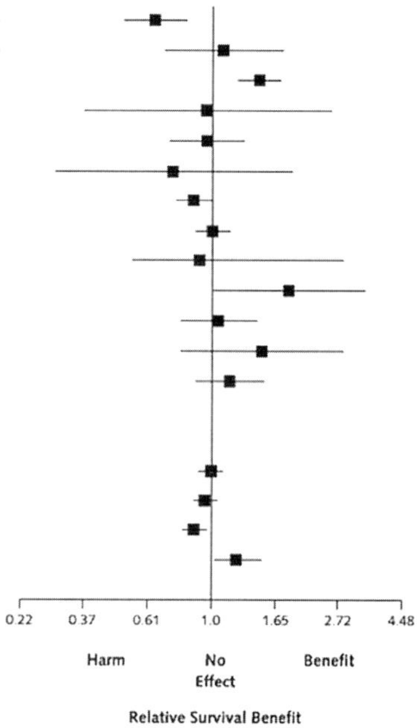

Figure 17.1 The effects of steroids on survival in sepsis: an analysis of clinical sepsis trials using high and low doses of corticosteroids. Relative survival benefits (closed boxes) are shown with 95% confidence intervals (horizontal lines). The meta-analysis of 13 trials (nine trials pre-1989 and four trials post-1997) demonstrates significant variability with no overall improvement in relative survival benefit (fixed-effects estimate, 1.01 (95% CI, 0.94 to 1.09). The effect of steroids in the trials published before 1989 (high-dose steroids) compared with those published after 1997 (low-dose steroids) differed significantly ($P = 0.02$). One significant outlier trial was removed and the remaining, eight pre-1989 steroid trials had a consistent harmful effect on survival (fixed-effects estimate, 0.89 (CI, 0.82 to 0.97)). In the four trials published after 1997, the effect of low-dose steroids on the relative survival benefit was consistently beneficial (fixed-effects estimate, 1.23 (CI, 1.01 to 1.50)). (Reproduced from US government work [34].)

than heart failure, he implored his colleagues to cease their empiric treatment with digitalis in febrile patients and instead focus on the restoration of effective blood volume [41].

By the 1950s, novel vasopressors were beginning to be identified [42–44]. In 1948, Goldenberg and colleagues showed that noradrenaline increased blood pressure by increased peripheral resistance, slowed pulse and decreased cardiac output [45]. Treatment of septic shock patients with levarterenol, an isomer of norepinephrine, and metaraminol, an α_1-adrenergic receptor agonist, increased blood pressure, decreased pulse rate and improved patient appearance [31, 46]. By the 1960s, however, it was recognized that increased peripheral resistance with vasopressor use would compromise effective blood flow to vital organs. In addition, the effects of different vasopressors in patients with different categories of shock were compared. Angiotensin increased peripheral resistance, lowered cardiac index and decreased urine flow compared to levarterenol or metaraminol [47] and thus its clinical use was discontinued. Dopamine increased cardiac output, renal blood flow and sodium excretion in patients with congestive heart failure suggesting a potential benefit over other vasopressors [48, 49]. In 1966, MacCannell noted cardiovascular improvement and increased urine output in shock patients who received dopamine after being treated with hydration and metaraminol [50].

In the 1960s, many experts felt a systematic approach was needed to decrease the high mortality in septic shock because much of the treatment at the time was empiric. Max Harry Weil started a shock research unit at the Los Angeles County/University of Southern California Medical Center and used his clinical observations to propose a bedside management algorithm of shock, regardless of cause. He termed this algorithm "Ventilation, Infusion, Pump" or VIP [51]. He reasoned that failure of gas exchange was the most frequent cause of death in shock [52, 53], and therefore ventilation was the first step in managing shock followed by the infusion of fluids or blood until an elevated central venous pressure suggested that cardiac preload was adequate. Lastly, if the central venous pressure rose to greater than 5 cm H_2O after fluid challenge it suggested that limited cardiac reserve was present, and treatment with digoxin, isoproterenol or phlebotomy was indicated. Weil believed this early systemic approach to shock resuscitation was needed before pharmacologic treatment with pressors (P) or surgical (S) interventions (termed PS) were instituted.

> *"Trial and error, at a time when the patient's likelihood of survival is small, has proven statistically unsound because the effect of the error is too costly... Systematic management in which the VIP precedes the PS though seemingly consuming greater time, provides essential guides for professional judgment and, therefore, greater assurance of ultimate success."* [51].

The widespread use of hemodynamic monitoring in the intensive care unit followed the development of the pulmonary artery catheter by Jeremy Swan and William Ganz for cardiac diagnostic procedures [54, 55], facilitating further characterization of cardiac performance in shock states. In the 1980s, William Shoemaker and others used hemodynamic monitoring to help direct therapies that would enhance flow and oxygen delivery to tissues in shock states [56–59]. Shoemaker *et al.* later studied the effects of achieving supranormal goals for cardiac output and oxygen delivery using fluids, blood and inotropes in high-risk general surgery patients. While their data suggested a survival advantage with this approach, the purported beneficial effects were not reproduced in other critically ill patients [60, 61].

17.5
Contemporary Developments in Sepsis and Non-infectious Systemic Inflammation: the Molecular Basis of Inflammation associated with Infection and Shock

In the early 1980s, a more detailed understanding of the pathogenesis and mechanisms associated with severe infections and injury began to emerge. Endotoxin, a principal bacterial mediator of shock, was a major component of many septic shock models. This discovery was facilitated by the development of methods to measure endotoxin in biological fluids [62] and the availability of purified endotoxin preparations. Jacob Fine postulated an important role for gut-derived endotoxin leaking into the circulation and contributing to the development of shock states [63]. Measurement of endotoxin in blood suggested that it contributed to the pathogenesis of both severe Gram-positive and Gram-negative infections [64]. However, models of septic shock showed that while sufficient to cause shock, endotoxin was not a necessary component of septic shock states since low or non-endotoxin producing organisms or non-endotoxin bacterial components were sufficient to cause inflammatory responses and shock [65].

Janeway and Medzhitov developed many fundamental concepts that defined host and pathogen interactions [66, 67]. Innate immune responses are germ-line encoded reactions of the host to non-self molecules on microbial pathogens. These molecules termed pathogen-associated molecular patterns, were microbial class (i.e. endotoxin, exotoxin, mannans, petidoglycans, bacterial DNA) rather than pathogen specific. Host receptors in the circulation (i.e. C-reactive protein, serum amyloid A, complement, mannose binding proteins) and receptors present on the cell surface (i.e. Toll-like receptors, CD14, mannose receptor, complement receptors) recognize and bind microbial products resulting in the activation of barrier (i.e. vascular endothelium, lung and gut epithelium) and inflammatory cells (i.e. tissue and circulating myeloid cells). An immediate cascade of inflammatory events ensues: endothelial cell activation with activation of coagulation, kinin and fibrinolytic pathways, as well as myeloid cell activation with the release of pleiotropic inflammatory molecules (i.e. cytokines and chemokines) that amplify inflammatory responses. The clinical manifestations of these systemic inflammatory responses include fever, leukocytosis, tachycardia, and tachypnea. The American College of Chest Physicians/Society of Critical Care Medicine Consensus Conference clarified the terminology used to describe this syndrome and standardized definitions for sepsis and organ failure [68]. This clinical syndrome was termed the Systemic Inflammatory Response Syndrome (SIRS) and if evidence of infection was present, it was called sepsis. If sufficiently intense, the cascade of inflammation culminates in the development of hypotension, with systemic inflammation and tissue injury [68]. Septic patients with organ injury were considered to have severe sepsis and if persistently hypotensive, were designated as having septic shock [68]. This terminology has been important for developing consistency in trial development and potentially helping clinicians recognize infections at an earlier stage. However, while pragmatic, the broad nature of these definitions provides little insight into the clinical differences that exist within heterogeneous patient populations that develop severe infections, tissue injury and shock.

The discovery of tumor necrosis factor (TNF) heralded the beginning of a new era in understanding the molecules and mechanisms that underlie host–pathogen interactions [69]. Beutler *et al.* showed that passive immunization against cachectin/tumor necrosis factor protected mice from the lethal effect of endotoxin [70]. The prophylactic effect was dose-dependent and most effective when antiserum was administered prior to endotoxin injection. Similar beneficial effects were observed in primates treated with anti-TNF antibodies prior to lethal *E. coli* bacterial infusion. Notably, in primates complete protection against shock, vital organ dysfunction, persistent stress hormone release and death was conferred by administration of antibodies 2 h before the bacterial infusion [71]. Such models provided the proof of principle that the inflammatory cascade could be modulated with beneficial effects. However, the issue of timing with respect to the onset of infection remained clinically problematic (see below).

17.6 Endotoxin as a Therapeutic Target

The pivotal role of endotoxin in the pathogenesis of sepsis was suggested by the identical appearance of shock syndromes that occurred with systemic administration of endotoxin to animals. With the predominance of Gram-negative infections that were present in hospitalized patients during the period 1950–1970s, this was recognized as an important target in the adjunctive treatment of sepsis. A. Braude and E. Ziegler demonstrated the potential efficacy of antiserum against Gram-negative pathogens first in animals and then in humans. Using an approach that was pioneered earlier in the century, these investigators immunized firemen with a strain of bacteria (*E. coli* J5), and then harvested their plasma. The immune plasma was administered to patients with Gram-negative sepsis. While the initial results were promising [72], later trials did not show similar beneficial effects [73]. Because of the practical difficulty of using plasma from immunized volunteers, monoclonal antibodies (HA1A, E5) directed against the core structure of endotoxin (Lipid A) were developed. The initial multicenter HA1A trial showed beneficial effects in septic patients [74] but because of irregularities noted in the trial results due to changes in the primary endpoints, the developers were required to conduct another clinical trial [75]. This later trial demonstrated that the antibody had no beneficial effects and was associated with worse outcomes than placebo [76]. Another endotoxin monoclonal antibody, E5, underwent three clinical trials in patients with septic shock and sepsis without shock, yet none was successful [77–79]. Notably, neither of these antibodies had reproducible neutralizing activity *in vitro* [80] (Figure 17.2).

Other approaches to the inhibition of the effects of endotoxin were also unsuccessful. Bactericidal permeability-increasing protein was evaluated in children with systemic meningococcal disease, a fulminant infection associated with high levels of circulating endotoxin, but did not improve outcome [81] (Figure 17.2). This was likely due in part to logistical delays of up to 6 h that occurred prior to patient entry into the trial and the rapid demise of the most severely ill patients within hours of hospital admission, leaving less ill patients to be enrolled in the study [82]. Two approaches are currently being evaluated in clinical trials to assess the effects of inhibiting the endotoxin receptor, Toll-like receptor 4 (TLR4). One approach uses a cyclohexene derivative (TAK-242) that inhibits TLR4 signal transduction and the other utilizes a synthetic lipid A analog, eritoran tetrasodium (E5564) that blocks the binding of endotoxin to TLR4 (http://clinicaltrials.gov NCT 00143611 and NCT 00334828 respectively).

17.7 Immunotherapy Revisited

Treating sepsis with immune serum from patients or previously immunized animals was pioneered at the turn of the 20th century, but has recently been the subject of renewed interest. Three recent meta-analyses however

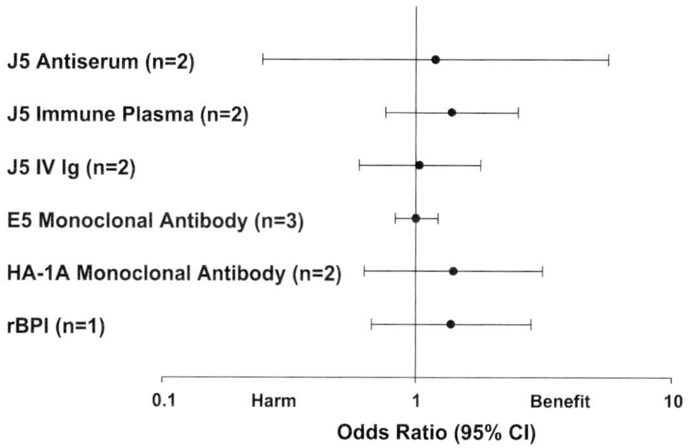

Figure 17.2 Summary of 12 clinical trials in sepsis that have used six different agents to neutralize the effects of endotoxin. Each agent with the number of reported clinical trials is shown. The circles represent the odds ratio of survival with the lines showing the 95% confidence intervals. (Data for figure compiled from [73] and [81].) rBPI, recombinant bactericidal permeability increasing protein.

reached different conclusions regarding the efficacy of this approach as an adjuvant therapy in critically ill patients with sepsis [83–85]. The most recent meta-analysis analyzed 20 randomized trials and concluded that polyclonal immunoglobulin may decrease the risk of death in sepsis [85]. However, the authors note several limitations in these trials; most had methodological limitations and were performed before modern intensive care management strategies were standard [85]. Thus, firm conclusions regarding the efficacy of this approach await new trials.

17.8 Anti-inflammatory Approaches to Sepsis and Systemic Inflammation

During the period 1980–1990s, numerous clinical trials of adjuvant therapies for septic shock were instituted. Several trials showed that broad suppression of host inflammatory responses with high doses (i.e. grams) of corticosteroids for 1 to 2 days was harmful [86–88]. In parallel, major developments in molecular medicine described both the triggers and the mediators of acute inflammation [89, 90]. These discoveries suggested that the promise of mediator-specific interventions would limit the toxicities seen with high doses of steroids. Multiple animal models demonstrated that inhibitors or antagonists of these molecules (i.e. TNF, IL-1, platelet activating factor, bradykinin) significantly improved survival in septic animals [91].

Several categories of non-glucocorticoid anti-inflammatory agents were evaluated in clinical trials in sepsis and septic shock. These included IL-1 receptor antagonist, anti-TNF antibodies and TNF soluble receptors, anti-prostaglandin, anti-bradykinin, anti-platelet factor antagonist, and phospholipase inhibitors [92–94]. While preclinical data suggested potential benefit from these agents, surprisingly none was clearly beneficial in sepsis trials. A meta-analysis of 27 trials of mediator-specific anti-inflammatory therapies in sepsis showed a small but significant overall improvement in survival (odds ratio of survival, 1.09; 95% confidence interval 1.01–1.18, $p = 0.03$) with an absolute risk reduction of 2% and a relative risk reduction of 7% [92] (Figure 17.3). This modest effect suggests that if future trials are conducted as before, enrollment of several thousand patients would be required to demonstrate benefit. However, the majority of these past trials did not target patients with the highest risk of death and this may have undermined the power of these trials to demonstrate beneficial effects.

17.9 Resurrecting the Use of Corticosteroids in Sepsis

In the 1990s, investigators began to reassess the role of corticosteroids in sepsis because of increased recognition that inflammatory injury is a major factor in

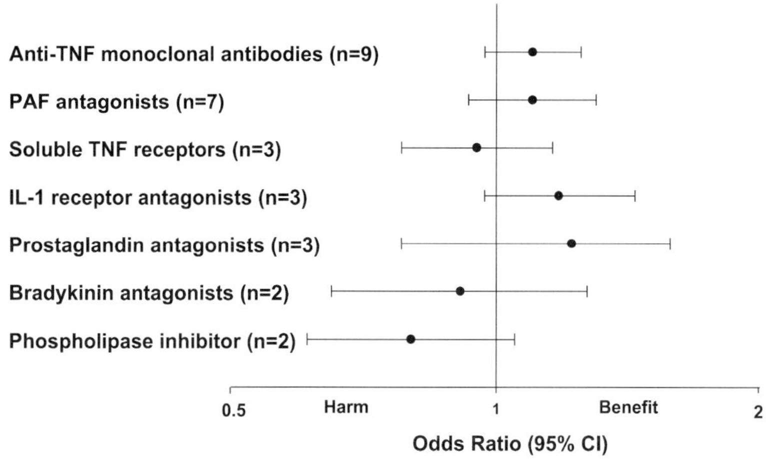

Figure 17.3 Summary of 29 clinical trials in sepsis evaluating seven different types of mediator-specific anti-inflammatory therapies. Each class of agent with the number of reported clinical trials is shown. The circles represent the odds ratio of survival with the lines showing the 95% confidence intervals. (Data for figure compiled from [92–94].)

the development of organ failure and death in sepsis. A meta-analysis showed that the use of high dose steroids (i.e. using a total median dose of 23 975 mg of hydrocortisone equivalents) for 1 day, was associated with significant harm possibly due to greater immunosuppression and greater risk of secondary infections [34] (Figure 17.1). Due in part to a new understanding of relative adrenal insufficiency and its association with mortality, five studies were performed in the 1990s to evaluate the role of smaller stress doses of steroids (i.e. 300 mg day of hydrocortisone) in sepsis and septic shock. These studies used 80-fold less corticosteroids (total median dose 1209 mg hydrocortisone) for a longer period of time (median 6 days) with a tapering schedule and found significant beneficial effects on shock reversal and mortality [34] (Figure 17.1). Thus, while the contribution of adrenal function to potential benefit in clinical sepsis has been recognized since the 1930s [27], it took almost 75 years to determine the dose of steroids that may benefit patients with severe sepsis.

17.10
The Role of Anti-coagulant Therapy in Sepsis

Microthrombi and fibrin deposition in damaged organs are common sequelae of systemic inflammation associated with injury or infection [95]. This results from endothelium activation by inflammatory mediators with the release of procoagulant and fibrinolytic factors [96]. Coagulation factors such as thrombin are pro-inflammatory and can accelerate the development of tissue injury [96].

Thirteen clinical trials using either antithrombin III, tissue factor pathway inhibitor or activated protein C (drotrecogin alfa activated) have been conducted to determine whether targeting this portion of acute systemic inflammation would benefit patients with severe infections [97] (Figure 17.4). All of these agents have an increased risk of bleeding associated with their use.

Only activated protein C has been approved for limited use in sepsis. On the basis of a single Phase III trial, the US Food and Drug Administration approved the use of activated protein C for patients with septic shock and a high severity of illness score (APACHE II score ≥ 25) [98]. This approval was somewhat controversial because of changes in the study protocol, and questions arose regarding its efficacy in patient subgroups and potential serious bleeding risks [99]. At the time of drug approval, there was no precedent for using the APACHE II system to select patients for the use of an anti-inflammatory agent in sepsis. The retrospective application of this score to the trial results, the absence of validation of this approach and the recognized inter-observer and intra-observer variability in calculating the APACHE II score raised serious concerns about this agent in sepsis [99]. Post-marketing trials of activated protein C in pediatric patients [100] and in patients with less severe sepsis (APACHE II < 25) were stopped prematurely because of lack of efficacy and resulted in further restrictions on its use (i.e. surgical patients with single organ failure) [101, 102]. In randomized trials comprising 5000 patients to date, only 17% have derived benefit from this agent (Figure 17.5). Of greater concern is the fact that the purported benefit of activated protein C has not been reproduced in subsequent trials and with greater use outside of the clinical trial setting, higher rates of serious bleeding and intracerebral hemorrhage

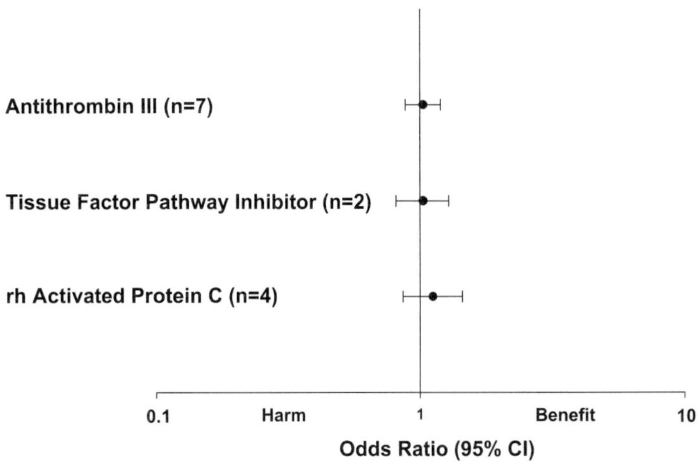

Figure 17.4 Summary of 13 clinical trials in sepsis evaluating three different types of anti-coagulant therapies. Each class of agent with the number of reported clinical trials is shown. The circles represent the odds ratio of survival with the lines showing the 95% confidence intervals. (Data for figure compiled from reference [96, 97].)

Trial (APACHE II Score)	Number of Patients (Nonsurvivors / Total)	
	Placebo	rhAPC
PROWESS		
(3-19)	26 / 215	33 / 218
(20-24)	57 / 222	49 / 218
(25-29)	58 / 162	48 / 204
(30-53)	118 / 241	80 / 210
ADDRESS		
(<25)	184 / 1147	197 / 1168
(≥25)	39 / 159	49 / 165
RESOLVE	36 / 198	34 / 201
Phase II**	4 / 15	5 / 15

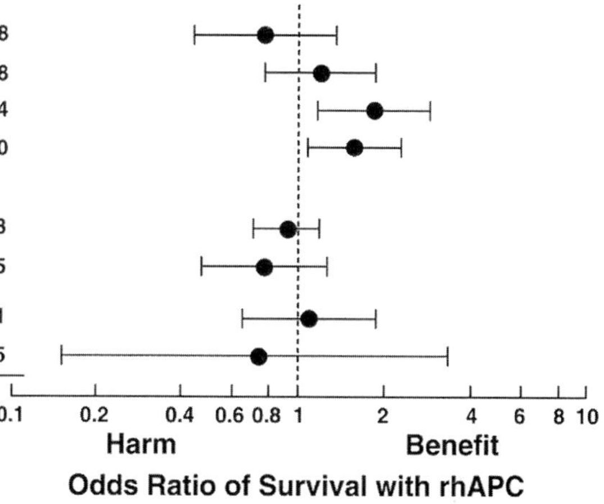

Figure 17.5 Summary of four clinical trials in sepsis evaluating activated protein C. The odds ratio of survival (95% confidence interval) for the effects of recombinant human activated protein C (rhAPC) versus placebo in the PROWESS, ADDRESS, RESOLVE, and phase II trials based on 28-day survival data. Trials were further categorized based on Acute Physiology and Chronic Health Evaluation (APACHE) II scores when available. (Reproduced from US government work [101].)

have been reported [103]. A new post-marketing trial of activated protein C compared to placebo is now being conducted (http://clinicaltrials.gov NCT 00279214).

17.11 Immunostimulatory Therapies for Sepsis and Shock

Patients with shock from any cause will develop a state of immunosuppression that is manifested by decreased host responsiveness to microbial antigens. Whole blood from patients with severe trauma, septic or cardiogenic shock has a diminished capacity to produce pro-inflammatory cytokines in response to endotoxin or intact bacteria [104–106]. Cell-mediated immunity is depressed and monocytes from patients with septic, traumatic injury or cardiogenic shock show decreased responsiveness to secondary stimuli (i.e. endotoxin, anti-CD3/CD28) with impaired Th1 cytokine secretion, T cell proliferative responses, and decreased expression of HLA-DR molecules [107–109]. This has been postulated to be an adaptive response to limit tissue damage [110]. However, the state of relative immunosuppression is associated with an increased risk for secondary infections. The etiology is multifactorial in nature and includes the effects of anti-inflammatory cytokines, extracellular ubiquitin and sex hormones [106, 110–112].

Immuno-enhancing therapies for the treatment of sepsis, including colony-stimulating factors and interferon gamma, have undergone limited evaluation in clinical trials. Granulocyte colony-stimulating factor (G-CSF) has been developed for use in patients with depressed immunity, neutropenia and high risk of infection due to cancer therapies or stem cell transplantation. Neutrophil number and function are enhanced by G-CSF that may also help to improve microbial clearance. In addition, G-CSF has some anti-inflammatory effects. While theoretically appealing, G-CSF failed to improve outcome in trials of patients with sepsis due to community-acquired or hospital-acquired pneumonia [113, 114]. This may be due in part to unique site and pathogen-specific effects of G-CSF. For example, G-CSF has beneficial or harmful effects in animal models of sepsis depending on the site of infection [115] and the bacterial species (*S. aureus* versus *E. coli* pneumonia) [116].

Granulocyte-macrophage colony stimulating factor (GM-CSF) enhances both the production and function of neutrophils and monocyte/macrophages. Trials of GM-CSF in premature neonates have been equivocal regarding benefit [117]. Limited data from small trials in adult patients with sepsis show no overall beneficial effects [118, 119].

Interferon gamma is a well-known immunostimulant that has been used for the prevention of infections in patients with congenital immunodeficiencies (i.e. chronic granulomatous disease). Interferon gamma enhances the antimicrobial function of both neutrophils and monocytes, yet when used as an immuno-adjuvant in clinical trials of traumatically injured or burns patients, has not shown reproducible differences in infection rates compared to placebo [120–122].

17.12
Contemporary Therapy for Infection and Shock

In patients with presumed septic shock, the prompt administration of antibiotics is essential. The choice of antibiotics is often empiric and broad-based targeting both the species and antibiotic sensitivity of the likely pathogen. The importance of appropriate antibiotic therapy has been recognized for many years. In 1980, Kreger *et al.* showed that appropriate antibiotics were associated with decreased mortality with nonfatal, ultimately fatal and rapidly fatal underlying diseases [123]. Recently, the time from the onset of hypotension due to sepsis and the initiation of antibiotic therapy was shown to be a critical determinant of survival [124]. In a large retrospective cohort study of septic patients, each hour of delay in antimicrobial administration was associated with a 7.6% decrease in survival [124]. These data confirm the critical importance of early appropriate antibiotics in determining outcome in septic shock.

In shock due to infection or injury, adequate methods to reverse the hypotension and tissue hypoperfusion must be provided in a timely fashion. Early trials of saline versus albumin solutions in trauma were equivocal regarding the benefit of either solution [125]. A recent large prospective cohort study of crystalloid compared to albumin solutions for the resuscitation of critically ill patients found no difference in 28-day mortality, days of mechanical ventilation or days of renal replacement [126].

If adequate volume resuscitation does not restore mean arterial pressure to greater than 60–65 mmHg, vasopressor therapy is indicated. Norepinephrine or dopamine are considered the vasopressors of choice and there is little data showing improved outcome with either agent [127]. Recently, there has been renewed interest in the vasopressor effect of vasopressin, although this was first described in 1895 [128]. Vasopressin deficiency has been noted in septic shock [129–131] and several groups have reported improved blood pressure, urine output and decreased vasopressor requirement [130, 132, 133] with vasopressin. This has prompted empiric use of low-dose vasopressin in conjunction with norepinehrine or dopamine in septic shock. Preliminary results of a multicenter trial found no difference in 28-day mortality with the combination of vasopressin and norepinephrine compared to norepinephrine alone in septic shock [134].

17.13
The Future of Therapies for Sepsis and Non-infectious Systemic Inflammation

During the last 50 years, substantial progress has been made in decreasing mortality in critically ill patients with severe infections. In 1964, a National Academy of Sciences workshop was held to address the excess mortality (>66%) in patients who developed septic shock [135]. An analysis of discharge data of septic patients in the United States from 1979 to 2000 showed that while the incidence of sepsis and the number of sepsis-related deaths increased, the overall mortality rate among patients with sepsis fell from 28 to 18% [136]. The development of organ failure associated with sepsis and its inflammatory injury also increased during this 22-year period and the number of failing organs had an additive effect on mortality; approximately 15% of patients without organ failure died, whereas 70% of patients with three or more failing organs died [136]. The improved outcome from sepsis during this time period may be due to recognition of the systemic consequences of serious infections, prompt resuscitation of shock and the use of appropriate antimicrobial agents. However, multiple challenges remain in the treatment of inflammation due to infection or injury. These include integrating advances in knowledge about inflammatory mechanisms into new therapies, using new biomarkers to characterize the high-risk patient who might benefit from such therapies, and developing a better understanding of the role of host and pathogen interactions.

Knowledge of the components and complex interactions of signaling pathways in the initiation and regulation of inflammation is rapidly emerging. Strategies to modulate these pathways must simultaneously address limiting maladaptive inflammatory responses that lead to organ injury while still maintaining an adequate response to eradicate an infection [137, 138]. The complexity of manipulating signaling pathways of inflammation during a serious life-threatening infection is daunting. At least five levels of negative regulators have been described for the TLR alone including extracellular decoy receptors, intracellular inhibitors, membrane-bound suppressors, degradation of the primary TLR and TLR-induced apoptosis [139]. Similarly, a novel gene family, CATERPILLER or NOD-LRR, has recently been described that has a similar role in sensing microbial products and regulating innate and adaptive inflammatory responses [140]. This gene family appears to be linked to several rare human immunodeficiencies and studies of mice deficient in genes from this family suggest that these products play a pivotal role in the control of inflammation and may represent potential therapeutic targets [140, 141]. However, the therapeutic challenge for future sepsis therapies may relate less to identifying new targets than in reassessing assumptions about final common pathways and patient selection in this heterogeneous disorder.

A meta-analysis of pre-clinical and clinical trials provides insight into the paradox of preclinical success and clinical trial failures of mediator-specific anti-inflammatory therapies in sepsis [91]. Twenty-two clinical trials conducted between 1986 and 2000, evaluated different anti-inflammatory approaches to sepsis (i.e. anti-TNF, anti-IL-1, platelet factor antagonists, bradykinin and prostaglandin antagonists). A meta-analysis of 95 preclinical experiments cited by these 22 trials showed that regardless of the type of challenge, the timing of the dose of treatment, the species studied, or duration of observation, the odds ratio of survival with all mediator-specific anti-inflammatory therapies diminished as the risk of death (control mortality rate studied) decreased [91] (Figure 17.6). An analysis of the clinical trials showed that anti-inflammatory agents were significantly more efficacious in septic patients with higher risk of death and were harmful in patients at low risk. This analysis was then tested prospectively in a large series of animal studies using differing doses and sites of bacteria challenge to produce a full range of risk of death. The anti-inflammatory agents were beneficial at higher control mortality rates and were less effective at moderate control mortality rates. The observation that the efficacy of mediator-specific anti-inflammatory agents is dependent on the risk of death helps to explain the apparent contradiction of preclinical success and subsequent failure to translate preclinical findings into successful clinical trials [91]. This analysis suggests that agents that failed in clinical trials in the past should be reassessed, specifically targeting high-risk patients rather than a heterogeneous population with variable severity of illness.

Developing biomarkers of severity of illness may assist in targeting patients who will benefit from agents that modify the host inflammatory response. The high-throughput technologies of functional genomics and proteomics hold

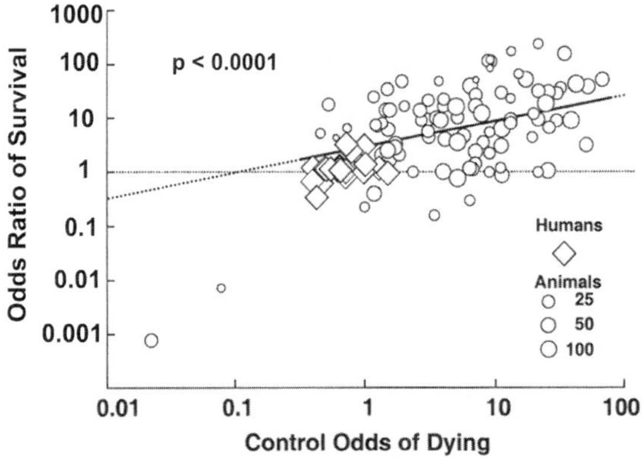

Figure 17.6 Relationship between the severity of illness and treatment effects of mediator-specific anti-inflammatory agents in preclinical and clinical trials of sepsis. The relationship between the control odds of dying (risk of death) on the x-axis and the odds ratio of survival (y-axis) in 95 experiments (open circles) from 38 published animal studies cited in 22 clinical trials (open diamonds) of mediator-specific anti-inflammatory agents in sepsis. The line is a weighted linear regression that shows that the treatment effect of these agents is significantly related to severity of illness as represented by the control odds of dying ($p < 0.0001$); these agents are beneficial in populations with a high severity of illness at a high risk of death and have minimal effect or are harmful in populations that are less severely ill with a low risk of death. (Reproduced from US government work [91].)

considerable promise for this task. This approach was pioneered in cancer biology and continues to be developed at a rapid pace [142]. Applying gene or protein profiling to characterize inflammation associated with infection, however, presents several challenges. Well-characterized phenotypes of the septic patient are necessary to address heterogeneity caused by factors such as co-morbid conditions, the presence of organ failure, as well as the type and site of infection. In contrast to most malignancies where mortality occurs over months or years, the mortality associated with serious infections occurs within days or weeks, suggesting that many dynamic changes in gene and protein expression profiles will occur during this period. This approach is made more complex when considering the large number of pathogens and the wide range of host cells from different tissues that are potential targets [143]. Exploring transcriptional and protein expression profiles in both the pathogen and host-target cells during the course of an infection may provide new insight into mechanisms of disease as well as identify both unique and common pathways activated by infection [144]. Characterizing the most relevant and predictive gene and protein profiles during sepsis is a major challenge that will require the integration of multiple disciplines including critical care, genomics, proteomics, biostatistics, bioinformatics, computational biology, genetics, and systems biology to fully exploit the potential of these technologies [145]. The magnitude of this undertaking will require cooperation between academic centers throughout the world to develop networks, acquire funding and share samples and data collections. A prototype of this type of investigation is the ongoing 10-year multicenter study of genomic and proteomic markers in blood to predict outcomes and new targets for therapies for the inflammation associated with traumatic injury and infection (http://www.gluegrant.org/).

References

1. Cohen, M.L. (**1998**) Candidate bacterial conditions. *Bull World Health Organ.*, **76**(Suppl 2), 61–63.
2. Castiglione, A. (**1941**) *A History of Medicine*, Knopf, New York, p. 159.
3. Karger, B., Sudhues, H. and Brinkmann, B. (**2001**) Arrow wounds: major stimulus in the history of surgery. *World J. Surg.*, **25**, 1550–5.
4. Levine, J.M. (**1992**) Historical notes on pressure ulcers: the cure of Ambrose Pare. *Decubitus*, **5**, 23–6.
5. Wyklicky, H. and Skopec, M. (**1983**) Ignaz Philipp Semmelweis, the prophet of bacteriology. *Infect. Control.*, **4**, 367–370.
6. Wangensteen, O. and Wangensteen, S.D. (**1978**) *The Rise of Surgery From Empiric Craft to Scientific Discipline*, University of Minnesota Press, Minneapolis, pp. 414–52.
7. Lesky, E. (**1976**) *The Vienna Medical School of the 19th Century*, The Johns Hopkins University Press, Baltimore, pp. 124–86.
8. Newsom, S.W. (**2003**) Pioneers in infection control-Joseph Lister. *J. Hosp. Infect.*, **55**, 246–253.
9. Porter, R. (**1997**) *The Greatest Benefit to Mankind-A Medical History of Humanity*, W.W. Norton and Company, New York, pp. 276–7.
10. Porter, R. (**1997**) *The Greatest Benefit to Mankind-A Medical History of Humanity*, W.W. Norton and Company, New York, pp. 433–4.
11. Behring, E. and Kitasato, S. (**1890**) Über das Zustandehommen der Diphterie-Immunität und der Tetanus-Immunität bei Thieren. *Dtsch. Med. Wochenschr.*, **16**, 1113–1114.
12. Molz, G. (**1970**) [Variations in child mortality in the past 100 years]. *Helv Paediatr Acta*, **25**, 1–12.
13. Grundbacher, F.J. (**1992**) Behring's discovery of diphtheria and tetanus antitoxins. *Immunol Today*, **13**, 188–190.
14. Porter, R. (**1997**) *The Greatest Benefit to Mankind – A Medical History of Humanity*, W.W. Norton and Company, New York, pp. 438–9.
15. Porter, R. (**1997**) *The Greatest Benefit to Mankind – A Medical History of Humanity*, W.W. Norton and Company, New York, p. 452.
16. Bendiner, E. (**1989**) Alexander Fleming: player with microbes. *Hosp. Pract. (Off Ed).*, **24**, 283–285, 290 293–286 passim.
17. Chain, E., Florey, H.W., Adelaide, M.B., Gardner, A.D., Heatley, N.G., Jennings, M.A., Orr-Ewing, J. and Sanders, A.G. (**1940**) Penicillin as a chemotherapeutic agent. *Lancet*, **24**, 226–231.
18. Ligon, B.L. (**2004**) Penicillin: its discovery and early development. *Semin. Pediatr. Infect. Dis.*, **15**, 52–57.
19. Martin, W.J. and Nichols, D.R. (**1956**) Bacteremic shock. *Mayo. Clin. Proc.*, **31**, 333–340.
20. Ezzo, J.A. and Knight, W.A., Jr. (**1957**) Bacterial shock; a clinical analysis of thirty-eight cases. *AMA Arch. Intern. Med.*, **99**, 701–707.
21. Porter, R. (**1997**) *The Greatest Benefit to Mankind – A Medical History of Humanity*, W.W. Norton and Company, New York, pp. 70–1.
22. Weissmann, G. (**1988**) *Inflammation: Basic Principles and Clinical Correlates*,

Raven Press, Ltd., New York, p. 5.
23. Waterhouse, R. (**1911**) Case of suprarenal apoplexy. *Lancet*, **1**, 577–578.
24. Friderichsen, C. (**1918**) Nebennierenapoplexie bei kleinen Kindern. *Jahrb Kinderh*, **87**, 109–125.
25. Liljestrand, G. (**1950**) The Nobel Prize in Physiology or Medicine 1950 (http://nobelprize.org/nobel-prizes/medicine/laureates/1950/press.html).
26. Rich, A.R. (**1944**) A peculiar type of adrenal cortical damage associated with acute infections and its possible relation to circulatory collapse. *Bull Johns Hopkins Hosp.*, **74**, 1–15.
27. Najib-Farah (**1938**) Enteric fever treated with suprarenal cortex extract and vitamin C intravenously. *Lancet*, 777–779.
28. Perla, D. and Marmorston, J. (**1940**) Suprarenal cortical hormone and salt in the treatment of pneumonia and other severe infections. *Endocrinology*, **27**, 367–374.
29. Kass, E.H. and Finland, M. (**1953**) Adrenocortical hormones in infection and immunity. *Annu. Rev. Microbiol.*, **7**, 361–388.
30. Weil, M.H. and Spink, W.W. (**1958**) The shock syndrome associated with bacteremia due to Gram-negative bacilli. *AMA Arch. Intern. Med.*, **101**, 184–193.
31. Weil, M.H. and Spink, W.W. (**1955**) Clinical studies on a vasopressor agent: metaraminol (aramine). I. Observations in normotensive subjects. *Am. J. Med. Sci.*, **229**, 661–669.
32. Melby, J.C., Bossenmaier, I.C., Egdahl, R.H. and Spink, W.W. (**1959**) Suppression by cortisol of increased serum-transaminase induced by endotoxin. *Lancet*, **1**, 441–444.
33. Lillehei, R.C. and Maclean, L.D. (**1959**) Physiological approach to successful treatment of endotoxin shock in the experimental animal. *AMA Arch. Surg.*, **78**, 464–471.
34. Minneci, P.C., Deans, K.J., Banks, S.M., Eichacker, P.Q. and Natanson, C. (**2004**) Meta-analysis: the effect of steroids on survival and shock during sepsis depends on the dose. *Ann. Intern. Med.*, **141**, 47–56.
35. Laennec, R.T. (**1831**) *Traité de l'Auscultation Mediate et des Maladies des Poumons et du Coeur*, Vol. 3, Brosson and Chaudé, Paris.
36. Romberg, E., Pässler, H., Bruhns, C. and Müller, W. (**1899**) Untersuchungen über die allgemeine Pathologie and Therapie der Kreislaufstörung bie acuten Infectionskrankheiten. *Arch. Klin. Med.*, **64**, 652–714.
37. Janeway, T. (**1907**) Some common misconceptions in the pathological physiology of the circulation and their practical significance. *New York State J. Med.*, **85**, 193.
38. Dale, H.H. and Laidlaw, P. (**1919**) Histamine shock, *J. Physiol.*, **52**, 355.
39. Brooks, B. and Blalock, A. (**1934**) Shock with particular reference to that due to hemorrhage and trauma to muscles. *Ann. Surg.*, **100**, 728–733.
40. Atchley, D.W. (**1930**) Medical shock. *JAMA*, **95**, 385–389.
41. Warfield, L.M. (**1936**) The treatment of circulatory failure. *JAMA*, **106**, 892–895.
42. Holton, P. (**1949**) Noradrenaline in tumours of the adrenal medulla. *J. Physiol.*, **108**, 525–529.
43. Gaddum, J.H., Peart, W.S. and Vogt, M. (**1949**) The estimation of adrenaline and allied substances in blood. *J. Physiol.*, **108**, 467–481.
44. Von Euler, U.S., Hamberg, U. and Purkhold, A. (**1949**) Noradrenaline and adrenaline in the suprarenals of the guinea-pig. *Experientia*, **5**, 451.
45. Barcroft, H. and Konzett, H. (**1949**) On the actions of noradrenaline, adrenaline and isopropyl noradrenaline on the arterial blood pressure, heart rate and muscle blood flow in man. *J. Physiol.*, **110**, 194–204.
46. Braude, A.I., Williams, D., Siemenski, J. and Murphy, R. (**1953**) Shock-like state due to transfusion of blood contaminated with Gram-negative bacilli; successful

treatment with antibiotics and arterenol. *AMA Arch. Intern. Med.*, **92**, 75–84.

47. Udhoji, V.N. and Weil, M.H. (**1964**) Circulatory effects of angiotensin, levarterenol and metaraminol in the treatment of shock. *N. Engl. J. Med.*, **270**, 501–505.
48. McDonald, R.H., Jr., Goldberg, L.I., McNay, J.L. and Tuttle, E.P., Jr. (**1964**) Effect of dopamine in man: augmentation of sodium excretion, glomerular filtration rate, and renal plasma flow. *J. Clin. Invest.*, **43**, 1116–1124.
49. Goldberg, L.I., McDonald, R.H., Jr. and Zimmerman, A.M. (**1963**) Sodium diuresis produced by dopamine in patients with congestive heart failure. *N. Engl. J. Med.*, **268**, 1060–1064.
50. MacCannell, K.L., McNay, J.L., Meyer, M.B. and Goldberg, L.I. (**1966**) Dopamine in the treatment of hypotension and shock. *N. Engl. J. Med.*, **275**, 1389–1398.
51. Weil, M.H. and Shubin, H. (**1969**) The "VIP" approach to the bedside management of shock. *JAMA*, **207**, 337–340.
52. Mackenzie, G.J., Flenley, D.C., Taylor, S.H., McDonald, A.H., Staunton, H.P. and Donald, K.W. (**1964**) Circulatory and respiratory studies in myocardial infarction and cardiogenic shock. *Lancet*, **22**, 825–832.
53. Foley, F.D., Moncrief, J.A. and Mason, A.D., Jr. (**1968**) Pathology of the lung in fatally burned patients. *Ann. Surg.*, **167**, 251–264.
54. Swan, H.J., Ganz, W., Forrester, J., Marcus, H., Diamond, G. and Chonette, D. (**1970**) Catheterization of the heart in man with use of a flow-directed balloon-tipped catheter. *N. Engl. J. Med.*, **283**, 447–451.
55. Ganz, W., Donoso, R., Marcus, H.S., Forrester, J.S. and Swan, H.J. (**1971**) A new technique for measurement of cardiac output by thermodilution in man. *Am. J. Cardiol.*, **27**, 392–396.
56. Shoemaker, W.C., Montgomery, E.S., Kaplan, E. and Elwyn, D.H. (**1973**) Physiologic patterns in surviving and nonsurviving shock patients. Use of sequential cardiorespiratory variables in defining criteria for therapeutic goals and early warning of death. *Arch. Surg.*, **106**, 630–636.
57. Shoemaker, W.C., Chang, P., Czer, L., Bland, R., Shabot, M.M. and State, D. (**1979**) Cardiorespiratory monitoring in postoperative patients: I. Prediction of outcome and severity of illness. *Crit. Care Med.*, **7**, 237–242.
58. Bland, R.D., Shoemaker, W.C., Abraham, E. and Cobo, J.C. (**1985**) Hemodynamic and oxygen transport patterns in surviving and nonsurviving postoperative patients. *Crit. Care Med.*, **13**, 85–90.
59. Hankeln, K.B., Senker, R., Schwarten, J.U., Beez, M.G., Engel, H.J. and Laniewsky, P. (**1987**) Evaluation of prognostic indices based on hemodynamic and oxygen transport variables in shock patients with adult respiratory distress syndrome. *Crit. Care Med.*, **15**, 1–7.
60. Shoemaker, W.C., Appel, P.L., Kram, H.B., Waxman, K. and Lee, T.S. (**1988**) Prospective trial of supranormal values of survivors as therapeutic goals in high-risk surgical patients. *Chest*, **94**, 1176–1186.
61. Hayes, M.A., Timmins, A.C., Yau, E.H., Palazzo, M., Hinds, C.J. and Watson, D. (**1994**) Elevation of systemic oxygen delivery in the treatment of critically ill patients. *N. Engl. J. Med.*, **330**, 1717–1722.
62. Hurley, J. and Levin, J. (**1999**) *Endotoxin in Health and Disease*, Marcel Dekker, Inc., New York.
63. Woodruff, P.W., O'Carroll, D.I., Koizumi, S. and Fine, J. (**1973**) Role of the intestinal flora in major trauma. *J. Infect. Dis.*, **128**(Suppl), 290–294.
64. Danner, R.L., Elin, R.J., Hosseini, J.M., Wesley, R.A., Reilly, J.M. and Parillo, J.E. (**1991**) Endotoxemia in human septic shock. *Chest*, **99**, 169–175.
65. Natanson, C., Danner, R.L., Elin, R.J., Hosseini, J.M., Peart, K.W., Banks, S.M.,

MacVittie, T.J., Walker, R.I. and Parrillo, J.E. (**1989**) Role of endotoxemia in cardiovascular dysfunction and mortality. *Escherichia coli* and *Staphylococcus aureus* challenges in a canine model of human septic shock. *J. Clin. Invest.*, **83**, 243–251.

66. Medzhitov, R. and Janeway, C., Jr. (**2000**) Innate immunity. *N. Engl. J. Med.*, **343**, 338–344.
67. Medzhitov, R., Preston-Hurlburt, P. and Janeway, C.A., Jr. (**1997**) A human homologue of the Drosophila Toll protein signals activation of adaptive immunity. *Nature*, **388**, 394–397.
68. (**1992**) American College of Chest Physicians/Society of Critical Care Medicine Consensus Conference: definitions for sepsis and organ failure and guidelines for the use of innovative therapies in sepsis. *Crit. Care Med.*, **20**, 864–874.
69. Carswell, E.A., Old, L.J., Kassel, R.L., Green, S., Fiore, N. and Williamson, B. (**1975**) An endotoxin-induced serum factor that causes necrosis of tumors. *Proc. Natl. Acad. Sci. USA*, **72**, 3666–3670.
70. Beutler, B., Milsark, I.W. and Cerami, A.C. (**1985**) Passive immunization against cachectin/tumor necrosis factor protects mice from lethal effect of endotoxin. *Science*, **229**, 869–871.
71. Tracey, K.J., Fong, Y., Hesse, D.G., Manogue, K.R., Lee, A.T., Kuo, G.C., Lowry, S.F. and Cerami, A. (**1987**) Anti-cachectin/TNF monoclonal antibodies prevent septic shock during lethal bacteraemia. *Nature*, **330**, 662–664.
72. Ziegler, E.J., McCutchan, J.A., Fierer, J., Glauser, M.P., Sadoff, J.C., Douglas, H. and Braude, A.I. (**1982**) Treatment of Gram-negative bacteremia and shock with human antiserum to a mutant *Escherichia coli*. *N. Engl. J. Med.*, **307**, 1225–1230.
73. Freeman, B.D. and Natanson, C. (**2000**) Anti-inflammatory therapies in sepsis and septic shock. *Expert. Opin. Investig. Drugs*, **9**, 1651–1663.
74. Ziegler, E.J., Fisher, C.J., Jr., Sprung, C.L., Straube, R.C., Sadoff, J.C., Foulke, G.E., Wortel, C.H., Fink, M.P., *et al.* (**1991**) Treatment of Gram-negative bacteremia and septic shock with HA-1A human monoclonal antibody against endotoxin. A randomized, double-blind, placebo-controlled trial. The HA-1A Sepsis Study Group. *N. Engl. J. Med.*, **324**, 429–436.
75. Warren, H.S., Danner, R.L. and Munford, R.S. (**1992**) Anti-endotoxin monoclonal antibodies. *N. Engl. J. Med.*, **326**, 1153–1157.
76. McCloskey, R.V., Straube, R.C., Sanders, C., Smith, S.M. and Smith, C.R. (**1994**) Treatment of septic shock with human monoclonal antibody HA-1A. A randomized, double-blind, placebo-controlled trial. CHESS Trial Study Group. *Ann. Intern. Med.*, **121**, 1–5.
77. Angus, D.C., Birmingham, M.C., Balk, R.A., Scannon, P.J., Collins, D., Kruse, J.A., Graham, D.R., Dedhia, H.V., *et al.* (**2000**) E5 murine monoclonal antiendotoxin antibody in Gram-negative sepsis: a randomized controlled trial. E5 Study Investigators. *JAMA*, **283**, 1723–1730.
78. Bone, R.C., Balk, R.A., Fein, A.M., Perl, T.M., Wenzel, R.P., Reines, H.D., Quenzer, R.W., Iberti, T.J., *et al.* (**1995**) A second large controlled clinical study of E5, a monoclonal antibody to endotoxin: results of a prospective, multicenter, randomized, controlled trial. The E5 Sepsis Study Group. *Crit. Care Med.*, **23**, 994–1006.
79. Greenman, R.L., Schein, R.M., Martin, M.A., Wenzel, R.P., MacIntyre, N.R., Emmanuel, G., Chmel, H., Kohler, R.B., *et al.* (**1991**) A controlled clinical trial of E5 murine monoclonal IgM antibody to endotoxin in the treatment of Gram-negative sepsis. The XOMA Sepsis Study Group, *JAMA*, **266**, 1097–1102.

80. Hoffman, W.D. and Natanson, C. (**1993**) Endotoxin in septic shock. *Anesth. Analg.*, **77**, 613–624.
81. Levin, M., Quint, P.A., Goldstein, B., Barton, P., Bradley, J.S., Shemie, S.D., Yeh, T., Kim, S.S., *et al.* (**2000**) Recombinant bactericidal/permeability-increasing protein (rBPI21) as adjunctive treatment for children with severe meningococcal sepsis: a randomised trial. rBPI21 Meningococcal Sepsis Study Group. *Lancet*, **356**, 961–967.
82. van Deuren, M. and Brandtzaeg, P. (**2000**) Parents' and GPs' key role in diagnosis of meningococcal septicaemia. *Lancet*, **356**, 954–955.
83. Alejandria, M.M., Lansang, M.A., Dans, L.F. and Mantaring, J.B. (**2002**) Intravenous immunoglobulin for treating sepsis and septic shock. *Cochrane Database Syst. Rev.*
84. Pildal, J. and Gotzsche, P.C. (**2004**) Polyclonal immunoglobulin for treatment of bacterial sepsis: a systematic review. *Clin. Infect. Dis.*, **39**, 38–46.
85. Turgeon, A.F., Hutton, B., Fergusson, D.A., McIntyre, L., Tinmouth, A.A., Cameron, D.W. and Hebert, P.C. (**2007**) Meta-analysis: intravenous immunoglobulin in critically ill adult patients with sepsis. *Ann. Intern. Med.*, **146**, 193–203.
86. The Veterans Administration Systemic Sepsis Cooperative Study Group (**1987**) Effect of high-dose glucocorticoid therapy on mortality in patients with clinical signs of systemic sepsis. *N. Engl. J. Med.*, **317**, 659–665.
87. Bone, R.C., Fisher, C.J., Jr., Clemmer, T.P., Slotman, G.J., Metz, C.A. and Balk, R.A. (**1987**) A controlled clinical trial of high-dose methylprednisolone in the treatment of severe sepsis and septic shock. *N. Engl. J. Med.*, **317**, 653–658.
88. Luce, J.M., Montgomery, A.B., Marks, J.D., Turner, J., Metz, C.A. and Murray, J.F. (**1988**) Ineffectiveness of high-dose methylprednisolone in preventing parenchymal lung injury and improving mortality in patients with septic shock. *Am. Rev. Resp. Dis.*, **138**, 62–68.
89. Cerami, A., Tracey, K.J., Lowry, S.F. and Beutler, B. (**1987**) Cachectin: a pluripotent hormone released during the host response to invasion. *Recent. Prog. Horm. Res.*, **43**, 99–112.
90. Dinarello, C.A. and Wolff, S.M. (**1993**) The role of interleukin-1 in disease. *N. Engl. J. Med.*, **328**, 106–113.
91. Eichacker, P.Q., Parent, C., Kalil, A., Esposito, C., Cui, X., Banks, S.M., Gerstenberger, E.P., Fitz, Y., *et al.* (**2002**) Risk and the efficacy of antiinflammatory agents: retrospective and confirmatory studies of sepsis. *Am. J. Resp. Crit. Care Med.*, **166**, 1197–1205.
92. Deans, K.J., Haley, M., Natanson, C., Eichacker, P.Q. and Minneci, P.C. (**2005**) Novel therapies for sepsis: a review. *J. Trauma*, **58**, 867–874.
93. Abraham, E., Naum, C., Bandi, V., Gervich, D., Lowry, S.F., Wunderink, R., Schein, R.M., Macias, W., Skerjanec, S., Dmitrienko, A., Farid, N., Forgue, S.T. and Jiang, F. (**2003**) Efficacy and safety of LY315920Na/S-5920, a selective inhibitor of 14-kDa group IIA secretory phospholipase A2, in patients with suspected sepsis and organ failure. *Crit. Care Med.*, **31**, 718–728.
94. Zeiher, B.G., Steingrub, J., Laterre, P.F., Dmitrienko, A., Fukiishi, Y. and Abraham, E. (**2005**) LY315920NA/S-5920, a selective inhibitor of group IIA secretory phospholipase A2, fails to improve clinical outcome for patients with severe sepsis. *Crit. Care Med.*, **33**, 1741–1748.
95. McGovern, V.J. (**1984**) Shock revisited. *Pathol. Annual*, **19**(Pt 1), 15–36.
96. Ware, L.B., Camerer, E., Welty-Wolf, K., Schultz, M.J. and Matthay, M.A. (**2006**) Bench to bedside: targeting coagulation and

fibrinolysis in acute lung injury. *Am. J. Phys.*, **291**, L307–311.

97. Minneci, P.C., Deans, K.J., Cui, X., Banks, S.M., Natanson, C. and Eichacker, P.Q. (**2006**) Antithrombotic therapies for sepsis: a need for more studies. *Crit. Care Med.*, **34**, 538–541.
98. Bernard, G.R., Vincent, J.L., Laterre, P.F., LaRosa, S.P., Dhainaut, J.F., Lopez-Rodriguez, A., Steingrub, J.S., Garber, G.E., *et al.* (**2001**) Efficacy and safety of recombinant human activated protein C for severe sepsis. *N. Engl. J. Med.*, **344**, 699–709.
99. Warren, H.S., Suffredini, A.F., Eichacker, P.Q. and Munford, R.S. (**2002**) Risks and benefits of activated protein C treatment for severe sepsis. *N. Engl. J. Med.*, **347**, 1027–1030.
100. Nadel, S., Goldstein, B., Williams, M.D., Dalton, H., Peters, M., Macias, W.L., Abd-Allah, S.A., Levy, H., Angle, R., *et al.* (**2007**) Drotrecogin alfa (activated) in children with severe sepsis: a multicentre phase III randomised controlled trial. *Lancet*, **369**, 836–843.
101. Eichacker, P.Q., Danner, R.L., Suffredini, A.F., Cui, X. and Natanson, C. (**2005**) Reassessing recombinant human activated protein C for sepsis: time for a new randomized controlled trial. *Crit. Care Med.*, **33**, 2426–2428.
102. Marti-Carvajal, A., Salanti, G. and Cardona, A. (**2007**) Human recombinant activated protein C for severe sepsis, *Cochrane Database Syst. Rev.*
103. Eichacker, P.Q. and Natanson, C. (**2007**) Increasing evidence that the risks of rhAPC may outweigh its benefits. *Intensive Care Med.*, **33**, 396–399.
104. de Werra, I., Zanetti, G., Jaccard, C., Chiolero, R., Schaller, M.D., Yersin, B., Glauser, M.P., Calandra, T. and Heumann, D. (**2001**) CD14 expression on monocytes and TNF alpha production in patients with septic shock, cardiogenic shock or bacterial pneumonia. *Swiss Med. Wkly.*, **131**, 35–40.
105. Ertel, W., Kremer, J.P., Kenney, J., Steckholzer, U., Jarrar, D., Trentz, O. and Schildberg, F.W. (**1995**) Downregulation of proinflammatory cytokine release in whole blood from septic patients. *Blood*, **85**, 1341–1347.
106. Majetschak, M., Krehmeier, U., Bardenheuer, M., Denz, C., Quintel, M., Voggenreiter, G. and Obertacke, U. (**2003**) Extracellular ubiquitin inhibits the TNF-alpha response to endotoxin in peripheral blood mononuclear cells and regulates endotoxin hyporesponsiveness in critical illness. *Blood*, **101**, 1882–1890.
107. Adib-Conquy, M., Moine, P., Asehnoune, K., Edouard, A., Espevik, T., Miyake, K., Werts, C. and Cavaillon, J.M. (**2003**) Toll-like receptor-mediated tumor necrosis factor and interleukin-10 production differ during systemic inflammation. *Am. J. Resp. Crit. Care Med.*, **168**, 158–164.
108. Decker, D., Schondorf, M., Bidlingmaier, F., Hirner, A. and von Ruecker, A.A. (**1996**) Surgical stress induces a shift in the type-1/type-2 T-helper cell balance, suggesting down-regulation of cell-mediated and up-regulation of antibody-mediated immunity commensurate to the trauma. *Surgery*, **119**, 316–325.
109. Docke, W.D., Randow, F., Syrbe, U., Krausch, D., Asadullah, K., Reinke, P., Volk, H.D. and Kox, W. (**1997**) Monocyte deactivation in septic patients: restoration by IFN-gamma treatment. *Nat. Med.*, **3**, 678–681.
110. Munford, R.S. and Pugin, J. (**2001**) Normal responses to injury prevent systemic inflammation and can be immunosuppressive. *Am. J. Resp. Crit. Care Med.*, **163**, 316–321.
111. Angele, M.K. and Chaudry, I.H. (**2005**) Surgical trauma and immunosuppression: pathophysiology and potential immunomodulatory approaches.

Langenbecks Arch. Surg., **390**, 333–341.

112. Angele, M.K. and Faist, E. (**2002**) Clinical review: immunodepression in the surgical patient and increased susceptibility to infection. *Crit. Care*, **6**, 298–305.
113. Nelson, S., Belknap, S.M., Carlson, R.W., Dale, D., DeBoisblanc, B., Farkas, S., Fotheringham, N., Ho, H., *et al.* (**1998**) A randomized controlled trial of filgrastim as an adjunct to antibiotics for treatment of hospitalized patients with community-acquired pneumonia. CAP Study Group. *J. Infect. Dis.*, **178**, 1075–1080.
114. Root, R.K., Lodato, R.F., Patrick, W., Cade, J.F., Fotheringham, N., Milwee, S., Vincent, J.L., Torres, A., *et al.* (**2003**) Multicenter, double-blind, placebo-controlled study of the use of filgrastim in patients hospitalized with pneumonia and severe sepsis. *Crit. Care Med.*, **31**, 367–373.
115. Sevransky, J.E., Parent, C., Cui, X., Karzai, W., Fitz, Y., Banks, S.M., Gerstenberger, E., Danner, R.L., Natanson, C. and Eichacker, P.Q. (**2004**) Granulocyte colony-stimulating factor has differing effects comparing intravascular versus extravascular models of sepsis. *J. Trauma*, **57**, 618–625.
116. Karzai, W., von Specht, B.U., Parent, C., Haberstroh, J., Wollersen, K., Natanson, C., Banks, S.M. and Eichacker, P.Q. (**1999**) G-CSF during Escherichia coli versus Staphylococcus aureus pneumonia in rats has fundamentally different and opposite effects. *Am. J. Resp. Crit. Care Med.*, **159**, 1377–1382.
117. Carr, R., Modi, N. and Dore, C. (**2003**) G-CSF and GM-CSF for treating or preventing neonatal infections. *Cochrane Database Syst. Rev.*
118. Presneill, J.J., Harris, T., Stewart, A.G., Cade, J.F. and Wilson, J.W. (**2002**) A randomized phase II trial of granulocyte-macrophage colony-stimulating factor therapy in severe sepsis with respiratory dysfunction. *Am. J. Resp. Crit. Care Med.*, **166**, 138–143.
119. Rosenbloom, A.J., Linden, P.K., Dorrance, A., Penkosky, N., Cohen-Melamed, M.H. and Pinsky, M.R. (**2005**) Effect of granulocyte-monocyte colony-stimulating factor therapy on leukocyte function and clearance of serious infection in nonneutropenic patients. *Chest*, **127**, 2139–2150.
120. Dries, D.J., Jurkovich, G.J., Maier, R.V., Clemmer, T.P., Struve, S.N., Weigelt, J.A., Stanford, G.G., Herr, D.L., *et al.* (**1994**) Effect of interferon gamma on infection-related death in patients with severe injuries. A randomized, double-blind, placebo-controlled trial. *Arch. Surg.*, **129**, 1031–1041; discussion 1042.
121. Polk, H.C., Jr., Cheadle, W.G., Livingston, D.H., Rodriguez, J.L., Starko, K.M., Izu, A.E., Jaffe, H.S. and Sonnenfeld, G. (**1992**) A randomized prospective clinical trial to determine the efficacy of interferon-gamma in severely injured patients. *Am. J. Surg.*, **163**, 191–196.
122. Wasserman, D., Ioannovich, J.D., Hinzmann, R.D., Deichsel, G. and Steinmann, G.G. (**1998**) Interferon-gamma in the prevention of severe burn-related infections: a European phase III multicenter trial. The Severe Burns Study Group. *Crit. Care Med.*, **26**, 434–439.
123. Kreger, B.E., Craven, D.E. and McCabe, W.R. (**1980**) Gram-negative bacteremia. IV. Re-evaluation of clinical features and treatment in 612 patients. *Am. J. Med.*, **68**, 344–355.
124. Kumar, A., Roberts, D., Wood, K.E., Light, B., Parrillo, J.E., Sharma, S., Suppes, R., Feinstein, D., *et al.* (**2006**) Duration of hypotension before initiation of effective antimicrobial therapy is the critical determinant of

survival in human septic shock. *Crit. Care Med.*, **34**, 1589–1596.
125. Cochrane Injuries Group Albumin Reviewers (**1998**). Human albumin administration in critically ill patients: systematic review of randomised controlled trials. *BMJ*, **317**, 235–240.
126. Finfer, S., Bellomo, R., Boyce, N., French, J., Myburgh, J. and Norton, R. (**2004**) A comparison of albumin and saline for fluid resuscitation in the intensive care unit. *N. Engl. J. Med.*, **350**, 2247–2256.
127. Hollenberg, S.M., Ahrens, T.S., Annane, D., Astiz, M.E., Chalfin, D.B., Dasta, J.F., Heard, S.O., Martin, C., *et al.* (**2004**) Practice parameters for hemodynamic support of sepsis in adult patients: 2004 update. *Crit. Care Med.*, **32**, 1928–1948.
128. Olivier, G. and Schafer, E.A. (**1895**) On the physiological action of extracts of pituitary body and certain other glandular organs. *J. Physiol.*, **18**, 277–279.
129. Landry, D.W., Levin, H.R., Gallant, E.M., Ashton, R.C., Jr., Seo, S., D'Alessandro, D., Oz, M.C. and Oliver, J.A. (**1997**) Vasopressin deficiency contributes to the vasodilation of septic shock. *Circulation*, **95**, 1122–1125.
130. Patel, B.M., Chittock, D.R., Russell, J.A. and Walley, K.R. (**2002**) Beneficial effects of short-term vasopressin infusion during severe septic shock. *Anesthesiology*, **96**, 576–582.
131. Sharshar, T., Blanchard, A., Paillard, M., Raphael, J.C., Gajdos, P. and Annane, D. (**2003**) Circulating vasopressin levels in septic shock. *Crit. Care Med.*, **31**, 1752–1758.
132. Holmes, C.L., Walley, K.R., Chittock, D.R., Lehman, T. and Russell, J.A. (**2001**) The effects of vasopressin on hemodynamics and renal function in severe septic shock: a case series. *Intensive Care Med.*, **27**, 1416–1421.
133. Malay, M.B., Ashton, R.C., Jr., Landry, D.W. and Townsend, R.N. (**1999**) Low-dose vasopressin in the treatment of vasodilatory septic shock. *J. Trauma*, **47**, 699–703; discussion 703-695.
134. Russell, J.A. (**2007**) VASST: A multicenter randomized trial of vasopressin in septic shock. Presented at *American Thoracic Society International Conference*, San Francisco, CA, May 20, 2007.
135. National Academy of Sciences-National Research Council (**1965**) *Proceedings of a Workshop: Septic Shock* (eds J.G. Strawitz, and N. Grossblatt), Washington, DC.
136. Martin, G.S., Mannino, D.M., Eaton, S. and Moss, M. (**2003**) The epidemiology of sepsis in the United States from 1979 through 2000. *N. Engl. J. Med.*, **348**, 1546–1554.
137. O'Neill, L.A. (**2006**) Targeting signal transduction as a strategy to treat inflammatory diseases. *Nature. Rev. Drug Discov.*, **5**, 549–563.
138. Serhan, C.N., Brain, S.D., Buckley, C.D., Gilroy, D.W., Haslett, C., O'Neill, L.A., Perretti, M., Rossi, A.G. and Wallace, J.L. (**2007**) Resolution of inflammation: state of the art, definitions and terms. *FASEB J*, **21**, 325–332.
139. Liew, F.Y., Xu, D., Brint, E.K. and O'Neill, L.A. (**2005**) Negative regulation of toll-like receptor-mediated immune responses. *Nat. Rev. Immunol.*, **5**, 446–458.
140. Ting, J.P., Kastner, D.L. and Hoffman, H.M. (**2006**) CATERPILLERs, pyrin and hereditary immunological disorders. *Nat. Rev. Immunol.*, **6**, 183–195.
141. Mariathasan, S. and Monack, D.M. (**2007**) Inflammasome adaptors and sensors: intracellular regulators of infection and inflammation. *Nat. Rev. Immunol.*, **7**, 31–40.
142. Staudt, L.M. and Dave, S. (**2005**) The biology of human lymphoid malignancies revealed by gene expression profiling. *Adv. Immunol.*, **87**, 163–208.

143. Hossain, H., Tchatalbachev, S. and Chakraborty, T. (**2006**) Host gene expression profiling in pathogen-host interactions. *Curr. Opin. Immunol.*, **18**, 422–429.

144. Waddell, S.J., Butcher, P.D. and Stoker, N.G. (**2007**) RNA profiling in host-pathogen interactions. *Curr. Opin. Microbiol.*, **10**, 297–302.

145. Cobb, J.P., Mindrinos, M.N., Miller-Graziano, C., Calvano, S.E., Baker, H.V., Xiao, W., Laudanski, K., Brownstein, B.H., *et al.* (**2005**) Application of genome-wide expression analysis to human health and disease. *Proc. Natl. Acad. Sci. USA*, **102**, 4801–4806.

Index

a

acetyltransferase 188
acinetobacter 28
Acinetobacter baumannii 79
activated partial thromboplastin time (aPTT) 7, 253
activated protein C (APC) 187
activating agent 355–356
acute lung injury (ALI) 60, 67, 165
– during severe falciparum malaria (SFM) 162
Acute Physiology and Chronic Health Evaluation (APACHE) II score 31, 241, 253
acute renal failure (ARF) 66, 165, 282
– during severe falciparum malaria (SFM) 162
acute respiratory distress syndrome (ARDS) 60, 67, 236
adherens junction (AJ) 281
adrenocorticotropin hormone (ACTH) 301
adrenomedullin (AM) 220
Akt 219–220
Ala896Gly 328
altered immune status
– cytokines 351–354
– leukocytes 350–351
altered TLR-dependent TNF production 357–358
Alzheimer's disease 209
AMB lipid complex (ABLC) 153
amphotericin B 147, 153
amphotericin B deoxycholate 153, 155
anaphylatoxins C3a and C5a 186
angiopoietins 191
angiotensin converting enzyme (ACE)-2 191–192
anidulafungin 155
animal models, of sepsis 383–384
anti-fibrinolytic activity, effects 69
anti-inflammatory cytokines 190, 260
anti-inflammatory mediator release effects 46
antimicrobial peptides (AMPs) 94, 95
antithrombin 255
apolipoprotein (apo) deficient mice 190
apoptosis
– adrenomedullin (AM) in treatment 220
– Akt in treatment 220
– anti-apoptotic effects of Bcl-2 217–218
– anti-apoptotic therapies 213–222
– biochemistry 209–210
– caspases inhibitor therapy 218
– clearance of apoptotic cells 220–221
– levels in experimental septic mice and septic patients 211
– milk fat globule epidermal growth factor 8 (MFG-E8) 221–222
– in MODS 280
– overview 206–207
– pathological role in sepsis 210–212
– pathways 207–209
– protease inhibitor therapy 219
– role of Bid protein 219
– targeting of apoptotic pathway 214–215
– therapies for target interventions 216
apoptosis-inducing factor (AIF) 208
arachidonic acid release 192
Arg753Gln 328
artemether 163
artemotil 164

Sepsis and Non-infectious Systemic Inflammation. Edited by J.-M. Cavaillon and C. Adrie

ISBN: 978-3-527-31935-0

Aspergillus fumigatus 361
Asp299Gly 150
Australian Aborigines 27
azoles 152–156
azotemia 282

b

Bacillus fragilis 315, 381
Bacillus subtilis 116
bacteremia 376
–models, of sepsis 378, 379
bactericidal/permeability-increasing protein (BPI) 93
barrier function alterations, of the intestinal epithelium 282
B-cell CCl/lymphoma 2 (Bcl-2) family207
BH3-only proteins 207
Bid protein 219
bradykinin 47

c

Caenorhabiditis elegans 209
calcitonin gene-related peptide (CGRP) 192
Candida albicans 146–151, 154
Candida endocarditis 151
Candida endophtalmitis 151
Candida glabrata 146–148, 153–154
Candida guilliermondii 154
Candida krusei 148, 153
Candida parapsilosis 146–148, 154
Candida parapsilosis fungemia 147
Candida rugosa 154
Candida sp. 27
Candida spp. bloodstream infections (CBSIs) 146–147
Candida spp. infection 316
Candida tropicalis 146–147, 154
candidemia
–complications and prognosis 151–152
–diagnosis 150–151
–epidemiology, in ICU 145–148
–pathogenesis of *Candida* spp. 149–150
–treatment 152–156
candidiasis 150
cardiocirculatory dysfunction
–cardiac dysfunction 64
–extrinsic factors 65
–intrinsic factors 64, 65
–vascular dysfunction 65–66
cardiomyocyte 242
caspase-8 210
caspases 209–210
–inhibitors 218
caspofungin 154
cathecholaminergic nerve terminals 305
cathecolamines 65
cathelicidins 94
C4b binding protein (C4bBP) 257
CD163 190
cell culture 354
cerebral damage and dysfunction, in sepsis 67–68
cholestasis 69
cholestatic jaundice 282
circulating markers, for systemic inflammation characterization 49–50
claudins 281
clindamycin 164
clinical evidence, of an infection site 3
coagulase-negative staphylococci (CNS) 134
coagulation abnormalities, in sepsis 187
–anticoagulants and restoration of pathways 266–268
–cross-talk with inflammation 260–261
–diagnosis of 261–265
–disseminated intravascular coagulation (DIC) 251–252
–impairment of 255–258
–incidence 253
–influence of polymorphisms of genes coding 329
–inhibition of endogenous fibrinolysis 258
–initiation of 254–255
–laboratory tests 262
–markers for 263
–pathogenetic pathways 254–261
–plasma and platelet substitution therapy 266
–platelet count and coagulation factors in patients 263–264
–regulatory role of cytokines 259–260
–relevance 251–252
–tests for fibrin formation and fibrin degradation 262–263
–treatment 264–265
coagulopathy 61
comorbidities, in sepsis 14
comparmentalization 360
complement system, role during sepsis 182–186

coronary artery bypass grafting (CABG) technique 48
C1q inhibitor 186
corticotrophin releasing hormone (CRH) 44
C-reactive protein 194
critical micellar concentration (CMC) 91–92
cyclooxygenase (COX) 236
cyclooxygenase inhibitors 188
cytochrome a_1a_3 inhibition 63
cytokines 45, 47, 324–328
–altered *ex vivo* 354–356
–cell culture of 354
–neural control of 304–306
–impaired production of 353
–neuro-endocrine mediation of 301–303
–production of 355
–stimulation of the nervous system 300
cytopathic hypoxia 62, 63, 279–280
cytosolic cytochrome c 208

d

damage associated molecular patterns (DAMPs) 193
death receptor pathway 207–209
death-inducing signaling complex (DISC) 216
dehydroepiandrosterone (DHEA) 302
dendritic cells 353, 354
desensitizing agents, in plasma 358
disseminated intravascular coagulation (DIC) 68–69
D-N-acetylglucosamine (GlcNAc) 86, 117
D-N-acetylmuramic acid (MurNAc) 86
doxycycline 164

e

Early Goal Directed Therapy trial 30
EBI3-deficient mice 180
electroncephalography 68
endothelial activation 62
endothelial cell functions and their changes, in SIRS and sepsis
–procoagulant properties of 232–233
–regulation of coagulation 232–233
–regulation of leukocyte adhesion/migration and vascular permeability 233–234
–regulation of the microcirculation/vasomotor properties 234, 239
–therapeutic approaches 239–243
endothelial dysfunction 62
endothelial injury 61
endothelial protein C receptor (EPCR) 261
endothelial-leukocyte adhesion molecule 1 (ELAM-1) 301
endothelin (ET) 238–239
endotoxemia 376
endotoxic conformation principle 93
endotoxin 181, 376–378
endotoxin tolerance 356, 361
Enterococcus sp. 27
environmental risk factors, for sepsis 28–29
epidemiology, of sepsis 13–15
epidermal growth factor (EGF)-like repeats 257
epithelial dysfunctions, in MODS 281–283
erythropoietin 45
Escherichia coli 79, 180, 187, 317, 354, 378–381
E-selectin 61, 233
E3 ubiquitin-protein ligase 91
exogenous peritonitis models, of sepsis 379–383
extracorporeal circulation (ECC) 48

f

Factor Xa 260
Factor XIII 187
FADD-like IL-1β-converting enzyme (FLICE) 207
Fas associated death domain (FADD) 207
fever 44–45
fibrin 260
fibrinolysis, in sepsis 258
flagellins 86
flavoprotein 210
fluconazole 147–148, 152–153, 155
flucytosine 153
fluid resuscitation 165
fluorescein isothiocyanate-labeled dextran (FD40) 286
fungi, in sepsis 145–156

g

genetic polymorphisms 160, 323–334
genitourinary infections 29
Glasgow Coma Scale 70, 158
GlcNAc, *see* D-N-acetylgluco-samine
glomerular thrombin deposition 282
glucocorticoids modulate cytokines 301

gluco-corticosteroids 42
glucosamine 82
glycosaminoglycans 256
G-protein coupled receptors 260
GNB-derived lipoproteins 84
Gram-negative bacteremia 28
Gram-positive infections 27
Gram-positive organisms 28
granulocyte-macrophage colony-stimulating factor (GM-CSF) 45
gut-associated lymphoid tissues (GALT) 211

h

H. influenzae 82
halofantrine 164
health care 30–31, 34
health parameters, linked to sepsis 322–353
heat shock proteins 88, 90
hematologic dysfunction 68, 69
heme oxygenase (HO)-1 191
heparin sulfate 49
heparins 240
hepatic dysfunction 69
hepatobiliary barrier function, in MODS 286–287
HIF, *see* hypoxia-inducible factor
Hickman Catheters 146
histidin rich proteins (HRP) 159
HIV patients, with sepsis 29–30
HLA 133
HMGB family of DNA-binding proteins 288–289
hydroxy methyl glutaryl-CoA (HMG-CoA) reductase inhibitors 242–243
hyperthermia 3
hypothalamic–growth hormone axis 303
hypothalamic–pituitary–gonadal (HPG) axis 302–303
hypothalamic–pituitary–thyroid (HPT) axis 303
hypothermia 3
hypovolemia 47
hypoxemia 67, 238
hypoxia 69
hypoxia-inducible factor (HIF)-1 280

i

IAP, *see* inhibitor of apoptosis
ibuprofen 188
ICAM, *see* intracellular adhesion molecule
immune dysregulation, during sepsis 178
immunoglobulin-like receptors 233
immunological (or inflammation, or humoral) hypothesis, in SFM 159
immunosuppressive mediators 359
inducible NO synthase (iNOS) 189
Infection Probability Score (IPS) 9
infections
–genetic polymorphisms 323–334
–infectious agent and site of infection 315–317
–non-genetic parameters 315
–parameters linked to health care 322–323
–parameters linked to patients 317–321
infectious etiology 47–48
inflammatory markers 61
inflammatory mediators, in sepsis
–anti-inflammatory mediators 190–192
–cellular markers of stress and soluble cell surface markers 189–190
–chemokines 177–182
–complement system 182–186
–cytokines 177–182
–disorders of coagulation and fibrinolysis 186–187
–growth factors 177–182
–lipid mediators 188, 189
–lipoproteins and lipopolysaccharide inhibitors 190
–neuromediators 192
–nitric oxide (NO) 189
–in non-infectious SIRS 192, 195
–proteases 187, 188
–transcription factor 183
inflammatory reflex
–afferent arm of 306
–cholinergic anti-inflammatory pathway 306–309
–clinical implications 309
inhibitors of apoptosis (IAPs) 208
innate immune receptors 328
integrin family 233
interferon-γ IFNγ 180, 181, 283
interleukin-1β (IL-1β) 179, 180
interleukin-2β (IL-2β) 45
interleukin-6 (IL-6) 191, 259
interleukin-10 (IL-10) 260
interleukin-12 (IL-12) 180
interleukin-15 (IL-15) 180
interleukin-18 (IL-18) 180
interleukin-27 (IL-27) mRNA 180

interleukin (IL)-1 receptor associated kinase (IRAK)-M 357
interleukin-33 (IL-33) receptor (sST2) 180
interleukin-1-like cytokines 180
interleukin-1 receptor-associated kinase (IRAK) family 90–91
International Classification of Disease (ICD) 20
international sepsis forum 9
interstitial pulmonary edema 67
intracellular cell adhesion molecule (ICAM) 233, 301
intravenous immunoglobulin (IVIG) therapy 133
isoprenoid geranylgeranylpyrophosphate (GGPP) 242
itraconazole 152, 153

k

Kawasaki disease 131
Klebsiella pneumoniae 190, 315
Kupffer cells 46

l

L-mevalonate 242
L. monocytogenes 116
lactate dehydrogenase (LDH) 289
lactic acidosis 162
lactoferrin 94
LBP, *see* LPS-binding protein
Legionella pneumophila 84, 86
leptin 192
leucocytosis 45–46
leukocytes
 –altered function of circulating 350
 –surface markers 348–350
leukotrienes 188
lipid A 83
5′-lipooxygenase 188
lipopolysaccharides (LPS) 44, 61, 81–84, 87–96, 180
 –activated TLR4/MD-2-signaling system 87–91
 –antagonistic compounds 97
 –antimicrobial peptides and proteins 94–95
 –biological activity 93–97
 –inhibitors 190
 –neutralization of 95–96
 –physico-chemical properties 91–93
 –pulmonary collectins 96–97
 –scheme 83
lipoproteins 84, 113–115
 –activating domains 85
 –activation of TLR2-dependent signaling 120–121, 123
 –inhibitors 190
liposomal AMB (LAMB) 153
LPS-binding protein (LBP) 190
LPS-neutralizing peptides 94–95
LPXTG-type proteins 116
lymphocytes 351–352

m

macrophage colony stimulating factor (M-CSF) 253
macrophage inflammatory protein-2 (CXCL2) 181
macrophage migration inhibitory factor (MIF) 181
major histocompatibility complex (MHC) 123–124
mass spectrometry techniques 8
matrix metalloproteinases (MMPs) 188
mefloquine 164
mesenteric lymph nodes (MLN) 285
MHC, *see* major histocompatibility complex
micafungin 155
microvascular dysfunction 60, 61
milk fat globule epidermal growth factor 8 (MFG-E8) 221–222
monocyte chemotactic protein (MCP)-1 (CCL2) 181
monocytes 352–353
morbidity, due to sepsis 33
mortality, in sepsis 14, 23–24
multi-organ dysfunction syndrome (MODS) 194
 –cellular apoptosis in the pathogenesis of 280
 –cross-talk between gut epithelial cells and intestinal bacteria 289–290
 –cytopathic hypoxia 279–280
 –epithelial dysfunctions 281–283
 –hepatobiliary barrier function 286–287
 –HMGB family of DNA-binding proteins 288
 –iNOS expression 285–286
 –systemic inflammation 287
 –TJ protein expression and function 283–285
Multiple Organ Dysfunction Score (MODS) 57, 58, 253
murein lipoprotein (MLP) 81
MurNAc, *see* D-N-acetylmuramic acid 116
Mycoplasma 123

MyD88 123, 357
myelopoiesis 46
myocardial depression 64

n

N. meningitidis 82
neuromediators 192
neuropeptides 304
neuropeptide Y (NPY) 192
neutrophils 48, 354
NF-κB expression 213
N^G-monomethyl-L-arginine (L-NMMA) 235
nitric oxide (NO) 64, 189, 234–236, 241–242
NK-lysin 94–95
N-methyl-D-aspartate (NMDA) 214
non-infectious etiology 48–49
NOS inhibitor 241
NOS 234, 243
NP-40 insoluble occludin 288
NSAIDs 237
nucleotide-binding oligomerization domain (Nod)-like receptors (NLRs) 80

o

occludin 284, 285
oliguria 282
organ dysfunction 20, 23
- –brain dysfunction 67–68
- –cardiocirculatory dysfunction 64–66
- –classification 57–59
- –coagulopathy 61
- –concept and scores used in ICU patients 69–70
- –cytopathic hypoxia 62, 63
- –endothelial dysfunction 61–63
- –epidemiology 59–60
- –hematologic dysfunction 68–69
- –hepatic dysfunction 69
- –inflammatory cytokines 63, 64
- –microvascular dysfunction 60, 61
- –pathophysiology 60–64
- –renal dysfunction 66
- –respiratory dysfunction 66, 67

organum vasculosum laminae terminalis (OVLT) 44

p

P. falciparum asexual parasitemia 157
P. falciparum Erythrocyte Membrane Protein-1 (PFEMP-1) 159
pancreatitis 60
parasite-induced sepsis. *See* severe falciparum malaria (SFM)
parasitized red blood cells (pRBC) 159
pathogen-associated molecular patterns (PAMPs) 149–150, 193
pathogen-associated molecular pattern (PAMP) molecules, of gram-negative bacteria
- –cell wall-associated 82
- –classes 80–81
- –flagellins 86
- –lipid bilayers 80
- –lipopolysaccharides (LPS) 81–84
- –lipoproteins of 84
- –of mammalian innate immunity 81
- –mDAP-type PGs 86–87
- –NLRs and RLRs 80
- –substructures 80

pathogen-associated molecular pattern (PAMP) molecules, of gram-positive bacteria
- –cell wall and major wall-associated 112
- –clustered embedding of staphylococci 112
- –lipoproteins 113–115, 120–123
- –lipoteichoic acids (LTA) and wall teichoic acids (WTA) 117–118, 120
- –of mammalian innate immunity 111
- –murein structure of 114–116
- –peptidoglycan matrix of 111–117
- –structure 110–113
- –superantigens 123–134
- –therapeutic concepts 134, 135

pattern recognition receptors (PRRs) 80–81
peripheral nervous system 304
peroxynitrite 63
PG-associated lipoprotein (PAL) 81
phosphatidylinositide 3′-OH kinase (PI3K)/c-Akt pathway 219
phosphodiesterases (PDEs) 235
phospholipase A2 (PLA2) 188
P13K, *see* phosphatidylinositide 3′-OH-kinase
PIRO concept 59, 70
plasma level of procalcitonin (PCT) measurement 50
plasminogen activator inhibitor, type 1 (PAI-1) 61, 187, 258
Plasmodium falciparum malaria 157
platelet-activating factor (PAF) 67, 188
P. mirabilis 381

Pneumocystis carinii 29–30
pneumonia 20
poly (ADP-ribose) polymerase (PARP-1) hypothesis 63
posaconazole 153–154
procalcitonin 194
procaspase-8 207
prostacyclin (PGI2) 188, 236–238
prostanoids (PGE2), in the brain 44, 237
proteases 187–188
–inhibitors 219, 260
protein C 253
Protein C Worldwide Evaluation in Severe Sepsis (PROWESS) trial 240–241
prothrombin time (PT) 253
P-selectin 233
Pseudomonas aeruginosa 28, 46, 79, 87, 94, 180, 315, 379
pyruvate dehydrogenase inhibition 63

q
quinine 164

r
RBC sequestration 159
recombinant activated Protein C (rhaPC) 213–214
recombinant human activated Protein C (rhaPC) 213–214
reimbursements, in United States 20
renal vasoconstriction 282
reprogrammation phenomenon 354
–clinical implication of 362
–leukocyte 358–360
–reversal 361–362
respiratory infections 28
retinoic acid-inducible gene (RIG)-I-like receptors (RLRs) 80
rifins 159
Ring Erythrocyte Surface Antigen (RESA Ag) 159

s
S. aureus 116, 119, 129, 131
S. bovis 381
S. enterica sv. Typhimurium 86
S. pyogenes 134
Salmonella spp 190
Salmonella typhimurium 180
sepsis
–age-adjusted population mortality rate 23–24
–animal models 383–384
–bacteremia models 378–379
–care 31–34
–case illustration 43
–cecal ligation and perforation (CLP) model 380–383
–in children 30
–comorbidities 14
–costs for treatment 30–31
–definition 376
–developing *vs* developed world 22
–discrepancies in data 22
–endotoxin models 376–378
–environmental factors 28–29
–epidemiology 13–19
–exogenous peritonitis models 379–383
–female *vs* male 24–27
–French ICUs cohort study 14
–hidden 21–22
–host response 41–42
–incidence and prevalence 13–14, 23–24
–in ICU patient 42
–malignancy 29
–markers 5–7
–mortality in 14, 23–24
–natural evolution of 316
–outcome other than mortality 33
–patient parameters 318–321
–PAMPS 41–42
–pathways 42
–in patients with HIV 29–30
–*a posteriori*
–prevalence studies 20–21
–problems with definitions of 14–20
–profile 7
–regional differences 22–23
–reimbursements in United States 20
–SIRS criteria 43–44
–site of primary infection 27
–syndrome diagnosis 42–43
–from systemic inflammation 7
–treatment 398ff.
–type of organism 27–28
sepsis-associated delirium (SAD) 68
septicemia 27
Sequential Organ Failure Assessment (SOFA) 58
sequestration hypothesis, in SFM 159
serum proteins 92
severe falciparum malaria (SFM)
–acute lung injury 162

severe falciparum malaria (SFM) *(continued)*
- –acute renal failure (ARF) 162
- –anemia and hypoglycemia 163
- –bacterial co-infections in 162
- –caring 164–165
- –clinical aspects 162
- –definitions 157–158
- –epidemiology 156–157
- –genetic polymorphisms 160
- –immunological (or inflammation, or humoral) hypothesis 159
- –lactic acidosis 162
- –management of 163–166
- –manifestations in adults 158
- –pathophysiology 158–161
- –at risk population 161
- –role of hemostasis and platelets 160
- –sequestration hypothesis 159
- –symptoms of uncomplicated falciparum malaria 161
- –thrombocytopenia 162

severe sepsis 3, 7
- –diagnosis of 20
- –mortality of ICU-acquired 14

SFM, *see* severe falcipasum malaria
Shigella flexneri 87
SIGIRR molecules 358
Simplified Acute Physiology Score (SAPS II) 157
single nucleotide polymorphism (SNPs) 323
SIRS, *see* systemic inflammatory response syndrome
small-angle X-ray diffraction (SAXS) measurements 92–93
SMEZ alleles 130
staphylococcal enterotoxin-like proteins (SEls) 130
staphylococcal enterotoxin-like toxins (SET) 130
staphylococcal enterotoxin (SE) 124
staphylococcal superantigen-like (SSL) proteins 130
Staphylococcus aureus 28, 79, 123, 131–132, 134, 146, 181, 317, 379
statins 242–243
steatosis 69
stem cell factor (SCF) 45
steroid therapy 165
stochastic events, concept of 317–318
streptococcal pyrogenic exotoxin (SPE) subfamilies 124
streptococcal superantigens 133
streptococcal toxic shock syndrome (STSS) 133–134
Streptococcus dysgalactiae 123
Streptococcus equi 123
Streptococcus pneumoniae 316, 379
Streptococcus pyogenes 123, 316
superantigens
- –and allelic variation 130
- –antigen presentation and pathogenic effects of 129
- –definition 123–124
- –involvement in sepsis 131–134
- –on pathogenicity islands 130
- –properties 126–128
- –staphylococcal enterotoxin-like proteins (SEls) 130
- –staphylococcal superantigen-like (SSL) proteins 130
- –structure 124–129

suppressor of cytokine signaling-1 (SOCS-1) 123
surviving sepsis campaign 42
sympathetic nervous system 304–306
systemic inflammatory disorder 316
systemic inflammatory response syndrome (SIRS) 3–5
- –clinical manifestations 44–47
- –epidemiology 13
- –incidence and prevalence 13

t

tachycardia 3, 45
tachypnea 3, 47
teicoplanin 116
therapeutic concepts, in Gram-positive septic diseases 132
thrombin 260
thrombin signaling, in endothelial cells 186
thrombin-activatable fibrinolysis inhibitor (TAFI) 187, 257
thrombocytopenia 68
thrombomodulin (TM) 61, 257
thromboxane 47, 188, 236–238
TIRAP (Mal) 123
tissue factor pathway inhibitor (TFPi) 61, 255
tissue-type plasminogen activator (t-PA) 189, 258
TJ protein expression and function, in MODS 283–285
TLR-4 pathway, of the immune cell 49

TLR4/MD-2 complex 87–91
TNF receptor apoptosis-mediating protein (TRAMP) 207
TNF receptor I associated death domain (TRADD) 207
Toll/IL-1R homology (TIR) domain-containing adapter proteins 89–91
Toll-like receptors (TLRs) 80
toxic shock syndrome (TSS) 123, 129–131
TRAF6-deubiquitinating enzyme A20 123
transient organic psychosis 20
trauma 60
treatment, of sepsis
- –anti-coagulant therapy 401–403
- –anti-inflammatory agents 400, 406
- –antibiotic therapy 404
- –biomarkers, development of 406–408
- –CATERPILLER or NOD-LRR genes 406
- –corticosteroids 400–401
- –developments 397–398
- –endotoxin 398–399
- –immunotherapy 399–404
- –infection, historical background 391–393
- –inflammation, historical background 393–394
- –initiation and regulation of inflammation 406
- –shock, historical background 394–396

TRIAD3A 91
triazoles 152–153
triggering receptor expression on myeloid cells (TREM)-1 189–190
tumor necrosis factor receptor 1 (TNFR1), 207
tumor necrosis factor receptor α (TNFRα) 253
tumor necrosis factor-α (TNFα) 179–180
TXA_2 synthesis inhibitors 237

u
urokinase-type plasminogen activator (u-PA) 258

v
vancomycin 116
vascular cell adhesion molecule 1 (VCAM-1) 301
vascular endothelial growth factor (VEGF) 182
vasoactive intestinal peptide 304
von Willebrand factor (vWF) 61
voriconazole 153–154

y
Yersinia 123

z
ZO-1 staining 285